Introduction to
ARITHMETIC GROUPS

Dave Witte Morris

Dave Witte Morris
University of Lethbridge

http://people.uleth.ca/~dave.morris/

Version 1.0 of April 2015

published by
Deductive Press
http://deductivepress.ca/

ISBN: 978-0-9865716-0-2 (paperback)
ISBN: 978-0-9865716-1-9 (hardcover)

A PDF file of this book is available for download from
http://arxiv.org/src/math/0106063/anc/

The Latex source files are also available from
http://arxiv.org/abs/math/0106063

Cover photo © Depositphotos.com/jbryson. Used with permission.

to my father

E. Kendall Witte
1926 — 2013

List of Chapters

i

Contents

Part III. Important Concepts

Appendices

Preface

This is a book about arithmetic subgroups of semisimple Lie groups, which means that we will discuss the group $SL(n, \mathbb{Z})$, and certain of its subgroups. By definition, the subject matter combines algebra (groups of matrices) with number theory (properties of the integers). However, it also has important applications in geometry. In particular, arithmetic groups arise in classical differential geometry as the fundamental groups of locally symmetric spaces. (See Chapters 1 and 2 for an elaboration of this line of motivation.) They also provide important examples and test cases in geometric group theory.

My intention in this text is to give a fairly gentle introduction to several of the main methods and theorems in the subject. There is no attempt to be encyclopedic, and proofs are usually only sketched, or only carried out for an illustrative special case. Readers with sufficient background will learn much more from [Ma] and [PR] (written by the masters) than they can find here.

The book assumes knowledge of algebra, analysis, and topology that might be taught in a first-year graduate course, plus some acquaintance with Lie groups. (Appendix A quickly recounts the essential Lie theory, and Appendix B lists the required facts from graduate courses.) Some individual proofs and examples assume additional background (but may be skipped).

Generally speaking, the chapters are fairly independent of each other (and they all have their own bibliographies), so there is no need to read the book linearly. To facilitate making a plan of study, the bottom of each chapter's first page states the main prerequisites that are not in appendices A and B. Individual chapters (or, sometimes, sections) could be assigned for reading in a course or presented in a seminar. (The book has been released into the public domain, so feel free to make copies for such purposes.) Notes at the end of each chapter have suggestions for further reading. (Many of the subjects have been given book-length treatments.) Several topics (such as amenability and Kazhdan's property (T)) are of interest well beyond the theory of arithmetic groups.

Although this is a long book, some very important topics have been omitted. In particular, there is almost no discussion of the cohomology of arithmetic groups, even though it is a subject with a long history and continues to be a very active field. (See the lecture notes of Borel [B2] for a recent survey.) Also, there is no mention at all of automorphic forms. (Recent introductions to this subject include [De] and [SS].)

Among the other books on arithmetic groups, the authoritative monographs of Margulis [Ma] and Platonov-Rapinchuk [PR] have already been mentioned. They are essential references, but would be difficult reading for my intended audience. Some works at a level more comparable to this book include:

[B1] This classic gives an explanation of reduction theory (discussed here in Chapter 19) and some of its important consequences.

[Hu] This exposition covers reduction theory (at a more elementary level than [B1]), adeles, ideles, and fundamentals of the Congruence Subgroup Property (mentioned here in Remark 17.1.3(4)).

[Ji] This extensive survey touches on many more topics than are covered here (or even in [Ma] and [PR]), with 60 pages of references.

[MR] This monograph thoroughly discusses arithmetic subgroups of the groups $SL(2, \mathbb{R})$ and $SL(2, \mathbb{C})$.

[Ra] This is an essential reference (along with [Ma] and [PR]). It is the standard reference for basic properties of lattices in Lie groups (covered here in Chapter 4). It also has proofs of the Godement Criterion (discussed here in Section 5.3), the existence of both cocompact and noncocompact arithmetic subgroups (discussed here in Section 18.7), and reduction theory for arithmetic groups of $\mathbb{Q}$-rank one (discussed here in Chapter 19). It also includes several topics not covered here, such as cohomology vanishing theorems, and lattices in non-semisimple Lie groups.

[Su] This textbook provides an elementary introduction to the Congruence Subgroup Property (which has only a brief mention here in Remark 17.1.3(4)).

[Zi] After developing the necessary prerequisites in ergodic theory and representation theory, this monograph provides proofs of three major theorems of Margulis: Superrigidity, Arithmeticity, and Normal Subgroups (discussed here in Chapters 16 and 17). It also proves a generalization of the superrigidity theorem that applies to "Borel cocycles."

Dave Morris
April 2015

References

[B1] A. Borel: *Introduction aux Groupes Arithmétiques.*
      Hermann, Paris, 1969. MR 0244260

[B2] A. Borel: Introduction to the cohomology of arithmetic groups,
      in L. Ji et al., eds.: *Lie Groups and Automorphic Forms.*
      American Mathematical Society, Providence, RI, 2006, pp. 51–86.
      ISBN 978-0-8218-4198-3, MR 2272919

[De] A. Deitmar: *Automorphic Forms.*
      Springer, London, 2013.   ISBN 978-1-4471-4434-2, MR 2977413

[Hu] J. E. Humphreys: *Arithmetic Groups.*
      Springer, Berlin, 1980.   ISBN 3-540-09972-7, MR 0584623

[Ji] L. Ji: *Arithmetic Groups and Their Generalizations.*
      American Mathematical Society, Providence, RI, 2008.
      ISBN 978-0-8218-4675-9, MR 2410298

[MR] C. Maclachlan & A. Reid: *The Arithmetic of Hyperbolic 3-Manifolds.*
      Springer, New York, 2003.   ISBN 0-387-98386-4, MR 1937957

[Ma] G. A. Margulis: *Discrete Subgroups of Semisimple Lie Groups.*
      Springer, Berlin, 1991.   ISBN 3-540-12179-X, MR 1090825

[PR] V. Platonov & A. Rapinchuk: *Algebraic Groups and Number Theory.*
      Academic Press, Boston, 1994.   ISBN 0-12-558180-7, MR 1278263

[Ra] M. S. Raghunathan: *Discrete Subgroups of Lie Groups.*
      Springer, New York, 1972.   ISBN 0-387-05749-8, MR 0507234

[SS] P. Sarnak & F. Shahidi, eds.: *Automorphic Forms and Applications.*
      American Mathematical Society, Providence, RI, 2007.
      ISBN 978-0-8218-2873-1, MR 2331351

[Su] B. Sury: *The Congruence Subgroup Problem.*
      Hindustan Book Agency, New Delhi, 2003.
      ISBN 81-85931-38-0, MR 1978430

[Zi] R. J. Zimmer: *Ergodic Theory and Semisimple Groups.*
      Birkhäuser, Boston, 1984.   ISBN 3-7643-3184-4, MR 0776417

Acknowledgments

In writing this book, I have received major help from discussions with G. A. Margulis, M. S. Raghunathan, S. G. Dani, T. N. Venkataramana, Gopal Prasad, Andrei Rapinchuk, Robert J. Zimmer, Scot Adams, and Benson Farb. I have also benefited from the comments and suggestions of many other colleagues, including Ian Agol, Marc Burger, Indira Chatterji, Yves Cornulier, Alessandra Iozzi, Anders Karlsson, Sean Keel, Nicolas Monod, Hee Oh, Alan Reid, Yehuda Shalom, Hugh Thompson, Shmuel Weinberger, Barak Weiss, and Kevin Whyte. They may note that some of the remarks they made to me have been reproduced here almost verbatim.

I would like to thank É. Ghys, D. Gaboriau and the other members of the École Normale Supérieure of Lyon for the stimulating conversations and invitation to speak that were the impetus for this work, and B. Farb, A. Eskin, and other members of the University of Chicago mathematics department for encouraging me to write this introductory text, and for the opportunity to lecture from it to a stimulating audience on numerous occasions.

I am also grateful to the University of Chicago, the École Normale Supérieure of Lyon, the University of Bielefeld, the Isaac Newton Institute for Mathematical Sciences, the University of Michigan, the Tata Institute for Fundamental Research, the University of Western Australia, and the Mathematical Sciences Research Institute, for their warm hospitality while I worked on various parts of the book. The preparation of this manuscript was partially supported by research grants from the National Science Foundation of the USA and the National Science and Engineering Research Council of Canada.

I also thank my wife, Joy Morris, for her patience and encouragement during the (too many!) years this project was underway.

Part I

Introduction

Chapter 1

What is a
Locally Symmetric Space?

In this chapter, we give a geometric introduction to the notion of a symmetric space or a locally symmetric space, and explain the central role played by simple Lie groups and their lattice subgroups. (Since geometers are the target audience here, we assume familiarity with differential geometry that will not be needed in other parts of the book.) This material is not a prerequisite for reading any of the later chapters, except Chapter 2; it is intended to provide a geometric motivation for the study of lattices in semisimple Lie groups. Since arithmetic subgroups are the primary examples of lattices, this also motivates the main topic of the rest of the book.

§1.1. Symmetric spaces

Recall that a ***Riemannian manifold*** is a smooth manifold M, together with the choice of an inner product $\langle \cdot \mid \cdot \rangle_x$ on the tangent space $T_x M$, for each $x \in M$, such that $\langle \cdot \mid \cdot \rangle_x$ varies smoothly as x varies. The nicest Riemannian manifolds are homogeneous. This means that every point looks exactly like every other point:

Main prerequisites for this chapter: understanding of geodesics, and other concepts of Differential Geometry.

(1.1.1) **Definition.** A Riemannian manifold X is a ***homogeneous space*** if its isometry group $\mathrm{Isom}(X)$ acts transitively. That is, for every $x, y \in X$, there is an isometry ϕ of X, such that $\phi(x) = y$.

(1.1.2) **Notation.** We use G° to denote the identity component of the group G.

(1.1.3) **Examples.** Here are some elementary examples of (simply connected) homogeneous spaces.

1) The round sphere $S^n = \{ x \in \mathbb{R}^{n+1} \mid \|x\| = 1 \}$. Rotations are the only orientation-preserving isometries of S^n, so we have $\mathrm{Isom}(S^n)^\circ = \mathrm{SO}(n+1)$. Any point on S^n can be rotated to any other point, so S^n is homogeneous.

2) Euclidean space $\mathbb{R}^n$. Every orientation-preserving isometry of $\mathbb{R}^n$ is a combination of a translation and a rotation, and this implies that $\mathrm{Isom}(\mathbb{R}^n)^\circ = \mathrm{SO}(n) \ltimes \mathbb{R}^n$. Any point in $\mathbb{R}^n$ can be translated to any other point, so $\mathbb{R}^n$ is homogeneous.

3) The hyperbolic plane $\mathfrak{H}^2 = \{ z \in \mathbb{C} \mid \mathrm{Im}\, z > 0 \}$, where the inner product on $T_z\mathfrak{H}^2$ is given by
$$\langle u \mid v \rangle_{\mathfrak{H}^2} = \frac{1}{4(\mathrm{Im}\, z)^2} \langle u \mid v \rangle_{\mathbb{R}^2}.$$
It is not difficult to show that
$$\mathrm{Isom}(\mathfrak{H}^2)^\circ \text{ is isomorphic to } \mathrm{PSL}(2, \mathbb{R})^\circ = \mathrm{SL}(2, \mathbb{R})/\{\pm 1\},$$
by noting that $\mathrm{SL}(2, \mathbb{R})$ acts on $\mathfrak{H}^2$ by linear-fractional transformations $z \mapsto (az + b)/(cz + d)$, and confirming, by calculation, that these linear-fractional transformations preserve the hyperbolic metric.

4) Hyperbolic space $\mathfrak{H}^n = \{ x \in \mathbb{R}^n \mid x_n > 0 \}$, where the inner product on $T_x\mathfrak{H}^n$ is given by
$$\langle u \mid v \rangle_{\mathfrak{H}^n} = \frac{1}{4x_n^2} \langle u \mid v \rangle_{\mathbb{R}^n}.$$
It is not difficult to see that $\mathfrak{H}^n$ is homogeneous (see Exercise 1). One can also show that that the group $\mathrm{Isom}(\mathfrak{H}^n)^\circ$ is isomorphic to $\mathrm{SO}(1, n)^\circ$ (see Exercise 4).

5) A cartesian product of any combination of the above (see Exercise 6).

(1.1.4) **Definitions.** Let $\phi \colon X \to X$.

1) We say that ϕ is ***involutive*** (or that ϕ is an ***involution***) if $\phi^2 = \mathrm{Id}$.

2) A ***fixed point*** of ϕ is a point $p \in X$, such that $\phi(p) = p$.

3) A fixed point p of ϕ is ***isolated*** if there is a neighborhood U of p, such that p is the only fixed point of ϕ that is contained in U.

Besides an isometry taking x to y, each of the above spaces also has a nice involutive isometry that fixes x.

1) Define $\phi_1 \colon S^n \to S^n$ by

$$\phi_1(x_1, \ldots, x_{n+1}) = (-x_1, \ldots, -x_n, x_{n+1}).$$

Then ϕ_1 is an isometry of S^n, such that ϕ_1 has only two fixed points: namely, e_{n+1} and $-e_{n+1}$, where $e_{n+1} = (0, 0, \ldots, 0, 1)$. Therefore, e_{n+1} is an isolated fixed point of ϕ_1.

2) Define $\phi_2 \colon \mathbb{R}^n \to \mathbb{R}^n$ by $\phi_2(x) = -x$. Then ϕ_2 is an isometry of $\mathbb{R}^n$, such that 0 is the only fixed point of ϕ_2.

3) Define $\phi_3 \colon \mathfrak{H}^2 \to \mathfrak{H}^2$ by $\phi_3(z) = -1/z$. Then i is the only fixed point of ϕ_3.

4) There are involutive isometries of $\mathfrak{H}^n$ that have a unique fixed point (see Exercise 3), but they are somewhat difficult to describe in the upper-half-space model that we are using.

The existence of such an isometry is the additional condition that is required to be a symmetric space.

(1.1.5) **Definition.** A Riemannian manifold X is a *symmetric space* if

1) X is connected,

2) X is homogeneous, and

3) there is an involutive isometry ϕ of X, such that ϕ has at least one isolated fixed point.

(1.1.6) *Remark.* If X is a symmetric space, then all points of X are essentially the same, so, for each $x \in X$ (not only for *some* $x \in X$), there is an isometry ϕ of X, such that $\phi^2 = \mathrm{Id}$ and x is an isolated fixed point of ϕ (see Exercise 9). Conversely, if Condition (3) is replaced with this stronger assumption, then Condition (2) can be omitted (see Exercise 10).

We constructed examples of involutive isometries of S^n, $\mathbb{R}^n$, and $\mathfrak{H}^n$ that have an isolated fixed point. The following proposition shows that no choice was involved: the involutive isometry with a given isolated fixed point p is unique, if it exists. Furthermore, in the exponential coordinates at p, the involution must simply be the map $x \mapsto -x$.

(1.1.7) **Proposition.** *Suppose ϕ is an involutive isometry of a Riemmanian manifold X, and suppose p is an isolated fixed point of ϕ. Then*

1) $d\phi_p = -\,\mathrm{Id}$, *and*

2) *for every geodesic y with $y(0) = p$, we have $\phi(y(t)) = y(-t)$, for all $t \in \mathbb{R}$.*

Proof. (1) From the Chain Rule, and the fact that $\phi(p) = p$, we have

$$d(\phi^2)_p = d\phi_{\phi(p)} \circ d\phi_p = (d\phi_p)^2.$$

Also, because $\phi^2 = \mathrm{Id}$, we know that $d(\phi^2)_p = d\,\mathrm{Id}_p = \mathrm{Id}$. We conclude that $(d\phi_p)^2 = \mathrm{Id}$; hence, the linear transformation $d\phi_p\colon T_pX \to T_pX$ satisfies the polynomial equation $x^2 - 1 = 0$.

Suppose $d\phi_p \neq -\mathrm{Id}$. (This will lead to a contradiction.) Since the polynomial $x^2 - 1$ has no repeated roots, we know that $d\phi_p$ is diagonalizable. Furthermore, because 1 and -1 are the only roots of $x^2 - 1$, we know that 1 and -1 are the only possible eigenvalues of $d\phi_p$. Therefore, because $d\phi_p \neq -\mathrm{Id}$, we conclude that 1 is an eigenvalue; so we may choose some nonzero $v \in T_pX$, such that $d\phi_p(v) = v$. Let γ be the geodesic with $\gamma(0) = p$ and $\gamma'(0) = v$. Then, because ϕ is an isometry, we know that $\phi \circ \gamma$ is also a geodesic. We have

$$(\phi \circ \gamma)(0) = \phi(\gamma(0)) = \phi(p) = p = \gamma(0)$$

and

$$(\phi \circ \gamma)'(0) = d\phi_{\gamma(0)}(\gamma'(0)) = d\phi_p(v) = v = \gamma'(0).$$

Since every geodesic is uniquely determined by prescribing its initial position and its initial velocity, we conclude that $\phi \circ \gamma = \gamma$. Therefore, $\phi(\gamma(t)) = \gamma(t)$, so $\gamma(t)$ is a fixed point of ϕ, for every t. This contradicts the fact that the fixed point $p = \gamma(0)$ is isolated.

(2) Define $\overline{\gamma}(t) = \gamma(-t)$, so $\overline{\gamma}$ is a geodesic. Because ϕ is an isometry, we know that $\phi \circ \gamma$ is also a geodesic. We have

$$(\phi \circ \gamma)(0) = \phi(\gamma(0)) = \phi(p) = p = \overline{\gamma}(0)$$

and, from (1),

$$(\phi \circ \gamma)'(0) = d\phi_{\gamma(0)}(\gamma'(0)) = -\gamma'(0) = \overline{\gamma}'(0).$$

Since a geodesic is uniquely determined by prescribing its initial position and its initial velocity, we conclude that $\phi \circ \gamma = \overline{\gamma}$, as desired.   □

(1.1.8) **Definition.** Let M be a Riemannian manifold, and let $p \in M$. It is a basic fact of differential geometry that there is a neighborhood V of 0 in T_pM, such that the exponential map $\exp_p$ maps V diffeomorphically onto a neighborhood U of p in M. By making V smaller, we may assume it is:

- *symmetric* (that is, $-V = V$), and
- *star-shaped* (that is, $tV \subseteq V$, for $0 \leq t < 1$).

The **geodesic symmetry** at p is the diffeomorphism τ of U that is defined by

$$\tau(\exp_p(v)) = \exp_p(-v),$$

for all $v \in V$.

In other words, for each geodesic γ in M, such that $\gamma(0) = p$, and for all $t \in \mathbb{R}$, such that $t\gamma'(0) \in V$, we have $\tau(\gamma(t)) = \gamma(-t)$.

Note. The geodesic symmetry τ is a local diffeomorphism, but, for most manifolds M, it is *not* a local isometry (cf. Remark 1.3.2).

In this terminology, the preceding proposition shows that if an involutive isometry ϕ has a certain point p as an isolated fixed point, then, locally, ϕ must agree with the geodesic symmetry at p. This has the following easy consequence, which is the motivation for the term **symmetric space**.

(1.1.9) **Corollary.** *A connected Riemannian manifold M is a symmetric space if and only if, for each $p \in M$, the geodesic symmetry at p extends to an isometry of M.*

Exercises for §1.1.

#1. Show that $\mathfrak{H}^n$ is homogeneous.

[*Hint:* For any $t \in \mathbb{R}^+$, the dilation $x \mapsto tx$ is an isometry of $\mathfrak{H}^n$. Also, for any $v \in \mathbb{R}^{n-1}$, the translation $x \mapsto x + v$ is an isometry of $\mathfrak{H}^n$.]

#2. Let $B^n = \{ x \in \mathbb{R}^n \mid \|x\| < 1 \}$ be the open unit ball in $\mathbb{R}^n$, equip $T_x B^n$ with the inner product

$$\langle u \mid v \rangle_{B^n} = \frac{1}{(1 - \|x\|^2)^2} \langle u \mid v \rangle_{\mathbb{R}^n},$$

and let $e_n = (0, 0, \ldots, 0, 1) \in \mathbb{R}^n$. Show that the map $\phi : B^n \to \mathfrak{H}^n$ defined by

$$\phi(x) = \frac{x + e_n}{\|x + e_n\|^2} - \frac{1}{2} e_n$$

is an isometry from B_n onto $\mathfrak{H}^n$. (In geometric terms, ϕ is obtained by composing a translation with the inversion centered at the south pole of B^n.)

#3. Show that $x \mapsto -x$ is an isometry of B^n (with respect to the Riemannian metric $\langle \cdot \mid \cdot \rangle_{B^n}$ defined in Exercise 2).

#4. For $u, v \in \mathbb{R}^{n+1}$, define

$$\langle u \mid v \rangle_{1,n} = u_0 v_0 - \sum_{j=1}^{n} u_j v_j.$$

(Note that, for convenience, we start our numbering of the coordinates at 0, rather than at 1.) Let

$$X_{1,n}^+ = \{ x \in \mathbb{R}^{n+1} \mid \langle x \mid x \rangle_{1,n} = 1, \ x_0 > 0 \},$$

so $X_{1,n}^+$ is one sheet of a 2-sheeted hyperboloid. Equip $T_x X_{1,n}^+$ with the inner product obtained by restricting $\langle \cdot \mid \cdot \rangle_{1,n}$ to this subspace.

a) Show that the bijection $\psi : B^n \to X_{1,n}^+$ defined by

$$\psi(x) = \frac{1}{1 - \|x\|^2} (1, x)$$

is an isometry. (Note that this implies that the restriction of $\langle \cdot \mid \cdot \rangle_{1,n}$ to $T_x X_{1,n}^+$ is positive definite, even though $\langle \cdot \mid \cdot \rangle_{1,n}$ is not positive definite on all of $\mathbb{R}^{n+1}$.)

b) Show $SO(1, n)^\circ$ acts transitively on $X_{1,n}^+$ by isometries.

#5. For $G = SO(1, n)^\circ = \text{Isom}(\mathfrak{H}^n)^\circ$, show there is some $p \in \mathfrak{H}^n$, such that $\text{Stab}_G(p) = SO(n)$.

[*Hint:* This is easy in the hyperboloid model $X_{1,n}^+$.]

#6. Show that if $X_1, X_2, \ldots, X_n$ are homogeneous spaces, then the cartesian product $X_1 \times X_2 \times \cdots \times X_n$ is also homogeneous.

#7. Show that every homogeneous space is **geodesically complete**. That is, for every geodesic segment $y: (-\epsilon, \epsilon) \to X$, there is a doubly-infinite geodesic $\overline{y}: \mathbb{R} \to X$, such that $\overline{y}(t) = y(t)$ for all $t \in (-\epsilon, \epsilon)$.

#8. Show that if $X_1, \ldots, X_n$ are symmetric spaces, then the cartesian product $X_1 \times X_2 \times \cdots \times X_n$ is also a symmetric space.

#9. Show that if X is a symmetric space, then, for each $x \in X$, there is an isometry ϕ of X, such that $\phi^2 = \text{Id}$ and x is an isolated fixed point of ϕ.

#10. Let X be a connected Riemannian manifold, and assume, for each $x \in X$, that there is an isometry ϕ of X, such that $\phi^2 = \text{Id}$ and x is an isolated fixed point of ϕ. Show that X is homogenous, and conclude that X is a symmetric space.

#11. Show that the real projective space $\mathbb{R}P^n$ (with the metric that makes its universal cover a round sphere) has an involutive isometry ϕ, such that ϕ has both an isolated fixed point, and a fixed point that is not isolated. Is $\mathbb{R}P^n$ a symmetric space?

§1.2. How to construct a symmetric space

In this section, we describe how Lie groups are used to construct symmetric spaces. Let us begin by recalling the well-known group-theoretic structure of any homogeneous space.

Suppose X is a connected homogeneous space, and let $G = \text{Isom}(X)^\circ$. Because $\text{Isom}(X)$ is transitive on X, and X is connected, we see that G is transitive on X (see Exercise 1), so we may identify X with the coset space G/K, where K is the stabilizer of some point in X. Note that K is compact (see Exercise 2).

Conversely, if K is any compact subgroup of any Lie group G, then there is a G-invariant Riemannian metric on G/K (see Exercise 4), so G/K (with this metric) is a homogeneous space. (For any closed subgroup H of G, the group G acts transitively on the manifold G/H, by diffeomorphisms. However, when H is not compact, G usually does not act by isometries of any Riemannian metric on G/H, so there is no reason to expect G/H to be a homogeneous space in the sense of Definition 1.1.1.)

(1.2.1) **Example.**
1) For $X = S^n$, we have $G = SO(n + 1)$, and we may let
$$K = \text{Stab}_G(e_{n+1}) = SO(n), \quad \text{so} \quad S^n = SO(n + 1)/SO(n).$$
Note that, letting σ be the diagonal matrix
$$\sigma = \text{diag}(-1, -1, \ldots, -1, 1),$$
we have $\sigma^2 = \text{Id}$, and $K = C_G(\sigma)$ is the centralizer of σ in G.
2) For $X = \mathbb{R}^n$, we have $G = SO(n) \ltimes \mathbb{R}^n$, and we may let
$$K = \text{Stab}_G(0) = SO(n), \quad \text{so} \quad \mathbb{R}^n = (SO(n) \ltimes \mathbb{R}^n)/SO(n).$$
Note that the map $\sigma: (k, v) \mapsto (k, -v)$ is an automorphism of G, such that $\sigma^2 = \text{Id}$, and
$$C_G(\sigma) = \{g \in G \mid \sigma(g) = g\} = K.$$
3) For $X = \mathfrak{H}^2$, we have $G \approx SL(2, \mathbb{R})$, and we may let
$$K = \text{Stab}_G(i) \approx SO(2), \quad \text{so} \quad \mathfrak{H}^2 = SL(2, \mathbb{R})/SO(2).$$
4) For $X = \mathfrak{H}^n$, we have $G = SO(1, n)^\circ$, and we may take $K = SO(n)$ (see Exercise 1.1#5). Note that, for $\sigma = \text{diag}(1, -1, -1, \ldots, -1)$, we have $\sigma^2 = \text{Id}$, and $K = C_G(\sigma)$.

Therefore, in each of these cases, there is an automorphism σ of G, such that K is the centralizer of σ. (In other words, $K = \{k \in G \mid \sigma(k) = k\}$ is the set of fixed points of σ in G.) The following proposition shows, in general, that a slightly weaker condition makes G/K symmetric, not just homogeneous.

(1.2.2) **Proposition.** *Let*
- *G be a connected Lie group,*
- *K be a compact subgroup of G, and*
- *σ be an involutive automorphism of G, such that K is an open subgroup of $C_G(\sigma)$.*

Then G/K can be given the structure of a symmetric space, such that the map $\tau(gK) = \sigma(g)K$ is an involutive isometry of G/K with eK as an isolated fixed point.

Proof. To simplify the proof slightly, let us assume that $K = C_G(\sigma)$ (see Exercise 5).

Because K is compact, we know there is a G-invariant Riemannian metric on G/K (see Exercise 4). Then, because $\langle \tau \rangle$ is finite, and normalizes G, it is not difficult to see that we may assume this metric is also τ-invariant (see Exercise 6). (This conclusion can also be reached by letting $G^+ = \langle \sigma \rangle \ltimes G$ and $K^+ = \langle \sigma \rangle \times K$, so K^+ is a compact subgroup of G^+, such that $G^+/K^+ = G/K$.) Therefore, τ is an involutive isometry of G/K.

Suppose gK is a fixed point of τ, with $g \approx e$. Then $\sigma(g) \in gK$, so we may write $\sigma(g) = gk$, for some $k \in K$. Since σ centralizes k (and σ is an

automorphism), we have
$$\sigma^2(g) = \sigma(\sigma(g)) = \sigma(gk) = \sigma(g)\,\sigma(k) = (gk)(k) = gk^2.$$
On the other hand, we know $\sigma^2(g) = g$ (because σ is involutive), so we conclude that $k^2 = e$.

Since $g \approx e$, and $\sigma(e) = e$, we have $\sigma(g) \approx g$, so $k = g^{-1}\sigma(g) \approx e$. Since $k^2 = e$, we conclude that $k = e$. (There is a neighborhood U of e in G, such that, for every $u \in U \smallsetminus \{e\}$, we have $u^2 \neq e$.) Therefore $\sigma(g) = gk = ge = g$, so $g \in C_G(\sigma) = K$; hence, $gK = eK$. $\square$

Conversely, for any symmetric space X, there exist G, K, and σ as in Proposition 1.2.2, such that X is isometric to G/K (see Exercise 7).

(1.2.3) Example. Let $G = \mathrm{SL}(n, \mathbb{R})$, $K = \mathrm{SO}(n)$, and define $\sigma(g) = (g^{-1})^T$ (the transpose-inverse). Then $\sigma^2 = 1$ and $C_G(\sigma) = K$, so the theorem implies that G/K is a symmetric space. Let us describe this space somewhat more concretely.

Recall that any real symmetric matrix A can be diagonalized over $\mathbb{R}$. In particular, all of its eigenvalues are real. If all the eigenvalues of A are strictly positive, then we say that A is **positive definite**.

Let
$$X = \{\, A \in \mathrm{SL}(n, \mathbb{R}) \mid A \text{ is symmetric and positive definite} \,\},$$
and define $\alpha \colon G \times X \to X$ by $\alpha(g, x) = gxg^T$. Then:

a) α defines an action of G on X; i.e., we have $\alpha(gh, x) = \alpha(g, \alpha(h, x))$ for all $g, h \in G$ and $x \in X$.

b) This action is transitive, and we have $K = \mathrm{Stab}_G(\mathrm{Id})$, so X may be identified with G/K.

c) $T_{\mathrm{Id}}X = \{\, u \in \mathrm{Mat}_{n \times n}(\mathbb{R}) \mid u \text{ is symmetric and } \mathrm{trace}(u) = 0 \,\}$. (By definition, we have $X \subseteq \mathrm{SL}(n, \mathbb{R})$. The condition $\mathrm{trace}(u) = 0$ is obtained by differentiating the restriction $\det(A) = 1$.)

d) The inner product $\langle u \mid v \rangle = \mathrm{trace}(uv)$ on $T_{\mathrm{Id}}X$ is K-invariant, so it may be extended to a G-invariant Riemannian metric on X.

e) The map $\tau \colon X \to X$, defined by $\tau(A) = A^{-1}$, is an involutive isometry of X, such that $\tau(\alpha(g, x)) = \sigma(g)\,\tau(x)$ for all $g \in G$ and $x \in X$.

(1.2.4) Example. Other examples of symmetric spaces are:

1) $\mathrm{SL}(n, \mathbb{C}) / \mathrm{SU}(n)$, and

2) $\mathrm{SO}(p, q)^\circ / (\mathrm{SO}(p) \times \mathrm{SO}(q))$.

These are special cases of a consequence of Proposition 1.2.2 that will be stated after we introduce some terminology.

(1.2.5) Definitions.

1) A symmetric space X is **irreducible** if its universal cover is not isometric to any nontrivial product $X_1 \times X_2$.

2) A Riemannian manifold is **flat** if its curvature tensor is identically zero, or, equivalently, if every point in X has a neighborhood that is isometric to an open subset of the Euclidean space $\mathbb{R}^n$.

(1.2.6) **Proposition.** *Let G be a connected, noncompact, simple Lie group with finite center. Then G has a maximal compact subgroup K (which is unique up to conjugacy), and G/K is a simply connected, noncompact, irreducible symmetric space. Furthermore, G/K has non-positive sectional curvature and is not flat.*

Conversely, any noncompact, non-flat, irreducible symmetric space is of the form G/K, where G is a connected, noncompact, simple Lie group with trivial center, and K is a maximal compact subgroup of G.

(1.2.7) *Remark.* Let K be a compact subgroup of a connected, simple Lie group G with finite center, such that G/K is a symmetric space (cf. Proposition 1.2.2). Proposition 1.2.6 shows that if G is not compact, then K must be a maximal compact subgroup of G, which is essentially unique.

On the other hand, if G is compact, then the subgroup K may not be unique, and may not be maximal. For example, both $\mathrm{SO}(n)/\mathrm{SO}(n-1)$ and $\mathrm{SO}(n)/\{e\}$ are symmetric spaces. The former is a round sphere, which has already been mentioned. The latter is a special case of the fact that every connected, compact Lie group is a symmetric space (see Exercise 10).

É. Cartan obtained a complete list of all the symmetric spaces (both compact and noncompact) by finding all of the simple Lie groups G (see Theorem A2.7), and determining, for each of them, which compact subgroups K can arise in Proposition 1.2.2.

Exercises for §1.2.

#1. Suppose a topological group G acts transitively (and continuously) on a connected topological space M. Show that the identity component G° is transitive on M.

#2. Let $\{g_n\}$ be a sequence of isometries of a connected, complete Riemannian manifold M, and assume there exists $p \in M$, such that $g_n p = p$ for all n.) Show there is a subsequence $\{g_{n_k}\}$ of $\{g_n\}$ that converges uniformly on compact subsets of M. (That is, there is some isometry g of M, such that, for every $\epsilon > 0$ and every compact subset C of M, there exists k_0, such that $d(g_{n_k}c, gc) < \epsilon$ for all $c \in C$ and all $k > k_0$.)

[*Hint:* This is a special case of the Arzelà-Ascoli Theorem. For each $c \in C$, the sequence $\{g_n c\}$ is bounded, and therefore has a convergent subsequence. By Cantor diagonalization, there is a subsequence that works for all c in a countable, dense subset of C.]

#3. Let K be a compact group, and let $\rho\colon K \to \mathrm{GL}(n, \mathbb{R})$ be a continuous homomorphism. Show that there is a K-invariant inner product $\langle \cdot \mid \cdot \rangle_K$ on $\mathbb{R}^n$; that is, such that $\langle \rho(k)u \mid \rho(k)v \rangle_K = \langle u \mid v \rangle_K$ for all $k \in K$ and all $u, v \in \mathbb{R}^n$.
[*Hint:* Define $\langle u \mid v \rangle_K = \int_K \langle \rho(k)u \mid \rho(k)v \rangle\, d\mu(k)$, where μ is Haar measure on K.]

#4. Let K be a compact subgroup of a Lie group G. Use Exercise 3 to show that there is a G-invariant Riemannian metric on G/K.
[*Hint:* A G-invariant Riemannian metric on G/K is determined by the inner product it assigns to the tangent space $T_{eK}(G/K)$.]

#5. Complete the proof of Proposition 1.2.2, by removing the simplifying assumption that $K = C_G(\sigma)$.

#6. Let F be a finite group of diffeomorphisms (not necessarily isometries) of a Riemannian manifold $(M, \langle \cdot \mid \cdot \rangle_x)$. Define a new inner product $\langle \cdot \mid \cdot \rangle'_x$ on each tangent space $T_x M$ by
$$\langle u \mid v \rangle'_x = \sum_{f \in F} \langle df_x(u) \mid df_x(v) \rangle_{f(x)}.$$
a) Show that the Riemannian metric $\langle \cdot \mid \cdot \rangle'$ on M is F-invariant.
b) Show that if G is a group of isometries of $(M, \langle \cdot \mid \cdot \rangle_x)$, and G is normalized by F, then $\langle \cdot \mid \cdot \rangle'$ is G-invariant.

#7. For any symmetric space X, show that there exist G, K, and σ as in Proposition 1.2.2, such that X is isometric to G/K.
[*Hint:* Suppose τ is an involutive isometry of X with an isolated fixed point p. Let $G = \mathrm{Isom}(X)^\circ$ and $K = \mathrm{Stab}_G(p)$. Define $\sigma(g) = \tau g \tau$. Show $K \subset C_G(\sigma)$ and, using the fact that p is isolated, show that K contains the identity component of $C_G(\sigma)$.]

#8. Verify assertions (a), (b), (c), (d), and (e) of Example 1.2.3.
[*Hint:* To prove transitivity in (b), you may assume that every symmetric matrix A is diagonalizable by an *orthogonal* matrix. That is, there exists g, such that gAg^{-1} is diagonal and $gg^T = \mathrm{Id}$. Note that every positive-definite diagonal matrix has a square root that is also a diagonal matrix.]

#9. Show that if X is a connected homogeneous space, then $\mathrm{Isom}(X)$ has only finitely many connected components.
[*Hint:* Every component of $G = \mathrm{Isom}(X)$ intersects the compact group $\mathrm{Stab}_g(x)$.]

#10. Show that if G is compact, then there is a G-invariant Riemannian metric on G that makes G a symmetric space.
[*Hint:* The involutive isometry is $g \mapsto g^{-1}$.]

§1.3. Locally symmetric spaces

The gist of the following definition is that a locally symmetric space is a Riemannian manifold that is locally isometric to a symmetric space; that is, every point has a neighborhood that is isometric to an open subset of some symmetric space.

(1.3.1) **Definition.** A complete Riemannian manifold M is ***locally symmetric*** if its universal cover is a symmetric space. In other words, there is a symmetric space X, and a group Γ of isometries of X, such that

   1) Γ acts freely and properly discontinuously on X, and

   2) M is isometric to $\Gamma \backslash X$.

(1.3.2) *Remark.* At every point of a symmetric space, the geodesic symmetry $\gamma(t) \mapsto \gamma(-t)$ extends to an isometry of the entire manifold (see Corollary 1.1.9). In a locally symmetric space, the geodesic symmetry τ at each point is an isometry on its domain, but it may not be possible to extend τ to an isometry that is well-defined on the entire manifold; that is, the geodesic symmetry is only a *local* isometry. That is the origin of the term ***locally symmetric***.

(1.3.3) **Example.** Define $g \colon \mathfrak{H}^2 \to \mathfrak{H}^2$ by $g(z) = z + 1$, let $\Gamma = \langle g \rangle$, and let $M = \Gamma \backslash \mathfrak{H}^2$. Then (obviously) M is locally symmetric.

   However, M is not symmetric. We provide several different geometric proofs of this fact, in order to illustrate the important distinction between symmetric spaces and locally symmetric spaces. (It can also be proved group-theoretically (see Exercise 2).) The manifold M is a cusp:

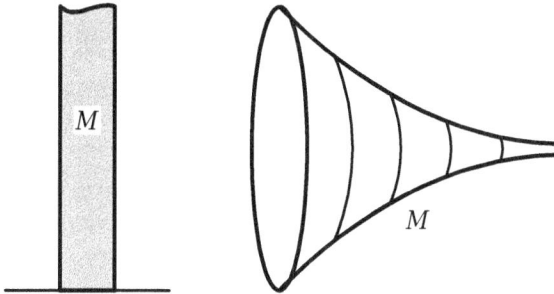

   1) Any point far out in the cusp lies on a short loop that is not null-homotopic, but points at the other end do not lie on such a loop. Therefore, M is not homogeneous, so it cannot be symmetric.

   2) The geodesic symmetry performs a 180° rotation. Therefore, if it is a well-defined diffeomorphism of M, it must interchange the two ends of the cusp. However, one end is thin, and the other end is (very!) wide, so no isometry can interchange these two ends. Hence, the geodesic symmetry (at any point) is not an isometry, so M is not symmetric.

   3) Let us show, directly, that the geodesic symmetry at some point $p \in \mathfrak{H}^2$ does not factor through to a well-defined map on $\Gamma \backslash \mathfrak{H}^2 = M$.

- Let $x = -1 + i$ and $y = 1 + i$, and let $p \in i\mathbb{R}$ be the midpoint of the geodesic segment joining x and y:

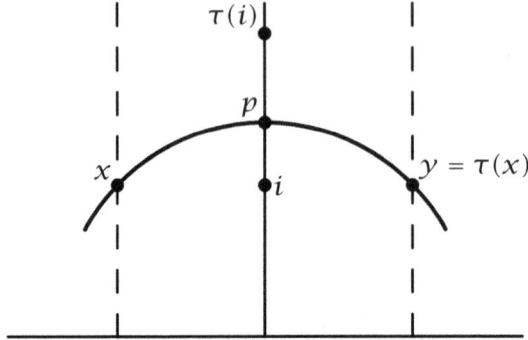

- Let τ be the geodesic symmetry at p. Then $\tau(x) = y = 1 + i$.
- Because the imaginary axis is a geodesic, we have $\tau(i) = ai$, for some $a > 1$.
- Now $i = x + 1 = g(x)$, so x and i represent the same point in M. However, $\tau(i) - \tau(x) = -1 + (a-1)i$ is not an integer (it is not even real), so $\tau(x)$ and $\tau(i)$ do **not** represent the same point in M. Therefore, τ does not factor through to a well-defined map on M.

(1.3.4) *Remarks.*

1) Some authors do not require M to be complete in their definition of a locally symmetric space. This would allow the universal cover of M to be an open subset of a symmetric space, instead of the entire symmetric space.

2) A more intrinsic (but more advanced) definition is that a complete, connected Riemannian manifold M is **locally symmetric** if and only if the curvature tensor of M is invariant under all parallel translations, and M is complete.

Any complete, connected manifold of constant negative curvature is a locally symmetric space, because the universal cover of such a manifold is $\mathfrak{H}^n$ (after normalizing the curvature to be -1). As a generalization of this, we are interested in locally symmetric spaces M whose universal cover $\widetilde{M}$ is of **noncompact type**, with no flat factors; that is, such that each irreducible factor of $\widetilde{M}$ is noncompact (and not flat). From Proposition 1.2.6, we see, in this case, that $\widetilde{M}$ can be written in the form $\widetilde{M} = G/K$, where $G = G_1 \times \cdots \times G_n$ is a product of noncompact simple Lie groups, and K is a maximal compact subgroup of G. We have $M = \Gamma \backslash \widetilde{M}$, for some discrete subgroup Γ of $\mathrm{Isom}(\widetilde{M})$. We know that $\mathrm{Isom}(\widetilde{M})$ has only finitely many connected components (see Exercise 1.2#9), so, if we replace M with an appropriate finite cover, we can arrange that $\Gamma \subset \mathrm{Isom}(\widetilde{M})^\circ = G$.

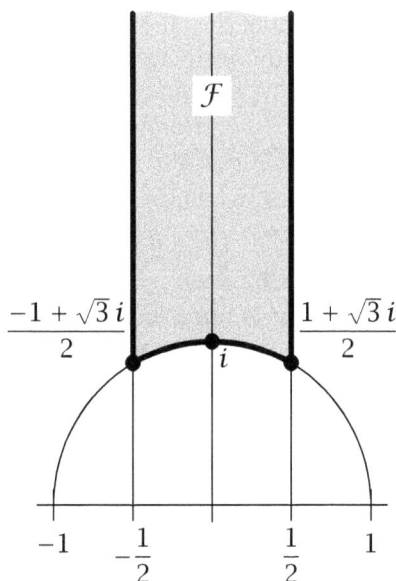

FIGURE 1.3A. A fundamental domain $\mathcal{F}$ for $\mathrm{SL}(2, \mathbb{Z})$ in $\mathrm{SL}(2, \mathbb{R})$.

Then
$$M = \Gamma \backslash G / K, \text{ and } \Gamma \text{ is a discrete subgroup of } G.$$

A topologist may like M to be compact, but it turns out that a very interesting theory is obtained by making the weaker assumption that M has finite volume. Hence, the subgroup Γ should be chosen so that $\Gamma \backslash G / K$ has finite volume. Because $\Gamma \backslash G$ is a principal K-bundle over $\Gamma \backslash G / K$, and K has finite measure, it is not difficult to see, from Fubini's Theorem, that $\Gamma \backslash G$ has finite volume (see Exercise 6). This leads to the following definition.

(1.3.5) **Definition.** A subgroup Γ of G is a *lattice* in G if Γ is discrete and $\Gamma \backslash G$ has finite volume (which respect to the Haar measure on G).

(1.3.6) **Example.** If Γ is discrete and $\Gamma \backslash G$ is compact, then Γ is a lattice in G, because any compact Riemannian manifold obviously has finite volume.

(1.3.7) **Example.** $\mathrm{SL}(2, \mathbb{Z})$ is a lattice in $\mathrm{SL}(2, \mathbb{R})$. To see this, let
$$\mathcal{F} = \{ z \in \mathfrak{H}^2 \mid |z| \geq 1 \text{ and } -1/2 \leq \operatorname{Re} z \leq 1/2 \} \qquad (1.3.8)$$
(see Figure 1.3A). It is well known (though not obvious) that $\mathcal{F}$ is a fundamental domain for the action of $\mathrm{SL}(2, \mathbb{Z})$ on $\mathfrak{H}^2$ (see Exercises 7 and 8); it therefore suffices to show that $\mathcal{F}$ has finite volume, or, more precisely, finite hyperbolic area.

The hyperbolic area dA of an infinitesimal rectangle is the product of its hyperbolic length and its hyperbolic width. If the Euclidean length is dx and the Euclidean width is dy, and the rectangle is located at the point $x + iy$, then, by definition of the hyperbolic metric, the hyperbolic length is $(dx)/(2y)$ and the hyperbolic width is $(dy)/(2y)$. Therefore,

$$dA = \frac{dx\,dy}{4y^2}.$$

Since $\operatorname{Im} z \geq \sqrt{3}/2$ for all $z \in \mathcal{F}$, we have

$$\operatorname{vol}(\mathcal{F}) = \int_{x+iy \in \mathcal{F}} dA \leq \int_{\sqrt{3}/2}^{\infty} \int_{-1/2}^{1/2} \frac{dx\,dy}{4y^2} = \frac{1}{4} \int_{\sqrt{3}/2}^{\infty} \frac{1}{y^2}\,dy < \infty.$$

Unfortunately, $\mathrm{SL}(2,\mathbb{Z}) \backslash \mathfrak{H}^2$ is not a locally symmetric space, because $\mathrm{SL}(2,\mathbb{Z})$ does not act freely on $\mathfrak{H}^2$ (so the quotient space is not a Riemann-ian manifold). However, there are finite-index subgroups of $\mathrm{SL}(2,\mathbb{Z})$ that do act freely (cf. Theorem 4.8.2), and these provide interesting locally symmetric spaces.

Calculations similar to (but more complicated than) Example 1.3.7 show:

- $\mathrm{SL}(n,\mathbb{Z})$ is a lattice in $\mathrm{SL}(n,\mathbb{R})$, and
- $\mathrm{SO}(p,q) \cap \mathrm{SL}(n,\mathbb{Z})$ is a lattice in $\mathrm{SO}(p,q)$.

As in the example of $\mathrm{SL}(2,\mathbb{Z}) \backslash \mathfrak{H}^2$, the hard part is to find a fundamental domain for $\Gamma \backslash G$ (or an appropriate approximation of a fundamental do-main); then it is not difficult to see that its volume is finite. These are special cases of the following general theorem, which implies that every simple Lie group has a lattice.

(1.3.9) **Theorem** (Arithmetic subgroups are lattices (see Theorem 5.1.11)). *Assume*

- $G = G_1 \times \cdots \times G_m$ *is a product of simple Lie groups,*
- $G \subseteq \mathrm{SL}(\ell,\mathbb{R})$, *and*
- $G \cap \mathrm{SL}(\ell,\mathbb{Q})$ *is dense in* G.

Then $G_{\mathbb{Z}} = G \cap \mathrm{SL}(\ell,\mathbb{Z})$ *is a lattice in* G.

Lattices constructed by taking the integer points of G in this way are said to be **arithmetic** (see Definition 5.1.19). (For most simple Lie groups, these are the only lattices (see Theorem 5.2.1).) When ℓ is large, there is more than one way to embed G in $\mathrm{SL}(\ell,\mathbb{R})$, and we will see that different embeddings can lead to quite different intersections with $\mathrm{SL}(\ell,\mathbb{Z})$. In particular, if G is a noncompact, simple Lie group, then:

- By taking an appropriate embedding of G in some $\mathrm{SL}(\ell,\mathbb{R})$, we will construct a lattice Γ in G, such that $\Gamma \backslash G$ is **not** compact (see Corol-lary 5.1.17).
- By taking a different embedding, we will construct a different lattice Γ', such that $\Gamma' \backslash G$ is compact (see Theorem 18.7.1).

We will also see that algebraic properties of Γ influence the geometry of the corresponding locally symmetric space M. In particular, the structure of Γ determines whether M is compact or not. (For example, the "Godement Criterion" (5.3.3) implies that M is compact if and only if every element of Γ is a diagonalizable matrix over $\mathbb{C}$.) Much more generally, the following important theorem implies that every geometric property of M is faithfully reflected in some group-theoretic property of Γ.

(1.3.10) **Theorem** (Mostow Rigidity Theorem (see Chapter 15)). *Let M_1 and M_2 be finite-volume locally symmetric spaces (not both 2-dimensional), such that*

- *the universal covers of M_1 and M_2 are neither compact, nor flat, nor reducible, and*
- *the volumes of M_1 and M_2 are normalized (i.e., $\mathrm{vol}\, M_1 = \mathrm{vol}\, M_2 = 1$).*

If $\pi_1(M_1) \cong \pi_1(M_2)$, then M_1 is isometric to M_2.

In fact, every homotopy equivalence is homotopic to an isometry.

The theorem implies that locally symmetric spaces have no nontrivial deformations, which is why it is called a "rigidity" theorem:

(1.3.11) **Corollary.** *Let $\{g_t\}$ be a continuous family of Riemannian metrics on a manifold M with $\dim M > 2$, such that, for each t:*

- *(M, g_t) is a finite-volume locally symmetric space whose universal cover is neither compact, nor flat, nor reducible, and*
- *$\mathrm{vol}(M, g_t) = 1$.*

Then (M, g_t) is isometric to (M, g_0), for every t.

(1.3.12) **Definition.** A locally symmetric space is ***irreducible*** if no *finite* cover of M can be written as a nontrivial cartesian product $M_1 \times M_2$.

It is important to note that the universal cover of an irreducible locally symmetric space need not be an irreducible symmetric space. In other words, there can be lattices in $G_1 \times \cdots \times G_n$ that are not of the form $\Gamma_1 \times \cdots \times \Gamma_n$ (see Example 5.5.3).

(1.3.13) *Remark.* Theorem 1.3.10 (and the corollary) can be generalized to the case where only M_1, rather than the universal cover of M_1, is irreducible. However, this requires the hypotheses to be strengthened: it suffices to assume that no irreducible factor of M_1 or M_2 is either compact or flat or 2-dimensional. Furthermore, the conclusion needs to be weakened: rather than simply multiplying by a single scalar to normalize the volume, there can be a different scalar on each irreducible factor of the universal cover.

Exercises for §1.3.

#1. Let
- X be a simply connected symmetric space,
- $\Gamma \backslash X$ be a locally symmetric space whose universal cover is X (so Γ is a discrete group of isometries that acts freely and properly discontinuously on X), and
- τ be an isometry of X.

Show that if τ factors through to a well-defined map on $\Gamma \backslash X$, then τ normalizes Γ (that is, $\tau \gamma \tau^{-1} \in \Gamma$, for every $\gamma \in \Gamma$).

#2. Define $g \colon \mathfrak{H}^2 \to \mathfrak{H}^2$ by $g(z) = z + 1$.
   a) Show the geodesic symmetry τ at i is given by $\tau(z) = -1/z$.
   b) Show that τ does not normalize $\langle g \rangle$.
   c) Conclude that τ does not factor through to a well-defined map on $\langle g \rangle \backslash \mathfrak{H}^2$.

#3. Let
- X be a simply connected symmetric space, and
- $\Gamma \backslash X$ be a locally symmetric space whose universal cover is X (so Γ is a discrete group of isometries that acts freely and properly discontinuously on X).

Show that X is homogeneous if and only if the normalizer $\mathcal{N}_G(\Gamma)$ is transitive on X, where $G = \mathrm{Isom}(X)$.

#4. Let $M = \Gamma \backslash G / K$ be a locally symmetric space, and assume that G has no compact factors. Show that if $\mathcal{N}_G(\Gamma)/\Gamma$ is finite, then $\mathrm{Isom}(M)$ is finite.

#5. Show that if K is any compact subgroup of a Lie group G, then there is a unique (up to a scalar multiple) G-invariant Borel measure ν on G/K, such that $\nu(C) < \infty$, for every compact subset C of G/K.

#6. Let
- K be a compact subgroup of a Lie group G, and
- Γ be a discrete subgroup of G that acts freely on G/K.

Show that $\Gamma \backslash G$ has finite volume if and only if $\Gamma \backslash G / K$ has finite volume.

#7. Let $\Gamma = \mathrm{SL}(2, \mathbb{Z})$, and define $\mathcal{F} \subset \mathfrak{H}^2$ as in (1.3.8). Show, for each $p \in \mathfrak{H}^2$, that there is some $\gamma \in \Gamma$ with $\gamma(p) \in \mathcal{F}$.
[*Hint:* If $\mathrm{Im}\, \gamma(p) \leq \mathrm{Im}\, p$ for all $\gamma \in \Gamma$, and $-1/2 \leq \mathrm{Re}\, p \leq 1/2$, then $p \in \mathcal{F}$.]

#8. Let $\Gamma = \mathrm{SL}(2, \mathbb{Z})$, and define $\mathcal{F} \subset \mathfrak{H}^2$ as in (1.3.8). Show, for $z, w \in \mathcal{F}$, that if there exists $\gamma \in \Gamma$ with $\gamma(z) = w$, then either $z = w$ or $z, w \in \partial \mathcal{F}$.
[*Hint:* Assume $\mathrm{Im}\, w \leq z$. Then $|\gamma_{2,1} z + \gamma_{2,2}| \leq 1$. Hence $|\gamma_{2,1}| \in \{0, 1\}$. If $|\gamma_{2,1}| = 1$ and $\gamma_{2,2} \neq 0$, then $|\mathrm{Re}\, z| = 1/2$, so $z \in \partial \mathcal{F}$. If $|\gamma_{2,1}| = 1$ and $\gamma_{2,2} = 0$, then

$w = (az - 1)/z$. Since $|\text{Re}(1/z)| \leq |\text{Re}\,z| \leq 1/2$, and $|\text{Re}\,w| \leq 1/2$, we see that either $\text{Re}\,z = 1/2$ or $w = -1/z$.]

Notes

Either of Helgason's books [2, 3] is a good reference for the geometric material on symmetric spaces and locally symmetric spaces, the connection with simple Lie groups, and much more. Lattices are the main topic of Raghunathan's book [8].

Theorem 1.3.9 is a result of Borel and Harish-Chandra [1] that will be proved in Chapters 7 and 19.

Theorem 1.3.10 combines work of Mostow [5], Prasad [7], and Margulis [4]. We will discuss it in Chapter 15.

Example 1.3.7 appears in many number theory texts, including [9, §7.1.2, pp. 77–79]. Our hints for Exercises 1.3#7 and 1.3#8 are taken from [6, Prop. 4.4, pp. 181–182].

References

[1] A. Borel and Harish-Chandra: Arithmetic subgroups of algebraic groups, *Ann. Math.* (2) 75 (1962) 485–535. MR 0147566, http://dx.doi.org/10.2307/1970210

[2] S. Helgason: *Differential Geometry and Symmetric Spaces.* American Mathematical Society, Providence, RI, 1962. ISBN 0-8218-2735-9, MR 1834454

[3] S. Helgason: *Differential Geometry, Lie Groups, and Symmetric Spaces.* Academic Press, New York, 1978. MR 0514561

[4] G. A. Margulis: Discrete groups of motions of manifolds of non-positive curvature, *Amer. Math. Soc. Translations* 109 (1977) 33–45. MR 0492072

[5] G. D. Mostow: *Strong Rigidity of Locally Symmetric Spaces.* Princeton Univ. Press, Princeton, 1973. MR 0385004

[6] V. Platonov and A. Rapinchuk: *Algebraic Groups and Number Theory.* Academic Press, Boston, 1994. ISBN 0-12-558180-7, MR 1278263

[7] G. Prasad: Strong rigidity of $\mathbb{Q}$-rank 1 lattices, *Invent. Math.* 21 (1973) 255–286. MR 0385005, http://eudml.org/doc/142232

[8] M. S. Raghunathan: *Discrete Subgroups of Lie Groups.* Springer, New York, 1972. ISBN 0-387-05749-8, MR 0507234

[9] J.-P. Serre: *A Course in Arithmetic,* Springer, New York 1973. MR 0344216

Chapter 2

Geometric Meaning of $\mathbb{R}$-rank and $\mathbb{Q}$-rank

This chapter, like the previous one, is motivational. It is not a prerequisite for later chapters.

§2.1. Rank and real rank

Let X be a symmetric space (see Definition 1.1.5). For example, X could be a Euclidean space $\mathbb{R}^n$, or a round sphere S^n, or a hyperbolic space $\mathfrak{H}^n$, or a product of any combination of these.

As is well known, the rank of X is a natural number that describes part of the geometry of X, namely, the dimension of a maximal flat.

(2.1.1) **Definition.** A *flat* in X is a connected, totally geodesic, flat submanifold of X.

(2.1.2) **Definition.** rank X is the largest natural number r, such that X contains an r-dimensional flat.

Let us assume that X has no flat factors. (That is, the universal cover of X is not isometric to a product of the form $Y \times \mathbb{R}^n$. Mostly, we will be interested in the case where X also does not have any compact factors.)

Main prerequisites for this chapter: locally symmetric spaces (Chapter 1) and other differential geometry.

21

Let $G = \text{Isom}(X)^\circ$. Then G acts transitively on X, and there is a compact subgroup K of G, such that $X = G/K$. Because X has no flat factors, G is a connected, semisimple, real Lie group with trivial center (see §1.2). (We remark that G is isomorphic to a closed subgroup of $\text{SL}(\ell, \mathbb{R})$, for some ℓ.)

The real rank can be understood similarly. It is an invariant of G that is defined algebraically (see Chapter 8), but it has the following geometric interpretation.

(2.1.3) **Theorem.** $\text{rank}_{\mathbb{R}} G$ is the largest natural number r, such that X contains a closed, **simply connected**, r-dimensional flat.

(2.1.4) **Warning.** By *closed*, we simply mean that the flat contains all of its accumulation points, not that it is compact. (A closed, simply connected flat is homeomorphic to some Euclidean space $\mathbb{R}^r$.)

For example, if X is compact, then every closed, totally geodesic, flat subspace of X must be a torus, not $\mathbb{R}^n$, so $\text{rank}_{\mathbb{R}} G = 0$. On the other hand, if X is not compact, then X has unbounded geodesics (for example, if X is irreducible, then every geodesic goes to infinity), so $\text{rank}_{\mathbb{R}} G \geq 1$. Hence:
$$\text{rank}_{\mathbb{R}} G = 0 \qquad \Leftrightarrow \qquad X \text{ is compact.}$$
Thus, there is a huge difference between $\text{rank}_{\mathbb{R}} G = 0$ and $\text{rank}_{\mathbb{R}} G > 0$, because no one would mistake a compact space for a noncompact one.

(2.1.5) *Remark.* $\text{rank}_{\mathbb{R}} G = \text{rank} X$ if and only if X has no compact factors.

There is also an important difference between $\text{rank}_{\mathbb{R}} G = 1$ and $\text{rank}_{\mathbb{R}} G > 1$. The following proposition is an important example of this.

(2.1.6) **Definition.** X is **two-point homogeneous** if, whenever (x_1, x_2) and (y_1, y_2) are two pairs of points in X with $d(x_1, x_2) = d(y_1, y_2)$, there is an isometry g of X with $g(x_1) = y_1$ and $g(x_2) = y_2$.

If $\text{rank}_{\mathbb{R}} G > 1$, then there exist maximal flats H_1 and H_2 that intersect nontrivially. On the other hand, there also exist some pairs x_1, x_2, such that $\{x_1, x_2\}$ is not contained in the intersection of any two (distinct) maximal flats. This establishes one direction of the following result.

(2.1.7) **Proposition.** *Assume X is noncompact and irreducible. The symmetric space X is two-point homogeneous if and only if $\text{rank}_{\mathbb{R}} G = 1$.*

The following is an infinitesimal version of this result.

(2.1.8) **Proposition.** *Assume X is noncompact and irreducible. The action of G on the set of unit tangent vectors of X is transitive iff $\text{rank}_{\mathbb{R}} G = 1$.*

(2.1.9) **Corollary.** $\text{rank}_{\mathbb{R}} \text{SO}(1, n) = 1$.

Proof. For $G = \text{SO}(1, n)$, we have $X = \mathfrak{H}^n$. The stabilizer $\text{SO}(n)$ of a point in $\mathfrak{H}^n$ acts transitively on the unit tangent vectors at that point. So G acts transitively on the unit tangent vectors of X. $\qquad\qquad\square$

More generally, it can be shown that $\mathrm{rank}_{\mathbb{R}}(SO(m,n)) = \min\{m,n\}$. Also, $\mathrm{rank}_{\mathbb{R}}(SL(n,\mathbb{R})) = n-1$. Although they may not be obvious geometrically, these real ranks are easy to calculate from the algebraic definition that will be given in Chapter 8.

(2.1.10) *Remark.* For every r, there is a difference between $\mathrm{rank}_{\mathbb{R}} G = r$ and $\mathrm{rank}_{\mathbb{R}} G > r$, but this difference is less important as r grows larger: the three main cases are $\mathrm{rank}_{\mathbb{R}} G = 0$, $\mathrm{rank}_{\mathbb{R}} G = 1$, and $\mathrm{rank}_{\mathbb{R}} G \geq 2$. (This is analogous to the situation with smoothness assumptions: countless theorems require a function to be C^0 or C^1 or C^2, but far fewer theorems require a function to be, say, C^7, rather than only C^6.)

Exercises for §2.1.

#1. Show $\mathrm{rank}_{\mathbb{R}}(G_1 \times G_2) = \mathrm{rank}_{\mathbb{R}} G_1 + \mathrm{rank}_{\mathbb{R}} G_2$.

#2. Assume $\mathrm{rank}_{\mathbb{R}} G = 1$. Show X is irreducible if and only if X has no compact factors.

#3. Show that if X is reducible, then X is *not* two-point homogeneous. (Do not assume the fact about maximal flats that was mentioned, without proof, before Proposition 2.1.7.)

§2.2. Q-rank

Now let $\Gamma \backslash X$ be a locally symmetric space modeled on X, and assume that $\Gamma \backslash X$ has finite volume. Hence, Γ is a (torsion-free) discrete subgroup of G, such that $\Gamma \backslash G$ has finite volume; in short, Γ is a **lattice** in G.

The real rank depends only on X, so it is not affected by the choice of a particular lattice Γ. We now describe an analogous algebraically defined invariant, $\mathrm{rank}_{\mathbb{Q}} \Gamma$, that does depend on Γ, and therefore distinguishes between some of the various locally homogeneous spaces that are modeled on X. We will mention some of the geometric implications of $\mathbb{Q}$-rank, leaving a more detailed discussion to later chapters.

(2.2.1) **Theorem** (see Subsections 19.3(iii) and 19.3(iv)).

1) $\mathrm{rank}_{\mathbb{Q}} \Gamma$ *is the largest natural number r, such that some finite cover of $\Gamma \backslash X$ contains a closed, simply connected, r-dimensional flat.*

2) $\mathrm{rank}_{\mathbb{Q}} \Gamma$ *is the smallest natural number r, for which there exists collection of finitely many closed, r-dimensional flats, such that all of $\Gamma \backslash X$ is within a bounded distance of the union of these flats.*

(2.2.2) *Remark.* It is clear from Theorem 2.2.1(1) that $\mathrm{rank}_{\mathbb{Q}} \Gamma$ always exists (and is finite). Furthermore, $0 \leq \mathrm{rank}_{\mathbb{Q}} \Gamma \leq \mathrm{rank}_{\mathbb{R}} G$. Although not so obvious, it can be shown that the extreme values are always attained: there are lattices Γ_c and Γ_s in G with $\mathrm{rank}_{\mathbb{Q}} \Gamma_c = 0$ and $\mathrm{rank}_{\mathbb{Q}} \Gamma_s = \mathrm{rank}_{\mathbb{R}} G$ (see Theorem 18.7.1 and Exercise 9.1#7). So it is perhaps surprising that there may be gaps in between. (For example, if $G \cong \mathrm{SO}(2, n)$, with $n \geq 5$, and n is odd, then $\mathrm{rank}_{\mathbb{R}}\, G = 2$, but Corollary 18.6.2 shows there does not exist a lattice Γ in G, such that $\mathrm{rank}_{\mathbb{Q}}\, \Gamma = 1$.)

(2.2.3) **Example** (see Example 9.1.5). From the algebraic definition, which will appear in Chapter 9, it is easy to calculate
$$\mathrm{rank}_{\mathbb{Q}}(\mathrm{SO}(m, n)_{\mathbb{Z}}) = \min\{m, n\} = \mathrm{rank}_{\mathbb{R}}(\mathrm{SO}(m, n))$$
and
$$\mathrm{rank}_{\mathbb{Q}}(\mathrm{SL}(n, \mathbb{Z})) = n - 1 = \mathrm{rank}_{\mathbb{R}}(\mathrm{SL}(n, \mathbb{R})).$$

As for the real rank, the biggest difference is between spaces where the invariant is zero and those where it is nonzero, because this is again the distinction between a compact space and a noncompact one:

(2.2.4) **Theorem** (see Exercise 9.1#5). $\mathrm{rank}_{\mathbb{Q}}\, \Gamma = 0$ *iff* $\Gamma \backslash X$ *is compact.*

Theorem 2.2.1(2) implies that the $\mathbb{Q}$-rank of Γ is directly reflected in the large-scale geometry of $\Gamma \backslash X$, as described by the asymptotic cone of $\Gamma \backslash X$. Intuitively, the asymptotic cone of a metric space is obtained by looking at it from a large distance. For example, if $\Gamma \backslash X$ is compact, then, as we move farther away, the manifold appears smaller and smaller (see the illustration below). In the limit, the manifold shrinks to a point.

$$(2.2.5)$$

An intuitive understanding is entirely sufficient for our purposes here, but, for the interested reader, we provide a more formal definition.

(2.2.6) **Definition.** The *asymptotic cone* of a metric space (M, d) is the limit space
$$\lim_{\epsilon \to 0^+} ((M, \epsilon d), p),$$
if the limit exists. Here, p is an arbitrary (but fixed!) point of M, and the limit is with respect to Gromov's Hausdorff distance. (Roughly speaking, a large ball around p in $(M, \epsilon d)$ is δ-close to being isometric to a large ball around a certain (fixed) point p_0 in the limit space (M_0, d_0).)

(2.2.7) **Examples.**
1) If $\Gamma \backslash X$ is compact, then the asymptotic cone of $\Gamma \backslash X$ is a point, as is illustrated in (2.2.5). This point is a 0-dimensional simplicial complex, which is a geometric manifestation of the fact that $\mathrm{rank}_{\mathbb{Q}}\, \Gamma = 0$.
2) If $\mathrm{rank}_{\mathbb{R}}\, G = 1$, and $\Gamma \backslash X$ is not compact, then, as is well known, $\Gamma \backslash X$ has finitely many cusps. The asymptotic cone of a cusp is a ray, so the asymptotic cone of $\Gamma \backslash X$ is a "star" of finitely many rays emanating from a single vertex (see Figure 2.2A). Therefore,

the asymptotic cone of $\Gamma \backslash X$ is a 1-dimensional simplicial complex. This manifests the fact that $\mathrm{rank}_{\mathbb{Q}}\, \Gamma = 1$.

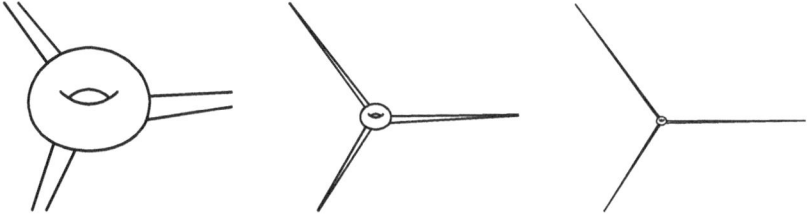

FIGURE 2.2A. Looking at a manifold with cusps from farther and farther away.

(2.2.8) **Theorem** (see Remark 19.3.9). *The asymptotic cone of $\Gamma \backslash X$ is a simplicial complex whose dimension is $\mathrm{rank}_{\mathbb{Q}}\, \Gamma$.*

(2.2.9) **Example.** Let $G = \mathrm{SL}(3,\mathbb{R})$ and $\Gamma = \mathrm{SL}(3,\mathbb{Z})$. From Theorem 2.2.8, we see that the asymptotic cone of $\Gamma \backslash G/K$ is a 2-dimensional simplicial complex. In fact, it turns out to be (isometric to) the sector

$$\left\{ (x,y) \in \mathbb{R}^2 \;\middle|\; 0 \le y \le \frac{\sqrt{3}}{2} x \right\}.$$

(It is not a coincidence that this sector is a Weyl chamber of the Lie algebra $\mathfrak{sl}(3,\mathbb{R})$.)

(2.2.10) *Remarks.*

1) If $\mathrm{rank}_{\mathbb{Q}}\, \Gamma = 1$, then the asymptotic cone of $\Gamma \backslash X$ is a star of finitely many rays emanating from the origin (cf. Example 2.2.7(2)). Note that this intersects the unit sphere in finitely many points.

2) In general, if $\mathrm{rank}_{\mathbb{Q}}\, \Gamma = k$, then the unit sphere contains a certain simplicial complex $\mathcal{T}_\Gamma$ of dimension $k-1$, such that the asymptotic cone of $\Gamma \backslash X$ is the union of all the rays emanating from the origin that pass through $\mathcal{T}_\Gamma$.

3) For $\Gamma = \mathrm{SL}(3,\mathbb{Z})$, the simplicial complex $\mathcal{T}_\Gamma$ is a single edge (cf. Example 2.2.9). In general, the Tits building $\mathcal{T}_G$ is a certain simplicial complex defined from the parabolic $\mathbb{Q}$-subgroups of G, and $\mathcal{T}_\Gamma$ can be obtained from $\mathcal{T}_G$ by modding out the action of Γ.

4) The asymptotic cone is also known as "tangent cone at infinity."

(2.2.11) *Remark.* Although we will not prove this, the $\mathbb{Q}$-rank is directly reflected in the cohomology of $\Gamma \backslash X$. Namely, let c be the cohomological dimension of $\Gamma \backslash X$. Because $\Gamma \backslash X$ is a manifold of dimension $\dim X$, we have $c = \dim X$ if and only if $\Gamma \backslash X$ is compact. So the deficiency $\dim X - c$ is, in some sense, a measure of how far $\Gamma \backslash X$ is from being compact. This measure is precisely $\mathrm{rank}_{\mathbb{Q}}\, \Gamma$ (if X has no compact factors).

Exercises for §2.2.

#1. Theorem 2.2.4 states that if $\Gamma \backslash X$ is compact, then $\mathrm{rank}_{\mathbb{Q}}\,\Gamma = 0$.
   a) Prove this directly from Theorem 2.2.1(1).
   b) Prove this directly from Theorem 2.2.1(2).

Notes

Helgason's book [4] has a thorough treatment of rank and ℝ-rank. Theorem 2.2.1(2) was proved by B. Weiss [8].

Theorem 2.2.4 was proved for arithmetic lattices by Borel and Harish-Chandra [1] and, independently, by Mostow and Tamagawa [7]. For non-arithmetic lattices, this theorem is part of the *definition* of ℚ-rank.

A more precise version of Theorem 2.2.8 (providing a description of the geometry of the simplicial complex) was proved by Hattori [3]. Proofs also appear in [5] and [6].

Remark 2.2.11 is due to Borel and Serre [2].

References

[1] A. Borel and Harish-Chandra: Arithmetic subgroups of algebraic groups, *Ann. Math.* (2) 75 (1962) 485–535. MR 0147566, http://dx.doi.org/10.2307/1970210

[2] A. Borel and J.-P. Serre: Corners and arithmetic groups, *Comment. Math. Helvetici* 48 (1973) 436–491. MR 0387495, http://dx.doi.org/10.5169/seals-37166

[3] T. Hattori: Asymptotic geometry of arithmetic quotients of symmetric spaces. *Math. Z.* 222 (1996) 247–277. MR 1429337, http://eudml.org/doc/174884

[4] S. Helgason: *Differential Geometry, Lie Groups, and Symmetric Spaces.* Academic Press, New York, 1978. MR 0514561

[5] L. Ji and R. MacPherson: Geometry of compactifications of locally symmetric spaces, *Ann. Inst. Fourier (Grenoble)* 52 (2002) 457–559. MR 1906482, http://eudml.org/doc/115986

[6] E. Leuzinger: Tits geometry, arithmetic groups, and the proof of a conjecture of Siegel, *J. Lie Theory* 14 (2004) 317–338. MR 2066859, http://www.emis.de/journals/JLT/vol.14_no.2/6.html

[7] G. D. Mostow and T. Tamagawa: On the compactness of arithmetically defined homogeneous spaces, *Ann. Math.* 76 (1962) 446–463. MR 0141672, http://dx.doi.org/10.2307/1970368

[8] B. Weiss: Divergent trajectories and ℚ-rank, *Israel J. Math.* 152 (2006), 221–227. MR 2214461, http://dx.doi.org/10.1007/BF02771984

Chapter 3

Brief Summary

This book is about arithmetic subgroups, and other lattices, in semisimple Lie groups. Given a lattice Γ in a semisimple Lie group G, we will investigate both the algebraic structure of Γ, and properties of the corresponding homogeneous space G/Γ. We will also study the close relationship between G and Γ. For example, we will see that G is essentially the only semisimple group in which Γ can be embedded as a lattice ("Mostow Rigidity Theorem"), and, conversely, we will usually be able to make a list of all the lattices in G ("Margulis Arithmeticity Theorem").

This chapter provides a very compressed outline of the material in this book. To help keep it brief, let us assume, for the remainder of the chapter, that

G is a noncompact, simple Lie group, and Γ is a lattice in G.

This means (see Definition 4.1.9):

- Γ is a discrete subgroup of G, and

- the homogeneous space G/Γ has finite volume (with respect to the Haar measure on G).

(If G/Γ is compact, which is a very important special case, we say Γ is *cocompact*.)

Main prerequisites for this chapter: none for most of the non-proof material.

Part I. Introduction

All three chapters in this part of the book are entirely optional; none of the material will be needed later (although some examples and remarks do refer back to it). Chapters 1 and 2 provide geometric motivation for the study of arithmetic groups, by explaining the connection with locally symmetric spaces. The present chapter (Chapter 3) is a highly condensed version of the entire book.

Part II. Fundamentals

This part of the book presents definitions and other foundational material for the study of arithmetic groups.

Chapter 4. Basic Properties of Lattices. This chapter presents a few important definitions, including the notions of *lattice subgroups*, *commensurable subgroups*, and *irreducible lattices*. It also proves a number of fundamental algebraic and geometric consequences of the assumption that Γ is a lattice, including the following.

(4.4.4) Recall that an element u of $\mathrm{SL}(n, \mathbb{R})$ is *unipotent* if its characteristic polynomial is $(x - 1)^n$ (or, in other words, its only eigenvalue is 1). If G/Γ is compact, then Γ does not have any nontrivial unipotent elements. This is proved by combining the Jacobson-Morosov Lemma (A5.8) with the observation that if a sequence $c_i\Gamma$ leaves all compact sets, and U is a precompact set in G, then, after passing to a subsequence, the sets $Uc_1\Gamma, Uc_2\Gamma, \ldots$ are all disjoint. However, G/Γ has finite volume, so it cannot have infinitely many disjoint open sets that all have the same volume.

(4.5#11) (*Borel Density Theorem*) Γ is not contained in any connected, proper, closed subgroup of G. Assuming that G/Γ is compact, the key to proving this is to note that if $\rho\colon G \to \mathrm{GL}(m, \mathbb{R})$ is any continuous homomorphism, u is any unipotent element of G, and $v \in \mathbb{R}^m$, then the coordinates of the vector $\rho(u^k)v$ are polynomial functions of k. However, if G/Γ is compact, and v happens to be $\rho(\Gamma)$-invariant, then the coordinates are all bounded. Since every bounded polynomial is constant, we conclude that every $\rho(\Gamma)$-invariant vector is $\rho(G)$-invariant. From this, the desired conclusion follows by looking at the action of G on exterior powers of its Lie algebra.

(4.7.10) Γ is finitely presented. When G/Γ is compact, this follows from the fact that the fundamental group of any compact manifold is finitely presented. For the noncompact case, it follows from the existence of a nice fundamental domain for the action of Γ on G (which will be explained in Chapter 19).

(4.8.2) (*Selberg's Lemma*) Γ has a torsion-free subgroup of finite index. For example, if $\Gamma = \mathrm{SL}(3, \mathbb{Z})$, then the desired torsion-free subgroup can be obtained by choosing any prime $p \geq 3$, and taking the matrices in Γ that are congruent to the identity matrix, modulo p.

(4.8#9) Γ is residually finite. For example, if $\Gamma = \mathrm{SL}(3, \mathbb{Z})$, then no nontrivial element of Γ is in the intersection of the finite-index subgroups used in the preceding paragraph's proof of Selberg's Lemma.

(4.9.2) (*Tits Alternative*) Γ contains a nonabelian free subgroup. This is proved by using the *Ping-Pong Lemma* (4.9.6), which, roughly speaking, states that if homeomorphism a contracts all of the space toward one point, and homeomorphism b contracts all of the space toward a different point, then the group generated by a and b is free.

(4.10.3) (*Moore Ergodicity Theorem*) If H is any noncompact, closed subgroup of G, then every real-valued, H-invariant, measurable function on G/Γ is constant (a.e.). The general case will be proved in Section 11.2, but suppose, for example, that $G = \mathrm{SL}(2, \mathbb{R})$, $H = \{a^s\}$ is the group of diagonal matrices, and f is an H-invariant function that, for simplicity, we assume is uniformly continuous. If we let $\{u^t\}$ be the group of upper-triangular matrices with 1's on the diagonal, then we have

$$u^t \cdot f = u^t a^s \cdot f = a^s u^{e^{-s}t} \cdot f \xrightarrow{s \to \infty} a^s u^0 \cdot f = f,$$

so f is invariant under $\{u^t\}$. Similarly, it is also invariant under the group of lower-triangular matrices. So f is G-invariant, and therefore constant.

Chapter 5. What is an Arithmetic Group? Roughly speaking, an *arithmetic subgroup* $G_{\mathbb{Z}}$ of G is obtained by embedding G in some $\mathrm{SL}(\ell, \mathbb{R})$, and taking the resulting set of integer points of G. That is, $G_{\mathbb{Z}}$ is the intersection of G with $\mathrm{SL}(\ell, \mathbb{Z})$. However, in order for $G_{\mathbb{Z}}$ to be called an arithmetic subgroup, the embedding $G \hookrightarrow \mathrm{SL}(\ell, \mathbb{Z})$ is required to satisfy a certain technical condition ("defined over $\mathbb{Q}$") (see Definition 5.1.2).

(5.1.11) Every arithmetic subgroup of G is a lattice in G. This fundamental fact will be proved in Chapters 7 and 19.

(5.2.1) (*Margulis Arithmeticity Theorem*) Conversely, if G is neither $\mathrm{SO}(1, n)$ nor $\mathrm{SU}(1, n)$, then every lattice in G is an arithmetic subgroup. Therefore, in most cases, "arithmetic subgroup" is synonymous with "lattice." This amazing theorem will be proved in Section 16.3.

It is a hugely important result. The definition of "lattice" is quite abstract, but a fairly explicit list of all the lattices in G can be obtained by combining this theorem with the classification of arithmetic subgroups that will be given in Chapter 18.

(5.3.1) (*Godement Compactness Criterion*) $G/G_{\mathbb{Z}}$ is compact if and only if the identity element is the only unipotent element of $G_{\mathbb{Z}}$. The direction ($\Rightarrow$) is very elementary and was proved in the previous chapter (see 4.4.4). The converse uses the same main idea, combined with the simple observation that if a polynomial has integer coefficients, and all of its roots are close to 1, then all of its roots are exactly equal to 1.

(5.5) The embedding of G in $\mathrm{SL}(\ell, \mathbb{R})$ is not at all unique, and different embeddings can yield quite different arithmetic subgroups $G_{\mathbb{Z}}$. One very important method of constructing non-obvious embeddings is called *Restriction of Scalars*. It starts by choosing a field F that is a finite extension of $\mathbb{Q}$. If we think of F as a vector space over $\mathbb{Q}$, then it can be identified with some $\mathbb{Q}^n$, in such a way that the ring $\mathcal{O}$ of algebraic integers of F is identified with $\mathbb{Z}^n$. This implies that the group $G_{\mathcal{O}}$ is isomorphic to $G'_{\mathbb{Z}}$, where G' is a semisimple group that has G as one of its factors. Therefore, this method allows arithmetic subgroups to be constructed not only from ordinary integers, but also from algebraic integers.

Chapter 6. Examples of Arithmetic Groups. This chapter explains how to construct many arithmetic subgroups of $\mathrm{SL}(2, \mathbb{R})$, $\mathrm{SO}(1, n)$, and $\mathrm{SL}(n, \mathbb{R})$, by using unitary groups and quaternion algebras (and other division algebras). (Restriction of scalars is also used for some of the cocompact ones.) It will be proved in Chapter 18 that these fairly simple constructions actually produce all of the arithmetic subgroups of these groups.

(6.5) There exist non-arithmetic lattices in $\mathrm{SO}(1, n)$ for every n. This was proved by M. Gromov and I. Piatetski-Shapiro. They "glued together" two arithmetic lattices to create a "hybrid" lattice that is not arithmetic.

Chapter 7. $\mathrm{SL}(n, \mathbb{Z})$ is a lattice in $\mathrm{SL}(n, \mathbb{R})$. This chapter explains two different proofs of the fundamental fact (already mentioned in Theorem 5.1.11) that $G_{\mathbb{Z}}$ is a lattice in G, in the illustrative special case where $G = \mathrm{SL}(n, \mathbb{R})$ and $G_{\mathbb{Z}} = \mathrm{SL}(n, \mathbb{Z})$.

The first proof is quite short and elementary, and is presented fairly completely. It constructs a nice set that is (approximately) a fundamental domain for the action of Γ on G. The key notion is that of a *Siegel set*. We begin with the Iwasawa decomposition $G = KAN$.

- $K = \mathrm{SO}(n)$ is a maximal compact subgroup of G.
- The group A of diagonal matrices in G is isomorphic to $\mathbb{R}^{n-1}$, so we can think of it as a real vector space. Under this identification, the "simple roots" are linear functionals $\alpha_1, \ldots, \alpha_{n-1}$ on A. Choose any $t \in \mathbb{R}$, and let
$$A_t = \{ a \in A \mid \alpha_i(a) \geq t \text{ for all } i \},$$
so A_t is a polyhedral cone in A.
- N is the group of upper-triangular matrices with 1's on the diagonal.

- Choose any compact subset N_0 of N.

Then the product $\mathfrak{S} = N_0 \, A_t \, K$ is a Siegel set (see Section 7.2). It depends on the choice of t and N_0.

A straightforward calculation shows that every Siegel set has finite volume (see Proposition 7.2.5). It is also not terribly difficult to find a Siegel set $\mathfrak{S}$ with the property that $G_{\mathbb{Z}} \cdot \mathfrak{S} = G$ (see Theorem 7.3.1). This implies that $G/G_{\mathbb{Z}}$ has finite volume, so $G_{\mathbb{Z}}$ is a lattice in G, or, in other words, $\mathrm{SL}(n, \mathbb{Z})$ is a lattice in $\mathrm{SL}(n, \mathbb{R})$.

Unfortunately, some difficulties arise when generalizing this method to other groups, because it is more difficult to use Siegel sets to construct an appropriate fundamental domain in the general case. The main ideas will be explained in Chapter 19.

So we also present a different proof that is much easier to generalize (see Section 7.4). Namely, the general case is quite easy to prove if one accepts the following key fact that was proved by Margulis: If

- u^t is any unipotent 1-parameter subgroup of $\mathrm{SL}(n, \mathbb{R})$, and

- $x \in \mathrm{SL}(n, \mathbb{R})/\mathrm{SL}(n, \mathbb{Z})$,

then there is a compact subset C of $\mathrm{SL}(n, \mathbb{R})/\mathrm{SL}(n, \mathbb{Z})$, and some $\epsilon > 0$, such that at least $\epsilon\%$ of the orbit $\{u^t x\}_{t \in \mathbb{R}}$ is in the set C (see Theorem 7.4.7).

Part III. Important Concepts

This part of the book explores several fundamental ideas that are important not only for their applications to arithmetic groups, but much more generally.

Chapter 8. Real rank. This chapter defines the real rank of G, which is an important invariant in the study of semisimple Lie groups. It also describes some consequences of assuming that the real rank is at least two, and presents the definition and basic structure of the minimal parabolic subgroups of G.

Chapter 9. $\mathbb{Q}$-rank. This chapter, unlike the others in this part of the book, discusses a topic that is primarily of interest in the theory of arithmetic groups (and related algebraic groups). Largely parallel to Chapter 8, it defines the $\mathbb{Q}$-rank of Γ, describes some consequences of assuming that the $\mathbb{Q}$-rank is at least two, and presents the definition and basic structure of the minimal parabolic $\mathbb{Q}$-subgroups of G.

Chapter 10. Quasi-isometries. Any finite generating set S for Γ yields a metric d_S on Γ: the distance from x to y is the minimal number of elements of S that need to be multiplied together to obtain $x^{-1}y$. Unfortunately, this "word metric" is not canonical, because it depends on the choice of the generating set S. However, it is well-defined up to a bounded factor, so, to get a geometric object that is uniquely determined by Γ, we consider two metric spaces to be equivalent (or *quasi-isometric*) if there is a map between them that only distorts distances by a bounded factor (see Definition 10.1.3).

(10.1.7) Some quasi-isometries arise from cocompact actions: it is not difficult to see that if Γ acts cocompactly, by isometries on a (nice) space X, then there is a quasi-isometry from Γ to X. Thus, for example, any cocompact lattice in $\mathrm{SO}(1, n)$ is quasi-isometric to the hyperbolic space $\mathfrak{H}^n$.

(10.2) Γ is *Gromov hyperbolic* if and only if $\mathrm{rank}_{\mathbb{R}} G = 1$ and Γ is compact, except that all lattices in $\mathrm{SL}(2, \mathbb{R})$ are hyperbolic, not only the cocompact ones. One direction is a consequence of the well-known fact that $\mathbb{Z} \times \mathbb{Z}$ is not contained in any hyperbolic group. The other direction (for the cocompact case) is a special case of the fact that the fundamental group of any closed manifold of strictly negative sectional curvature is hyperbolic.

Chapter 11. Unitary representations. This chapter presents some basic concepts in the theory of unitary representations, the study of group actions on Hilbert spaces. The Moore Ergodicity Theorem (4.10.3) is proved in Section 11.2, and the "induced representations" defined in Section 11.3 will be used in Section 13.4 to prove that Γ has Kazhdan's Property (T) if $\mathrm{rank}_{\mathbb{R}} G \geq 2$.

(11.2.2) (*Decay of matrix coefficients*) If π is a continuous homomorphism from G to the unitary group of a Hilbert space $\mathcal{H}$, then

$$\lim_{\|g\| \to \infty} \langle \pi(g)\phi \mid \psi \rangle = 0, \text{ for all } \phi, \psi \in \mathcal{H}.$$

This yields the Moore Ergodicity Theorem (4.10.3) as an easy corollary, and the proof is based on the existence of $a \in G$ and (unipotent) subgroups U^+ and U^- of G, such that $\langle U^+, U^- \rangle = G$ and $a^n u a^{-n} \to e$ as $n \to \infty$ (or $-\infty$), for all $u \in U^+$ (or U^-, respectively).

(11.4.2) Every unitary representation of any compact Lie group is a direct sum of finite-dimensional, irreducible unitary representations.

(11.5.3) Every unitary representation of any abelian Lie group is a direct integral of one-dimensional unitary representations.

(11.6.4) Every unitary representation of G is a direct integral of irreducible unitary representations.

Chapter 12. Amenable Groups. Amenability is such a fundamental notion that it has very many quite different definitions, all of which determine exactly the same class of groups (see Definition 12.1.3 and Theorem 12.3.1). One useful choice is that a group Λ is *amenable* if every continuous action of Λ on a compact, metric space has a finite, invariant measure.

(12.4.2) The fact that the lattice Γ contains a nonabelian free subgroup (see Corollary 4.9.2) implies that it is not amenable. This is because subgroups of amenable groups are amenable (see Proposition 12.2.8), and free groups do have actions (such as the actions described in the Ping-Pong Lemma (4.9.6)) that do not have a finite, invariant measure.

Even so, amenability plays an important role in the study of Γ, through the following observation:

(12.6.1) (*Furstenberg Lemma*) If P is an amenable subgroup of G, and we have a continuous action of Γ on some compact, metric space X, then there exists a measurable, Γ-equivariant map from G/P to the space $\mathrm{Prob}(X)$ of measures μ on X, such that $\mu(X) = 1$. To prove this, let $\mathcal{F}$ be the set of measurable, Γ-equivariant maps from G to $\mathrm{Prob}(X)$. With an appropriate weak topology, this is a compact, metrizable space, and P acts on it by translation on the right. Since P is amenable, there is a P-invariant, finite measure μ on $\mathcal{F}$. The barycenter of this measure is a fixed point of P in $\mathcal{F}$, and this fixed point is a function on G that factors through to a well-defined Γ-equivariant map from G/P to $\mathrm{Prob}(X)$.

Chapter 13. Kazhdan's Property (T). To say Γ has Kazhdan's property (T) means that if a unitary representation of Γ does not have any (nonzero) vectors that are fixed by Γ, then it does not *almost-invariant vectors*, that is, vectors that are moved only a small distance by the elements of any given finite subset of Γ (see Definition 13.1.1).

(13.1.5) Kazhdan's property (T) is, in a certain sense, the antithesis of amenability: a discrete group cannot have both properties unless it is finite. This is because the regular representation of any amenable group has almost-invariant vectors.

(13.1.7) Every discrete group with Kazhdan's property (T) is finitely generated. To see this, let $\mathcal{H} = \bigoplus_F \mathcal{L}^2(\Lambda/F)$, where F ranges over all the finitely generated subgroups of Λ. Then, by construction, every finite subset of Λ fixes some nonzero vector in $\mathcal{H}$.

(13.2.4) G has Kazhdan's property (T), unless G is either $\mathrm{SO}(1,n)$ or $\mathrm{SU}(1,n)$. To prove this for $G = \mathrm{SL}(3,\mathbb{R})$, first note that the semidirect product $\mathrm{SL}(2,\mathbb{R}) \ltimes \mathbb{R}^2$ can be embedded in G. Also note that there are elements a and b of $\mathrm{SL}(2,\mathbb{R})$, such that, if Q is any of the 4 quadrants of $\mathbb{R}^2$, then either aQ or bQ is disjoint from Q (except for the 0 vector).

Applying this to the Pontryagin dual of $\mathbb{R}^2$ implies that if a representation of the semidirect product $SL(2,\mathbb{R}) \ltimes \mathbb{R}^2$ has almost-invariant vectors, then it must have a nonzero vector that is invariant under $\mathbb{R}^2$. This vector must be invariant under all of $SL(3,\mathbb{R})$, by a generalization of the Moore Ergodicity Theorem that is called the *Mautner phenomenon* (11.2.8).

(13.4.1) Γ has Kazhdan's property (T), unless G is either $SO(1,n)$ or $SU(1,n)$. Any unitary representation π of Γ can be "induced" to a representation π_Γ^G of G. If π has almost-invariant vectors, then the induced representation has almost-invariant vectors, and, since G has Kazhdan's property (T), this implies that π_Γ^G has G-invariant vectors. Any such vector must come from a Γ-invariant vector in π.

(13.5.4) A group has Kazhdan's property (T) if and only if every action of the group by (affine) isometries on any Hilbert space has a fixed point. This is not at all obvious, but here is the proof of one direction.

Suppose Γ does not have Kazhdan's property (T), so there exists a unitary representation of Γ on some Hilbert space $\mathcal{H}$ that has almost-invariant vectors, but does not have invariant vectors. Choose an increasing chain $F_1 \subseteq F_2 \subseteq \cdots$ of finite subsets whose union is all of Γ. Since $\mathcal{H}$ has almost-invariant vectors, there exists a unit vector $v_n \in \mathcal{H}$, such that $\|f v_n - v_n\| < 1/2^n$ for all $f \in F_n$. Now, define $\alpha \colon \Gamma \to \mathcal{H}^\infty$ by

$$\alpha(g)_n = n(g v_n - v_n).$$

Then α is a 1-cocycle, so defining $g * v = gv + \alpha(g)$ yields an action of Γ on the Hilbert space $\mathcal{H}^\infty$. Since $\mathcal{H}$ has no nonzero invariant vectors, it is not difficult to see that α is an unbounded function on Γ, so α is not a coboundary. This implies that the corresponding action on $\mathcal{H}^\infty$ has no fixed points.

Chapter 14. Ergodic Theory. *Ergodic Theory* can be defined as the measure-theoretic study of group actions. In this category, the analogue of the transitive actions are the so-called *ergodic* actions, for which every measurable, invariant function is constant (a.e.) (see Definition 14.2.1).

(14.3.2) (*Pointwise Ergodic Theorem*) If $\mathbb{Z}$ acts ergodically on X, with finite invariant measure, and f is any L^1-function on X, then the average of f on almost every $\mathbb{Z}$-orbit is equal to the average of f on the entire space X.

(14.4.3) Every measure-preserving action of G can be measurably decomposed into a union of ergodic actions.

(14.5.10) If the action of G on a space X is ergodic, with a finite, invariant measure, then the action of G on $X \times X$ is also ergodic.

Part IV. Major Results

Here are some of the major theorems in the theory of arithmetic groups.

Chapter 15. Mostow Rigidity Theorem.

(15.1.2) (*Mostow Rigidity Theorem*) Suppose Γ_i is a lattice in G_i, for $i = 1, 2$, and $\varphi: \Gamma_1 \to \Gamma_2$. If G_i has trivial center and no compact factors, and is not PSL$(2, \mathbb{R})$, then φ extends to an isomorphism $\overline{\varphi}: G_1 \to G_2$.

In most cases, the desired conclusion is a consequence of the Margulis Superrigidity Theorem, which will be discussed in Chapter 16. However, a different proof is needed when $G_1 = G_2 = \text{SO}(1, n)$ (and some other cases). Assuming that the lattices are cocompact, the proof uses the fact (mentioned in Proposition 10.1.7) that Γ_1 and Γ_2 are quasi-isometric to $\mathfrak{H}^n$. Comparing the two embeddings yields a quasi-isometry φ from $\mathfrak{H}^n$ to itself. By proving that this quasi-isometry induces a map on the boundary that is conformal (i.e., preserves angles), it is shown that the two embeddings are conjugate by an isometry of $\mathfrak{H}^n$.

(15.3.6) Mostow's theorem does not apply to PSL$(2, \mathbb{R})$: in this group, there are uncountably many lattices that are isomorphic to each other, but are not conjugate. This follows from the fact that there are uncountably many different right-angled hexagons in the hyperbolic plane $\mathfrak{H}^2$. A compact surface of genus g can be constructed by gluing $4g - 4$ of these hexagons together, in such a way that the fundamental group is a cocompact lattice in PSL$(2, \mathbb{R})$. The uncountably many different hexagons yield uncountably many non-conjugate lattices.

(15.4.1) From Mostow's Theorem, we know that lattices in two different groups G_1 and G_2 cannot be isomorphic. In fact, the lattices cannot even be quasi-isometric. Some ideas in the proof of this fact are similar to the argument of Mostow's theorem, but we omit the details.

Chapter 16. Margulis Superrigidity Theorem.

(16.1) (*Margulis Superrigidity Theorem*) Suppose $\rho: \Gamma \to \text{GL}(n, \mathbb{R})$ is a homomorphism. If G is neither SO$(1, n)$ nor SU$(1, n)$, and mild hypotheses are satisfied, then ρ extends to a homomorphism $\overline{\rho}: G \to \text{GL}(n, \mathbb{R})$.

Assuming $\text{rank}_{\mathbb{R}} G \geq 2$, a proof is presented in Section 16.5. Start by letting H be the Zariski closure of $\rho(\Gamma)$, and let Q be a parabolic subgroup of H. Furstenberg's Lemma (12.6.1) provides a Γ-equivariant map $\psi: G/P \to \text{Prob}(\mathbb{RP}^n)$. By using "proximality," ψ can be promoted to a map $\hat{\psi}: G/A \to \mathbb{R}^n$ (where A is a maximal $\mathbb{R}$-split torus of G). Thus, we have an A-invariant (measurable) section of the flat vector bundle over G/Γ that is associated to φ. Since G is generated by the centralizers of nontrivial, connected subgroups of A, this implies there is a finite-dimensional, G-invariant space of sections of the bundle, from which it follows that φ has the desired extension to a homomorphism defined on all of G.

(16.2.1) This theorem of Margulis is a strengthening of the Mostow Rigidity Theorem (15.1.2), because the homomorphism ρ is not required to be an isomorphism. (On the other hand, Mostow's theorem applies to the groups $SO(1, n)$ and $SU(1, n)$, which are not allowed in the superrigidity theorem.)

(16.2.3) In geometric terms, the superrigidity theorem implies (under mild hypotheses) that flat vector bundles over G/Γ become trivial on a finite cover.

(16.3) (*Margulis Arithmeticity Theorem*) If G is neither $SO(1, n)$ nor $SU(1, n)$, then the superrigidity theorem implies that every lattice in G is an arithmetic subgroup (as was stated without proof in Theorem 5.2.1).

The basic idea of the proof is that if there is some $\rho(\gamma)$ with a matrix entry that is transcendental, then composing ρ with arbitrary elements of the Galois group $\mathrm{Gal}(\mathbb{C}/\mathbb{Q})$ would result in uncountably many different n-dimensional representations of Γ. Since G has only finitely many representations of each dimension, this would contradict superrigidity. Thus, we conclude that $\rho(\Gamma) \subseteq GL(n, \overline{\mathbb{Q}})$. By using a p-adic version of the superrigidity theorem, $\overline{\mathbb{Q}}$ can be replaced with $\mathbb{Z}$.

(16.8) For groups of real rank one, the proof of superrigidity described in Section 16.5 does not apply, because A does not have any nontrivial, proper subgroups. Instead, a more geometric approach is used (but only a brief sketch will be provided). Let X and Y be the symmetric spaces associated to G and H, respectively, where H is the Zariski closure of $\rho(\Gamma)$. By minimizing a certain energy functional, one can show there is a harmonic Γ-equivariant map $\psi \colon X \to Y$. Then, by using the geometry of X and Y, it can be shown that this harmonic map must be a totally geodesic embedding. This provides an embedding of the isometry group of X in the isometry group of Y. In other words, an embedding of G in H.

Chapter 17. Normal Subgroups of Γ.

(17.1.1) If $\mathrm{rank}_{\mathbb{R}} G \geq 2$, then Γ is almost simple. More precisely, every normal subgroup of Γ either is finite, or has finite index. This is proved by showing that if N is any infinite, normal subgroup of Γ, then the quotient Γ/N is amenable. Since Γ/N has Kazhdan's property (T) (because we saw in Proposition 13.4.1 that Γ has this property), this implies Γ/N is finite.

(17.2.1) On the other hand, if $\mathrm{rank}_{\mathbb{R}} G = 1$, then Γ is very far from being simple — there are many, many infinite normal subgroups of Γ. In fact, Γ is "SQ-universal," which means that if Λ is any finitely generated group, then there is a normal subgroup N of Γ, such that Λ is isomorphic to a subgroup of Γ/N (see Theorem 17.2.5).

Chapter 18. Arithmetic Subgroups of Classical Groups. The main result of this chapter is the table on page 380 that provides a list of all of the arithmetic subgroups of G (unless G is either an exceptional group or a group whose complexification $G_{\mathbb{C}}$ is isogenous to $SO(8, \mathbb{C})$). Inspection of the list establishes several results that were stated without proof in previous chapters.

(18.4) It was stated without proof in Section 6.8 that every arithmetic subgroup of $SL(n, \mathbb{R})$ is either a special linear group or a unitary group (if we allow division algebras in the construction). The proof of this fact is based on a calculation of the group cohomology of Galois groups (or *Galois cohomology*, for short). To introduce this method in a simpler setting, it is first proved that the only $\mathbb{R}$-forms of the complex Lie group $SL(n, \mathbb{C})$ are $SL(n, \mathbb{R})$, $SL(n/2, \mathbb{H})$, and $SU(k, \ell)$ (see Section 18.3).

(18.5) The same methods show that all the $\mathbb{Q}$-forms of any classical group G are classical groups (except that there is a problem when G is a real form of $SO(8, \mathbb{C})$ (see Remark 18.5.10)). However, we do not provide the calculations.

(18.7.4) We say that a semisimple group $H = G_1 \times \cdots \times G_r$ is *isotypic* if all the simple factors of $H_{\mathbb{C}}$ are isogenous to each other. A theorem of Borel and Harder (18.7.3) on Galois cohomology implies that if H is isotypic, then it has an arithmetic subgroup that is *irreducible*: it is not commensurable to a nontrivial direct product $\Gamma_1 \times \Gamma_2$. (The converse follows from the Margulis Arithmeticity Theorem unless H is either $SO(1, n) \times K$ or $SU(1, n) \times K$.)

Chapter 19. Construction of a Coarse Fundamental Domain. This chapter presents some of the main ideas involved in the construction of a nice subset of G that approximates a fundamental domain for G/Γ (when Γ is an arithmetic subgroup). This generalizes the construction for $\Gamma = SL(n, \mathbb{Z})$ that was explained in Chapter 7.

As in Chapter 7, the key notion is that of a *Siegel set*. The main difference is that, instead of the maximal $\mathbb{R}$-split torus A, we must work with a subtorus T of A that is $\mathbb{Q}$-split, not merely $\mathbb{R}$-split:

- K is a maximal compact subgroup of G (same as before),
- S is a maximal $\mathbb{Q}$-split torus in G,
- S_t is a sector in S,
- P is a minimal parabolic $\mathbb{Q}$-subgroup of G that contains S, and
- P_0 is a compact subset of P.

Then KS_tP_0 is a *Siegel set* for Γ in G (see Definition 19.1.2).

It may not be possible to find a Siegel set $\mathfrak{S}$, such that $\mathfrak{S} \cdot \Gamma = G$ (see Example 19.2.1). (When $\dim S = 1$, this is because each Siegel set can only cover one cusp, and G/Γ may have several cusps.) However, there is always a finite union of (translates of) Siegel sets that will suffice (see Theorem 19.2.2).

The existence of a nice set $\mathcal{F}$, such that $\mathcal{F} \cdot \Gamma = G$, has important consequences, such as the fact that Γ is finitely presented (see Subsection 19.3(i)). This fact was stated in Theorem 4.7.10, but could only be proved for the cocompact case there.

Chapter 20. Ratner's Theorems on Unipotent Flows. If $\{a^t\}$ is any 1-parameter group of diagonal matrices in G, then there are $\{a^t\}$-orbits in G/Γ that have bad closures: the closure is a fractal. M. Ratner proved that if a subgroup V is generated by 1-parameter unipotent subgroups, then it is much better behaved: the closure of every V-orbit is a C^∞ submanifold of G/Γ (see Theorem 20.1.3).

This theorem has important consequences in geometry and number theory. As a sample application in the theory of arithmetic groups, we mention that if Γ_1 and Γ_2 are any two lattices in G, then the subset $\Gamma_1 \Gamma_2$ of G is either discrete or dense (see Corollary 20.2.6). This is proved by letting $\Gamma = \Gamma_1 \times \Gamma_2$ in $G \times G$, and letting V be the diagonal embedding of G in the same group.

Ratner proved that the actions of 1-parameter unipotent subgroups on G/Γ also have nice measurable properties: every finite, invariant probability measure is the Haar measure on a closed orbit of some subgroup of G (see Theorem 20.3.4), and every dense orbit is uniformly distributed (see Theorem 20.3.3).

We will not prove Ratner's theorems, but some of the ideas in the proof will be described. One of the main ingredients is called "shearing" (see Section 20.4). For example, suppose $G = \mathrm{SL}(2, \mathbb{R})$ and $V = \{u^t\}$ is a 1-parameter unipotent subgroup. Then the key point is that if x and y are two nearby points in G/Γ (and are not on the same $\{u^t\}$-orbit), then the fastest relative motion between the two points is along the V-orbits. More precisely, there is some t, such that $u^t x$ is close to either $u^{t+1} y$ or $u^{t-1} y$.

Appendices

The main text is followed by three appendices. The first two (appendices A and B) recall some facts that are used in the main text. The third (Appendix C) defines the notion of *S-arithmetic group*, and quickly summarizes how the results on arithmetic groups extend to this more general setting.

Part II

Fundamentals

Chapter 4

Basic Properties of Lattices

This book is about lattices in semisimple Lie groups (with emphasis on the "arithmetic" ones).

> (4.0.0) **Standing Assumptions.** Throughout this book:
> 1) *G* **is a linear, semisimple Lie group** (see Appendix A1 for an explanation of these terms), with only finitely many connected components, and
> 2) Γ **is a lattice in** *G* (see Definition 4.1.9).
>
> Similar restrictions apply to the symbols G_1, G_2, G', Γ_1, Γ_2, Γ', etc.

(4.0.1) *Remark.* Without losing any of the main ideas, it may be assumed, throughout, that *G* is either $\mathrm{SL}(n, \mathbb{R})$ or $\mathrm{SO}(m, n)$ (or a product of these groups), but it is best if the reader is also acquainted with the other "classical groups," such as unitary groups and symplectic groups (see Definition A2.1).

Three definitions in this chapter are very important: lattice subgroups (4.1.9), commensurable subgroups (4.2.1), and irreducible lattices

Main prerequisites for this chapter: none.

(4.3.1 and 4.3.3). The rest of the material in this chapter may not be essential for a first reading, and can be referred back to when necessary. However, if the reader has no prior experience with lattices, then the basic properties discussed in Section 4.1 will probably be helpful.

§4.1. Definition

(4.1.1) Lemma. *If Λ is a discrete subgroup of G, then there is a **strict fundamental domain** for G/Λ in G. That is, there is a Borel subset $\mathcal{F}$ of G, such that the natural map $\mathcal{F} \to G/\Lambda$, defined by $g \mapsto g\Lambda$, is bijective.*

Proof. Since Λ is discrete, there is a nonempty, open subset U of G, such that $(U^{-1} U) \cap \Lambda = \{e\}$. Since G is second countable (or, if you prefer, since G is σ-compact), there is a sequence $\{g_n\}$ of elements of G, such that $\bigcup_{n=1}^{\infty} g_n U = G$. Let

$$\mathcal{F} = \bigcup_{n=1}^{\infty} \left(g_n U \smallsetminus \bigcup_{i<n} g_i U \Lambda \right).$$

Then $\mathcal{F}$ is obviously Borel, and it is a strict fundamental domain for G/Λ (see Exercise 2). □

(4.1.2) *Remark.*

1) The above lemma is stated for the space G/Λ of left cosets of Λ, but, in some situations, it is more natural to work with the space $\Lambda\backslash G$ of right cosets. In this book, we will feel free to use whichever is most convenient at a particular time, and leave it to the reader to translate between the two, by using the fact that the function $g\Lambda \mapsto \Lambda g^{-1}$ is a homeomorphism from G/Λ to $\Lambda\backslash G$ (see Exercises 3 and 4). Our choice will usually be determined by the preference for most mathematicians to write their actions on the left. (Therefore, if G is acting, then we will tend to use G/Λ, but if we are thinking of Λ as acting on G, then we usually consider the quotient $\Lambda\backslash G$.)

2) Definitions in the literature vary somewhat, but saying that a subset $\mathcal{F}$ of G is a **fundamental domain** for G/Λ typically means:
   (a) $\mathcal{F}\Lambda = G$,
   (b) $\mathcal{F}$ is a closed set that is nice: its interior $\mathring{\mathcal{F}}$ is dense in $\mathcal{F}$, and its boundary $\mathcal{F} \smallsetminus \mathring{\mathcal{F}}$ has measure 0, and
   (c) $\mathcal{F}\lambda \cap \mathring{\mathcal{F}} = \varnothing$, for all nonidentity $\lambda \in \Lambda$.
   It is not difficult to see that if $\mathcal{F}$ is a fundamental domain, then it has a Borel subset $\mathcal{F}'$, such that $\mathcal{F}'$ is a strict fundamental domain and $\mathcal{F} \smallsetminus \mathcal{F}'$ has measure 0. This means that, for many purposes (such as calculating integrals), it suffices to have a fundamental domain, rather than finding a set that is precisely a strict fundamental domain.

(4.1.3) **Proposition.** *Let Λ be a discrete subgroup of G, and let μ be Haar measure on G. There is a unique (up to a scalar multiple) σ-finite, G-invariant Borel measure ν on G/Λ. More precisely:*

1) *For any strict fundamental domain $\mathcal{F}$, the measure ν can be defined by*

$$\nu(A/\Lambda) = \mu(A \cap \mathcal{F}), \tag{4.1.4}$$

for every Borel set A in G, such that $A\Lambda = A$.

2) *Conversely, for $A \subseteq G$, we have*

$$\mu(A) = \int_{G/\Lambda} \#(A \cap x\Lambda) \, d\nu(x\Lambda). \tag{4.1.5}$$

Proof. See Exercises 7 and 8 for (1) and (2). The uniqueness of ν follows from (2) and the uniqueness of the Haar measure μ. $\qquad\square$

(4.1.6) *Remark.* We always assume that the G-invariant measure ν on G/Λ is normalized so that (4.1.4) and (4.1.5) hold.

(4.1.7) **Corollary.** *Let Λ be a discrete subgroup of G, and let $\phi\colon G \to G/\Lambda$ be the natural quotient map $\phi(g) = g\Lambda$. If A is a Borel subset of G, such that the restriction $\phi|_A$ is injective, then $\nu(\phi(A)) = \mu(A)$.*

(4.1.8) *Remarks.*

1) The Haar measure μ on G is given by a smooth volume form, so the associated measure ν on G/Λ is also given by a volume form. Therefore, we say that G/Λ has **finite volume** if $\nu(G/\Lambda) < \infty$.

2) The assumption that Λ is discrete cannot be eliminated from Proposition 4.1.3. However, a G-invariant measure on G/Λ can be constructed under the weaker assumption that Λ is closed and unimodular (see Exercise 9).

(4.1.9) **Definition.** A subgroup Γ of G is a **lattice** in G if

- Γ is a discrete subgroup of G, and
- G/Γ has finite volume.

(4.1.10) *Remark.* The definition is not vacuous: we will explain in Corollary 5.1.16 that G does have at least one lattice (in fact, infinitely many), although part of the proof will be postponed to Chapter 7.

(4.1.11) **Proposition.** *Let Λ be a discrete subgroup of G, and let μ be Haar measure on G. The following are equivalent:*

1) *Λ is a lattice in G.*

2) *There is a strict fundamental domain $\mathcal{F}$ for G/Λ, with $\mu(\mathcal{F}) < \infty$.*

3) *There is a strict fundamental domain $\mathcal{F}'$ for $\Lambda\backslash G$, with $\mu(\mathcal{F}') < \infty$.*

4) *There is a Borel subset C of G, such that $C\Lambda = G$ and $\mu(C) < \infty$.*

Proof. (1 ⇔ 2) From Equation (4.1.4), we have $v(G/\Lambda) = \mu(\mathcal{F})$. Therefore, G/Λ has finite volume if and only if $\mu(\mathcal{F}) < \infty$.

(2 ⇔ 3) If $\mathcal{F}$ is any strict fundamental domain for G/Λ, then $\mathcal{F}^{-1}$ is a strict fundamental domain for $\Lambda \backslash G$ (see Exercise 4). Since G is unimodular, we have $\mu(\mathcal{F}^{-1}) = \mu(\mathcal{F})$ (see Exercise 5).

(2 ⇒ 4) Obvious.

(4 ⇒ 1) We have $C \cap x\Lambda \neq \emptyset$, for every $x \in G$, so, from (4.1.5), we see that

$$v(G/\Lambda) = \int_{G/\Lambda} 1 \, dv(x\Lambda) \leq \int_{G/\Lambda} \#(C \cap x\Lambda) \, dv(\Lambda x) = \mu(C) < \infty. \quad \square$$

(4.1.12) Example. As mentioned in Example 1.3.7, $\mathrm{SL}(2, \mathbb{Z})$ is a lattice in $\mathrm{SL}(2, \mathbb{R})$.

(4.1.13) Definition. A closed subgroup Λ of G is **cocompact** (or **uniform**) if G/Λ is compact.

(4.1.14) Corollary.

1) *Every cocompact, discrete subgroup of G is a lattice.*

2) *Every finite-index subgroup of a lattice is a lattice.*

Proof. Exercises 12 and 13. $\square$

(4.1.15) Remark. Lattices in G are our main interest, but we will occasionally encounter lattices in Lie groups H that are not semisimple. If H is unimodular, then all of the above results remain valid with H in the place of G. In contrast, if H is not unimodular, then Proposition 4.1.3 may fail: there may exist a discrete subgroup Λ, such that there is no H-invariant Borel measure on H/Λ. Instead, there is sometimes only a semi-invariant measure v:

$$v(hA) = \Delta(h) \, v(A),$$

where Δ is the modular function of H (see Exercise 14).

For completeness, let us specifically state the following concrete generalization of Definition 4.1.9 (cf. 4.1.11).

(4.1.16) Definition. A subgroup Λ of a Lie group H is a **lattice** in H if

- Λ is a discrete subgroup of H, and
- there is an H-invariant measure v on H/Λ, such that $v(H/\Lambda) < \infty$.

(4.1.17) Example. $\mathbb{Z}^n$ is a cocompact lattice in $\mathbb{R}^n$.

(4.1.18) Proposition. *If a Lie group H has a lattice, then H is unimodular.*

Proof. Let $\mathcal{F}$ be a strict fundamental domain for H/Λ. The proof of Proposition 4.1.3 shows $v(A/\Lambda) = v(A \cap \mathcal{F})$, for every Borel set A in G, such that $A\Lambda = A$. Then Exercise 14 implies $\nu(hA/\Lambda) = \Delta(h)\,\nu(A/\Lambda)$. In particular, we see that $\nu(H/\Lambda) = \Delta(h)\,\nu(H/\Lambda)$, by letting $A = H$ (and noting that $hH = H$). Since $\nu(H/\Lambda) < \infty$, this implies $\Delta(h) = 1$, as desired. $\square$

Exercises for §4.1. Recall that (in accordance with the Standing Assumptions (4.0.0)), Γ is a lattice in G, and G is a semisimple Lie group.

#1. Show that Γ is finite if and only if G is compact.

#2. Complete the proof of Lemma 4.1.1; that is, show that $\mathcal{F}$ is a strict fundamental domain.

#3. Define $f\colon G/\Lambda \to \Lambda\backslash G$ by $f(g\Lambda) = \Lambda g^{-1}$. Show that f is a homeomorphism.

#4. Show Lemma 4.1.1 easily implies an analogous statement that applies to right cosets. More precisely, show that if
   - Λ is a discrete subgroup of G,
   - $\mathcal{F}$ is a strict fundamental domain for G/Λ, and
   - $\mathcal{F}^{-1} = \{\, x^{-1} \mid x \in \mathcal{F} \,\}$,
   then the natural map $\mathcal{F}^{-1} \to \Lambda\backslash G$, defined by $g \mapsto \Lambda g$, is bijective.

#5. Show that $\mu(A^{-1}) = \mu(A)$ for every Borel subset A of G.

   [*Hint:* Defining $\mu'(A) = \mu(A^{-1})$ yields a G-invariant measure on G. The uniqueness of Haar measure implies $\mu' = \mu$. Where did you use the fact that G is unimodular?]

#6. Let
   - Λ be a discrete subgroup of G,
   - $\mathcal{F}$ and $\mathcal{F}'$ be strict fundamental domains for G/Λ,
   - μ be Haar measure on G, and
   - A be a Borel subset of G.
   Show:
   a) For each $g \in G$, there is a unique $\lambda \in \Lambda$, such that $g\lambda \in \mathcal{F}$.
   b) For each $\lambda \in \Lambda$, if we let $A_\lambda = \{\, a \in A \mid a\lambda \in \mathcal{F} \,\}$, then A_λ is Borel, and A is the disjoint union of the sets $\{\, A_\lambda \mid \lambda \in \Lambda \,\}$.
   c) $\mu(\mathcal{F}) = \mu(\mathcal{F}')$.
   d) If $A\Lambda = A$, then $\mu(A \cap \mathcal{F}) = \mu(A \cap \mathcal{F}')$.

#7. Show, for every Haar measure μ on G, that the Borel measure ν defined in Proposition 4.1.3(1) is G-invariant.

   [*Hint:* For any $g \in G$, the set $g\mathcal{F}$ is a strict fundamental domain. From Exercise 6(d), we know that ν is independent of the choice of the strict fundamental domain $\mathcal{F}$.]

#8. If Λ is a discrete subgroup of G, and ν is a σ-finite, G-invariant Borel measure on G/Λ, show that the Borel measure μ defined in Proposition 4.1.3(2) is G-invariant.

#9. Let H be a closed subgroup of G. Show that there is a σ-finite, G-invariant Borel measure ν on G/H if and only if H is unimodular.

[*Hint:* (⇒) For a left Haar measure ρ on H, define a left Haar measure μ on G by

$$\mu(A) = \int_{G/H} \rho(x^{-1}A \cap H)\, d\nu(xH).$$

Then $\mu(A) = \Delta_H(h)\,\mu(Ah)$ for $h \in H$, where Δ_H is the modular function of H. Since G is unimodular, we must have $\Delta_H \equiv 1$.]

#10. Show that if Λ is a discrete subgroup of G that contains Γ, then Λ is a lattice in G, and Γ has finite index in Λ.

[*Hint:* Let $\mathcal{F}$ be a strict fundamental domain for G/Λ, and let F be a set of coset representatives for Γ in Λ. Then $\mathcal{F} \cdot F$ is a strict fundamental domain for G/Γ, and therefore has finite measure.]

#11. Let Λ be a discrete subgroup of G. Show that a subset A of G/Λ is precompact if and only if there is a compact subset C of G, such that $A \subseteq C\Lambda/\Lambda$.

[*Hint:* (⇐) The continuous image of a compact set is compact. (⇒) Let $\mathcal{U}$ be a cover of G by precompact, open sets.]

#12. Prove Corollary 4.1.14(1).

[*Hint:* Exercise 11 and Proposition 4.1.11(4).]

#13. Prove Corollary 4.1.14(2).

[*Hint:* Proposition 4.1.11. A finite union of sets of finite measure has finite measure.]

#14. Let
- H be a Lie group,
- Λ be a discrete subgroup of H,
- μ be the right Haar measure on H, and
- $\mathcal{F}$ be a strict fundamental domain for H/Λ.

Define a σ Borel measure ν on H/Λ by $\nu(A/\Lambda) = \mu(A \cap \mathcal{F})$, for every Borel set A in H, such that $A\Lambda = A$. Show $\nu(hA/\Lambda) = \Delta(h)\,\nu(A/\Lambda)$, where Δ is the modular function of H.

[*Hint:* Cf. Exercise 7.]

#15. Show that every discrete, cocompact subgroup of every Lie group is a lattice.

[*Hint:* Define ν as in Exercise 14. Since $\nu(H/\Lambda) < \infty$ (why?), we must have $\Delta(h) = 1$.]

§4.2. Commensurability and isogeny

We usually wish to ignore the minor differences that come from passing to a finite-index subgroup. The following definition describes the resulting equivalence relation.

(4.2.1) **Definition.** We say that two subgroups Λ_1 and Λ_2 of a group H are *commensurable* if $\Lambda_1 \cap \Lambda_2$ is a finite-index subgroup of both Λ_1 and Λ_2. This is an equivalence relation on the collection of all subgroups of H (see Exercise 1).

(4.2.2) Examples.

1) Two cyclic subgroups $a\mathbb{Z}$ and $b\mathbb{Z}$ of $\mathbb{R}$ are commensurable if and only if a is a nonzero rational multiple of b; therefore, commensurability of subgroups generalizes the classical notion of commensurability of real numbers.

2) It is easy to show that every subgroup commensurable to a lattice is itself a lattice. (For example, this follows from Corollary 4.1.14(2) and Exercise 4.1#10.)

The analogous notion for Lie groups (with finite center and finitely many connected components) is called "isogeny:"

(4.2.3) Definitions.

1) G_1 is *isogenous* to G_2 if some finite cover of $(G_1)^\circ$ is isomorphic to some finite cover of $(G_2)^\circ$. This is an equivalence relation.

2) A (continuous) homomorphism $\varphi\colon G_1 \to G_2$ is an *isogeny* if it is an isomorphism modulo finite groups. More precisely:
   - the kernel of φ is finite, and
   - the image of φ has finite index in G_2.

(4.2.4) Remark. The following are equivalent:

1) G_1 is isogenous to G_2.

2) $\mathrm{Ad}(G_1)^\circ \cong \mathrm{Ad}(G_2)^\circ$.

3) G_1 and G_2 are *locally isomorphic*, that is, the Lie algebras $\mathfrak{g}_1$ and $\mathfrak{g}_2$ are isomorphic.

4) There is an isogeny from some finite cover of $(G_1)^\circ$ to G_2.

The normalizer of a subgroup is very important in group theory. Because we are ignoring finite groups, the following definition is natural in our context.

(4.2.5) Definition. An element g of G *commensurates* Γ if $g\Gamma g^{-1}$ is commensurable to Γ. Let

$$\mathrm{Comm}_G(\Gamma) = \{\, g \in G \mid g \text{ commensurates } \Gamma \,\}.$$

This is called the *commensurator* of Γ.

(4.2.6) Remark. The commensurator of Γ is sometimes much larger than the normalizer of Γ. For example, let $G = \mathrm{SL}(n, \mathbb{R})$ and $\Gamma = \mathrm{SL}(n, \mathbb{Z})$. Then $\mathcal{N}_G(\Gamma)$ is commensurable to Γ (see Corollary 4.5.5), but $\mathrm{Comm}_G(\Gamma)$ contains $\mathrm{SL}(n, \mathbb{Q})$(see Exercise 4.8#11), so $\mathrm{Comm}_G(\Gamma)$ is dense in G, even though $\mathcal{N}_G(\Gamma)$ is discrete. Therefore, in this example (and, more generally, whenever Γ is "arithmetic"), $\mathcal{N}_G(\Gamma)$ has infinite index in $\mathrm{Comm}_G(\Gamma)$.

On the other hand, if $G = \mathrm{SO}(1, n)$, then it is known that there are examples in which Γ, $\mathcal{N}_G(\Gamma)$, and $\mathrm{Comm}_G(\Gamma)$ are commensurable to each other (see Exercise 5.2#3 and Corollary 6.5.16).

(4.2.7) **Definition.** We say that two groups Λ_1 and Λ_2 are *abstractly commensurable* if some finite-index subgroup of Λ_1 is isomorphic to some finite-index subgroup of Λ_2.

Note that if Λ_1 and Λ_2 are commensurable, then they are abstractly commensurable, but not conversely.

Exercises for §4.2.

#1. Verify that commensurability is an equivalence relation.

#2. If Γ_1 is commensurable to Γ_2, show $\mathrm{Comm}_G(\Gamma_1) = \mathrm{Comm}_G(\Gamma_2)$.

§4.3. Irreducible lattices

Note that $\Gamma_1 \times \Gamma_2$ is a lattice in $G_1 \times G_2$. A lattice that can be decomposed as a product of this type is said to be *reducible*.

(4.3.1) **Definition.** Γ is *irreducible* if ΓN is dense in G, for every noncompact, closed, normal subgroup N of G° (and Γ is infinite, or, equivalently, G is not compact).

(4.3.2) **Example.** If G is simple (and not compact), then every lattice in G is irreducible. Conversely, if G is not simple, then not every lattice in G is irreducible. To see this, assume, for simplicity, that G is connected and has trivial center (and is not compact). Then we may write G as a nontrivial direct product $G = G_1 \times G_2$, where each of G_1 and G_2 is semisimple. If we let Γ_i be any lattice in G_i, for $i = 1, 2$, then $\Gamma_1 \times \Gamma_2$ is a reducible lattice in G.

The following proposition shows (under mild assumptions) that every lattice is commensurable to a product of irreducible lattices. Therefore, the preceding example provides essentially the only way to construct reducible lattices, so most questions about lattices can be reduced to the irreducible case. We postpone the proof, because it relies on some results from later in this chapter.

(4.3.3) **Proposition** (see proof on page 56). *Assume*
- *G has trivial center, and*
- *Γ projects densely into the maximal compact factor of G.*

Then there is a direct-product decomposition $G = G_1 \times \cdots \times G_r$, such that Γ is commensurable to $\Gamma_1 \times \cdots \times \Gamma_r$, where $\Gamma_i = \Gamma \cap G_i$, and Γ_i is an irreducible lattice in G_i, for each i.

For readers familiar with locally symmetric spaces, these results can be restated in the following geometric terms.

(4.3.4) **Definition.** Recall that a locally symmetric space $\Gamma \backslash X$ is *irreducible* if there do not exist (nontrivial) locally symmetric spaces $\Gamma_1 \backslash X_1$ and $\Gamma_2 \backslash X_2$, such that the product $(\Gamma_1 \backslash X_1) \times (\Gamma_2 \backslash X_2)$ finitely covers $\Gamma \backslash X$.

The following is obvious by induction on $\dim X$.

(4.3.5) **Proposition.** *There exist locally symmetric spaces $\Gamma_1 \backslash X_1, \dots, \Gamma_r \backslash X_r$ that are irreducible, such that the product $(\Gamma_1 \backslash X_1) \times \cdots \times (\Gamma_r \backslash X_r)$ finitely covers $\Gamma \backslash X$.*

The following is a restatement of Proposition 4.3.3 (in the special case where G has no compact factors).

(4.3.6) **Proposition.** *Let M be an irreducible locally symmetric space, such that the universal cover X of M has no compact factors, and no flat factors. For any nontrivial cartesian product decomposition $X = X_1 \times X_2$ of X, the image of X_1 is dense in M.*

We will see in Example 5.5.3 that $SL(2, \mathbb{R}) \times SL(2, \mathbb{R})$ has an irreducible lattice (for example, a lattice isomorphic to $SL(2, \mathbb{Z}[\sqrt{2}])$). More generally, Corollary 18.7.4 shows that G has an irreducible lattice if all the simple factors of the "complexification" of G are isogenous to each other. The converse is proved in Theorem 5.6.2, under the additional assumption that G has no compact factors.

Exercises for §4.3.

#1. Show that if Γ is irreducible, then Γ projects densely into the maximal compact factor of G.

§4.4. Unbounded subsets of $\Gamma \backslash G$

Geometrically, looking at the fundamental domain described in Example 1.3.7 makes it clear that the sequence $\{ni\}$ tends to ∞ in $SL(2, \mathbb{Z}) \backslash \mathfrak{H}^2$. In this section, we give an algebraic criterion that determines whether or not a sequence tends to ∞ in G/Γ, without any need for a fundamental domain.

Recall that the *injectivity radius* of a Riemannian manifold X is the maximal $r \geq 0$, such that, for every $x \in X$, the exponential map is a diffeomorphism on the open ball of radius r around x. If X is compact, then the injectivity radius is nonzero. The following proposition shows that the converse holds in the special case where $X = \Gamma \backslash G/K$ is locally symmetric of finite volume.

(4.4.1) **Proposition.** *For $g \in G$, define $\phi_g \colon G \to G/\Gamma$ by $\phi_g(x) = xg\Gamma$. The homogeneous space G/Γ is compact if and only if there is a nonempty, open* subset U of G, such that, for every $g \in G$, the restriction $\phi_g|_U$ of ϕ_g to U is injective.

Proof. ($\Rightarrow$) Define $\phi \colon G \to G/\Gamma$ by $\phi(x) = x\Gamma$. Then ϕ is a covering map, so, for each $p \in G/\Gamma$, there is a connected neighborhood V_p of p, such that the restriction of ϕ to each component of $\phi^{-1}(V_p)$ is a diffeomorphism onto V_p. Since $\{V_p \mid p \in G/\Gamma\}$ is an open cover of G/Γ, and G/Γ is compact, there is a connected neighborhood U of e in G, such that, for each $p \in G/\Gamma$, there is some $p' \in G/\Gamma$, with $Up \subseteq V_{p'}$ (see Exercise 1). Then $\phi_g|_U$ is injective, for each $g \in G$.

($\Leftarrow$) We prove the contrapositive. Let U be any nonempty, precompact, open subset of G. (We wish to show, for some $g \in G$, that $\phi_g|_U$ is not injective.) If C is any compact subset of G/Γ, then, because G/Γ is not compact, we have

$$(G/\Gamma) \smallsetminus (U^{-1}C) \neq \varnothing.$$

Hence, we may inductively construct a sequence $\{g_n\}$ of elements of G, such that the open sets $\phi_{g_1}(U), \phi_{g_2}(U), \ldots$ are pairwise disjoint. Since G/Γ has finite volume, these sets cannot all have the same volume, so, for some n, the restriction $\phi_{g_n}|_U$ is not injective (see Corollary 4.1.7). $\square$

Let us restate this geometric result in algebraic terms.

(4.4.2) Notation. For elements a and b of a group H, and subsets A and B of H, let

$$^b a = bab^{-1}, \qquad\qquad\qquad {}^B a = \{\, {}^b a \mid b \in B \,\},$$

$$^b A = \{\, {}^b a \mid a \in A \,\}, \qquad\qquad {}^B A = \{\, {}^b a \mid a \in A, b \in B \,\}.$$

(4.4.3) Corollary. G/Γ is compact if and only if the identity element e is **not** an accumulation point of ${}^G\Gamma$.

Proof. We have

$$\phi_g|_U \text{ is injective} \quad \Leftrightarrow \quad \nexists u_1, u_2 \in U, \ u_1 g\Gamma = u_2 g\Gamma \text{ and } u_1 \neq u_2$$

$$\Leftrightarrow \quad {}^g\Gamma \cap (U^{-1}U) = \{e\}. \qquad\qquad \square$$

This has the following interesting consequence.

(4.4.4) Corollary. If Γ has a nontrivial, unipotent element, then G/Γ is **not** compact.

Proof. If u is a nontrivial, unipotent element of Γ, then, from the Jacobson-Morosov Lemma (A5.8), we know there is a continuous homomorphism $\phi \colon \mathrm{SL}(2,\mathbb{R}) \to G$, with $\phi \begin{bmatrix} 1 & 1 \\ 0 & 1 \end{bmatrix} = u$. Let $a = \phi \begin{bmatrix} 1/2 & 0 \\ 0 & 2 \end{bmatrix} \in G$. Then

$$a^n u a^{-n} = \phi \left(\begin{bmatrix} 2^{-n} & 0 \\ 0 & 2^n \end{bmatrix} \begin{bmatrix} 1 & 1 \\ 0 & 1 \end{bmatrix} \begin{bmatrix} 2^n & 0 \\ 0 & 2^{-n} \end{bmatrix} \right)$$

$$= \phi \left(\begin{bmatrix} 1 & 2^{-2n} \\ 0 & 1 \end{bmatrix} \right) \to \phi \left(\begin{bmatrix} 1 & 0 \\ 0 & 1 \end{bmatrix} \right) = e.$$

Therefore, e is an accumulation point of ${}^G u$, so Corollary 4.4.3 implies that G/Γ is not compact. $\square$

(4.4.5) *Remarks.*

1) If G has no compact factors, then the converse of Corollary 4.4.4 is true. However, we will prove this only in the special case where Γ is "arithmetic" (see Section 5.3).

2) In general (without any assumption on compact factors), it can be shown that G/Γ is compact if and only if every element of Γ is semisimple (see Exercise 5.3#6).

The proofs of Proposition 4.4.1 and Corollary 4.4.3 establish the following more general version of those results.

(4.4.6) **Proposition** (see Exercise 2). *Let Λ be a lattice in a Lie group H, and let C be a subset of H. The image of C in H/Λ is precompact if and only if the identity element e is **not** an accumulation point of ${}^C \Lambda$.*

The following is a similar elementary result that applies to the important special case where $G = \mathrm{SL}(\ell, \mathbb{R})$ and $\Gamma = \mathrm{SL}(\ell, \mathbb{Z})$, without relying on the fact that $\mathrm{SL}(\ell, \mathbb{Z})$ is a lattice.

(4.4.7) **Proposition** (Mahler Compactness Criterion). *Let $C \subseteq \mathrm{SL}(\ell, \mathbb{R})$. The image of C in $\mathrm{SL}(\ell, \mathbb{R}) / \mathrm{SL}(\ell, \mathbb{Z})$ is precompact if and only if 0 is **not** an accumulation point of*

$$C\mathbb{Z}^\ell = \{ cv \mid c \in C, v \in \mathbb{Z}^\ell \}.$$

Proof. ($\Rightarrow$) Since the image of C in $\mathrm{SL}(\ell, \mathbb{R}) / \mathrm{SL}(\ell, \mathbb{Z})$ is precompact, there is a compact subset C_0 of G, such that $C \subseteq C_0 \mathrm{SL}(\ell, \mathbb{Z})$ (see Exercise 4.1#11). There is no harm in assuming that $C = C_0 \mathrm{SL}(\ell, \mathbb{Z})$ (by enlarging C). Then $C(\mathbb{Z}^\ell \smallsetminus \{0\}) = C_0(\mathbb{Z}^\ell \smallsetminus \{0\})$ is closed (since $\mathbb{Z}^\ell \smallsetminus \{0\}$, being discrete, is closed and C_0 is compact), so $C(\mathbb{Z}^\ell \smallsetminus \{0\})$ contains all of its accumulation points. In addition, since 0 is fixed by every element of C, we know that $0 \notin C(\mathbb{Z}^\ell \smallsetminus \{0\})$. Therefore, 0 is not an accumulation point of $C(\mathbb{Z}^\ell \smallsetminus \{0\})$.

($\Leftarrow$) To simplify the notation (while retaining the main ideas), let us assume $\ell = 2$ (see Exercise 6). Suppose $\{g_n\}$ is a sequence of elements of $\mathrm{SL}(2, \mathbb{R})$, such that 0 is **not** an accumulation point of $\bigcup_{n=1}^\infty g_n \mathbb{Z}^2$. We wish to show there is a sequence $\{y_n\}$ of elements of $\mathrm{SL}(2, \mathbb{Z})$, such that $\{g_n y_n\}$ has a convergent subsequence.

For each n, let

- $v_n \in \mathbb{Z}^2 \smallsetminus \{0\}$, such that $\|g_n v_n\|$ is minimal,
- $\pi_n \colon \mathbb{R}^2 \to \mathbb{R}g_n v_n$ and $\pi_n^\perp \colon \mathbb{R}^2 \to (\mathbb{R}g_n v_n)^\perp$ be the orthogonal projections, and
- $w_n \in \mathbb{Z}^2 \smallsetminus \mathbb{R}v_n$, such that $\|\pi_n^\perp(g_n w_n)\|$ is minimal.

By replacing w_n with $w_n + kv_n$, for an appropriately chosen $k \in \mathbb{Z}$, we may assume $\|\pi_n(g_n w_n)\| \le \|g_n v_n\|/2$. Then, since the minimality of $\|g_n v_n\|$ implies $\|g_n v_n\| \le \|g_n w_n\|$, we have

$$\|g_n v_n\| \le \|\pi_n^\perp(g_n w_n)\| + \|\pi_n(g_n w_n)\| \le \|\pi_n^\perp(g_n w_n)\| + \frac{\|g_n v_n\|}{2},$$

so

$$\|\pi_n^\perp(g_n w_n)\| \ge \frac{\|g_n v_n\|}{2}. \qquad (4.4.8)$$

Let C be the convex hull of $\{0, v_n, w_n\}$ and (thinking of v_n and w_n as column vectors) let $y_n = [v_n \; w_n] \in \mathrm{Mat}_{2\times2}(\mathbb{Z})$. From the minimality of $\|g_n v_n\|$ and $\|\pi_n^\perp(g_n w_n)\|$, we see that $C \cap \mathbb{Z}^2 = \{0, v_n, w_n\}$ (see Exercise 7), so $\det y_n = \pm 1$ (see Exercise 8). Therefore, perhaps after replacing w_n with $-w_n$, we have $y_n \in \mathrm{SL}(2, \mathbb{Z})$. Since $y_n\left[\begin{smallmatrix}1\\0\end{smallmatrix}\right] = v_n$ and $y_n\left[\begin{smallmatrix}0\\1\end{smallmatrix}\right] = w_n$, we may assume, by replacing g_n with $g_n y_n$, that

$$v_n = \left[\begin{smallmatrix}1\\0\end{smallmatrix}\right] \qquad \text{and} \qquad w_n = \left[\begin{smallmatrix}0\\1\end{smallmatrix}\right].$$

Note that

$$\|\pi_n^\perp(g_n w_n)\| \cdot \|g_n v_n\| = \det g_n = 1. \qquad (4.4.9)$$

By combining this with (4.4.8), we see that $\{g_n v_n\}$ is a bounded sequence, so, by passing to a subsequence, we may assume $g_n v_n$ converges to some vector v. By assumption, we have $v \ne 0$.

Now, from (4.4.9), and the fact that $\|g_n v_n\| \to \|v\|$ is bounded away from 0, we see that $\|\pi_n^\perp(g_n w_n)\|$ is bounded. Because $\|\pi_n(g_n w_n)\|$ is also bounded, we conclude that $\|g_n w_n\|$ is bounded. Hence, by passing to a subsequence, we may assume $g_n w_n$ converges to some vector w. From (4.4.8), we know that $\|\pi_n^\perp(g_n w_n)\| \not\to 0$, so $w \notin \mathbb{R}v$.

Since $v \ne 0$ and $w \notin \mathbb{R}v$, there is some $g \in \mathrm{GL}(\ell, \mathbb{R})$ with $g\left[\begin{smallmatrix}1\\0\end{smallmatrix}\right] = v$ and $g\left[\begin{smallmatrix}0\\1\end{smallmatrix}\right] = w$. We have

$$g_n\left[\begin{smallmatrix}1\\0\end{smallmatrix}\right] = g_n v_n \to v = g\left[\begin{smallmatrix}1\\0\end{smallmatrix}\right]$$

and, similarly, $g_n\left[\begin{smallmatrix}0\\1\end{smallmatrix}\right] \to g\left[\begin{smallmatrix}0\\1\end{smallmatrix}\right]$, so $g_n x \to gx$ for all $x \in \mathbb{R}^2$. Therefore, $g_n \to g$, as desired. $\qquad \square$

Exercises for §4.4.

#1. Suppose a Lie group H acts continuously on a compact topological space M, and $\mathcal{V}$ is an open cover of M. Show that there is a neighborhood U of e in H, such that, for each $m \in M$, there is some $V \in \mathcal{V}$ with $Um \subseteq V$.

[*Hint:* This is analogous to the fact that every open cover of a compact metric space has a "Lebesgue number." Each $m \in M$ is contained in some $V_m \in \mathcal{V}$. Choose V'_m containing m, and a neighborhood U'_m of e, such that $U'_m V'_m \subseteq V_m$. Cover M with finitely many V'_m.]

#2. Prove Proposition 4.4.6.

[*Hint:* See the proofs of Proposition 4.4.1 and Corollary 4.4.3.]

#3. Use Proposition 4.4.6 to show that if H is a closed subgroup of G, such that H/Γ is a lattice in H, then the natural inclusion map $H/(\Gamma \cap H) \hookrightarrow G/\Gamma$ is proper.

[*Hint:* It suffices to show that if C is a subset of H, such that the image of C in G/Γ is precompact, then the image of C in $H/(\Gamma \cap H)$ is also precompact.]

#4. Let $G = \mathrm{SL}(2, \mathbb{R})$, $\Gamma = \mathrm{SL}(2, \mathbb{Z})$, and A be the subgroup consisting of all the diagonal matrices in G. Show that the natural inclusion map $A/(\Gamma \cap A) \hookrightarrow G/\Gamma$ is proper, but $\Gamma \cap A$ is *not* a lattice in A. (Therefore, the converse of Exercise 3 does not hold.)

#5. Let $G = \mathrm{SL}(3, \mathbb{R})$, $\Lambda = \mathrm{SL}(3, \mathbb{Z})$, and $a = \mathrm{diag}(1/2, 1, 2) \in H$. Show that $a^n \Lambda \to \infty$ in G/Λ as $n \to \infty$. That is, show, for each compact subset C of G/Λ, that, for all sufficiently large n, we have $a^n \Lambda \notin C$. (For the purposes of this exercise, do *not* assume that Λ is a lattice in G.)

#6. Prove Proposition 4.4.7($\Leftarrow$) without assuming $\ell = 2$.

[*Hint:* Extend the definition of v_n and w_n to an inductive construction of vectors $u_{1,n}, \ldots, u_{\ell,n} \in \mathbb{Z}^\ell$.]

#7. Suppose that v and w are linearly independent vectors in $\mathbb{R}^\ell$, and $x = av + bw$, with $a, b \geq 0$ and $a + b \leq 1$. Show that either
- $x \in \{v, w\}$, or
- $\|x\| < \|v\|$, or
- $x \notin \mathbb{R}v$ and $d(x, \mathbb{R}v) < d(w, \mathbb{R}v)$.

[*Hint:* Either $b = 1$, or $b = 0$, or $0 < b < 1$.]

#8. Let C be the convex hull of $\{0, v, w\}$, where v and w are linearly independent vectors in $\mathbb{Z}^2$. Show that if $C \cap \mathbb{Z}^2 = \{0, v, w\}$, then $\det[v \ w] = \pm 1$.

[*Hint:* Let P be the the convex hull of $\{0, v, w, v + w\}$, so $|\det[v \ w]|$ is the area of P. If this area is > 1, then the translates of P by elements of $\mathbb{Z}^2$ cannot be a tiling of $\mathbb{R}^2$.]

#9. Let H be a closed subgroup of G.
a) Show that if $\Gamma \cap H$ is a lattice in H, then $H\Gamma$ is closed in G.
b) Show that the converse holds if H is normal in G.

[*Hint:* (a) Exercise 3. (b) Since G/Γ is a bundle over $G/(H\Gamma)$ with fiber $H\Gamma/\Gamma$, Fubini's Theorem implies that $H\Gamma/\Gamma$ has finite volume. So the H-equivariantly homeomorphic space $H/(\Gamma \cap H)$ also has finite volume.]

#10. Suppose
- Λ is a non-cocompact lattice in a topological group H, and

- *H* has a compact, open subgroup *K*.

Show that Λ has a nontrivial element of finite order.

[*Hint:* Since *K* is compact and Λ is discrete, it suffices to show that some conjugate of Λ intersects *K* nontrivially.]

§4.5. Borel Density Theorem and some consequences

The results in this section require the minor assumption that Γ projects densely into the maximal compact factor of *G*. This hypothesis is automatically satisfied (vacuously) if *G* has no compact factors. Recall that G° denotes the identity component of *G* (see Notation 1.1.2).

(4.5.1) **Theorem** (Borel). *Assume*

- Γ *projects densely into the maximal compact factor of G,*
- *V is a finite-dimensional vector space over $\mathbb{R}$ or $\mathbb{C}$, and*
- $\rho\colon G \to \mathrm{GL}(V)$ *is a continuous homomorphism.*

Then:

1) *Every $\rho(\Gamma)$-invariant vector in V is $\rho(G^\circ)$-invariant.*
2) *Every $\rho(\Gamma)$-invariant subspace of V is $\rho(G^\circ)$-invariant.*

Proof. The proof is not difficult, but, in order to get to the applications more quickly, we will postpone it until Section 4.6. For now, to illustrate the main idea, let us just prove (1), in the special case where G/Γ is compact (and *G* is connected). Assume also that *G* has no compact factors (see Exercise 1); then *G* is generated by its unipotent elements (see Exercise 2), so it suffices to show that v is invariant under $\rho(u)$, for every nontrivial unipotent element *u* of *G*. Because $\rho(u)$ is unipotent (see Exercise 3), we know that $\rho(u^n)v$ is a polynomial function of *n* (see Exercise 4). However, because G/Γ is compact and $\rho(\Gamma)v = v$, we also know that $\rho(G)v$ is compact, so $\{\rho(u^n)v \mid n \in \mathbb{N}\}$ is bounded. Every bounded polynomial is constant, so we conclude that $\rho(u^n)v = v$ for all *n*; in particular, $\rho(u)v = \rho(u^1)v = v$, as desired. □

(4.5.2) **Corollary.** *Assume Γ projects densely into the maximal compact factor of G. If H is a connected, closed subgroup of G that is normalized by Γ, then H is normal in G°.*

Proof. The Lie algebra $\mathfrak{h}$ of *H* is a vector subspace of the Lie algebra $\mathfrak{g}$ of *G*. Also, because Γ normalizes *H*, we know that $\mathfrak{h}$ is invariant under $\mathrm{Ad}_G\Gamma$. From Theorem 4.5.1(2), we conclude that $\mathfrak{h}$ is invariant under $\mathrm{Ad}\,G^\circ$. Since *H* is connected, this implies that *H* is a normal subgroup of G°. □

(4.5.3) **Corollary.** *If* Γ *projects densely into the maximal compact factor of* G *(and* G *is connected), then* $C_G(\Gamma) = Z(G)$.

Proof. Recall that $G \subseteq SL(\ell, \mathbb{R})$, for some ℓ (see the Standing Assumptions (4.0.0)). Let $V = \text{Mat}_{\ell \times \ell}(\mathbb{R})$ be the vector space of all real $\ell \times \ell$ matrices, so $G \subseteq V$. For $g \in G$ and $v \in V$, define $\rho(g)v = gvg^{-1}$, so $\rho \colon G \to GL(V)$ is a continuous homomorphism. If $c \in C_G(\Gamma)$, then $\rho(\gamma)c = \gamma c \gamma^{-1} = c$ for every $\gamma \in \Gamma$, so Theorem 4.5.1(1) implies that $\rho(G)c = c$. Therefore $c \in Z(G)$. $\square$

(4.5.4) **Corollary.** *Assume* Γ *projects densely into the maximal compact factor of* G *(and* G *is connected). If* N *is a finite, normal subgroup of* Γ, *then* $N \subseteq Z(G)$.

Proof. The quotient $\Gamma / C_\Gamma(N)$ is finite, because it embeds in the finite group $\text{Aut}(N)$, so $C_\Gamma(N)$ is a lattice in G (see Corollary 4.1.14(2)). Then, because $N \subseteq C_G(C_\Gamma(N))$, Corollary 4.5.3 implies $N \subseteq Z(G)$. $\square$

(4.5.5) **Corollary.** *If* Γ *projects densely into the maximal compact factor of* G, *then* Γ *has finite index in its normalizer* $\mathcal{N}_G(\Gamma)$.

Proof. By passing to a subgroup of finite index, we may assume G is connected. Because Γ is discrete, the identity component $\mathcal{N}_G(\Gamma)^\circ$ of $\mathcal{N}_G(\Gamma)$ must centralize Γ. So $\mathcal{N}_G(\Gamma)^\circ \subseteq C_G(\Gamma) = Z(G)$ is finite. On the other hand, $\mathcal{N}_G(\Gamma)^\circ$ is connected. Therefore, $\mathcal{N}_G(\Gamma)^\circ$ is trivial, so $\mathcal{N}_G(\Gamma)$ is discrete. Hence Γ has finite index in $\mathcal{N}_G(\Gamma)$ (see Exercise 4.1#10). $\square$

(4.5.6) **Corollary** (Borel Density Theorem). *If* Γ *projects densely into the maximal compact factor of* G *(and* G *is connected), then* Γ *is Zariski dense in* G. *That is, if* $Q \in \mathbb{R}[x_{1,1}, \dots, x_{\ell,\ell}]$ *is a polynomial function on* $\text{Mat}_{\ell \times \ell}(\mathbb{R})$, *such that* $Q(\Gamma) = 0$, *then* $Q(G) = 0$.

Proof. Let

$$\mathcal{Q} = \{ Q \in \mathbb{R}[x_{1,1}, \dots, x_{\ell,\ell}] \mid Q(\Gamma) = 0 \}.$$

From the definition of $\mathcal{Q}$, it is obvious that Γ is contained in the corresponding variety $\text{Var}(\mathcal{Q})$ (see Definition A4.1). Since $\text{Var}(\mathcal{Q})$ has only finitely many connected components (see Theorem A4.6), this implies that $\text{Var}(\mathcal{Q})^\circ$ is a connected subgroup of G that contains a finite-index subgroup of Γ. Hence Corollary 4.5.2 implies that $\text{Var}(\mathcal{Q})^\circ = G$ (see Exercise 11), so $G \subseteq \text{Var}(\mathcal{Q})$, as desired. $\square$

With the above results, we can now provide the proof that was postponed from page 48:

Proof of Proposition 4.3.3. We may assume Γ is reducible (otherwise, let $r = 1$). Hence, there is some noncompact, connected, closed, normal subgroup N of G, such that $N\Gamma$ is *not* dense in G; let H be the closure of $N\Gamma$, and let $H_1 = H^\circ$. Because $\Gamma \subset H$, we know that Γ normalizes H_1, so H_1 is a normal subgroup of G (see Corollary 4.5.2 and Exercise 4.3#1)).

Let $\Lambda_1 = H_1 \cap \Gamma$. By definition, H_1 is open in H, so the subgroup $H_1\Gamma$ is also open in H. It is therefore closed, so Λ_1 is a lattice in H_1 (see Exercise 4.4#9).

Because H_1 is normal in G and G is semisimple (with trivial center), there is a normal subgroup H_2 of G, such that $G = H_1 \times H_2$ (see Exercise A1#6). Let $\Lambda = H_1 \cap (H_2\Gamma)$ be the projection of Γ to H_1. Now Γ normalizes Λ_1, and H_2 centralizes Λ_1, so Λ must normalize Λ_1. Therefore Corollary 4.5.5 implies that Λ is discrete (hence closed), so $H_2\Gamma = \Lambda \times H_2$ is closed, so $\Lambda_2 = H_2 \cap \Gamma$ is a lattice in H_2 (see Exercise 4.4#9).

Because Λ_1 is a lattice in H_1 and Λ_2 is a lattice in H_2, we know that $\Lambda_1 \times \Lambda_2$ is a lattice in $H_1 \times H_2 = G$. Hence, $\Lambda_1 \times \Lambda_2$ has finite index in Γ (see Exercise 4.1#10).

By induction on $\dim G$, we may write

$$H_1 = G_1 \times \cdots \times G_s \text{ and } H_2 = G_{s+1} \times \cdots \times G_r,$$

so that $\Gamma \cap G_i$ is an irreducible lattice in G_i, for each i. $\square$

(4.5.7) *Remark.* For simplicity, the statement of Theorem 4.5.1 assumes that Γ projects densely into the maximal compact factor of G. Without this assumption, the proof of Theorem 4.5.1(1) establishes the weaker conclusion that v is $\rho(S)$-invariant, for every noncompact, simple factor S of G. This leads to alternate versions of the corollaries that make no assumption about the compact factor of G. For example, the analogue of Corollary 4.5.2 states that if H is a connected, closed subgroup of G that is normalized by Γ, then H is normalized by every noncompact, simple factor of G.

Exercises for §4.5.

#1. Prove 4.5.1(1), under the assumption that G/Γ is compact (but allowing G to have compact factors).

[*Hint:* The above proof shows that v is invariant under the the product N of all the noncompact factors of G. So it is invariant under the closure of $N\Gamma$, which is G.]

#2. Show that if G is connected, and has no compact factors, then it is generated by its unipotent elements.

[*Hint:* Consider each simple factor of G individually. The conjugates of a unipotent element are also unipotent.]

#3. Suppose $\rho: G \to \text{SL}(\ell, \mathbb{R})$ is a continuous homomorphism. Show that if u is unipotent, then $\rho(u)$ is unipotent.

[*Hint:* The Jacobson-Morosov Lemma (A5.8) allows you to assume $G = \mathrm{SL}(2, \mathbb{R})$ and $u = \begin{bmatrix} 1 & 1 \\ 0 & 1 \end{bmatrix}$. Then a sequence of conjugates of u converges to Id, so the characteristic polynomial of $\rho(u)$ is the same as the characteristic polynomial of Id.]

#4. Show that if u is a unipotent element of $\mathrm{SL}(\ell, \mathbb{R})$ and $v \in \mathbb{R}^\ell$, then each coordinate of the vector $u^n v$ is a polynomial function of n.

[*Hint:* Let $u = \mathrm{Id} + T$, where $T^{\ell+1} = 0$. Then $u^n v = (\mathrm{Id} + T)^n v = \sum_{k=0}^{\ell} \binom{n}{k} T^k v$.]

#5. Show that if G is not compact, then Γ is not abelian.

#6. Generalizing Exercise 5, show that if G is not compact, then Γ is not solvable.

#7. Strengthening Exercise 5, show that if G is not compact, then the commutator subgroup $[\Gamma, \Gamma]$ is infinite.

#8. Assume the hypotheses of Theorem 4.5.1, and that G is connected. For definiteness, assume that V is a real vector space. For any subgroup H of G, let $V[H]$ be the $\mathbb{R}$-span of $\{ \rho(h) \mid h \in H \}$ in $\mathrm{End}(V)$. Show that $V[\Gamma] = V[G]$.

#9. Show the identity component of $\mathcal{N}_G(\Gamma)$ is contained in the maximal compact factor of G.

[*Hint:* Apply Corollary 4.5.5 to G/K, where K is the maximal compact factor.]

#10. Show that if G is not compact, then Γ has an element that is not unipotent.

[*Hint:* Any unipotent element y of $\mathrm{SL}(\ell, \mathbb{R})$ satisfies the polynomial $(x-1)^\ell = 0$.]

#11. Assume G has no compact factors. Show that if H is a connected, closed subgroup of G that contains a finite-index subgroup of Γ, then $H = G°$.

[*Hint:* H is normalized by $\Gamma \cap H$, so $H \triangleleft G°$.]

#12. Assume G has trivial center and no compact factors. Show that Γ is reducible if and only if there is a finite-index subgroup Γ' of Γ such that Γ' is isomorphic to $A \times B$, for some infinite groups A and B. (Actually, although you do not need to prove it, there is no need to assume the center of G is trivial. This is because Γ has a subgroup of finite index that is torsion free (see Theorem 4.8.2), and therefore does not intersect the center of G.)

#13. Show that if Γ is irreducible, then $N \cap \Gamma$ is finite, for every connected, closed, normal subgroup N of G, such that G/N is not compact.

[*Hint:* See the proof of Proposition 4.3.3.]

#14. Let ρ_1 and ρ_2 be finite-dimensional, real representations of G. Assume G is connected, and has no compact factors. Show that if the restrictions $\rho_1|_\Gamma$ and $\rho_2|_\Gamma$ are isomorphic, then ρ_1 and ρ_2 are isomorphic.

[*Hint:* We are assuming $\rho_i : G \to \mathrm{GL}(n, \mathbb{R})$, for some n, and that there is some $A \in \mathrm{GL}(n, \mathbb{R})$, such that $\rho_1(g) = A \rho_2(g) A^{-1}$, for all $y \in \Gamma$. You wish to show there is some $A' \in \mathrm{GL}(n, \mathbb{R})$, such that the same condition holds for all $g \in G$, with A' in the place of A. The Borel Density Theorem implies that you may take $A' = A$.]

§4.6. Proof of the Borel Density Theorem

The proof of the Borel Density Theorem (4.5.1) is based on the contrast between two behaviors. On the one hand, if u is a unipotent matrix, and v is a vector that is not fixed by u, then some component of $u^n v$ is a nonconstant polynomial, and therefore tends to $\pm\infty$ with n. On the other hand, the following observation implies that if v is fixed by Γ, then some subsequence converges to a finite limit.

(4.6.1) **Lemma** (Poincaré Recurrence Theorem). *Let*

- *(X, d) be a metric space,*
- *$T: X \to X$ be a homeomorphism, and*
- *μ be a T-invariant measure on X, such that $\mu(X) < \infty$.*

Then, for almost every $x \in X$, there is a sequence $n_k \to \infty$, such that $T^{n_k} x \to x$.

Proof. Let
$$A_\epsilon = \{ a \in X \mid \forall m > 0, \ d(T^m x, x) > \epsilon \}.$$
It suffices to show $\mu(A_\epsilon) = 0$ for every ϵ.

Suppose $\mu(A_\epsilon) > 0$. Then we may choose a subset B of A_ϵ, such that $\mu(B) > 0$ and $\mathrm{diam}(B) < \epsilon$. Because the sets $B, T^{-1}B, T^{-2}B, \ldots$ all have the same measure, and $\mu(X) < \infty$, they cannot all be disjoint: there exists $m < n$, such that $T^{-m}B \cap T^{-n}B \neq \emptyset$. By applying T^n, we may assume $n = 0$. For $x \in T^{-m}B \cap B$, we have $T^m x \in B$ and $x \in B$, so
$$d(T^m x, x) \leq \mathrm{diam}(B) < \epsilon.$$
This contradicts the definition of A_ϵ. □

(4.6.2) *Remark.* Part (1) of Theorem 4.5.1 is a corollary of Part (2). Namely, if v is $\rho(\Gamma)$-invariant, then the 1-dimensional subspace $\mathbb{R}v$ (or $\mathbb{C}v$) is also invariant, so (2) implies that the subspace is $\rho(G)$-invariant. Since G has no nontrivial homomorphism to the abelian group $\mathbb{R}^\times$ (or $\mathbb{C}^\times$), this implies that the vector v is $\rho(G)$-invariant.

However, we will provide a direct proof of (1), since it is quite short (and a little more elementary than the proof of (2)).

Proof of Theorem 4.5.1(1). Suppose v is a vector in V that is fixed by $\rho(\Gamma)$. It suffices to show, for every unipotent $u \in G$, that v is fixed by $\rho(u)$.

Since u is $\rho(\Gamma)$-invariant, the map ρ induces a well-defined map $\overline{\rho}: G/\Gamma \to V$, defined by $\overline{\rho}(g\Gamma) = \rho(g)v$. Since $\overline{\rho}$ is G-equivariant, it pushes the G-invariant, finite measure ν on G/Γ to a $\rho(G)$-invariant, finite measure $\overline{\nu}$ on V. Therefore, Lemma 4.6.1 implies, for a.e. $g \in G$, that $\{\rho(u^n g)v\}$ has a convergent subsequence.

However, each component of the vector $\rho(u^n g)v$ is a polynomial function of n (see Exercise 4.5#4). Therefore, the preceding paragraph implies, for a.e. $g \in G$, that $\rho(u^n g)v$ is constant (independent of n). This means that $\rho(g)v$ is fixed by $\rho(u)$. Since this is true for a.e. g, we conclude, by continuity, that it is true for all g, including $g = e$. Hence, v is fixed by u. $\qquad\square$

To prepare for the proof of Theorem 4.5.1(2), we make a few observations about the action of G on the projectivization of V.

(4.6.3) **Proposition.** *Assume*
- *G has no compact factors,*
- *V is a finite-dimensional vector space over $\mathbb{R}$ or $\mathbb{C}$,*
- *$\rho\colon G \to \mathrm{GL}(V)$ is a continuous homomorphism, and*
- *μ is a $\rho(G)$-invariant measure on the projective space $\mathbb{P}(V)$.*

If $\mu(\mathbb{P}(V)) < \infty$, then μ is supported on the set of fixed points of $\rho(G^\circ)$.

Proof. We know that G° is generated by its unipotent elements (see Exercise 4.5#2), so it suffices to show that μ is supported on the set of fixed points of $\rho(u)$, for every unipotent element u of G.

Let
- u be a unipotent element of G,
- $T = \rho(u) - \mathrm{Id}$, and
- $v \in V \smallsetminus \{0\}$.

Then T is nilpotent (because $\rho(u)$ is unipotent (see Exercise 4.5#3)), so there is some integer $r \geq 0$, such that $T^r v \neq 0$, but $T^{r+1} v = 0$. We have
$$\rho(u) T^r v = (\mathrm{Id} + T)(T^r v) = T^r v + T^{r+1} v = T^r v + 0 = T^r v,$$
so $[T^r v] \in \mathbb{P}(V)$ is a fixed point for $\rho(u)$. Also, for each $n \in \mathbb{N}$, we have
$$\rho(u^n)[v] = \left[\sum_{k=0}^r \binom{n}{k} T^k v \right] = \left[\binom{n}{r}^{-1} \sum_{k=0}^r \binom{n}{k} T^k v \right] \to [T^r v]$$
(because, for $k < r$, we have $\binom{n}{k} / \binom{n}{r} \to 0$ as $n \to \infty$). Therefore, $\rho(u^n)[v]$ converges to a fixed point of $\rho(u)$, as $n \to \infty$.

The Poincaré Recurrence Theorem (4.6.1) implies, for μ-almost every $[v] \in \mathbb{P}(V)$, that there is a sequence $n_k \to \infty$, such that $\rho(u^{n_k})[v] \to [v]$. On the other hand, the preceding paragraph tells us that $\rho(u^{n_k})[v]$ converges to a fixed point of $\rho(u)$. Therefore, μ-almost every element of $\mathbb{P}(V)$ is a fixed point of $\rho(u)$. In other words, μ is supported on the set of fixed points of $\rho(u)$, as desired. $\qquad\square$

The assumption that G has no compact factors cannot be omitted from Proposition 4.6.3. For example, the usual Lebesgue measure is an $SO(n)$-invariant, finite measure on S^{n-1}, but $SO(n)$ has no fixed points on S^{n-1}. We can, however, make the following weaker statement.

(4.6.4) **Corollary.** *Assume*

- *V is a finite-dimensional vector space over $\mathbb{R}$ or $\mathbb{C}$,*
- *$\rho: G \to GL(V)$ is a continuous homomorphism, and*
- *μ is a $\rho(G)$-invariant measure on the projective space $\mathbb{P}(V)$.*

If $\mu(\mathbb{P}(V)) < \infty$, then there is a cocompact, closed, normal subgroup G' of G, such that μ is supported on the set of fixed points of $\rho(G')$.

Proof. Let K be the maximal connected, compact, normal subgroup of G, and write $G \approx G' \times K$, for some closed, normal subgroup G' of G. Then G' has no compact factors, so we may apply Proposition 4.6.3 to the restriction $\rho|_{G'}$. $\qquad\square$

It is now easy to prove the other part of Theorem 4.5.1:

Proof of Theorem 4.5.1(2). By passing to a subgroup of finite index, we may assume G is connected. For simplicity, let us also assume that G has no compact factors (see Exercise 1).

Suppose W is a subspace in V that is fixed by $\rho(\Gamma)$, and let $d = \dim W$. Note that ρ induces a continuous homomorphism $\hat{\rho}: G \to GL(\bigwedge^d V)$, and, since W is $\rho(\Gamma)$-invariant, the 1-dimensional subspace $\bigwedge^d W$ is $\hat{\rho}(\Gamma)$-invariant. Hence, $\hat{\rho}$ induces a well-defined map $\overline{\rho}: G/\Gamma \to \mathbb{P}(\bigwedge^d V)$, with $\overline{\rho}(e\Gamma) = [\bigwedge^d W]$. Then, since $\overline{\rho}$ is G-equivariant, it pushes the G-invariant, finite measure ν on G/Γ to a $\hat{\rho}(G)$-invariant, finite measure $\overline{\nu}$ on $\mathbb{P}(\bigwedge^d V)$. Then Proposition 4.6.3 tells us that $\overline{\rho}(G/\Gamma)$ is contained in the set of fixed points of $\overline{\rho}(G)$. In particular, $[\bigwedge^d W] = \overline{\rho}(e\Gamma)$ is fixed by $\hat{\rho}(G)$. This means that W is $\rho(G)$-invariant. $\qquad\square$

(4.6.5) *Remark.* The proofs of the two parts of Theorem 4.5.1 never use the fact that the lattice Γ is discrete. Therefore, Γ can be replaced with any closed subgroup H of G, such that there is a G-invariant, finite measure on G/H, and H projects densely into the maximal compact factor of G.

Exercises for §4.6.

#1. Complete the proof of Theorem 4.5.1(2), by removing the assumption that G has no compact factors.
[*Hint:* See the hint to Exercise 4.5#1.]

#2. Let H be a closed subgroup of G that projects densely into the maximal compact factor of G. Show that if there is a G-invariant, finite measure on G/H, then the identity component H° is a normal subgroup of G°.

[*Hint:* Remark 4.6.5 and the proof of Corollary 4.5.2.]

§4.7. Γ is finitely presented

(4.7.1) **Definitions.** Let Λ be a group.

1) Λ is *finitely generated* it has a finite generating set. That is, there is a finite subset of Λ that is not contained in any proper subgroup of Λ.

2) Λ is *finitely presented* it has a presentation with only finitely many generators and finitely many relations. In other words, there exist:
   - a finitely generated free group F,
   - a surjective homomorphism $\phi \colon F \to \Lambda$, and
   - a finite subset R of the kernel of ϕ,

   such that $\ker \phi$ is the smallest *normal* subgroup of F that contains R.

It is easy to see that every finitely presented group is finitely generated. However, the converse is not true.

In this section, we describe the proof that Γ is finitely presented. Much like the usual proof that the fundamental group of any compact manifold is finitely presented, it is based on the existence of a nice set that is close to being a fundamental domain for the action of Γ on G.

(4.7.2) **Definition.** Suppose Γ acts properly discontinuously on a topological space Y. A subset $\mathcal{F}$ of Y is a *coarse fundamental domain* for Γ if

1) $\Gamma \mathcal{F} = Y$, and
2) $\{ y \in \Gamma \mid y\mathcal{F} \cap \mathcal{F} \neq \varnothing \}$ is finite.

(4.7.3) **Other terminology.** Some authors call $\mathcal{F}$ a *fundamental set*, rather than a coarse fundamental domain.

The following general principle will be used to show that Γ is finitely generated:

(4.7.4) **Proposition.** *Suppose a discrete group Λ acts properly discontinuously on a topological space Y. If Y is connected, and Λ has a coarse fundamental domain $\mathcal{F}$ that is an open subset of Y, then Λ is finitely generated.*

Proof. Let $S = \{ s \in \Lambda \mid s\mathcal{F} \cap \mathcal{F} \neq \varnothing \}$. We know that S is finite (see Definition 4.7.2(2)), so it suffices to show that S generates Λ. Here is the idea: think of $\{ \lambda\mathcal{F} \mid \lambda \in \Lambda \}$ as a tiling of Y. The elements of S can move $\mathcal{F}$ to any adjacent tile, and Y is connected, so a composition of elements of S can move $\mathcal{F}$ to any tile. Therefore $\langle S \rangle$ is transitive on the set of tiles. Since S also contains the entire stabilizer of the tile $\mathcal{F}$, we conclude that $\langle S \rangle = \mathcal{F}$.

Now, here is the formal proof. Consider some $\lambda \in \Lambda$, such that $\lambda \mathcal{F} \cap \langle S \rangle \mathcal{F} \neq \varnothing$. This means there exists $s \in \langle S \rangle$, such that $\lambda \mathcal{F} \cap s \mathcal{F} \neq \varnothing$, so $s^{-1} \lambda \mathcal{F} \cap \mathcal{F} \neq \varnothing$. Therefore, by the definition of S, we have $s^{-1} \lambda \in S$, so $\lambda \in sS \subseteq \langle S \rangle$. Thus, we have shown that $(\Lambda \smallsetminus \langle S \rangle) \mathcal{F}$ is disjoint from $\langle S \rangle \mathcal{F}$.

However, both of these sets are open (since $\mathcal{F}$ is open), and their union is all of Y (since $\Lambda \mathcal{F} = Y$). Therefore, since Y is connected, the two sets cannot both be nonempty. Since $\langle S \rangle$ is obviously nonempty, we conclude that $\Lambda \smallsetminus \langle S \rangle = \varnothing$, so $\langle S \rangle = \Lambda$. $\qquad\square$

(4.7.5) **Corollary.** *If* $\Gamma \backslash G$ *is compact, then* Γ *is finitely generated.*

Proof. Since $\Gamma \backslash G$ is compact, there is a compact subset C of G, such that $\Gamma C = G$ (see Exercise 4.1#11). Let $\mathcal{F}$ be a precompact, open subset of G, such that $C \subseteq \mathcal{F}$. Because $C \subseteq \mathcal{F}$, we have $\Gamma \mathcal{F} = G$. Also, because $\mathcal{F}$ is precompact, and Γ acts properly discontinuously on G, we know that Condition 4.7.2(2) holds. Therefore, $\mathcal{F}$ is a coarse fundamental domain for Γ. By passing to a subgroup of finite index, we may assume G is connected (see Exercise 1), so Proposition 4.7.4 applies. $\qquad\square$

(4.7.6) **Example.** Let $\mathcal{F}$ be the closed unit square in $\mathbb{R}^2$, so $\mathcal{F}$ is a coarse fundamental domain for the usual action of $\mathbb{Z}^2$ on $\mathbb{R}^2$ by translations. Define S as in the proof of Proposition 4.7.4, so

$$S = \{ (m, n) \in \mathbb{Z}^2 \mid m, n \in \{-1, 0, 1\} \} = \{0, \pm a_1, \pm a_2, \pm a_3\},$$

where $a_1 = (1, 0)$, $a_2 = (0, 1)$, and $a_3 = (1, 1)$. Then S generates $\mathbb{Z}^2$; in fact, the subset $\{a_1, a_2\}$ is already a generating set.

Proposition 4.7.4 does not apply to this situation, because $\mathcal{F}$ is not open. We could enlarge $\mathcal{F}$ slightly, without changing S. Alternatively, the proposition can be proved under the weaker hypothesis that $\mathcal{F}$ is in the interior of $\bigcup_{s \in S} \mathcal{F}$ (see Exercise 5).

Note that $\mathbb{Z}^2$ has the presentation

$$\mathbb{Z}^2 = \langle x_1, x_2, x_3 \mid x_1 x_2 = x_3, \ x_2 x_1 = x_3 \rangle.$$

(More precisely, if F_3 is the free group on 3 generators x_1, x_2, x_3, then there is a surjective homomorphism $\phi \colon F_3 \to \mathbb{Z}^2$, defined by

$$\phi(x_1) = a_1, \qquad \phi(x_2) = a_2, \qquad \phi(x_3) = a_3,$$

and the kernel of ϕ is the smallest normal subgroup of F_3 that contains both $x_1 x_2 x_3^{-1}$ and $x_2 x_1 x_3^{-1}$.) Each of the relations in this presentation is of a very simple form, merely stating that the product of two elements of S is equal to another element of S. The proof of the following proposition shows that relations of this type suffice to define Λ in a very general situation.

(4.7.7) **Proposition.** *Suppose Λ acts properly discontinuously on a topological space Y. If*

- *Y is both connected and simply connected, and*
- *there is a coarse fundamental domain $\mathcal{F}$ for Λ that is a connected, open subset of Y,*

then Λ is finitely presented.

Proof. This is somewhat similar to the proof of Proposition 4.7.4, but is more elaborate. As before, let $S = \{\lambda \in \Lambda \mid \lambda\mathcal{F} \cap \mathcal{F} \neq \varnothing\}$. For each $s \in S$, define a formal symbol x_s, and let F be the free group on $\{x_s\}$. Finally, let

$$R = \{x_s x_t x_{st}^{-1} \mid s, t, st \in S\},$$

so R is a finite subset of F.

We have a homomorphism $\phi\colon F \to \Lambda$ determined by $\phi(x_s) = s$. From the proof of Proposition 4.7.4, we know that ϕ is surjective, and it is clear that $R \subseteq \ker\phi$. The main part of the proof is to show that $\ker\phi$ is the smallest normal subgroup of F that contains R. (Since R is finite, and $F/\ker\phi \cong \Lambda$, this implies that Λ is finitely presented, as desired.)

Let N be the smallest normal subgroup of F that contains R. (It is clear that $N \subseteq \ker(\phi)$; we wish to show $\ker(\phi) \subseteq N$.)

- Define an equivalence relation $\sim$ on $(F/N) \times \mathcal{F}$, by stipulating that $(fN, y) \sim (f'N, y')$ if and only if there exists $s \in S$, such that $x_s fN = f'N$ and $sy = y'$ (see Exercise 6).
- Let $\tilde{Y}$ be the quotient space $((F/N) \times \mathcal{F})/\sim$.
- Define a map $\psi\colon (F/N) \times \mathcal{F} \to Y$ by $\psi(fN, y) = \phi(f^{-1})y$. (Note that, because $N \subseteq \ker(\phi)$, the map ψ is well defined.)

Because

$$\psi(x_s fN, sy) = (\phi(f^{-1})s^{-1})(sy) = \psi(fN, y),$$

we see that ψ factors through to a well-defined map $\tilde{\psi}\colon \tilde{Y} \to Y$.

Let $\tilde{\mathcal{F}}$ be the image of $(\ker(\phi)/N) \times \mathcal{F}$ in $\tilde{Y}$. Then it is obvious, from the definition of ψ, that $\tilde{\psi}(\tilde{\mathcal{F}}) = \mathcal{F}$. In fact, it is not difficult to see that $\tilde{\psi}^{-1}(\mathcal{F}) = \tilde{\mathcal{F}}$ (see Exercise 7).

For each $f \in F$, the image $\mathcal{F}_f$ of $(fN) \times \mathcal{F}$ in $\tilde{Y}$ is open (see Exercise 8), and, for $f_1, f_2 \in \ker(\phi)$, one can show that $\mathcal{F}_{f_1} \cap \mathcal{F}_{f_2} = \varnothing$ if $f_1 \not\equiv f_2 \pmod{N}$ (cf. Exercise 9). Therefore, from the preceding paragraph, we see that $\tilde{\psi}$ is a covering map over $\mathcal{F}$. Since Y is covered by translates of $\mathcal{F}$ (and $\mathcal{F}$ is open) it follows that $\tilde{\psi}$ is a covering map.

Because $\mathcal{F}$ is connected, it is not difficult to see that $\tilde{Y}$ is connected (see Exercise 10). Since Y is simply connected, and $\tilde{\psi}$ is a covering map, this implies that $\tilde{\psi}$ is a homeomorphism. Hence, $\tilde{\psi}$ is injective, and it is easy to see that this implies $\ker(\phi) = N$, as desired.   □

(4.7.8) *Remark.* The assumption that $\mathcal{F}$ is connected can be replaced with the assumption that Y is locally connected. However, the proof is somewhat more complicated in this setting.

(4.7.9) **Corollary.** *If* $\Gamma \backslash G$ *is compact, then* Γ *is finitely presented.*

Proof. Let K be a maximal compact subgroup of G, so Γ acts properly discontinuously on G/K. Arguing as in the proof of Corollary 4.7.5, we see that Γ has a coarse fundamental domain $\mathcal{F}$ that is an open subset of G/K. From the "Iwasawa decomposition" $G = KAN$ (see Theorem 8.4.9), we see that G/K is connected and simply connected (see Exercise 8.4#11(b)). So Proposition 4.7.7 implies that Γ is finitely presented.   □

If $\Gamma \backslash G$ is not compact, then it is more difficult to prove that Γ is finitely presented (or even finitely generated).

(4.7.10) **Theorem.** Γ *is finitely presented.*

Idea of proof. It suffices to find a coarse fundamental domain for Γ that is a connected, open subset of G/K. Assume, without loss of generality, that Γ is irreducible.

In each of the following two cases, a coarse fundamental domain $\mathcal{F}$ can be constructed as the union of finitely many translates of "Siegel sets." (This will be discussed in Chapters 7 and 19.)

1) Γ is "arithmetic," as defined in (5.1.19), or
2) G has a simple factor of real rank one, or, more generally, we have $\mathrm{rank}_{\mathbb{Q}} \Gamma \leq 1$ (see Definitions 8.1.6 and 9.1.4).

The (amazing!) Margulis Arithmeticity Theorem (5.2.1) implies that these two cases are exhaustive, which completes the proof.   □

(4.7.11) *Remark.* It is not necessary to appeal to the Margulis Arithmeticity Theorem in order to prove only that Γ is finitely generated (and not that it is finitely presented). Namely, if (2) does not apply, then the real rank of every simple factor of G is at least two, so Kazhdan's Property (T) implies that Γ is finitely generated (see Proposition 13.1.7(3)).

(4.7.12) *Remark.* For $n \geq 2$, Γ is said to be of *type* F_n if there is a compact CW complex X, such that:

• the fundamental group $\pi_1(X)$ is isomorphic to Γ, and
• the homotopy group $\pi_k(X)$ is trivial for $2 \leq k < n$.

Since Γ is finitely presented (see Theorem 4.7.10), it is easy to see that Γ is of type F_2. (In fact, a group is of type F_2 if and only if it is finitely presented.) Borel and Serre proved the much stronger result that Γ is of type F_n for every n. (If G/Γ is compact, and Γ is torsion free, then one may let $X = \Gamma\backslash G/K$, where K is a maximal compact subgroup of G. (In other words, X is the locally symmetric space associated to Γ.) When G/Γ is not compact, but Γ is torsion free, then X is a certain space called the "Borel-Serre compactification" of $\Gamma\backslash G/K$.)

Exercises for §4.7.

#1. Show that if some finite-index subgroup of Λ is finitely generated, then Λ is finitely generated.

[*Hint:* If F is a finite set of coset representatives for $\langle S\rangle$ in Λ, then $S \cup F$ generates Λ.]

#2. Assume Λ is abstractly commensurable to Λ'. Show Λ is finitely generated if and only if Λ' is finitely generated.

[*Hint:* Exercise 1 is half of the proof. For the other half, suppose F is a finite set of coset representatives for the subgroup Λ' of Λ, and S is a finite generating set for Λ. For each $f \in F$ and $s \in S$, we have $fs \in \Lambda' f'$, for some $f'_{f,s} \in F$. Then $\{fs(f'_{f,s})^{-1} \mid f \in F, s \in S\}$ generates Λ'. Alternatively, it is easy to prove this topologically: If Λ is the fundamental group of a CW-complex Σ with only finitely many 1-cells, then Λ' is the fundamental group of a finite cover of Σ, which must also have only finitely many 1-cells.]

#3. Assume Λ is abstractly commensurable to Λ'. Show Λ is finitely presented if and only if Λ' is finitely presented.

[*Hint:* Suppose Λ' has finite index in Λ, and F is a set of coset representatives. Let S and S' be finite generating sets of Λ and Λ', respectively (see Exercise 2).
($\Leftarrow$) For each $s \in S \cup F$ and $f \in F$, there exist $g \in \Lambda'$ and $f' \in F$, such that $fs = gf'$. Adding these relations to a presentation of Λ' yields a presentation of Λ.
($\Rightarrow$) Proving this direction algebraically is somewhat more complicated, but there is an easy topological proof: If Λ is the fundamental group of a CW-complex Σ whose 2-skeleton is finite, then Λ' is the fundamental group of a finite cover of Σ, which must also have only finitely many 1-cells and 2-cells.]

#4. Suppose Λ is a discrete subgroup of a locally compact group H. Show that if H/Λ is compact, and H is **compactly generated** (that is, there is a compact subset C of H, such that $\langle C\rangle = H$), then Λ is finitely generated. (This provides an alternate proof of Proposition 4.7.4 that does not require G to be connected.)

[*Hint:* Assume $e \in C$. Choose a compact subset $\mathcal{F}$ of H, such that $\mathcal{F}\Lambda = H$ (and $e \in \mathcal{F}$), and let $S = \Lambda \cap (\mathcal{F}^{-1}C^{\pm 1}\mathcal{F})$. If $\lambda = c_1 c_2 \cdots c_n$ with $c_i \in C^{\pm 1}$, then $\lambda = \lambda_1 \cdots \lambda_n$, with $\lambda_i \in S$.]

#5. Prove Proposition 4.7.4, replacing the assumption that $\mathcal{F}$ is open with the weaker assumption that $\mathcal{F}$ is in the interior of $\bigcup_{s\in S} s\mathcal{F}$ (where S is as defined in the proof of Proposition 4.7.4).

#6. Show that the relation $\sim$ defined in the proof of Proposition 4.7.7 is an equivalence relation.

#7. In the notation of the proof of Proposition 4.7.7, show that if $\psi(fN, y) \in \mathcal{F}$, then $(fN, y) \sim (f'N, y')$, for some $f' \in \ker(\phi)$ and some $y' \in \mathcal{F}$.

[*Hint:* We have $\phi(f) \in S$, because $\phi(f^{-1})y \in \mathcal{F}$.]

#8. In the notation of the proof of Proposition 4.7.7, show that the inverse image of $\mathcal{F}_f$ in $(F/N) \times \mathcal{F}$ is

$$\bigcup_{s \in S} ((x_s fN/N) \times (\mathcal{F} \cap s\mathcal{F})),$$

which is open.

#9. In the notation of the proof of Proposition 4.7.7, show that if we have $\mathcal{F}_f \cap \mathcal{F}_e \neq \varnothing$ and $f \in \ker(\phi)$, then $f \in N$.

[*Hint:* If $(fN, y_1) \sim (N, y_2)$, then there is some $s \in S$ with $x_s N = fN$. Since $f \in \ker(\phi)$, we have $s = \phi(x_s) = \phi(f) = e$.]

#10. Show that the set $\tilde{Y}$ defined in the proof of Proposition 4.7.7 is connected.

[*Hint:* For $s_1, \ldots, s_r \in S$, define $\mathcal{F}_j = \{x_{s_j} \cdots x_{s_1} N\} \times \mathcal{F}$. Show there exist $a \in \mathcal{F}_j$ and $b \in \mathcal{F}_{j+1}$, such that $a \sim b$.]

#11. Assume Λ acts properly discontinuously on a topological space Y. Show that a Borel subset $\mathcal{F}$ of Y is a coarse fundamental domain for Λ if and only if
    a) $\mathcal{F}$ contains a strict fundamental domain $\mathcal{F}_0$ for Λ, and
    b) there is a finite subset F of Λ, such that $\mathcal{F} \subseteq F\mathcal{F}_0$.

§4.8. Γ has a torsion-free subgroup of finite index

(4.8.1) **Definition.** A group is ***torsion free*** if it has no nontrivial finite subgroups. Equivalently, the identity element e is the only element of finite order.

(4.8.2) **Theorem** (Selberg's Lemma). *Γ has a torsion-free subgroup of finite index.*

Proof. From the Standing Assumptions (4.0.0), we know $\Gamma \subseteq \mathrm{SL}(\ell, \mathbb{R})$, for some ℓ. Let us start with an illustrative special case.

Case 1. Assume $\Gamma = \mathrm{SL}(\ell, \mathbb{Z})$. For any positive integer n, the natural ring homomorphism $\mathbb{Z} \to \mathbb{Z}/n\mathbb{Z}$ induces a group homomorphism $\Gamma \to \mathrm{SL}(\ell, \mathbb{Z}/n\mathbb{Z})$ (see Exercise 2); let Γ_n be the kernel of this homomorphism. (This is called the ***principal congruence subgroup*** of $\mathrm{SL}(\ell, \mathbb{Z})$ of level n.) Since it is the kernel of a group homomorphism, we know that Γ_n is a normal subgroup of Γ. It is also not difficult to see that Γ_n has finite index in Γ (see Exercise 3). It therefore suffices to show that Γ_n is torsion free, for some n. In fact, Γ_n is torsion free whenever $n \geq 3$ (see Exercise 5), but, for simplicity, we will assume $n = p$ is an odd prime.

Given $y \in \Gamma_p \smallsetminus \{\mathrm{Id}\}$ and $k \in \mathbb{N} \smallsetminus \{0\}$, we wish to show that $y^k \neq \mathrm{Id}$. We may write

$$y = \mathrm{Id} + p^d T,$$

where

- $d \geq 1$,
- $T \in \mathrm{Mat}_{\ell \times \ell}(\mathbb{Z})$, and
- $p \nmid T$ (that is, not every matrix entry of T is divisible by p).

Also, we may assume k is prime (see Exercise 4). Therefore, either $p \nmid k$ or $p = k$.

Subcase 1.1. Assume $p \nmid k$. Noting that

$$(p^d T)^2 = p^{2d} T^2 \equiv 0 \pmod{p^{d+1}},$$

and using the Binomial Theorem, we see that

$$y^k = (\mathrm{Id} + p^d T)^k \equiv \mathrm{Id} + k(p^d T) \not\equiv \mathrm{Id} \pmod{p^{d+1}},$$

as desired.

Subcase 1.2. Assume $p = k$. Using the Binomial Theorem (and noting that $\binom{p}{i} p^{di}$ is divisible by p^{d+2} for $i > 1$ (see Exercise 1)), we have

$$y^k = y^p = (\mathrm{Id} + p^d T)^p \equiv \mathrm{Id} + p(p^d T) = \mathrm{Id} + p^{d+1} T \not\equiv \mathrm{Id} \pmod{p^{d+2}}.$$

Case 2. Assume $\Gamma \subseteq \mathrm{SL}(\ell, \mathbb{Z})$. From Case 1, we know there is a torsion-free, finite-index subgroup Γ_n of $\mathrm{SL}(\ell, \mathbb{Z})$. Then $\Gamma \cap \Gamma_n$ is a torsion-free subgroup of finite index in Γ.

Case 3. The general case. The proof is very similar to Case 1, with the addition of some commutative algebra (or algebraic number theory) to account for the more general setting.

We know that Γ is finitely generated (see Theorem 4.7.10), so there exist $a_1, \ldots, a_r \in \mathbb{C}$, such that every matrix entry of every element of Γ is contained in the ring $Z = \mathbb{Z}[a_1, \ldots, a_r]$ generated by $\{a_1, \ldots, a_r\}$ (see Exercise 7). Therefore, letting $\Lambda = \mathrm{SL}(\ell, Z)$, we have $\Gamma \subseteq \Lambda$.

Now let $\mathfrak{p}$ be a maximal ideal in Z. Then $Z/\mathfrak{p}$ is a field, so, because $Z/\mathfrak{p}$ is also known to be a finitely generated ring, it must be a finite field. Therefore, the kernel of the natural homomorphism $\Lambda \to \mathrm{SL}(\ell, Z/\mathfrak{p})$ has finite index in Λ. Basic facts of Algebraic Number Theory allow us to work with the prime ideal $\mathfrak{p}$ in very much the same way as we used the prime number p in Case 1. $\qquad\square$

(4.8.3) **Warning.** Our standing assumption that $G \subseteq \mathrm{SL}(\ell, \mathbb{R})$ is needed for Theorem 4.8.2. For example, the group $\mathrm{Sp}(4, \mathbb{R})$ has an 8-fold cover, which we call H. The inverse image of $\mathrm{Sp}(4, \mathbb{Z})$ in H is a lattice Λ in H. It can be shown that every finite-index subgroup of Λ contains an element of order 2, so no subgroup of finite index is torsion free. This does not contradict Theorem 4.8.2, because H is not linear: it has no faithful embedding in any $\text{SL}(\ell, \mathbb{R})$.

If $y^k = \text{Id}$, then every eigenvalue of y must be a kth root of unity. If, in addition, $y \neq \text{Id}$, then at least one of these roots of unity must be nontrivial. Therefore, the following is a strengthening of Theorem 4.8.2.

(4.8.4) **Theorem.** *There is a finite-index subgroup Γ' of Γ, such that no eigenvalue of any element of Γ' is a nontrivial root of unity.*

Proof. Assume $\Gamma = \text{SL}(\ell, \mathbb{Z})$. Let

- n be some (large) natural number,
- Γ_n be the principal congruence subgroup of Γ of level n,
- ω be a nontrivial k^{th} root of unity, for some k,
- y be an element of Γ_n, such that ω is an eigenvalue of y,
- $T = y - \text{Id}$,
- $Q(x)$ be the characteristic polynomial of T, and
- $\lambda = \omega - 1$, so λ is a nonzero eigenvalue of T.

Since $y \in \Gamma_n$, we know that $n \mid T$, so $Q(x) = x^\ell + nR(x)$, for some integral polynomial $R(x)$. Since $Q(\lambda) = 0$, we conclude that $\lambda^\ell = n\zeta$, for some $\zeta \in \mathbb{Z}[\lambda]$. Therefore, λ^ℓ is divisible by n, in the ring of algebraic integers.

The proof can be completed by noting that any particular nonzero algebraic integer is divisible by only finitely many natural numbers, and there are only finitely many roots of unity that satisfy a monic integral polynomial of degree ℓ. See Exercise 8 for a slightly different argument. $\square$

(4.8.5) *Remarks.*

1) The proof of Theorem 4.8.2 shows that Γ has nontrivial, proper, normal subgroups, so Γ is not simple. However, the normal subgroups constructed there all have finite index. In fact, it is often the case that every nontrivial, normal subgroup of Γ has finite index (see Theorem 17.1.1).

   Moreover, although it will not be proved in this book, it is often the case that all of the normal subgroups of finite index are close to being of the type constructed in the course of the proof. More precisely, the "Congruence Subgroup Property" asserts there is a constant C, such that if N is any finite-index, normal subgroup of Γ, then there is a principal congruence subgroup Γ' of Γ, such that $|\Gamma' : N'| < C$. This is not always true, but it has been proved for the lattices in many groups.

2) Arguing more carefully, one can obtain a finite-index subgroup Γ' with the stronger property that, for every $\gamma \in \Gamma'$, the multiplicative group generated by the (complex) eigenvalues of γ does not contain any nontrivial roots of unity. Such a subgroup is sometimes called **net**.

3) If F is any field of characteristic zero, then Theorem 4.8.2 remains valid (with the same proof) when Γ is replaced with any finitely generated subgroup Λ of $\mathrm{SL}(\ell, F)$.

Let us now present an alternate approach to the general case of Theorem 4.8.2. It requires only the Nullstellensatz, not Algebraic Number Theory.

Another proof of Theorem 4.8.2 (optional). Let

- Z be the subring of $\mathbb{C}$ generated by the matrix entries of the elements of Γ, and
- F be the quotient field of Z.

Because Γ is a finitely generated group (see Theorem 4.7.10), we know that Z is a finitely generated ring (see Exercise 7), so F is a finitely generated extension of $\mathbb{Q}$.

Step 1. We may assume that $F = \mathbb{Q}(x_1, \ldots, x_r)$ is a purely transcendental extension of $\mathbb{Q}$. Choose a subfield $L = \mathbb{Q}(x_1, \ldots, x_r)$ of F, such that

- L is a purely transcendental extension of $\mathbb{Q}$, and
- F is an algebraic extension of L.

Let d be the degree of F over L. Because F is finitely generated (and algebraic over L), we know that $d < \infty$. Therefore, we may identify F^ℓ with $L^{d\ell}$, so there is an embedding

$$\Gamma \subseteq \mathrm{SL}(\ell, F) \hookrightarrow \mathrm{SL}(d\ell, L).$$

Hence, by replacing F with L (and replacing ℓ with $d\ell$), we may assume that F is purely transcendental. (Identifying F^ℓ with $L^{d\ell}$ is the foundation of an important technique called "Restriction of Scalars" that will be introduced in Section 5.5.)

Step 2. If γ is any element of finite order in $\mathrm{SL}(\ell, F)$, then $\mathrm{trace}(\gamma) \in \mathbb{Z}$, and $|\mathrm{trace}(\gamma)| \leq \ell$. There is a positive integer k with $\gamma^k = \mathrm{Id}$, so every eigenvalue of γ is a kth root of unity. The trace of γ is the sum of these eigenvalues, and any root of unity is an algebraic integer, so we conclude that the trace of γ is an algebraic integer.

Since $\mathrm{trace}(\gamma)$ is the sum of the diagonal entries of γ, we know that $\mathrm{trace}(\gamma) \in F$. Since $\mathrm{trace}(\gamma)$ is algebraic, but F is a purely transcendental extension of $\mathbb{Q}$, this implies $\mathrm{trace}(\gamma) \in \mathbb{Q}$. Since $\mathrm{trace}(\gamma)$ is an algebraic integer, this implies $\mathrm{trace}(\gamma) \in \mathbb{Z}$.

Since trace(γ) is the sum of ℓ roots of unity, and every root of unity is on the unit circle, we see, from the triangle inequality, that $| \text{trace}(\gamma)| \le \ell$.

Step 3. There is a prime number $p > 2\ell$, such that $1/p \notin Z$. From the Nullstellensatz (B4.5), we know that there is a nontrivial homomorphism $\phi \colon Z \to \overline{\mathbb{Q}}$, where $\overline{\mathbb{Q}}$ is the algebraic closure of $\mathbb{Q}$ in $\mathbb{C}$. Replacing Z with $\phi(Z)$, let us assume that $Z \subset \overline{\mathbb{Q}}$. Thus, for each $z \in Z$, there is some nonzero integer n, such that nz is an algebraic integer. More precisely, because Z is finitely generated, there is an integer n, such that, for each $z \in Z$, there is some positive integer k, such that $n^k z$ is an algebraic integer. It suffices to choose p so that it is not a divisor of n.

Step 4. There is a finite field E of characteristic p, and a nontrivial homomorphism $\phi_p \colon Z \to E$. Because $1/p \notin Z$, there is a maximal ideal $\mathfrak{p}$ of Z, such that $p \in \mathfrak{p}$. Then $E = Z/\mathfrak{p}$ is a field of characteristic p. Because it is a finitely generated ring, E must be a finite extension of the prime field $\mathbb{Z}/p\mathbb{Z}$ (see Theorem B4.3), so E is finite.

Step 5. Let Λ be the kernel of the homomorphism $\hat{\phi}_p \colon \mathrm{SL}(\ell, Z) \to \mathrm{SL}(\ell, E)$ that is induced by ϕ_p. Then Λ is torsion free. Let γ be an element of finite order in Λ. Then
$$\text{trace}(\hat{\phi}_p(\gamma)) = \text{trace}(\text{Id}) = \ell \ (\text{mod } p),$$
so $p \mid (\ell - \text{trace}(\gamma))$ (since Step 2 tells us that $\text{trace}(\gamma) \in \mathbb{Z}$). Since we also know that $| \text{trace}(\gamma)| \le \ell$ and $p > 2\ell$, we conclude that $\text{trace}(\gamma) = \ell$. Since the ℓ eigenvalues of γ are roots of unity, and $\text{trace}(\gamma)$ is the sum of these eigenvalues, we conclude that 1 is the only eigenvalue of γ. Since $\gamma^k = \text{Id}$, we know that γ is elliptic (hence, diagonalizable over $\mathbb{C}$), so this implies $\gamma = \text{Id}$, as desired. $\square$

Exercises for §4.8.

#1. Show that if p is an odd prime, $d \ge 1$, and $2 \le i \le p$, then $\binom{p}{i} p^{di}$ is divisible by p^{d+2}.
   [*Hint:* If either $d > 1$ or $i > 2$, then $di \ge d + 2$.]

#2. Show that $\mathrm{SL}(\ell, \cdot)$ is a (covariant) functor from the category of rings with identity to the category of groups. That is, show:
   a) if A is any ring with identity, then $\mathrm{SL}(\ell, A)$ is a group,
   b) for every ring homomorphism $\phi \colon A \to B$ (with $\phi(1) = 1$), there is a group homomorphism $\phi_* \colon \mathrm{SL}(\ell, A) \to \mathrm{SL}(\ell, B)$, and
   c) if $\phi \colon A \to B$ and $\psi \colon B \to C$ are ring homomorphisms (with $\phi(1) = 1$ and $\psi(1) = 1$), then $(\psi \circ \phi)_* = \psi_* \circ \phi_*$.

#3. Show that if B is a finite ring with identity, then $\mathrm{SL}(\ell, B)$ is finite. Use this fact to show, for every positive integer n, that if Γ_n denotes the principal congruence subgroup of $SL(\ell, \mathbb{Z})$ of level n (cf. Case 1 of the proof of Theorem 4.8.2), then Γ_n has finite index in $SL(\ell, \mathbb{Z})$.

#4. Show that if Γ has a nontrivial element of finite order, then Γ has an element of prime order.

#5. Show the principal congruence subgroup Γ_n is torsion free if $n \geq 3$.
[*Hint:* Since $\Gamma_m \subseteq \Gamma_n$ whenever $n \mid m$, you may assume, without loss of generality, that n is either 4 or an odd prime.]

#6. In the notation of Case 1 of the proof of Theorem 4.8.2, show that Γ_2 is not torsion free. Where does your solution of Exercise 5 fail?

#7. Show that if Λ is a finitely generated subgroup of $SL(\ell, \mathbb{C})$, then there is a finitely generated subring B of $\mathbb{C}$, such that $\Lambda \subset SL(\ell, B)$.
[*Hint:* Let B be the subring of $\mathbb{C}$ generated by the matrix entries of the generators of Λ.]

#8. Suppose ω is a nontrivial root of unity, and $(\omega - 1)^\ell = n\zeta$, for some $n, \ell \in \mathbb{Z}^+$ and $\zeta \in \mathbb{Z} + \mathbb{Z}\omega + \cdots + \mathbb{Z}\omega^{\ell-1}$. Show $n < 2^{(\ell+1)!}$.
[*Hint:* Let F be the Galois closure of the field extension $\mathbb{Q}(\omega)$ of $\mathbb{Q}$ generated by ω, and define $N: F \to \mathbb{Q}$ by $N(x) = \prod_{\sigma \in \text{Gal}(F/\mathbb{Q})} \sigma(x)$. Then $N(\omega - 1)^\ell = n^d N(\zeta)$, and $|N(\omega - 1)| \leq 2^d \leq 2^{\ell!}$, where d is the degree of F over $\mathbb{Q}$.]

#9. Show that Γ is residually finite. That is, for every $\gamma \in \Gamma \smallsetminus \{e\}$, show that there is a finite-index, normal subgroup Γ' of Γ, such that $\gamma \notin \Gamma'$. (In particular, if Γ is infinite, then Γ is *not* a simple group.)

#10. Show there is a sequence $N_1, N_2, \ldots$ of subgroups of Γ, such that
   a) $N_1 \supset N_2 \supset \cdots$,
   b) each N_k is a finite-index, normal subgroup of Γ, and
   c) $N_1 \cap N_2 \cap \cdots = \{e\}$.
[*Hint:* Use Exercise 9.]

#11. Show that $SL(n, \mathbb{Q})$ commensurates $SL(n, \mathbb{Z})$.
[*Hint:* For each $g \in SL(n, \mathbb{Q})$, there is a principal congruence subgroup Γ_m of $SL(n, \mathbb{Z})$, such that $g^{-1}\Gamma_m g \subseteq SL(n, \mathbb{Z})$.]

#12. Show that if $\varphi: \Gamma \to GL(n, \mathbb{R})$ is a homomorphism, such that every element of $\varphi(\Gamma)$ has finite order, then $\varphi(\Gamma)$ is finite.
[*Hint:* Remark 4.8.5(3).]

§4.9. Γ has a nonabelian free subgroup

In this section, we describe the main ideas in the proof of the following important result.

(4.9.1) **Theorem** (Tits Alternative). *If Λ is a subgroup of $SL(\ell, \mathbb{R})$, then either*

   1) *Λ contains a nonabelian free group, or*
   2) *Λ has a solvable subgroup of finite index.*

Since Γ is not solvable when G is not compact (see Exercise 4.5#6), the following is an immediate corollary.

(4.9.2) Corollary. *If G is not compact, then Γ contains a nonabelian free group.*

(4.9.3) Definition. Let us say that a homeomorphism ϕ of a topological space M is (A^-, B, A^+)**-contracting** if A^-, B and A^+ are nonempty, disjoint, open subsets of M, such that $\phi(B \cup A^+) \subseteq A^+$ and $\phi^{-1}(B \cup A^-) \subseteq A^-$.

In a typical example, A^- and A^+ are small neighborhoods of points p^- and p^+, such that ϕ collapses a large open subset of M into A^+, and ϕ^{-1} collapses a large open subset of M into A^- (see Figure 4.9A).

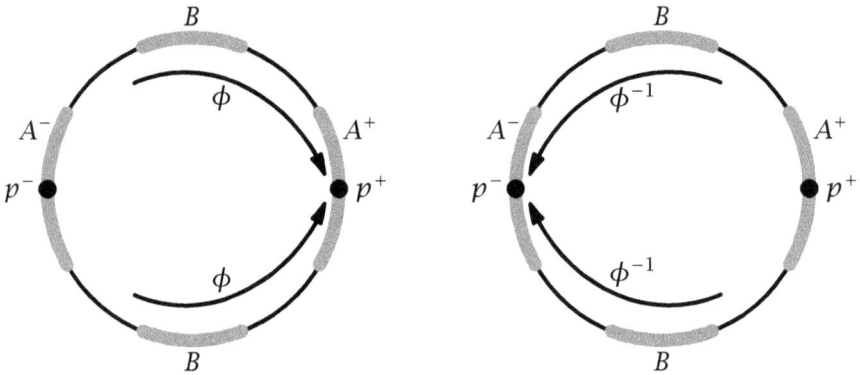

FIGURE 4.9A. A typical (A^-, B, A^+)-contracting homeomorphism of the circle.

(4.9.4) Example. Let

- M be the real projective line $\mathbb{P}(\mathbb{R}^2)$,
- $y = \begin{bmatrix} 2 & 0 \\ 0 & 1/2 \end{bmatrix} \in \mathrm{SL}(2, \mathbb{R})$,
- A^- be any (small) neighborhood of $p^- = [0:1]$ in $\mathbb{P}(\mathbb{R}^2)$,
- A^+ be any (small) neighborhood of $p^+ = [1:0]$ in $\mathbb{P}(\mathbb{R}^2)$, and
- B be any precompact, open subset of $\mathbb{P}(\mathbb{R}^2) \smallsetminus \{p^-, p^+\}$.

For any $(x, y) \in \mathbb{R}^2$ with $x \neq 0$, we have

$$y^n[x:y] = [2^n x : 2^{-n} y] = [1 : 2^{-2n} y/x] \to [1:0] = p^+ \text{ as } n \to \infty,$$

and the convergence is uniform on compact subsets. Similarly, we have $y^{-n}[x:y] \to p^-$ as $n \to \infty$. Hence, for sufficiently large n, the homeomorphism y^n is (A^-, B, A^+)-contracting on $\mathbb{P}(\mathbb{R}^2)$.

More generally, if y is any nontrivial, hyperbolic element of $SL(2, \mathbb{R})$, then y^n is (A^-, B, A^+)-contracting on $\mathbb{P}(\mathbb{R}^2)$, for some appropriate choice of A^-, B, and A^+ (see Exercise 1).

The following is easy to prove by induction on n.

(4.9.5) Lemma. *If ϕ is (A^-, B, A^+)-contracting, then*

1) $\phi^n(B) \subseteq A^+$ *for all $n > 0$,*
2) $\phi^n(B) \subseteq A^-$ *for all $n < 0$,*
3) $\phi^n(B) \subseteq A^- \cup A^+$ *for all $n \neq 0$.*

The following lemma is the key to the proof of Theorem 4.9.1.

(4.9.6) Lemma (Ping-Pong Lemma). *Suppose*

- *ϕ and ψ are homeomorphisms of a topological space M,*
- *A^-, A^+, B^-, and B^+ are nonempty, pairwise-disjoint, open subsets of M,*
- *ϕ is (A^-, B, A^+)-contracting, where $B = B^- \cup B^+$, and*
- *ψ is (B^-, A, B^+)-contracting, where $A = A^- \cup A^+$.*

Then ϕ and ψ have no nontrivial relations; so $\langle \phi, \psi \rangle$ is free.

Proof. Consider a word of the form $w = \phi^{m_1} \psi^{n_1} \ldots \phi^{m_k} \psi^{n_k}$, with each m_j and n_j nonzero. We wish to show $w \neq e$.

From Lemma 4.9.5(3), we have

$$\phi^{m_j}(B) \subseteq A \qquad \text{and} \qquad \psi^{n_j}(A) \subseteq B, \qquad \text{for } j = 1, 2, \ldots, k.$$

Therefore

$$\psi^{n_k}(A) \subseteq B,$$
$$\phi^{m_k} \psi^{n_k}(A) \subseteq A,$$
$$\psi^{n_{k-1}} \phi^{m_k} \psi^{n_k}(A) \subseteq B,$$
$$\phi^{m_{k-1}} \psi^{n_{k-1}} \phi^{m_k} \psi^{n_k}(A) \subseteq A,$$

and so on: points bounce back and forth between A and B. (Hence, the name of the lemma.) In the end, we see that $w(A) \subseteq A$.

Assume, for definiteness, that $m_1 > 0$. Then, by applying 4.9.5(1) in the last step, instead of 4.9.5(3), we obtain the more precise conclusion that $w(A) \subseteq A^+$. Since $A \not\subseteq A^+$ (recall that A^- is disjoint from A^+), we conclude that $w \neq e$, as desired. $\square$

(4.9.7) Corollary. *If y_1 and y_2 are two nontrivial hyperbolic elements of $SL(2, \mathbb{R})$ that have no common eigenvector, then, for sufficiently large $n \in \mathbb{Z}^+$, the group $\langle (y_1)^n, (y_2)^n \rangle$ is free.*

Proof. Let

- v_j and w_j be linearly independent eigenvectors of y_j, with eigenvalues λ_j and $1/\lambda_j$, such that $\lambda_j > 1$,

- A^+ and A^- be small neighborhoods of $[v_1]$ and $[w_1]$ in $\mathbb{P}(\mathbb{R}^2)$, and
- B^+ and B^- be small neighborhoods of $[v_2]$ and $[w_2]$ in $\mathbb{P}(\mathbb{R}^2)$.

By the same argument as in Example 4.9.4, we see that if n is sufficiently large, then

- $(\gamma_1)^n$ is $(A^-, B^- \cup B^+, A^+)$-contracting, and
- $(\gamma_2)^n$ is $(B^-, A^- \cup A^+, B^+)$-contracting

(see Exercise 1). Therefore, the Ping-Pong Lemma (4.9.6) implies that $\langle (\gamma_1)^n, (\gamma_2)^n \rangle$ is free.   □

We can now give a direct proof of Corollary 4.9.2, in the special case where $G = \mathrm{SL}(2, \mathbb{R})$.

(4.9.8) Corollary. *If* $G = \mathrm{SL}(2, \mathbb{R})$, *then* Γ *contains a nonabelian, free group.*

Proof. By passing to a subgroup of finite index, we may assume that Γ is torsion free (see Theorem 4.8.2). Hence, Γ has no elliptic elements. Not every element of Γ is unipotent (see Exercise 4.5#10), so we conclude that some nontrivial element γ_1 of Γ is hyperbolic.

Let v and w be linearly independent eigenvectors of γ_1. The Borel Density Theorem (4.5.6) implies that there is some $\gamma \in \Gamma$, such that $\{\gamma v, \gamma w\} \cap (\mathbb{R}v \cup \mathbb{R}w) = \varnothing$ (see Exercise 2). Let $\gamma_2 = \gamma \gamma_1 \gamma^{-1}$, so γ_2 is a hyperbolic element of Γ with eigenvectors γv and γw.

From Corollary 4.9.7, we conclude that $\langle (\gamma_1)^n, (\gamma_2)^n \rangle$ is a nonabelian, free subgroup of Γ, for some $n \in \mathbb{Z}^+$.   □

The same ideas work in general:

Idea of direct proof of Corollary 4.9.2. Assume $G \subseteq \mathrm{SL}(\ell, \mathbb{R})$. Choose some nontrivial, hyperbolic element γ_1 of Γ, and let $\lambda_1 \geq \lambda_2 \geq \cdots \geq \lambda_\ell$ be its eigenvalues. We may assume, without loss of generality, that $\lambda_1 > \lambda_2$. (If the eigenvalue λ_1 has multiplicity d, then we may pass to the dth exterior power $\wedge^d(\mathbb{R}^\ell)$, to obtain a representation in which the largest eigenvalue of γ_1 is simple.)

Let us assume that the smallest eigenvalue λ_ℓ is also simple; that is, $\lambda_\ell < \lambda_{\ell-1}$. (One can show that this is a generic condition in G, so it can be achieved by replacing γ_1 with some other element of Γ.)

Let v be an eigenvector corresponding to the eigenvalue λ_1 of γ_1, and let w be an eigenvector for the eigenvalue λ_ℓ. Assume, to simplify the notation, that all of the eigenspaces of γ_1 are orthogonal to each other. Then, for any $x \in \mathbb{R}^\ell \smallsetminus v^\perp$, we have $(\gamma_1)^n[x] \to [v]$ in $\mathbb{P}(\mathbb{R}^\ell)$, as $n \to \infty$ (see Exercise 3). Similarly, if $x \notin w^\perp$, then $(\gamma_1)^{-n}[x] \to [w]$.

We may assume, by replacing $\mathbb{R}^\ell$ with a minimal G-invariant subspace, that $\mathbb{R}^\ell$ has no nontrivial, proper, G-invariant subspaces. Then the

Borel Density Theorem implies that there exists $y \in \Gamma$, such that we have $\{yv, yw\} \cap (\mathbb{R}v \cup \mathbb{R}w) = \emptyset$.

Then, for any small neighborhoods A^-, A^+, B^-, and B^+ of $[v]$, $[w]$, $[yv]$, and $[yw]$, and any sufficiently large n, the Ping-Pong Lemma implies that the subgroup $\langle (\gamma_1)^n, (y\gamma_1 y^{-1})^n \rangle$ is free.   □

(4.9.9) *Remark.* The proof of Theorem 4.9.1 is similar, but involves additional complications.

1) In order to replace $\mathbb{R}^\ell$ with an irreducible subspace W, it is necessary to have $\dim W > 1$ (otherwise, there do not exist two linearly independent eigenvectors v and w). Unfortunately, the minimal Λ-invariant subspaces may be 1-dimensional. After modding these out, the minimal subspaces in the quotient may also be 1-dimensional, and so on. In this case, the group Λ consists entirely of upper-triangular matrices (after a change of basis), so Λ is solvable.

2) The subgroup Λ may not have any hyperbolic elements. Even worse, it may be the case that 1 is the absolute value of every eigenvalue of every element of Λ. (For example, Λ may be a subgroup of the compact group $SO(n)$, so that every element of Λ is elliptic.) In this case, the proof replaces the usual absolute value with an appropriate p-adic norm. Not all eigenvalues are roots of unity (cf. Theorem 4.8.4), so Algebraic Number Theory tells us that some element of Λ has an eigenvalue whose p-adic norm is greater than 1. The proof is completed by using this eigenvalue, and the corresponding eigenvector, just as we used λ_1 and the corresponding eigenvector v.

Exercises for §4.9.

#1. In the notation of the proof of Corollary 4.9.7, show that if A^-, A^+, B^-, and B^+ are disjoint, then, for all large n, the homeomorphism $(\gamma_1)^n$ is $(A^-, B^- \cup B^+, A^+)$-contracting on $\mathbb{P}(\mathbb{R}^2)$.

#2. Assume that G is irreducible in $SL(\ell, \mathbb{R})$ (see Definition A7.3), and that Γ projects densely into the maximal compact factor of G. If F is a finite subset of $\mathbb{R}^\ell \smallsetminus \{0\}$, and $\mathcal{W}$ is a finite set of proper subspaces of $\mathbb{R}^\ell$, show that there exists $y \in \Gamma$, such that

$$yF \cap \bigcup_{W \in \mathcal{W}} W = \emptyset.$$

[*Hint:* For $v \in F$ and $W \in \mathcal{W}$, the set $A_{v,W} = \{g \in G \mid gv \in W\}$ is Zariski closed, so $\bigcup_{v,W} A_{v,W}$ is Zariski closed. Apply the Borel Density Theorem and Exercise A4#7.]

#3. Let
  - y be a hyperbolic element of $SL(\ell, \mathbb{R})$,
  - $\lambda_1 > \lambda_2 \geq \cdots \geq \lambda_\ell$ be the eigenvalues of y,
  - v be an eigenvector of y corresponding to the eigenvalue λ_1, and
  - W be the sum of the other eigenspaces.

  Show that if $x \in \mathbb{P}(\mathbb{R}^\ell) \smallsetminus [W]$, then $y^n x \to [v]$ as $n \to \infty$. Furthermore, the convergence is uniform on compact subsets.

#4. (another version of the Ping-Pong Lemma) Suppose A and B are disjoint, nonempty subsets of M, such that $\phi^n(B) \subseteq A$ and $\psi^n(A) \subseteq B$, for every *nonzero* integer n. Show $\langle \phi, \psi \rangle$ is free.

  [*Hint:* If every m_j and n_j is nonzero, then $\phi^{m_1} \psi^{n_1} \cdots \phi^{m_k} \psi^{n_k} \phi^{m_{k+1}}(B) \subseteq A$.]

#5. Let $\Gamma' = \left\langle \begin{bmatrix} 1 & 2 \\ 0 & 1 \end{bmatrix}, \begin{bmatrix} 1 & 0 \\ 2 & 1 \end{bmatrix} \right\rangle$ be the ***Sanov subgroup*** of $SL(2, \mathbb{Z})$. Show that Γ' is a free subgroup of finite index in $SL(2, \mathbb{Z})$.

  [*Hint:* Exercise 4 implies Γ' is free. Matrices of the form $\begin{bmatrix} 4k+1 & 2\ell \\ 2m & 4n+1 \end{bmatrix}$ are in Γ'.]

#6. Generalizing Exercise 5, show that every torsion-free subgroup of $SL(2, \mathbb{Z})$ is a free group.

  [*Hint:* Let $\partial \mathcal{F}$ be the boundary of the usual fundamental domain for the action of $SL(2, \mathbb{Z})$ on the upper half plane $\mathfrak{H}^2$ (see Figure 1.3A). Then $\bigcup_{\gamma \in \Gamma} \gamma \cdot \partial \mathcal{F}$ is a contractible 1-dimensional simplicial complex; in other words, it is a tree. Γ acts properly on this tree, so any subgroup of Γ that acts freely must be a free group.]

#7. Show there is an irreducible lattice Γ in $SL(2, \mathbb{R}) \times SO(3)$, such that $\Gamma \cap SL(2, \mathbb{R})$ is infinite.

  [*Hint:* There is a free group F and a homomorphism $\phi \colon F \to SO(3)$, such that $\phi(F)$ is dense in $SO(3)$.]

§4.10. Moore Ergodicity Theorem

All mathematicians encounter situations in which they would like to prove that some function φ on some space X is constant. If $X = G/\Gamma$, this means that they would like to prove φ is G-invariant.

(4.10.1) **Definition.** Suppose f is a function on G/Γ, and H is a subgroup of G. We say that f is ***H-invariant*** if $f(hx) = f(x)$ for all $h \in H$ and $x \in G/\Gamma$.

The following fundamental result shows that it suffices to prove φ is invariant under a much smaller subgroup of G (if we make the very weak assumption that φ is measurable). It does not suffice to prove that φ is invariant under a compact subgroup, because it is easy to find a non-constant, continuous function on G/Γ that is invariant under any given compact subgroup of G (unless G itself is compact) (see Exercise 1), so the following result is optimal.

(4.10.2) **Theorem** (Moore Ergodicity Theorem). *Suppose*
- *G is connected and simple,*
- *H is a closed, noncompact subgroup of G, and*
- $\varphi \colon G/\Gamma \to \mathbb{C}$ *is H-invariant and measurable.*

Then φ is constant (a.e.).

Exercise 3 shows how to derive this theorem from the following more general result that replaces Γ with a discrete subgroup that need not be a lattice. (This generalization will be a crucial ingredient in Section 7.4's proof of the important fact that $SL(n,\mathbb{Z})$ is a lattice in $SL(n,\mathbb{R})$.) In this more general situation, we impose an $\mathscr{L}^p$-integrability hypothesis on φ, in order to compensate for the fact that G/Λ is not assumed to have finite measure (cf. Exercise 2).

(4.10.3) **Theorem.** *Suppose*
- *G is connected and simple,*
- *H is a closed, noncompact subgroup of G,*
- *Λ is a discrete subgroup of G, and*
- *φ is an H-invariant $\mathscr{L}^p$-function on G/Λ (with $1 \le p < \infty$).*

Then φ is constant (a.e.).

Idea of proof. To illustrate the key ingredient in the proof, let us consider only the special case where $G = SL(2,\mathbb{R})$ and H is the group of diagonal matrices. (A proof of the general case will be given in Section 11.2.) Let

$$a^t = \begin{bmatrix} e^t & 0 \\ 0 & e^{-t} \end{bmatrix} \in H \quad \text{and} \quad u \in \begin{bmatrix} 1 & 0 \\ * & 1 \end{bmatrix}.$$

Note that straightforward matrix multiplication (see Exercise 4) shows

$$\lim_{t\to\infty} a^t u a^{-t} = e. \tag{4.10.4}$$

For $g \in G$, define $g\varphi \colon G/\Lambda \to \mathbb{C}$ by $g\varphi(x) = \varphi(g^{-1}x)$. We are assuming that φ is a^t-invariant (which means $a^t\varphi = \varphi$), and the crux of the proof is the observation that we can use (4.10.4) to show that φ must also be u-invariant: we have

$$\|u\varphi - \varphi\|_p = \|a^t u\varphi - a^t\varphi\|_p \qquad \text{(Exercise 10)}$$
$$= \|(a^t u a^{-t})a^t\varphi - a^t\varphi\|_p \qquad \text{(inserting } a^{-t}a^t)$$
$$= \|(a^t u a^{-t})\varphi - \varphi\|_p \qquad \begin{pmatrix} a^t\varphi = \varphi \text{ because} \\ \varphi \text{ is H-invariant} \end{pmatrix}$$
$$\to \|e\varphi - \varphi\|_p \quad \text{as } t\to\infty \qquad \begin{pmatrix} (4.10.4) \text{ and} \\ \text{Exercise 11} \end{pmatrix}$$
$$= 0,$$

so $u\varphi = \varphi$ (a.e.).

Thus, from the fact that φ is H-invariant, we have shown that
$$\varphi \text{ must also be } \left[\begin{smallmatrix} 1 & 0 \\ * & 1 \end{smallmatrix}\right]\text{-invariant (a.e.).}$$
The same calculation, but with $t \to -\infty$, shows that
$$\varphi \text{ must also be } \left[\begin{smallmatrix} 1 & * \\ 0 & 1 \end{smallmatrix}\right]\text{-invariant (a.e.).}$$
Since $\left[\begin{smallmatrix} 1 & 0 \\ * & 1 \end{smallmatrix}\right]$ and $\left[\begin{smallmatrix} 1 & * \\ 0 & 1 \end{smallmatrix}\right]$ generate $SL(2, \mathbb{R}) = G$ (see Exercise 6), we conclude that φ is G-invariant (a.e.). Since G is transitive on G/Λ, this implies that φ is constant (a.e.) (see Exercise 7).   □

Exercises for §4.10.

#1. Show that if K is any compact subgroup of G, and G is not compact, then there is a continuous, K-invariant function on G/Γ that is not constant.

#2. Show there is a counterexample to Theorem 4.10.3 if we remove the assumption that the measurable function φ is $\mathscr{L}^p$.
   [*Hint:* It is easy to construct an counterexample by taking Λ to be trivial (or finite).]

#3. Derive Theorem 4.10.2 from Theorem 4.10.3.
   [*Hint:* If there is a nonconstant H-invariant function on G/Γ, then there is one that is bounded.]

#4. Verify Equation (4.10.4).

#5. Suppose
   - G and H are as in the Moore Ergodicity Theorem (4.10.2), and
   - X is an H-invariant, measurable subset of G/Γ.
   Show that either X has measure 0, or the complement of X has measure 0.

#6. Show $SL(2, \mathbb{R})$ is generated by the subgroups $\left\{\left[\begin{smallmatrix} 1 & 0 \\ * & 1 \end{smallmatrix}\right]\right\}$ and $\left\{\left[\begin{smallmatrix} 1 & * \\ 0 & 1 \end{smallmatrix}\right]\right\}$.
   [*Hint:* If a matrix can be reduced to the identity matrix by a sequence of elementary row operations, then it is a product of elementary matrices.]

#7. Suppose φ is a measurable function on G/Λ, and for each $g \in G$, we have $\varphi(gx) = \varphi(x)$ for a.e. $x \in G/\Lambda$. Show φ is constant (a.e.).
   [*Hint:* Use Fubini's Theorem to reverse the quantifiers.]

#8. Suppose G and H are as in the Moore Ergodicity Theorem (4.10.2). Show that Hx is dense in G/Γ, for a.e. $x \in G/\Gamma$.
   [*Hint:* Use Exercise 5. For any open subset $\mathcal{O}$ of G/Γ, the set $H\mathcal{O}$ is measurable (why?) and H-invariant.]

#9. Assume G is simple, and let H be a subgroup of G (not necessarily closed). Show that every (real-valued) H-invariant measurable function on G/Γ is constant (a.e.) if and only if the closure of H is not compact.

#10. Show that if Λ is a discrete subgroup of G, and μ is a G-invariant measure on G/Λ, then $\int_X g\varphi \, d\mu = \int_X \varphi \, d\mu$, for every $g \in G$ and measurable $\varphi \colon G/\Lambda \to \mathbb{C}$.

[*Hint:* Apply a change of variables, and use the fact that $g_*\mu = \mu$ (since μ is G-invariant).]

#11. Show that G acts continuously on $\mathscr{L}^p(G/\Lambda)$ if $1 \le p < \infty$. More precisely, show that if Λ is a discrete subgroup of G, and we define $\alpha \colon G \times \mathscr{L}^p(G/\Lambda) \to \mathscr{L}^p(G/\Lambda)$ by $\alpha(g, \varphi) = g\varphi$, then α is continuous.

[*Hint:* To show α is continuous in g, use Lusin's Theorem (B6.6) to approximate φ by a uniformly continuous function. Then use Exercise 10 (and the Triangle Inequality) to complete the proof.]

Notes

Raghunathan's book [16] is the standard reference for the basic properties of lattices. It contains almost all of the material in this chapter, except the Tits Alternative (Section 4.9) and the Moore Ergodicity Theorem (Section 4.10).

Remark 4.4.5 (the existence of unipotent elements in noncompact lattices) was proved by Kazhdan and Margulis [13]. Expositions can be found in [3] and [16, Cor. 11.13, p. 180].

The Borel Density Theorem (4.5.1) was proved by Borel [2]. It appears in [14, Thm. 2.4.4, p. 93], [16, Thm. 5.5, p. 79], and [21, Thm. 3.2.5, pp. 41–42]. Several authors have published generalizations or alternative proofs (for example, [6, 9, 20]).

Our presentation of Propositions 4.7.4 and 4.7.7 is based on [15, pp. 195–199]. A proof of Remark 4.7.8 can also be found there. A proof of Theorem 4.7.10 for the case where Γ is arithmetic can be found in [4] or [15, Thm. 4.2, p. 195]. For the case where $\mathrm{rank}_{\mathbb{Q}} \Gamma = 1$, see [10] or [16, Cor. 13.20, p. 210].

Borel and Serre [5, §11.1] proved Γ is of type F_n (see Remark 4.7.12). (We remark that there is no harm in assuming Γ is torsion free, since being of type F_n is invariant under passage to finite-index subgroups [11, Cor. 7.2.4, p. 170].)

Theorem 4.8.2 is proved in [16, Thm. 6.11, p. 93] and [4, Cor. 17.7, p. 119], in stronger forms that establish Remark 4.8.5(2,3). Our alternate proof of Theorem 4.8.2 is excerpted from the elementary proof in [1].

Warning 4.8.3 is due to P. Deligne [8]. See also [17].

For an introduction to the Congruence Subgroup Property, see [12, Chap. 6] or [18].

The Tits Alternative (4.9.1) was proved by Tits [19]. A nice introduction (and a proof of some special cases) can be found in [7].

See Section 14.2 for more on the Moore Ergodicity Theorem (4.10.2) and related results.

References

[1] R. C. Alperin: An elementary account of Selberg's Lemma, *L'Enseign. Math.* 33 (1987) 269–273. MR 0925989,
http://dx.doi.org/10.5169/seals-87896

[2] A. Borel: Density properties for certain subgroups of semi-simple groups without compact components, *Ann. Math.* 72 (1960) 179–188. MR 0123639, http://dx.doi.org/10.2307/1970150

[3] A. Borel: Sous-groupes discrets de groups semi-simples (d'après D. A. Kajdan et G. A. Margoulis), *Séminaire Bourbaki* 1968/1969, no. 358. *Springer Lecture Notes in Math.* 175 (1971) 199–215. MR 3077127,
http://www.numdam.org/item?id=SB_1968-1969__11__199_0

[4] A. Borel: *Introduction aux Groupes Arithmétiques.* Hermann, Paris, 1969. MR 0244260

[5] A. Borel and J.-P. Serre: Corners and arithmetic groups, *Comment. Math. Helv.* 48 (1973), 436–491. MR 0387495,
http://dx.doi.org/10.5169/seals-37166

[6] S. G. Dani: On ergodic quasi-invariant measures of group automorphism, *Israel J. Math.* 43 (1982) 62–74. MR 0728879,
http://dx.doi.org/10.1007/BF02761685

[7] P. de la Harpe: Free groups in linear groups, *L'Enseign. Math.* 29 (1983) 129–144. MR 0702736,
http://dx.doi.org/10.5169/seals-52975

[8] P. Deligne: Extensions centrales non résiduellement finies de groupes arithmétiques, *C. R. Acad. Sci. Paris Ser. A* 287 (1978), no. 4, 203–208. MR 0507760

[9] H. Furstenberg: A note on Borel's Density Theorem, *Proc. Amer. Math. Soc.* 55 (1976) 209–212. MR 0422497,
http://dx.doi.org/10.1090/S0002-9939-1976-0422497-X

[10] H. Garland and M. S. Raghunathan: Fundamental domains for lattices in ($\mathbb{R}$-)rank 1 semisimple Lie groups, *Ann. Math.* 92 (1970) 279–326. MR 0267041, http://dx.doi.org/10.2307/1970838

[11] R. Geoghegan: *Topological Methods in Group Theory.* Springer, New York, 2008. ISBN 978-0-387-74611-1, MR 2365352

[12] J. E. Humphreys: *Arithmetic Groups.* Springer, Berlin, 1980. ISBN 3-540-09972-7, MR 0584623

[13] D. Každan and G. A. Margulis: A proof of Selberg's hypothesis, *Math. USSR–Sbornik* 4 (1968), no. 1, 147–152. (Translated from *Mat. Sb. (N.S.)* 75 (117) (1968) 163–168. MR 0223487,
http://dx.doi.org/10.1070/SM1968v004n01ABEH002782

[14] G. A. Margulis: *Discrete Subgroups of Semisimple Lie Groups.* Springer, Berlin Heidelberg New York, 1991. ISBN 3-540-12179-X, MR 1090825

[15] V. Platonov and A. Rapinchuk: *Algebraic Groups and Number Theory.* Academic Press, Boston, 1994. ISBN 0-12-558180-7, MR 1278263

[16] M. S. Raghunathan: *Discrete Subgroups of Lie Groups.* Springer, New York, 1972. ISBN 0-387-05749-8, MR 0507234

[17] M. S. Raghunathan: Torsion in cocompact lattices in coverings of Spin(2, n), *Math. Ann.* 266 (1984), no. 4, 403–419. Correction *Math. Ann.* 303 (1995), no. 3, 575–578. MR 0735524, MR 1355004, http://resolver.sub.uni-goettingen.de/purl?GDZPPN002324733, http://eudml.org/doc/251794

[18] B. Sury: *The Congruence Subgroup Problem.* Hindustan Book Agency, New Delhi, 2003. ISBN 81-85931-38-0, MR 1978430

[19] J. Tits: Free subgroups in linear groups, *J. Algebra* 20 (1972) 250–270. MR 0286898, http://dx.doi.org/10.1016/0021-8693(72)90058-0

[20] D. Wigner: Un théorème de densité analytique pour les groupes semisimples, *Comment. Math. Helvetici* 62 (1987) 390–416. MR 0910168, http://dx.doi.org/10.5169/seals-47353

[21] R. J. Zimmer: *Ergodic Theory and Semisimple Groups.* Birkhäuser, Boston, 1984. ISBN 3-7643-3184-4, MR 0776417

Chapter 5

What is an Arithmetic Group?

SL($n, \mathbb{Z}$) is the most basic example of an "arithmetic group." We will see that, by definition, the other arithmetic groups are obtained by intersecting SL($n, \mathbb{Z}$) with some semisimple subgroup G of SL($n, \mathbb{R}$). More precisely, if G is a subgroup of SL($n, \mathbb{R}$) that satisfies certain technical conditions (to be explained in Section 5.1), then $G \cap$ SL($n, \mathbb{Z}$) (the group of "integer points" of G) is said to be an ***arithmetic subgroup*** of G. However, the official definition (5.1.19) also allows certain modifications of this subgroup to be called ***arithmetic***.

Different embeddings of G into SL($n, \mathbb{R}$) can yield different intersections with SL($n, \mathbb{Z}$), so G has many different arithmetic subgroups. (Examples can be found in Chapter 6.) Theorem 5.1.11 tells us that all of them are lattices in G. In particular, SL($n, \mathbb{Z}$) is a lattice in SL($n, \mathbb{R}$).

§5.1. Definition of arithmetic subgroups

We are assuming that G is a subgroup of SL($\ell, \mathbb{R}$) (see the Standing Assumptions (4.0.0)), and we are interested in $\Gamma = G \cap$ SL($\ell, \mathbb{Z}$), the set of

Recall: The Standing Assumptions (4.0.0 on page 41) are in effect, so, as always, Γ is a lattice in the semisimple Lie group $G \subseteq$ SL($\ell, \mathbb{R}$).

Main prerequisites for this chapter: Definitions 4.1.9 and 4.2.1 (lattice subgroups and commensurability).

"integer points" of G. However, in order for the integer points to form a lattice, G needs to be well-placed with respect to $SL(\ell, \mathbb{Z})$. (If we replace G by a conjugate under some terrible irrational matrix, perhaps $G \cap SL(\ell, \mathbb{Z})$ would become trivial (see Exercise 1).) The following proposition is an elementary illustration of this idea.

(5.1.1) Proposition. *The following are equivalent, for every subspace W of $\mathbb{R}^\ell$:*

1) *$W \cap \mathbb{Z}^\ell$ is a cocompact lattice in W.*

2) *W is spanned by $W \cap \mathbb{Z}^\ell$.*

3) *$W \cap \mathbb{Q}^\ell$ is dense in W.*

4) *W can be defined by a set of linear equations with coefficients in $\mathbb{Q}$.*

Proof. Let $k = \dim W$.

$(1 \Rightarrow 2)$ Let V be the $\mathbb{R}$-span of $W \cap \mathbb{Z}^\ell$. Then W/V, being a vector space over $\mathbb{R}$, is homeomorphic to $\mathbb{R}^d$, for some d. On the other hand, we know that $W \cap \mathbb{Z}^\ell \subset V$, and that $W/(W \cap \mathbb{Z}^\ell)$ is compact, so W/V is compact. Hence $d = 0$, so $V = W$.

$(2 \Rightarrow 1)$ Let $\{\varepsilon_1, \ldots, \varepsilon_k\}$ be the standard basis of $\mathbb{R}^k$. Because $W \cap \mathbb{Z}^\ell$ contains a basis of W, there is a linear isomorphism $T \colon \mathbb{R}^k \to W$, such that $T(\{\varepsilon_1, \ldots, \varepsilon_k\}) \subseteq W \cap \mathbb{Z}^\ell$. This implies that $T(\mathbb{Z}^k) \subseteq W \cap \mathbb{Z}^\ell$. Since $\mathbb{R}^k/\mathbb{Z}^k$ is compact, and T is continuous, we conclude that $W/(W \cap \mathbb{Z}^\ell)$ is compact.

$(2 \Rightarrow 3)$ As in the proof of $(2 \Rightarrow 1)$, there is a linear isomorphism $T \colon \mathbb{R}^k \to W$, such that $T(\mathbb{Z}^k) \subseteq W \cap \mathbb{Z}^\ell$. Then $T(\mathbb{Q}^k) \subseteq W \cap \mathbb{Q}^\ell$. Since $\mathbb{Q}^k$ is dense in $\mathbb{R}^k$, and T is continuous, we conclude that $T(\mathbb{Q}^k)$ is dense in W.

$(4 \Rightarrow 2)$ By assumption, W is the solution space of a system of linear equations whose coefficients belong to $\mathbb{Q}$. (Since $\mathbb{R}^\ell$ is finite dimensional, only finitely many of the equations are necessary.) Therefore, by elementary linear algebra (row reductions), we may find a basis for W that consists entirely of vectors in $\mathbb{Q}^\ell$. Multiplying by a scalar to clear the denominators, we may assume that the basis consists entirely of vectors in $\mathbb{Z}^\ell$.

$(3 \Rightarrow 4)$ Since $W \cap \mathbb{Q}^\ell$ is dense in W, we know that the orthogonal complement $W^\perp$ is defined by a set of linear equations with rational coefficients. (For each $w \in W \cap \mathbb{Q}^\ell$, we write the equation $w \cdot x = 0$.) Thus, from $(4 \Rightarrow 2)$, we conclude that there is a basis $v_1, \ldots, v_m$ of $W^\perp$, such that each $v_j \in \mathbb{Q}^\ell$. Then $W = (W^\perp)^\perp$ is defined by the system of equations $v_1 \cdot x = 0, \ldots, v_m \cdot x = 0$. $\square$

With the above proposition in mind, we make the following definition.

(5.1.2) **Definition** (cf. Definitions A4.1 and A4.7). Let H be a closed subgroup of $SL(\ell, \mathbb{R})$. We say H is ***defined over*** $\mathbb{Q}$ (or that H is a $\mathbb{Q}$-***subgroup***) if there is a subset $\mathcal{Q}$ of $\mathbb{Q}[x_{1,1}, \ldots, x_{\ell,\ell}]$, such that

- $Var(\mathcal{Q}) = \{ g \in SL(\ell, \mathbb{R}) \mid Q(g) = 0,\ \forall Q \in \mathcal{Q} \}$ is a subgroup of $SL(\ell, \mathbb{R})$,
- $H^\circ = Var(\mathcal{Q})^\circ$, and
- H has only finitely many components.

In other words, H is commensurable to the variety $Var(\mathcal{Q})$, for some set $\mathcal{Q}$ of $\mathbb{Q}$-polynomials.

(5.1.3) **Examples.**

1) $SL(\ell, \mathbb{R})$ is defined over $\mathbb{Q}$: let $\mathcal{Q} = \varnothing$.

2) If $n < \ell$, we may embed $SL(n, \mathbb{R})$ in the top left corner of $SL(\ell, \mathbb{R})$. This copy of $SL(n, \mathbb{R})$ is defined over $\mathbb{Q}$: let
$$\mathcal{Q} = \{ x_{i,j} - \delta_i^j \mid \max\{i, j\} > n \}.$$

3) For $A \in SL(\ell, \mathbb{Q})$, the group $SO_\ell(A; \mathbb{R}) = \{ g \in SL(\ell, \mathbb{R}) \mid gAg^T = A \}$ is defined over $\mathbb{Q}$: let
$$\mathcal{Q} = \left\{ \sum_{1 \le p,q \le m+n} x_{i,p} A_{p,q} x_{j,q} - A_{i,j} \;\middle|\; 1 \le i, j \le m + n \right\}.$$
In particular, $SO(m, n)$, under its usual embedding in $SL(m+n, \mathbb{R})$, is defined over $\mathbb{Q}$.

4) $SL(n, \mathbb{C})$, under its usual embedding in $SL(2n, \mathbb{R})$, is defined over $\mathbb{Q}$ (cf. Example A4.2(4)).

(5.1.4) *Remarks.*

1) There is always a subset $\mathcal{Q}$ of $\mathbb{R}[x_{1,1}, \ldots, x_{\ell,\ell}]$, such that G is commensurable to $Var(\mathcal{Q})$ (see Theorem A4.9); that is, G is ***defined over*** $\mathbb{R}$. However, it may not be possible to find a set $\mathcal{Q}$ that consists entirely of polynomials whose coefficients are rational, so G may not be defined over $\mathbb{Q}$.

2) If G is defined over $\mathbb{Q}$, then the set $\mathcal{Q}$ of Definition 5.1.2 can be chosen to be finite (because the ring $\mathbb{Q}[x_{1,1}, \ldots, x_{\ell,\ell}]$ is Noetherian).

(5.1.5) **Proposition.** *G is isogenous to a group that is defined over $\mathbb{Q}$.*

Proof. It is easy to handle direct products, so the crucial case is when G is simple. This is easy if G is classical. Indeed, the groups in Examples A2.3 and A2.4 are defined over $\mathbb{Q}$ (after identifying $SL(\ell, \mathbb{C})$ and $SL(\ell, \mathbb{H})$ with appropriate subgroups of $SL(2\ell, \mathbb{R})$ and $SL(4\ell, \mathbb{R})$, in a natural way).

The general case is not difficult for someone familiar with exceptional groups. Namely, since $\mathrm{Ad}\, G$ is a finite-index subgroup of $\mathrm{Aut}(\mathfrak{g})$, it suffices to find a basis of $\mathfrak{g}$, for which the structure constants of the Lie algebra are rational. We omit the details. $\qquad \square$

(5.1.6) **Notation.** For each subring $\mathcal{O}$ of $\mathbb{R}$ (containing 1), we construct $G_{\mathcal{O}} = G \cap \mathrm{SL}(n, \mathcal{O})$. That is, $G_{\mathcal{O}}$ is the subgroup consisting of the elements of G whose matrix entries all belong to $\mathcal{O}$.

(5.1.7) **Example.** Let $\phi \colon \mathrm{SL}(n, \mathbb{C}) \to \mathrm{SL}(2n, \mathbb{R})$ be the natural embedding. Then

$$\phi(\mathrm{SL}(n, \mathbb{C}))_{\mathbb{Q}} = \phi(\mathrm{SL}(n, \mathbb{Q}[i])).$$

Therefore, if we think of $\mathrm{SL}(n, \mathbb{C})$ as a Lie group over $\mathbb{R}$, then $\mathrm{SL}(n, \mathbb{Q}[i])$ represents the "$\mathbb{Q}$-points" of $\mathrm{SL}(n, \mathbb{C})$.

The following result provides an alternate point of view on being defined over $\mathbb{Q}$. It is the nonabelian version of $(3 \Leftrightarrow 4)$ of Proposition 5.1.1.

(5.1.8) **Proposition.** *Let H be a connected subgroup of $\mathrm{SL}(\ell, \mathbb{R})$ that is almost Zariski closed. The group H is defined over $\mathbb{Q}$ if and only if $H_{\mathbb{Q}}$ is dense in H.*

Proof. ($\Leftarrow$) Let $\mathcal{Q}_{\mathbb{C}} = \{ Q \in \mathbb{C}[x_{1,1}, \ldots, x_{\ell,\ell}] \mid Q(h) = 0, \ \forall h \in H \}$. Also, for $d \in \mathbb{N}$, let $\mathcal{Q}_{\mathbb{C}}^d = \{ Q \in \mathcal{Q}_{\mathbb{C}} \mid \deg Q \leq d \}$. Since $H_{\mathbb{Q}}$ is dense in H (and polynomials are continuous), it is clear that $\mathcal{Q}_{\mathbb{C}}^d$ is invariant under the Galois group $\mathrm{Gal}(\mathbb{C}/\mathbb{Q})$, so it is not difficult to see that $\mathcal{Q}_{\mathbb{C}}^d$ is spanned (as a vector space over $\mathbb{C}$) by a collection $\mathcal{Q}^d$ of polynomials with rational coefficients (see Exercise 2). Since H is almost Zariski closed, and polynomial rings are Noetherian, we have $H^{\circ} = \mathrm{Var}(\mathcal{Q}^d)^{\circ}$ for d sufficiently large. The polynomials in $\mathcal{Q}^d$ all have rational coefficients, so this implies that H is defined over $\mathbb{Q}$.

($\Rightarrow$) See Exercise 5.3#8 for a proof when G is simple and $G/G_{\mathbb{Z}}$ is not compact. The general case utilizes a fact from the theory of algebraic groups that will not be proved in this book (see Exercise 3). $\qquad\square$

(5.1.9) **Warning.** Proposition 5.1.8 requires the assumption that H is connected; there are subgroups H of $\mathrm{SL}(\ell, \mathbb{R})$, such that H is defined over $\mathbb{Q}$, but $H_{\mathbb{Q}}$ is not dense in H. For example, let

$$H = \{ h \in \mathrm{SO}(2) \mid h^8 = \mathrm{Id} \}.$$

(5.1.10) *Remark.* The Jacobson-Morosov Lemma (A5.8) has a relative version: if G is defined over $\mathbb{Q}$, and u is a nontrivial, unipotent element of $G_{\mathbb{Q}}$, then there is a (polynomial) homomorphism $\phi \colon \mathrm{SL}(2, \mathbb{R}) \to G$, such that $\phi\left(\left[\begin{smallmatrix} 1 & 1 \\ 0 & 1 \end{smallmatrix}\right]\right) = u$ and $\phi(\mathrm{SL}(2, \mathbb{Q})) \subseteq G_{\mathbb{Q}}$.

We now state a theorem of fundamental importance in the theory of lattices and arithmetic groups. It is a nonabelian analogue of the obvious fact that $\mathbb{Z}^{\ell}$ is a lattice in $\mathbb{R}^{\ell}$, and of $(4 \Rightarrow 1)$ of Proposition 5.1.1.

(5.1.11) **Major Theorem.** *If G is defined over $\mathbb{Q}$, then $G_{\mathbb{Z}}$ is a lattice in G.*

Proof. The statement of this theorem is more important than its proof, so, for most purposes, the reader could accept this fact as an axiom, without learning the proof. For those who do not want to take this on faith, a discussion of two different proofs can be found in Chapter 7 (with some additional details in Chapter 19).   □

(5.1.12) **Example.** Here are some standard cases of Theorem 5.1.11.

1) $SL(2, \mathbb{Z})$ is a lattice in $SL(2, \mathbb{R})$. (We proved this in Example 1.3.7.)

2) $SL(n, \mathbb{Z})$ is a lattice in $SL(n, \mathbb{R})$. (We will prove this in Chapter 7.)

3) $SO(m, n)_{\mathbb{Z}}$ is a lattice in $SO(m, n)$.

4) $SL(n, \mathbb{Z}[i])$ is a lattice in $SL(n, \mathbb{C})$ (cf. Example 5.1.7).

(5.1.13) **Example.** As an additional example, let
$$G = SO(7x_1^2 - x_2^2 - x_3^2; \mathbb{R}) \cong SO(1, 2).$$
Then Theorem 5.1.11 implies that $G_{\mathbb{Z}}$ is a lattice in G. This illustrates that the theorem is a highly nontrivial result. For example, in this case, it may not even be obvious to the reader that $G_{\mathbb{Z}}$ is infinite.

(5.1.14) **Warning.** Theorem 5.1.11 requires our standing assumption that G is semisimple; there are subgroups H of $SL(\ell, \mathbb{R})$, such that H is defined over $\mathbb{Q}$, but $H_{\mathbb{Z}}$ is not a lattice in H. For example, if H is the group of diagonal matrices in $SL(2, \mathbb{R})$, then $H_{\mathbb{Z}}$ is finite, not a lattice in H.

(5.1.15) *Remark.* The converse of Theorem 5.1.11 holds when G has no compact factors (see Exercise 5).

Combining Proposition 5.1.5 with Theorem 5.1.11 yields the following important conclusion:

(5.1.16) **Corollary.** *G has a lattice.*

In fact, a more careful look at the proof shows that if G is not compact, then the lattice we constructed is not cocompact:

(5.1.17) **Corollary.** *If G is not compact, then G has a noncocompact lattice.*

Proof. Assume that G is classical, which means it is one of the groups listed in Examples A2.3 and A2.4. As was mentioned in the proof of Proposition 5.1.5, each of these groups has an obvious $\mathbb{Q}$-form $G_{\mathbb{Q}}$, obtained by replacing $\mathbb{R}$ with $\mathbb{Q}$ (or replacing $\mathbb{C}$ with $\mathbb{Q}[i]$), in a natural way. Whenever G is noncompact, it is not difficult to see that $G_{\mathbb{Q}}$ has a nontrivial unipotent element (see Exercise 8), so Corollary 4.4.4 tells us that $G/G_{\mathbb{Z}}$ is not compact.   □

(5.1.18) *Remark.* We will show in Theorem 18.7.1 that G also has a co-compact lattice, and a special case that illustrates the main idea of the proof will be seen much earlier, in Example 5.5.4.

A lattice of the form $G_{\mathbb{Z}}$ is said to be *arithmetic*. However, for the following reasons, a somewhat more general class of lattices is also said to be arithmetic. The idea is that there are some obvious modifications of $G_{\mathbb{Z}}$ that are also lattices, and any subgroup that is obviously a lattice should be called arithmetic.

- If $\phi\colon G_1 \to G_2$ is an isomorphism, and Γ_1 is an arithmetic subgroup of G_1, then we wish to be able to say that $\phi(\Gamma_1)$ is an arithmetic subgroup in G_2.
- We wish to ignore compact groups; that is, modding out a compact subgroup should not affect arithmeticity. So we wish to be able to say that if K is a compact normal subgroup of G, and Γ is a lattice in G, then Γ is arithmetic if and only if $\Gamma K/K$ is an arithmetic subgroup of G/K
- Arithmeticity should be independent of commensurability.

The following formal definition implements these considerations.

(5.1.19) **Definition.** Γ is an ***arithmetic*** subgroup of G if and only if there exist

- a closed, connected, semisimple subgroup G' of some $\mathrm{SL}(n, \mathbb{R})$, such that G' is defined over $\mathbb{Q}$,
- compact normal subgroups K and K' of G° and G', respectively, and
- an isomorphism $\phi\colon G^\circ/K \to G'/K'$,

such that $\phi(\overline{\Gamma})$ is commensurable to $\overline{G'_{\mathbb{Z}}}$, where $\overline{\Gamma}$ and $\overline{G'_{\mathbb{Z}}}$ are the images of $\Gamma \cap G^\circ$ and $G'_{\mathbb{Z}}$ in G°/K and G'/K', respectively.

(5.1.20) *Remarks.*

1) If G has no compact factors, then it is obvious that the subgroup K in Definition 5.1.19 must be finite.
2) Corollary 5.3.2 will show that if G/Γ is not compact (and Γ is irreducible), then the annoying compact subgroups are not needed in Definition 5.1.19.
3) On the other hand, if Γ is cocompact, then a nontrivial (connected) compact group K' may be required (even if G has no compact factors). We will see many examples of this phenomenon, starting with Example 5.5.4.
4) Up to conjugacy, there are only countably many arithmetic lattices in G, because there are only countably many finite subsets of the polynomial ring $\mathbb{Q}[x_{1,1}, \ldots, x_{\ell,\ell}]$.

(5.1.21) **Other terminology.** Our definition of *arithmetic subgroup* assumes the perspective of Lie theory, where Γ is assumed to be embedded in some Lie group G. The theory of algebraic groups has a more strict definition, which requires Γ to be commensurable to $G_{\mathbb{Z}}$: arbitrary isomorphisms are not allowed, and compact subgroups cannot be ignored. At the other extreme, abstract group theory has a much looser definition, which completely ignores G: if an abstract group Λ is abstractly commensurable to a group that is arithmetic in our sense, then Λ is considered to be arithmetic.

Exercises for §5.1.

#1. Show that if G is connected, $G \subseteq \mathrm{SL}(\ell, \mathbb{R})$, and $-\mathrm{Id} \notin G$, then there exists $h \in \mathrm{SL}(\ell, \mathbb{R})$, such that $(h^{-1}Gh) \cap \mathrm{SL}(\ell, \mathbb{Z})$ is trivial.

[*Hint:* For each nontrivial $\gamma \in \mathrm{SL}(\ell, \mathbb{Z})$, let $X_\gamma = \{ h \in \mathrm{SL}(\ell, \mathbb{R}) \mid h\gamma h^{-1} \in G \}$. Then each X_γ is nowhere dense in $\mathrm{SL}(\ell, \mathbb{R})$ (see Exercise A4#4(b)).]

#2. Let W be a vector subspace of $\mathbb{C}^n$, for some n. Show that W is invariant under $\mathrm{Gal}(\mathbb{C}/\mathbb{Q})$ if and only if W is spanned by a set of vectors with rational coordinates.

[*Hint:* ($\Rightarrow$) Choose $w \in W \smallsetminus \{0\}$ with a minimal number of nonzero coordinates, and multiply by a scalar to assume at least one coordinate is a nonzero rational. Since $\sigma(w) - w \in W$ for all $\sigma \in \mathrm{Gal}(\mathbb{C}/\mathbb{Q})$, the minimality implies $w \in \mathbb{Q}^n$. Mod out w and induct on the dimension.]

#3. It can be shown that G° is *unirational*. This means there exists an open subset U of some $\mathbb{R}^n$, and a function $f \colon U \to G^\circ$, such that
   - $f(U)$ contains an open subset of G, and
   - each matrix entry of $f(x)$ is a rational function of x (that is, a quotient of two polynomials).

Furthermore, if G is defined over $\mathbb{Q}$, then f can be chosen to be defined over $\mathbb{Q}$ (that is, all of the coefficients of f are in $\mathbb{Q}$).

Assuming the above, show that $G_{\mathbb{Q}}$ is dense in G if G is connected and G is defined over $\mathbb{Q}$.

[*Hint:* Unirationality implies that $\overline{G_{\mathbb{Q}}}$ contains an open subset of G.]

#4. For H as in Warning 5.1.9, show that $H_{\mathbb{Q}}$ is not dense in H.

[*Hint:* H is finite, and $H_{\mathbb{Q}} \neq H$.]

#5. Show that if $G \subseteq \mathrm{SL}(\ell, \mathbb{R})$, G has no compact factors, and $G_{\mathbb{Z}}$ is a lattice in G, then G is defined over $\mathbb{Q}$.

[*Hint:* See the proof of Proposition 5.1.8($\Leftarrow$). Since $G_{\mathbb{Z}}$ is a lattice in G, the Borel Density Theorem (4.5.6) implies that $\mathcal{Q}_{\mathbb{C}}^d$ is invariant under the Galois group.]

#6. Show that if
   - $G \subseteq \mathrm{SL}(\ell, \mathbb{R})$, and
   - $G_{\mathbb{Z}}$ is Zariski dense in G,

then $G_{\mathbb{Z}}$ is a lattice in G.

[*Hint:* It suffices to show that G is defined over $\mathbb{Q}$.]

#7. Show that if
- G has no compact factors,
- Γ_1 and Γ_2 are arithmetic subgroups of G, and
- $\Gamma_1 \cap \Gamma_2$ is Zariski dense in G,

then Γ_1 is commensurable to Γ_2.

[*Hint:* Suppose $\phi_j \colon G \to H_j$ is an isomorphism, such that $\phi_j(\Gamma_j) = (H_j)_{\mathbb{Z}}$. Define $\phi \colon G \to H_1 \times H_2$ by $\phi(g) = (\phi_1(g), \phi_2(g))$. Then $\phi(G)_{\mathbb{Z}} = \phi(\Gamma_1 \cap \Gamma_2)$ is Zariski dense in $\phi(G)$, so $\Gamma_1 \cap \Gamma_2$ is a lattice in G (see Exercise 6). A similar (but slightly more complicated) argument applies if $\phi_j \colon G \to H_j/K_j$, where K_j is compact.]

#8. For each classical simple group G in Examples A2.3 and A2.4, let $G_{\mathbb{Q}}$ be the subgroup obtained by replacing $\mathbb{R}$ with $\mathbb{Q}$, $\mathbb{C}$ with $\mathbb{Q}[i]$, or $\mathbb{H}$ with $\mathbb{H}_{\mathbb{Q}} = \mathbb{Q} + \mathbb{Q}i + \mathbb{Q}j + \mathbb{Q}k$, as appropriate. Show that if G is not compact, then $G_{\mathbb{Q}}$ contains a nontrivial unipotent element.

[*Hint:* Show that $G_{\mathbb{Q}}$ contains a copy of either $\mathrm{SL}(2,\mathbb{Q})$, $\mathrm{SO}(1,2)_{\mathbb{Q}}$, or $\mathrm{SU}(1,1)_{\mathbb{Q}}$ (cf. Remark A2.6).]

§5.2. Margulis Arithmeticity Theorem

The following astonishing theorem shows that taking integer points is usually the only way to make a lattice. (See Section 16.3 for a sketch of the proof.)

(5.2.1) **Theorem** (Margulis Arithmeticity Theorem). *If*
- *G is not isogenous to $\mathrm{SO}(1,n) \times K$ or $\mathrm{SU}(1,n) \times K$, for any compact group K, and*
- *Γ is irreducible,*

then Γ is arithmetic.

(5.2.2) **Warning.** Unfortunately,
- $\mathrm{SL}(2,\mathbb{R})$ is isogenous to $\mathrm{SO}(1,2)$, and
- $\mathrm{SL}(2,\mathbb{C})$ is isogenous to $\mathrm{SO}(1,3)$,

so the arithmeticity theorem says nothing about the lattices in these two important groups.

(5.2.3) *Remark.* The conclusion of Theorem 5.2.1 can be strengthened: the subgroup K of Definition 5.1.19 can be taken to be finite. More precisely, if G and Γ are as in Theorem 5.2.1, and G is noncompact and has trivial center, then there exist
- a closed, connected, semisimple subgroup G' of some $\mathrm{SL}(\ell, \mathbb{R})$, such that G' is defined over $\mathbb{Q}$, and
- a surjective (continuous) homomorphism $\phi \colon G' \to G$,

such that
1) $\phi(G'_{\mathbb{Z}})$ is commensurable to Γ; and
2) the kernel of ϕ is compact.

(5.2.4) *Remarks.*

1) For any G, it is possible to give a reasonably complete description of the arithmetic subgroups of G (up to conjugacy and commensurability). Some examples are worked out in fair detail in Chapter 6. More generally, Theorem 18.5.3 (or the table on page 380) essentially provides a list of all the irreducible arithmetic subgroups of almost all of the classical groups. Thus, for most groups, the Margulis Arithmeticity Theorem provides a list of all the lattices in G.

2) Furthermore, knowing that Γ is arithmetic provides a foothold to use algebraic and number-theoretic techniques to explore the detailed structure of Γ. For example, we saw that it is easy to show Γ is torsion free if Γ is arithmetic (see Theorem 4.8.2). A more important example is that (apparently) the only known proof that every lattice is finitely presented (see Theorem 4.7.10) relies on the Margulis Arithmeticity Theorem.

3) It is known that there are nonarithmetic lattices in $\mathrm{SO}(1, n)$ for every n (see Corollary 6.5.16), but we do not yet have a theory that describes them all when $n \geq 3$. Also, nonarithmetic lattices have been constructed in $\mathrm{SU}(1, n)$ for $n \in \{1, 2, 3\}$, but (apparently) it is still not known whether they exist when $n \geq 4$.

(5.2.5) *Remark.* The subgroup

$$\mathrm{Comm}_G(\Gamma) = \{ g \in G \mid g\Gamma g^{-1} \text{ is commensurable to } \Gamma \}$$

is called the ***commensurator*** of Γ in G. It is easy to see that if G is defined over $\mathbb{Q}$, then $G_{\mathbb{Q}} \subseteq \mathrm{Comm}_G(G_{\mathbb{Z}})$ (cf. Exercise 4.8#11).

1) This implies that if Γ is arithmetic (and G is connected, with no compact factors), then $\mathrm{Comm}_G(G_{\mathbb{Z}})$ is dense in G (see Proposition 5.1.8). Margulis proved a converse. Namely, if G is connected and has no compact factors, then Γ **is arithmetic iff** $\mathrm{Comm}_G(\Gamma)$ **is dense in** G (see Theorem 16.3.3). This is known as the Commensurability Criterion for Arithmeticity.

2) In some cases, the commensurator of $G_{\mathbb{Z}}$ is much larger than $G_{\mathbb{Q}}$ (see Exercise 1). However, it was observed by Borel that this never happens when the "complexification" of G° has trivial center (and other minor conditions are satisfied) (see Exercise 4). (See Section 18.1 for an explanation of the complexification.)

Exercises for §5.2.

#1. Let $G = \mathrm{SL}(2, \mathbb{R})$ and $G_{\mathbb{Z}} = \mathrm{SL}(2, \mathbb{Z})$. Show $\mathrm{Comm}_G(G_{\mathbb{Z}})$ is *not* commensurable to $G_{\mathbb{Q}}$.

[*Hint:* The diagonal matrix $\mathrm{diag}(\sqrt{p}, 1/\sqrt{p})$ commensurates $G_{\mathbb{Z}}$, for all $p \in \mathbb{Z}^+$.]

#2. Show $\mathrm{Comm}_{\mathrm{SL}(3,\mathbb{R})}(\mathrm{SL}(3, \mathbb{Z}))$ is *not* commensurable to $\mathrm{SL}(3, \mathbb{Q})$.

#3. Show that if G is simple and Γ is not arithmetic, then Γ, $\mathcal{N}_G(\Gamma)$, and $\mathrm{Comm}_G(\Gamma)$ are commensurable to each other.

#4. (*requires some knowledge of algebraic groups*) Assume G is connected and $G_{\mathbb{Z}}$ is Zariski dense in G (cf. Corollary 4.5.6). The complexification $G \otimes \mathbb{C}$ is defined in Notation 18.1.3.

Show that if $Z(G \otimes \mathbb{C}) = \{e\}$, then $\mathrm{Comm}_G(G_{\mathbb{Z}}) = G_{\mathbb{Q}}$.

[*Hint:* For $g \in \mathrm{Comm}_G(G_{\mathbb{Z}})$, we know that $\mathrm{Ad}\,g$ is an automorphism of the Lie algebra $\mathfrak{g}$ that is defined over $\mathbb{Q}$, so $\mathrm{Ad}\,g \in (\mathrm{Ad}\,G)_{\mathbb{Q}}$. However, the assumptions imply that the adjoint representation is an isomorphism (and it is defined over $\mathbb{Q}$).]

#5. Show that the assumption $Z(G \otimes \mathbb{C}) = \{e\}$ cannot be replaced with the weaker assumption $Z(G) = \{e\}$ in Exercise 4.

[*Hint:* Any matrix in $\mathrm{GL}(3, \mathbb{Q})$ has a scalar multiple that is in $\mathrm{SL}(3, \mathbb{R})$, but $\mathrm{SL}(3, \mathbb{Q})$ has infinite index in $\mathrm{GL}(3, \mathbb{Q})$.]

§5.3. Unipotent elements of noncocompact lattices

The following result answers one of the most basic topological questions about the manifold $G/G_{\mathbb{Z}}$: is it compact?

(5.3.1) **Proposition** (Godement Compactness Criterion). *Assume that G is defined over $\mathbb{Q}$. The homogeneous space $G/G_{\mathbb{Z}}$ is compact if and only if $G_{\mathbb{Z}}$ has no nontrivial unipotent elements.*

Proof. ($\Rightarrow$) This is the easy direction (see Corollary 4.4.4).

($\Leftarrow$) We prove the contrapositive: suppose $G/G_{\mathbb{Z}}$ is not compact. (We wish to show that $G_{\mathbb{Z}}$ has a nontrivial unipotent element.) From Proposition 4.4.6 (and the fact that $G_{\mathbb{Z}}$ is a lattice in G (see Theorem 5.1.11)), we know that there exist nontrivial $y \in G_{\mathbb{Z}}$ and $g \in G$, such that ${}^g y \approx \mathrm{Id}$. Because the characteristic polynomial of a matrix is a continuous function of the matrix entries of the matrix, we conclude that the characteristic polynomial of ${}^g y$ is approximately $(x - 1)^\ell$ (the characteristic polynomial of Id). On the other hand, similar matrices have the same characteristic polynomial, so this means that the characteristic polynomial of y is approximately $(x - 1)^\ell$. Now all the coefficients of the characteristic polynomial of y are integers (because y is an integer matrix), so the only way this polynomial can be close to $(x - 1)^\ell$ is by being exactly equal to $(x - 1)^\ell$. Therefore, the characteristic polynomial of y is $(x - 1)^\ell$, so y is unipotent. $\qquad\square$

The following important consequence of the Godement Criterion tells us that there is often no need for compact subgroups in Definition 5.1.19, the definition of an arithmetic group:

(5.3.2) **Corollary.** *Assume*

- Γ *is an irreducible, arithmetic subgroup of G,*

- G/Γ is not compact, and
- G is connected and has no compact factors.

Then, perhaps after replacing G by an isogenous group, there is an embedding of G in some $\mathrm{SL}(\ell, \mathbb{R})$, such that

1) G is defined over $\mathbb{Q}$, and
2) Γ is commensurable to $G_{\mathbb{Z}}$.

Proof. From Definition 5.1.19 (and Remark 5.1.20(1)) we know that (up to isogeny and commensurability) there is a compact group K', such that we may embed $G' = G \times K'$ in some $\mathrm{SL}(\ell, \mathbb{R})$, such that G' is defined over $\mathbb{Q}$, and $\Gamma K' = G'_{\mathbb{Z}} K'$.

Let N be the almost-Zariski closure of the subgroup of G' generated by all of the unipotent elements of $G'_{\mathbb{Z}}$. Since G/Γ is not compact, the proposition implies N is infinite. However, K' has no unipotent elements (see Remark A5.2(2)), so $N \subseteq G$. Also, the definition of N implies that it is normalized by the Zariski closure of $G'_{\mathbb{Z}}$. Therefore, the Borel Density Theorem (4.5.7) implies that N is a normal subgroup of G.

Assume, for simplicity, that G is simple (see Exercise 1). Then the conclusion of the preceding paragraph tells us that $N = G$. Therefore, G is the almost-Zariski closure of a subset of $G'_{\mathbb{Z}}$, which implies that G is defined over $\mathbb{Q}$ (cf. Exercise 5.1#5). Hence, $G_{\mathbb{Z}}$ is a lattice in G, and it is easy to see that it is commensurable to Γ (see Exercise 2). $\square$

In the special case where Γ is arithmetic, the following result is an easy consequence of Proposition 5.3.1, but we will not prove the general case (which is more difficult). The assumption that G has no compact factors cannot be eliminated (see Exercise 3).

(5.3.3) **Theorem.** *Assume G has no compact factors. The homogeneous space G/Γ is compact if and only if Γ has no nontrivial unipotent elements.*

The above proof of Proposition 5.3.1 relies on the fact that $G_{\mathbb{Z}}$ is a lattice in G, which will not be proved until Chapter 7. The following result illustrates that the cocompactness of $G_{\mathbb{Z}}$ can sometimes be proved quite easily from the Mahler Compactness Criterion (4.4.7), without assuming that it is a lattice.

(5.3.4) **Proposition.** *If*

- $B(x, y)$ *is a symmetric, bilinear form on $\mathbb{Q}^{\ell}$, such that*
- $B(x, x) \neq 0$ *for all nonzero $x \in \mathbb{Q}^{\ell}$,*

then $\mathrm{SO}(B)_{\mathbb{Z}}$ is cocompact in $\mathrm{SO}(B)_{\mathbb{R}}$.

Proof. Let $G = \mathrm{SO}(B)$ and $\Gamma = \mathrm{SO}(B)_{\mathbb{Z}} = G_{\mathbb{Z}}$. (Our proof will not use the fact that Γ is a lattice in G.) Replacing B by an integer multiple to clear the denominators, we may assume $B(\mathbb{Z}^{\ell}, \mathbb{Z}^{\ell}) \subseteq \mathbb{Z}$.

Step 1. The image of G in $\mathrm{SL}(\ell, \mathbb{R}) / \mathrm{SL}(\ell, \mathbb{Z})$ is precompact. Let

- $\{g_n\}$ be a sequence of elements of G and
- $\{v_n\}$ be a sequence of elements of $\mathbb{Z}^\ell \smallsetminus \{0\}$.

Suppose that $g_n v_n \to 0$. (This will lead to a contradiction, so the desired conclusion follows from the Mahler Compactness Criterion (4.4.7).)

Since $B(v, v) \neq 0$ for all nonzero $v \in \mathbb{Z}^\ell$, and $B(\mathbb{Z}^\ell, \mathbb{Z}^\ell) \subseteq \mathbb{Z}$, we have $|B(v_n, v_n)| \geq 1$ for all n. Therefore

$$1 \leq |B(v_n, v_n)| = |B(g_n v_n, g_n v_n)| \to |B(0,0)| = 0.$$

This is a contradiction.

Step 2. The image of G in $\mathrm{SL}(\ell, \mathbb{R}) / \mathrm{SL}(\ell, \mathbb{Z})$ is closed. Suppose

$$g_n \gamma_n \to h \in \mathrm{SL}(\ell, \mathbb{R}), \quad \text{with } g_n \in G \text{ and } \gamma_n \in \mathrm{SL}(\ell, \mathbb{Z}).$$

We wish to show $h \in G\, \mathrm{SL}(\ell, \mathbb{Z})$.

Let $\{\varepsilon_1, \cdots, \varepsilon_\ell\}$ be the standard basis of $\mathbb{R}^\ell$ (so each $\varepsilon_j \in \mathbb{Z}^\ell$). Then

$$B(\gamma_n \varepsilon_j, \gamma_n \varepsilon_k) \in B(\mathbb{Z}^\ell, \mathbb{Z}^\ell) \subseteq \mathbb{Z}.$$

We also have

$$B(\gamma_n \varepsilon_j, \gamma_n \varepsilon_k) = B(g_n \gamma_n \varepsilon_j, g_n \gamma_n \varepsilon_k) \to B(h\varepsilon_j, h\varepsilon_k).$$

Since $\mathbb{Z}$ is discrete, we conclude that $B(\gamma_n \varepsilon_j, \gamma_n \varepsilon_k) = B(h\varepsilon_j, h\varepsilon_k)$ for any sufficiently large n. Therefore $h\gamma_n^{-1} \in \mathrm{SO}(B)$ (see Exercise 9), so we have $h \in G\gamma_n \subseteq G\, \mathrm{SL}(\ell, \mathbb{Z})$.

Step 3. Completion of the proof. Define $\phi \colon G/\Gamma \to \mathrm{SL}(\ell, \mathbb{R}) / \mathrm{SL}(\ell, \mathbb{Z})$ by $\phi(g\Gamma) = g\, \mathrm{SL}(\ell, \mathbb{Z})$. By combining Steps 1 and 2, we see that the image of ϕ is compact. Therefore, it suffices to show that ϕ is a homeomorphism onto its image.

Given a sequence $\{g_n\}$ in G, such that $\{\phi(g_n\Gamma)\}$ converges, we wish to show that $\{g_n\Gamma\}$ converges. There is a sequence $\{\gamma_n\}$ in $\mathrm{SL}(\ell, \mathbb{Z})$, and some $h \in G$, such that $g_n \gamma_n \to h$. The proof of Step 2 shows, for all large n, that $h \in G\gamma_n$. Then $\gamma_n \in Gh = G$ (and we know $\gamma_n \in \mathrm{SL}(\ell, \mathbb{Z})$), so $\gamma_n \in G_\mathbb{Z} = \Gamma$. Therefore, $\{g_n\Gamma\}$ converges (to $h\Gamma$), as desired. □

Exercises for §5.3.

#1. The proof of Corollary 5.3.2 assumes that G is simple. Eliminate this hypothesis.

[*Hint:* The proof shows that $N \cap \Gamma$ is a lattice in N. Since Γ is irreducible, this implies $N = G$.]

#2. At the end of the proof of Corollary 5.3.2, show that $G_\mathbb{Z}$ is commensurable to Γ.

[*Hint:* We know $G_\mathbb{Z}' K' = \Gamma K'$, and $G_\mathbb{Z}$ has finite index in $G_\mathbb{Z}'$ (see Exercise 4.1#10). Mod out K'.]

#3. Show there is a noncocompact lattice Γ in $\mathrm{SL}(2,\mathbb{R}) \times \mathrm{SO}(3)$, such that no nontrivial element of Γ is unipotent.

[*Hint:* $\mathrm{SL}(2,\mathbb{R})$ has a lattice Γ' that is free. Let Γ be the graph of a homomorphism from Γ' to $\mathrm{SO}(3)$.]

#4. Suppose $G \subseteq \mathrm{SL}(\ell,\mathbb{R})$ is defined over $\mathbb{Q}$.
   a) Show that if N is a closed, normal subgroup of G, and N is defined over $\mathbb{Q}$, then $G_{\mathbb{Z}}N$ is closed in G.
   b) Show that $G_{\mathbb{Z}}$ is irreducible if and only if no proper, closed, connected, normal subgroup of G is defined over $\mathbb{Q}$. (That is, if and only if G is $\mathbb{Q}$-*simple*.)
   c) Let H be the Zariski closure of the subgroup generated by the unipotent elements of $G_{\mathbb{Z}}$. Show that H is defined over $\mathbb{Q}$.

#5. Show that if every element of Γ is semisimple, then G/Γ is compact.

[*Hint:* There is no harm in assuming that G has no compact factors (why?), so Theorem 5.3.3 applies.]

#6. (*assumes some familiarity with reductive groups*) Prove the converse of Exercise 5.

[*Hint:* Let kau be the real Jordan decomposition of an element g of Γ. Since $C_G(ka)$ is reductive (see Exercise 8.2#2), the Jacobson-Morosov Lemma provides a subgroup L of $C_G(ka)$ that contains u and is isogenous to $\mathrm{SL}(2,\mathbb{R})$. So ka is in the closure of ${}^G g$. However, ${}^G g$ is closed, since Γ is discrete and cocompact. Therefore $g = ka$ is semisimple.]

#7. Assuming $\Gamma = G_{\mathbb{Z}}$ is arithmetic (and G is defined over $\mathbb{Q}$), prove the following are equivalent:
   a) $G/G_{\mathbb{Z}}$ is compact.
   b) $G_{\mathbb{Q}}$ has no nontrivial unipotent elements.
   c) Every element of $G_{\mathbb{Q}}$ is semisimple.
   d) Every element of Γ is semisimple.
   e) $G_{\mathbb{Q}}$ does not contain a subgroup isogenous to $\mathrm{SL}(2,\mathbb{Q})$. (More precisely, there does not exist a continuous homomorphism $\rho\colon \mathrm{SL}(2,\mathbb{R}) \to G$, such that $\rho(\mathrm{SL}(2,\mathbb{Q})) \subseteq G_{\mathbb{Q}}$.)

[*Hint:* $(b \Rightarrow c)$ Jordan decomposition. $(e \Rightarrow b)$ Jacobson-Morosov Lemma (5.1.10).]

#8. Show that $G_{\mathbb{Q}}$ is dense in G if G is defined over $\mathbb{Q}$, G is simple, and $G/G_{\mathbb{Z}}$ is not compact.

[*Hint:* The Godement Criterion implies that $G_{\mathbb{Q}}$ has a nontrivial unipotent element u. Write $u = \exp T = \sum_{k=0}^{\ell} T^k/k!$ (where $T \in \mathrm{Mat}_{\ell \times \ell}(\mathbb{Q})$ and $T^{\ell+1} = 0$). Then $\exp(rT) \in G_{\mathbb{Q}}$ for all $r \in \mathbb{Q}$, so the identity component of $\overline{G_{\mathbb{Q}}}$ is nontrivial. Combining Theorem 5.1.11 with the Borel Density Theorem (4.5.2) implies that $\overline{G_{\mathbb{Q}}} = G$.]

#9. Let $B(x,y)$ be a symmetric, bilinear form on $\mathbb{R}^{\ell}$, let $\{v_1, \cdots, v_{\ell}\}$ be a basis of $\mathbb{R}^{\ell}$, and let $y, h \in \mathrm{SL}(\ell,\mathbb{R})$. If $B(yv_j, yv_k) = B(hv_j, hv_k)$ for all j and k, show that $hy^{-1} \in \mathrm{SO}(B)$.

[*Hint:* $\{yv_1, \ldots, yv_{\ell}\}$ is a basis of $\mathbb{R}^{\ell}$.]

§5.4. How to make an arithmetic subgroup

The definition that (modulo commensurability, isogenies, and compact factors) an arithmetic subgroup must be the $\mathbb{Z}$-points of G has the virtue of being concrete. However, this concreteness imposes a certain lack of flexibility. (Essentially, we have limited ourselves to the standard basis of the vector space $\mathbb{R}^n$, ignoring the possibility that some other basis might be more convenient in some situations.) We now describe a more abstract viewpoint that makes the construction of general arithmetic lattices more transparent. (In particular, this approach will be used in §5.5.) The key point is that there are analogues of $\mathbb{Z}^\ell$ and $\mathbb{Q}^\ell$ in any real vector space, not just $\mathbb{R}^\ell$ (see Lemma 5.4.3(1)).

(5.4.1) **Definitions.** Let V be a real vector space.

1) A $\mathbb{Q}$-subspace $V_{\mathbb{Q}}$ of V is a $\mathbb{Q}$-*form* of V if the natural $\mathbb{R}$-linear map $V_{\mathbb{Q}} \otimes_{\mathbb{Q}} \mathbb{R} \to V$ is an isomorphism (see Exercise 1). (The map is defined by $v \otimes t \mapsto tv$.)

2) A polynomial f on V is *defined over* $\mathbb{Q}$ (with respect to the $\mathbb{Q}$-form $V_{\mathbb{Q}}$) if $f(V_{\mathbb{Q}}) \subseteq \mathbb{Q}$ (see Exercise 2).

3) A subgroup $\mathcal{L}$ of the additive group of $V_{\mathbb{Q}}$ is a $\mathbb{Z}$-*lattice* in $V_{\mathbb{Q}}$ if it is finitely generated and the natural $\mathbb{Q}$-linear map $\mathcal{L} \otimes_{\mathbb{Z}} \mathbb{Q} \to V_{\mathbb{Q}}$ is an isomorphism (see Exercise 3). (The map is defined by $v \otimes t \mapsto tv$.)

4) Each $\mathbb{Q}$-form $V_{\mathbb{Q}}$ of V yields a corresponding $\mathbb{Q}$-form of the real vector space $\mathrm{End}(V)$ by $\mathrm{End}(V)_{\mathbb{Q}} = \{ A \in \mathrm{End}(V) \mid A(V_{\mathbb{Q}}) \subseteq V_{\mathbb{Q}} \}$ (see Exercise 5).

5) A function Q on a real vector space W is a *polynomial* if for some (hence, every) $\mathbb{R}$-linear isomorphism $\phi \colon \mathbb{R}^\ell \cong W$, the composition $f \circ \phi$ is a polynomial function on $\mathbb{R}^\ell$.

6) A subgroup H of $\mathrm{SL}(V)$ is *defined over* $\mathbb{Q}$ (with respect to the $\mathbb{Q}$-form $V_{\mathbb{Q}}$) if there exists a set $\mathcal{Q}$ of polynomials on $\mathrm{End}(V)$, such that
   - every $Q \in \mathcal{Q}$ is defined over $\mathbb{Q}$ (with respect to the $\mathbb{Q}$-form $V_{\mathbb{Q}}$),
   - $\mathrm{Var}(\mathcal{Q}) = \{ g \in \mathrm{SL}(V) \mid Q(g) = 0 \text{ for all } Q \in \mathcal{Q} \}$ is a subgroup of $\mathrm{SL}(V)$, and
   - $\mathrm{Var}(\mathcal{Q})^\circ$ is a finite-index subgroup of H.

(5.4.2) *Remarks.*

1) Suppose $G \subseteq \mathrm{SL}(\ell, \mathbb{R})$, as usual. For the standard $\mathbb{Q}$-form $\mathbb{Q}^\ell$ of $\mathbb{R}^\ell$, it is easy to see that G is defined over $\mathbb{Q}$ in terms of Definition 5.4.1 if and only if it is defined over $\mathbb{Q}$ in terms of Definition 5.1.2.

2) Some authors simply call $\mathcal{L}$ a *lattice in* $V_{\mathbb{Q}}$, but this could cause confusion, because $\mathcal{L}$ is *not* a lattice in $V_{\mathbb{Q}}$, in the sense of Definition 4.1.9 (although it *is* a lattice in V).

A $\mathbb{Q}$-form $V_{\mathbb{Q}}$ and $\mathbb{Z}$-lattice $\mathcal{L}$ simply represent $\mathbb{Q}^{\ell}$ and $\mathbb{Z}^{\ell}$, under some identification of V with $\mathbb{R}^{\ell}$:

(5.4.3) Lemma. *Let V be an ℓ-dimensional real vector space.*

1) *If $V_{\mathbb{Q}}$ is a $\mathbb{Q}$-form of V, then there exists an $\mathbb{R}$-linear isomorphism $\phi: V \to \mathbb{R}^{\ell}$, such that $\phi(V_{\mathbb{Q}}) = \mathbb{Q}^{\ell}$. Furthermore, if $\mathcal{L}$ is any $\mathbb{Z}$-lattice in $V_{\mathbb{Q}}$, then ϕ may be chosen so that $\phi(\mathcal{L}) = \mathbb{Z}^{\ell}$.*

2) *A polynomial f on $\mathbb{R}^{\ell}$ is defined over $\mathbb{Q}$ (with respect to the standard $\mathbb{Q}$-form $\mathbb{Q}^{\ell}$) if and only if every coefficient of f is in $\mathbb{Q}$ (see Exercise 2).*

Also note that any two $\mathbb{Z}$-lattices in $V_{\mathbb{Q}}$ are commensurable:

(5.4.4) Lemma (see Exercise 6). *If $\mathcal{L}_1$ and $\mathcal{L}_2$ are two $\mathbb{Z}$-lattices in $V_{\mathbb{Q}}$, then there is some nonzero $p \in \mathbb{Z}$, such that $p\mathcal{L}_1 \subseteq \mathcal{L}_2$ and $p\mathcal{L}_2 \subseteq \mathcal{L}_1$.*

It is now easy to prove the following more abstract characterization of arithmetic subgroups (see Exercises 7 and 8).

(5.4.5) Proposition. *Suppose $G \subseteq \mathrm{GL}(V)$, and G is defined over $\mathbb{Q}$, with respect to the $\mathbb{Q}$-form $V_{\mathbb{Q}}$.*

1) *If $\mathcal{L}$ is any $\mathbb{Z}$-lattice in $V_{\mathbb{Q}}$, then*

$$G_{\mathcal{L}} = \{ g \in G \mid g\mathcal{L} = \mathcal{L} \}$$

is an arithmetic subgroup of G.

2) *If $\mathcal{L}_1$ and $\mathcal{L}_2$ are $\mathbb{Z}$-lattices in $V_{\mathbb{Q}}$, then $G_{\mathcal{L}_1}$ is commensurable to $G_{\mathcal{L}_1}$.*

From Proposition 5.4.5(2), we see that the arithmetic subgroup $G_{\mathcal{L}}$ is almost entirely determined by the $\mathbb{Q}$-form $V_{\mathbb{Q}}$; choosing a different $\mathbb{Z}$-lattice in $V_{\mathbb{Q}}$ will yield a commensurable arithmetic subgroup.

Exercises for §5.4.

#1. Show that a $\mathbb{Q}$-subspace $V_{\mathbb{Q}}$ of V is a $\mathbb{Q}$-form if an only if there is a subset $\mathcal{B}$ of $V_{\mathbb{Q}}$, such that $\mathcal{B}$ is both a $\mathbb{Q}$-basis of $V_{\mathbb{Q}}$ and an $\mathbb{R}$-basis of V.

#2. For the standard $\mathbb{Q}$-form $\mathbb{Q}^{\ell}$ of $\mathbb{R}^{\ell}$, show that a polynomial is defined over $\mathbb{Q}$ if and only if all of its coefficients are rational.

#3. Show that a subgroup $\mathcal{L}$ of $V_{\mathbb{Q}}$ is a $\mathbb{Z}$-lattice in $V_{\mathbb{Q}}$ if and only if there is a $\mathbb{Q}$-basis $\mathcal{B}$ of $V_{\mathbb{Q}}$, such that $\mathcal{L}$ is the additive abelian subgroup of $V_{\mathbb{Q}}$ generated by $\mathcal{B}$.

#4. Let V be a real vector space of dimension ℓ, and let $\mathcal{L}$ be a discrete subgroup of the additive group of V. Recall that the **rank** of an abelian group is the largest r, such that the group contains a copy of $\mathbb{Z}^r$.

   a) Show that $\mathcal{L}$ is a finitely generated, abelian group of rank $\leq \ell$, with equality if and only if the $\mathbb{R}$-span of $\mathcal{L}$ is V.

b) Show that if the rank of $\mathcal{L}$ is ℓ, then the $\mathbb{Q}$-span of $\mathcal{L}$ is a $\mathbb{Q}$-form of V, and $\mathcal{L}$ is a $\mathbb{Z}$-lattice in $V_{\mathbb{Q}}$.

[*Hint:* Induction on ℓ. For $\lambda \in \mathcal{L}$, show that the image of $\mathcal{L}$ in $V/\mathbb{R}\lambda$ is discrete.]

#5. Verify: if $V_{\mathbb{Q}}$ is a $\mathbb{Q}$-form of V, then $\mathrm{End}(V)_{\mathbb{Q}}$ is a $\mathbb{Q}$-form of $\mathrm{End}(V)$.

#6. Prove Lemma 5.4.4. Conclude that Λ_1 and Λ_2 are commensurable.

#7. Prove Proposition 5.4.5(1). [*Hint:* Use Lemma 5.4.3.]

#8. Prove Proposition 5.4.5(2). [*Hint:* Use Lemma 5.4.4.]

§5.5. Restriction of scalars

We know that $\mathrm{SL}(2, \mathbb{Z})$ is an arithmetic subgroup of $\mathrm{SL}(2, \mathbb{R})$. In this section, we explain that $\mathrm{SL}(2, \mathbb{Z}[\sqrt{2}])$ is an arithmetic subgroup of the group $\mathrm{SL}(2, \mathbb{R}) \times \mathrm{SL}(2, \mathbb{R})$ (see Example 5.5.3). More generally, recall that any finite extension of $\mathbb{Q}$ is called an ***algebraic number field***. We will see that if $\mathcal{O}$ is the ring of algebraic integers in any algebraic number field F, and G is defined over F, then $G_{\mathcal{O}}$ is an arithmetic subgroup of a certain group G' that is related to G.

(5.5.1) *Remark.* In practice, we do not require $\mathcal{O}$ to be the entire ring of algebraic integers in F: it suffices for the ring $\mathcal{O}$ to have finite index in the ring of integers (as an additive group); equivalently, the $\mathbb{Q}$-span of $\mathcal{O}$ should be all of F, or, in other words, the ring $\mathcal{O}$ should be a $\mathbb{Z}$-lattice in F. (A $\mathbb{Z}$-lattice in F that is also a subring is called an ***order*** in F.)

Any complex vector space can be thought of as a real vector space (of twice the dimension). Similarly, any complex Lie group can be thought of as a real group (of twice the dimension). Restriction of scalars is the generalization of this idea to any field extension F/L, not just $\mathbb{C}/\mathbb{R}$. This yields a general method to construct arithmetic subgroups.

(5.5.2) **Example.** Let
- $F = \mathbb{Q}[\sqrt{2}]$,
- $\mathcal{O} = \mathbb{Z}[\sqrt{2}]$, and
- σ be the nontrivial Galois automorphism of F,

and define a ring homomorphism $\Delta \colon F \to \mathbb{R}^2$ by $\Delta(x) = (x, \sigma(x))$.

It is easy to show that $\Delta(\mathcal{O})$ is discrete in $\mathbb{R}^2$. Namely, for $x \in \mathcal{O}$, the product of the coordinates of $\Delta(x)$ is the product $x \cdot \sigma(x)$ of all the Galois conjugates of x. This is the ***norm*** of the algebraic number x. Because x is an algebraic integer, its norm is an ordinary integer; hence, its norm is bounded away from 0. So it is impossible for both coordinates of $\Delta(x)$ to be small simultaneously.

More generally, if $\mathcal{O}$ is the ring of integers of any algebraic number field F, this same argument shows that if we let $\{\sigma_1, \ldots, \sigma_r\}$ be the set of all embeddings of $\mathcal{O}$ in $\mathbb{C}$, and define $\Delta\colon \mathcal{O} \to \mathbb{C}^r$ by

$$\Delta(x) = (\sigma_1(x), \ldots, \sigma_r(x)),$$

then $\Delta(\mathcal{O})$ is a *discrete* subring of $\mathbb{C}^r$.

Now Δ induces a homomorphism $\Delta_*\colon \mathrm{SL}(\ell, \mathcal{O}) \to \mathrm{SL}(\ell, \mathbb{C}^r)$ (because $\mathrm{SL}(\ell, \cdot)$ is a functor from the category of commutative rings to the category of groups). Furthermore, the group $\mathrm{SL}(\ell, \mathbb{C}^r)$ is naturally isomorphic to $\mathrm{SL}(\ell, \mathbb{C})^r$. Therefore, we have a homomorphism (again called Δ) from $\mathrm{SL}(\ell, \mathcal{O})$ to $\mathrm{SL}(\ell, \mathbb{C})^r$. Namely, for $y \in \mathrm{SL}(\ell, \mathcal{O})$, we let $\sigma_i(y) \in \mathrm{SL}(\ell, \mathbb{C})$ be obtained by applying σ_i to each entry of y, and then

$$\Delta(y) = (\sigma_1(y), \ldots, \sigma_r(y)).$$

Since $\Delta(\mathcal{O})$ is discrete in $\mathbb{C}^r$, it is obvious that the image of Δ_* is discrete in $\mathrm{SL}(\ell, \mathbb{C}^r)$, so $\Delta(\Gamma)$ is a discrete subgroup of $\mathrm{SL}(\ell, \mathbb{C})^r$, for any subgroup Γ of $\mathrm{SL}(\ell, \mathcal{O})$.

The main goal of this section is to show that if $\Gamma = G_{\mathcal{O}}$, and G is defined over F, then the discrete group $\Delta(\Gamma)$ is an arithmetic subgroup of a certain subgroup of $\mathrm{SL}(\ell, \mathbb{C})^r$.

To illustrate, let us show that $\mathrm{SL}(2, \mathbb{Z}[\sqrt{2}])$ is isomorphic to an arithmetic subgroup of $\mathrm{SL}(2, \mathbb{R}) \times \mathrm{SL}(2, \mathbb{R})$.

(5.5.3) **Example.** Let

- $\Gamma = \mathrm{SL}(2, \mathbb{Z}[\sqrt{2}])$,

- $G = \mathrm{SL}(2, \mathbb{R}) \times \mathrm{SL}(2, \mathbb{R})$, and

- σ be the conjugation on $\mathbb{Q}[\sqrt{2}]$ (so $\sigma(a + b\sqrt{2}) = a - b\sqrt{2}$, for $a, b \in \mathbb{Q}$),

and define $\Delta\colon \Gamma \to G$ by $\Delta(y) = (y, \sigma(y))$.

Then $\Delta(\Gamma)$ is an irreducible, arithmetic subgroup of G.

Proof. Let $F = \mathbb{Q}[\sqrt{2}]$ and $\mathcal{O} = \mathbb{Z}[\sqrt{2}]$. Then F is a 2-dimensional vector space over $\mathbb{Q}$, and $\mathcal{O}$ is a $\mathbb{Z}$-lattice in F.

Since $\{(1,1), (\sqrt{2}, -\sqrt{2})\}$ is both a $\mathbb{Q}$-basis of $\Delta(F)$ and an $\mathbb{R}$-basis of $\mathbb{R}^2$, we see that $\Delta(F)$ is a $\mathbb{Q}$-form of $\mathbb{R}^2$. Therefore,

$$\Delta(F^2) = \{ (u, \sigma(u)) \in F^4 \mid u \in F^2 \}$$

is a $\mathbb{Q}$-form of $\mathbb{R}^4$, and $\Delta(\mathcal{O}^2)$ is a $\mathbb{Z}$-lattice in $\Delta(F^2)$.

Now G is defined over $\mathbb{Q}$ (see Exercise 2), so $G_{\Delta(\mathcal{O}^2)}$ is an arithmetic subgroup of G. It is not difficult to see that $G_{\Delta(\mathcal{O}^2)} = \Delta(\Gamma)$ (see Exercise 3). Furthermore, because $\Delta(\Gamma) \cap (\mathrm{SL}(2, \mathbb{R}) \times e)$ is trivial, we see that the lattice $\Delta(\Gamma)$ must be irreducible in G (see Proposition 4.3.3). $\qquad\square$

More generally, the proof of Example 5.5.3 shows that if G is defined over $\mathbb{Q}$, then $G_{\mathbb{Z}[\sqrt{2}]}$ is isomorphic to an (irreducible) arithmetic subgroup of $G \times G$.

Here is another sample application of the method.

(5.5.4) Example. Let $G = \mathrm{SO}(x^2 + y^2 - \sqrt{2}z^2; \mathbb{R}) \cong \mathrm{SO}(1, 2)$. Then $G_{\mathbb{Z}[\sqrt{2}]}$ is a cocompact, arithmetic subgroup of G.

Proof. As above, let σ be the conjugation on $\mathbb{Q}[\sqrt{2}]$. Let $\Gamma = G_{\mathbb{Z}[\sqrt{2}]}$.

Let $K' = \mathrm{SO}(x^2 + y^2 + \sqrt{2}z^2) \cong \mathrm{SO}(3)$, so $\sigma(\Gamma) \subseteq K'$. (However, $\sigma(\Gamma) \not\subseteq G$.) Then, we may

$$\text{define } \Delta\colon \Gamma \to G \times K' \text{ by } \Delta(y) = (y, \sigma(y)).$$

Arguing as in the proof of Example 5.5.3 establishes that $\Delta(\Gamma)$ is an arithmetic subgroup of $G \times K'$. (See Exercise 4 for the technical point of verifying that $G \times K'$ is defined over $\mathbb{Q}$.) Since K' is compact, we see, by modding out K', that Γ is an arithmetic subgroup of G. (This type of example is the reason for including the compact normal subgroup K' in Definition 5.1.19.)

Let y be any nontrivial element of Γ. Since $\sigma(y) \in K'$, and compact groups have no nontrivial unipotent elements (see Remark A5.2(2)), we know that $\sigma(y)$ is not unipotent. Therefore, $\sigma(y)$ has some eigenvalue $\lambda \neq 1$. Hence, y has the eigenvalue $\sigma^{-1}(\lambda) \neq 1$, so y is not unipotent. Therefore, Godement's Criterion (5.3.1) implies that Γ is cocompact. Alternatively, this conclusion can easily be obtained directly from the Mahler Compactness Criterion (4.4.7) (see Exercise 6). □

Let us consider one more example before stating the general result.

(5.5.5) Example. Let

- $F = \mathbb{Q}[\sqrt[4]{2}]$,
- $\mathcal{O} = \mathbb{Z}[\sqrt[4]{2}]$,
- $\Gamma = \mathrm{SL}(2, \mathcal{O})$, and
- $G = \mathrm{SL}(2, \mathbb{R}) \times \mathrm{SL}(2, \mathbb{R}) \times \mathrm{SL}(2, \mathbb{C})$.

Then Γ is isomorphic to an irreducible, arithmetic subgroup of G.

Proof. For convenience, let $\alpha = \sqrt[4]{2}$. There are exactly 4 distinct embeddings $\sigma_0, \sigma_1, \sigma_2, \sigma_3$ of F in $\mathbb{C}$ (corresponding to the 4 roots of $x^4 - 2 = 0$); they are determined by:

$$\sigma_0(\alpha) = \alpha \text{ (so } \sigma_0 = \mathrm{Id}), \quad \sigma_1(\alpha) = -\alpha, \quad \sigma_2(\alpha) = i\alpha, \quad \text{and} \quad \sigma_3(\alpha) = -i\alpha.$$

Define $\Delta\colon F \to \mathbb{R} \oplus \mathbb{R} \oplus \mathbb{C}$ by $\Delta(x) = (x, \sigma_1(x), \sigma_2(x))$. Then, arguing much as before, we see that $\Delta(F^2)$ is a $\mathbb{Q}$-form of $\mathbb{R}^2 \oplus \mathbb{R}^2 \oplus \mathbb{C}^2$, G is defined over $\mathbb{Q}$, and $G_{\Delta(\mathcal{O}^2)} = \Delta(\Gamma)$. □

These examples illustrate all the ingredients of the general result that will be stated in Proposition 5.5.8 after the necessary definitions.

(5.5.6) **Definition.** Let F be an algebraic number field (or, in other words, let F be a finite extension of $\mathbb{Q}$).

1) Two distinct embeddings $\sigma_1, \sigma_2 \colon F \to \mathbb{C}$ are said to be **equivalent** if $\sigma_1(x) = \overline{\sigma_2(x)}$, for all $x \in F$ (where $\overline{z}$ denotes the usual complex conjugate of the complex number z).

2) A **place** of F is an equivalence class of embeddings in $\mathbb{C}$. Therefore, each place consists of either one or two embeddings of F:
   - a **real place** consists of only one embedding (with $\sigma(F) \subset \mathbb{R}$), but
   - a **complex place** consists of two embeddings (with $\sigma(F) \not\subset \mathbb{R}$).

3) We let $S^\infty = \{$ places of $F \}$, or, abusing notation, we assume that S^∞ is a set of embeddings, consisting of exactly one embedding from each place.

4) For $\sigma \in S^\infty$, we let
$$F_\sigma = \begin{cases} \mathbb{R} & \text{if } \sigma \text{ is real,} \\ \mathbb{C} & \text{if } \sigma \text{ is complex.} \end{cases}$$
Note that $\sigma(F)$ is dense in F_σ, so F_σ is often called the **completion** of F at the place σ.

5) For $Q \subset F_\sigma[x_{1,1}, \ldots, x_{\ell,\ell}]$, let
$$\mathrm{Var}_{F_\sigma}(Q) = \{\, g \in \mathrm{SL}(\ell, F_\sigma) \mid Q(g) = 0, \ \forall Q \in Q \,\}.$$
Thus, for $F_\sigma = \mathbb{R}$, we have $\mathrm{Var}_{\mathbb{R}}(Q) = \mathrm{Var}(Q)$, and $\mathrm{Var}_{\mathbb{C}}(Q)$ is analogous, using the field $\mathbb{C}$ in place of $\mathbb{R}$.

6) Suppose $G \subseteq \mathrm{SL}(\ell, \mathbb{R})$, and G is defined over F, so there is some subset Q of $F[x_{1,1}, \ldots, x_{\ell,\ell}]$, such that $G^\circ = \mathrm{Var}(Q)^\circ$. For each place σ of F, let
$$G^\sigma = \mathrm{Var}_{F_\sigma}(\sigma(Q))^\circ.$$
Then G^σ, the **Galois conjugate** of G by σ, is defined over $\sigma(F)$.

(5.5.7) **Other terminology.** Our definition requires places to be **infinite** (or **archimedean**); that is the reason for the superscript ∞ on S^∞. Other authors also allow places that are **finite** (or **nonarchimedean**, or **p-adic**). These additional places are of fundamental importance in number theory, and, therefore, in deeper aspects of the theory of arithmetic groups. For example, superrigidity at the finite places will play a crucial role in the proof of the Margulis Arithmeticity Theorem in Section 16.3. Finite places are also essential for the definition of the "S-arithmetic" groups discussed in Appendix C.

(5.5.8) **Proposition.** *If G is defined over an algebraic number field $F \subset \mathbb{R}$, and $\mathcal{O}$ is the ring of integers of F, then there is a finite-index subgroup $\dot{G}_\mathcal{O}$ of $G_\mathcal{O}$, such that*

$$\dot{G}_\mathcal{O} \text{ embeds as an arithmetic subgroup of } \prod_{\sigma \in S^\infty} G^\sigma,$$

via the natural embedding $\Delta\colon \gamma \mapsto (\sigma(\gamma))_{\sigma \in S^\infty}$
   Furthermore, if G is simple, then the lattice $\Delta(G_\mathcal{O})$ is irreducible.

(5.5.9) **Warning.** By our definition, G^σ is always connected, since it is the identity component of $\mathrm{Var}_{F_\sigma}(\sigma(\mathcal{Q}))$. If G is assumed to be Zariski closed (so it is equal to $\mathrm{Var}(\mathcal{Q})$, rather than merely being isogenous to it), then it is sometimes more convenient to define G^σ to be the entire variety $\mathrm{Var}_{F_\sigma}(\sigma(\mathcal{Q}))$, rather than merely the identity component. In particular, that would eliminate the need to pass to a finite-index subgroup $\dot{G}_\mathcal{O}$ in the statement of Proposition 5.5.8. Taking the best of both worlds, we will usually ignore the difference between $G_\mathcal{O}$ and $\dot{G}_\mathcal{O}$, and pretend that the map Δ of Proposition 5.5.8 is defined on all of $G_\mathcal{O}$. For example, the statements of Corollary 5.5.10 and Proposition 5.5.12 below omit the dots that should be in $\Delta(\dot{G}_\mathcal{O})$ and $\phi(\Delta(\dot{H}_\mathcal{O}))$.

The argument in the last paragraph of the proof of Example 5.5.4 shows the following:

(5.5.10) **Corollary.** *If G^σ is compact, for some $\sigma \in S^\infty$, then $\Delta(G_\mathcal{O})$ is cocompact.*

(5.5.11) *Remark.* Proposition 5.5.8 is stated only for real groups, but the same conclusions hold if

- $G \subseteq \mathrm{SL}(\ell, \mathbb{C})$,
- F is an algebraic number field, such that $F \not\subset \mathbb{R}$, and
- G is defined over F, as an algebraic group over $\mathbb{C}$; that is, there is a subset $\mathcal{Q}$ of $F[x_{1,1}, \ldots, x_{\ell,\ell}]$, such that $G^\circ = \mathrm{Var}_\mathbb{C}(\mathcal{Q})^\circ$ (see Notation 18.1.3).

For example, we have the following irreducible arithmetic lattices:

1) $\mathrm{SO}(n, \mathbb{Z}[i, \sqrt{2}])$ in $\mathrm{SO}(n, \mathbb{C}) \times \mathrm{SO}(n, \mathbb{C})$, and
2) $\mathrm{SO}\left(n, \mathbb{Z}\left[\sqrt{1 - \sqrt{2}}\right]\right)$ in $\mathrm{SO}(n, \mathbb{C}) \times \mathrm{SO}(n, \mathbb{R}) \times \mathrm{SO}(n, \mathbb{R})$.

The following converse shows that restriction of scalars is the only way to make a group of $\mathbb{Z}$-points that is irreducible.

(5.5.12) **Proposition.** *If $\Gamma = G_\mathbb{Z}$ is an irreducible lattice in G (and G is connected), then there exist*

1) *an algebraic number field F, with completion F_∞ ($= \mathbb{R}$ or $\mathbb{C}$),*
2) *a connected, simple subgroup H of $\mathrm{SL}(\ell, F_\infty)$, for some ℓ, such that H is defined over F (as an algebraic group over F_∞), and*

3) *an isogeny*

$$\phi\colon \prod_{\sigma \in S^\infty} H^\sigma \to G,$$

such that $\phi(\Delta(H_{\mathcal{O}}))$ is commensurable to Γ.

Proof. It is easier to work with the algebraically closed field $\mathbb{C}$, instead of $\mathbb{R}$, so, to avoid minor complications, let us assume that $G \subseteq SL(\ell, \mathbb{C})$ is defined over $\mathbb{Q}[i]$ (as an algebraic group over $\mathbb{C}$), and that $\Gamma = G_{\mathbb{Z}[i]}$. This assumption results in a loss of generality, but similar ideas apply in general.

Write $G = G_1 \times \cdots \times G_r$, where each G_i is simple. Let $H = G_1$. We remark that if $r = 1$, then the desired conclusion is obvious: let $F = \mathbb{Q}[i]$, and let ϕ be the identity map.

Let Σ be the Galois group of $\mathbb{C}$ over $\mathbb{Q}[i]$. Because G is defined over $\mathbb{Q}[i]$, we have $\sigma(G) = G$ for every $\sigma \in \Sigma$. Hence, σ must permute the simple factors $\{G_1, \ldots, G_r\}$.

We claim that Σ acts transitively on $\{G_1, \ldots, G_r\}$. To see this, suppose, for example, that $r = 5$, and that $\{G_1, G_2\}$ is invariant under Σ. Then $A = G_1 \times G_2$ is invariant under Σ, so A is defined over $\mathbb{Q}[i]$. Similarly, $A' = G_3 \times G_4 \times G_5$ is also defined over $\mathbb{Q}[i]$. Then $A_{\mathbb{Z}[i]}$ and $A'_{\mathbb{Z}[i]}$ are lattices in A and A', respectively, so $\Gamma = G_{\mathbb{Z}[i]} \approx A_{\mathbb{Z}[i]} \times A'_{\mathbb{Z}[i]}$ is reducible. This is a contradiction.

Let

$$\Sigma_1 = \{\, \sigma \in \Sigma \mid \sigma(G_1) = G_1 \,\}$$

be the stabilizer of G_1, and let

$$F = \{\, z \in \mathbb{C} \mid \sigma(z) = z, \ \forall \sigma \in \Sigma_1 \,\}$$

be the fixed field of Σ_1. Because Σ is transitive on a set of r elements, we know that Σ_1 is a subgroup of index r in Σ, so Galois Theory tells us that F is an extension of $\mathbb{Q}[i]$ of degree r.

Since Σ_1 is the Galois group of $\mathbb{C}$ over F, and $\sigma(G_1) = G_1$ for all $\sigma \in \Sigma_1$, we see that G_1 is defined over F.

Let $\sigma_1, \ldots, \sigma_r$ be coset representatives of Σ_1 in Σ. Then $\sigma_1|_F, \ldots, \sigma_r|_F$ are the r places of F and, after renumbering, we have $G_j = \sigma_j(G_1)$. So (with $H = G_1$), we have

$$\prod_{\sigma \in S^\infty} H^\sigma = H^{\sigma_1|_F} \times \cdots \times H^{\sigma_r|_F} = \sigma_1(G_1) \times \cdots \times \sigma_r(G_1) = G_1 \times \cdots \times G_r = G.$$

Let ϕ be the identity map.

For $h \in H_F$, let $\Delta'(h) = \prod_{j=1}^{r} \sigma_j(h)$. Then $\sigma(\Delta'(h)) = \Delta'(h)$ for all $\sigma \in \Sigma$, so $\Delta'(h) \in G_{\mathbb{Q}[i]}$. In fact, it is not difficult to see that $\Delta'(H_F) = G_{\mathbb{Q}[i]}$, and then one can verify that $\Delta'(H_{\mathcal{O}}) \approx G_{\mathbb{Z}[i]} = \Gamma$, so $\phi(\Delta(H_{\mathcal{O}}))$ is commensurable to Γ. $\square$

(5.5.13) *Remark.* Although it may not be clear from our proof, the group G' in Corollary 5.5.15 can be chosen to be "absolutely simple." This means that if $F \subset \mathbb{R}$, then the following three equivalent conditions must be true: G' remains simple over $\mathbb{C}$, $\mathfrak{g}' \otimes_{\mathbb{R}} \mathbb{C}$ is simple, and G' is not isogenous to any "complexification" $(G'')_{\mathbb{C}}$.

Combining Proposition 5.5.12 with Corollary 5.5.10 yields the following result.

(5.5.14) **Corollary.** *If $G_{\mathbb{Z}}$ is an irreducible lattice in G, and $G/G_{\mathbb{Z}}$ is not cocompact, then G has no compact factors.*

By combining Proposition 5.5.12 with Definition 5.1.19, we see that every irreducible arithmetic subgroup can be constructed by using restriction of scalars, and then modding out a compact subgroup:

(5.5.15) **Corollary.** *If Γ is an irreducible, arithmetic lattice in G (and G is connected), then there exist*

1) *an algebraic number field F, with completion F_∞ ($= \mathbb{R}$ or $\mathbb{C}$),*
2) *a connected, simple subgroup G' of $\mathrm{SL}(\ell, F_\infty)$, for some ℓ, such that G' is defined over F (as an algebraic group over F_∞), and*
3) *a continuous surjection*

$$\phi \colon \prod_{\sigma \in S^\infty} (G')^\sigma \to G,$$

with compact kernel,
such that $\phi(\Delta(G'_{\mathcal{O}}))$ is commensurable to Γ.

When G is simple, the restriction of ϕ to some simple factor of $\prod_{\sigma \in S^\infty} (G')^\sigma$ must be an isogeny, so the conclusion can be stated in the following much simpler form:

(5.5.16) **Corollary.** *If Γ is an arithmetic subgroup of G, and G is simple, then there exist*

1) *an algebraic number field F, with completion F_∞ ($= \mathbb{R}$ or $\mathbb{C}$),*
2) *a connected, simple subgroup G' of $\mathrm{SL}(\ell, F_\infty)$, for some ℓ, such that G' is defined over F (as an algebraic group over F_∞), and*
3) *an isogeny $\phi \colon G' \to G$,*

such that $\phi(G'_{\mathcal{O}})$ is commensurable to Γ.

However, we should point out that this result is of interest only when Γ is cocompact (or is reducible with at least one cocompact factor). This is because there is no need for restriction of scalars when the irreducible lattice Γ is not cocompact (see Corollary 5.3.2).

Exercises for §5.5.

#1. In the notation of the proof of Example 5.5.3, show, for the $\mathbb{Q}$-form $\Delta(F^2)$ of $\mathbb{R}^4$, that

$$\operatorname{End}(\mathbb{R}^4)_\mathbb{Q} = \left\{ \begin{bmatrix} A & B \\ \sigma(B) & \sigma(A) \end{bmatrix} \,\middle|\, A, B \in \operatorname{Mat}_{2\times 2}(F) \right\}.$$

[*Hint:* Since the F-span of $\Delta(F^2)$ is F^4, we have $\operatorname{End}(\mathbb{R}^4)_\mathbb{Q} \subseteq \operatorname{Mat}_{4\times4}(F)$. Thus, for any $T \in \operatorname{End}(\mathbb{R}^4)_\mathbb{Q}$, we may write $T = \begin{bmatrix} A & B \\ C & D \end{bmatrix}$, with $A, B, C, D \in \operatorname{Mat}_{2\times2}(F)$. Now use the fact that, for all $u \in F^2$, we have $T(u) = (v, \sigma(v))$, for some $v \in F^2$.]

#2. In the notation of the proof of Example 5.5.3, let

$$\mathcal{Q} = \left\{ x_{i,j+2} + x_{i+2,j}, \ x_{i,j+2} x_{i+2,j} \,\middle|\, 1 \le i, j \le 2 \right\}$$
$$\cup \left\{ \frac{1}{\sqrt{2}} \big((x_{1,1}x_{2,2} - x_{1,2}x_{2,1}) - (x_{3,3}x_{4,4} - x_{3,4}x_{4,3}) \big) \right\}.$$

a) Use the conclusion of Exercise 1 to show that each $Q \in \mathcal{Q}$ is defined over $\mathbb{Q}$.
b) Show that $\operatorname{Var}(\mathcal{Q})^\circ = \operatorname{SL}(2,\mathbb{R}) \times \operatorname{SL}(2,\mathbb{R})$.

#3. In the notation of the proof of Example 5.5.3, use Exercise 1 to show that $G_{\Delta(\mathcal{O}^2)} = \Delta(\Gamma)$.

#4. Let F, $\mathcal{O}$, σ, Δ be as in the proof of Example 5.5.3. If $G \subseteq \operatorname{SL}(\ell, \mathbb{R})$, and G is defined over F, show $G \times G$ is defined over $\mathbb{Q}$ (with respect to the $\mathbb{Q}$-form on $\operatorname{End}(\mathbb{R}^{2\ell})$ induced by the $\mathbb{Q}$-form $\Delta(F^\ell)$ on $\mathbb{R}^{2\ell}$).
[*Hint:* For each $Q \in \mathbb{Q}[x_{1,1},\ldots,x_{\ell,\ell}]$, let us define a corresponding polynomial $Q^+ \in \mathbb{Q}[x_{\ell+1,\ell+1},\ldots,x_{2\ell,2\ell}]$ by replacing every occurrence of each variable $x_{i,j}$ with $x_{\ell+i,\ell+j}$. For example, if $\ell = 2$, then
$$(x_{1,1}^2 + x_{1,2}x_{2,1} - 3x_{1,1}x_{2,2})^+ = x_{3,3}^2 + x_{3,4}x_{4,3} - 3x_{3,3}x_{4,4}.$$
Choose $\mathcal{Q}_0 \subset \mathbb{Q}[x_{1,1},\ldots,x_{\ell,\ell}]$ that defines G as a subgroup of $\operatorname{SL}(\ell,\mathbb{R})$, and let
$$\mathcal{Q}_1 = \{ Q + \sigma(Q^+),\ Q\sigma(Q^+) \mid Q \in \mathcal{Q}_0 \}.$$
A natural generalization of Exercise 2 shows that $\operatorname{SL}(\ell,\mathbb{R}) \times \operatorname{SL}(\ell,\mathbb{R})$ is defined over $\mathbb{Q}$: let $\mathcal{Q}_2$ be the corresponding set of $\mathbb{Q}$-polynomials. Now define $\mathcal{Q} = \mathcal{Q}_1 \cup \mathcal{Q}_2$.]

#5. Suppose $\mathcal{O}$ is the ring of integers of an algebraic number field F.
a) Show $\Delta(\mathcal{O})$ is discrete in $\bigoplus_{\sigma \in S^\infty} F_\sigma$.
b) Show $\Delta(F)$ is a $\mathbb{Q}$-form of $\bigoplus_{\sigma \in S^\infty} F_\sigma$.
c) Show $\Delta(\mathcal{O})$ is a $\mathbb{Z}$-lattice in $\Delta(F)$.

#6. Let
- $B(v,w) = v_1 w_1 + v_2 w_2 - \sqrt{2} v_3 w_3$, for $v, w \in \mathbb{R}^3$,
- $G = \operatorname{SO}(B)^\circ$,
- $G^* = G \times G^\sigma$,
- $\Gamma = G_{\mathbb{Z}[\sqrt{2}]}$, and
- $\Gamma^* = \Delta(\Gamma)$.

Show:

    a) The image of G^* in $\mathrm{SL}(6,\mathbb{R})/\mathrm{SL}(6,\mathbb{R})_{\Delta(\mathcal{O}^3)}$ is precompact (by using the Mahler Compactness Criterion).

    b) The image of G^* in $\mathrm{SL}(6,\mathbb{R})/\mathrm{SL}(6,\mathbb{R})_{\Delta(\mathcal{O}^3)}$ is closed.

    c) G^*/Γ^* is compact.

    d) G/Γ is compact (without using the fact that Γ is a lattice in G).

[*Hint:* This is similar to Proposition 5.3.4.]

#7. For any algebraic number field F, the $\mathbb{Q}$-form $\Delta(F^\ell)$ on $\bigoplus_{\sigma\in S^\infty}(F_\sigma)^\ell$ induces a natural $\mathbb{Q}$-form on $\mathrm{End}_{\mathbb{R}}(\bigoplus_{\sigma\in S^\infty}(F_\sigma)^\ell)$. Show the group $\prod_{\sigma\in S^\infty}\mathrm{SL}(\ell,F_\sigma)$ is defined over $\mathbb{Q}$, with respect to this $\mathbb{Q}$-form.

[*Hint:* This is a generalization of Exercise 2. That proof is based on the elementary symmetric functions of two variables: $P_1(a_1,a_2)=a_1+a_2$ and $P_2(a_1,a_2)=a_1a_2$. For the general case, use symmetric functions of d variables, where d is the degree of F over $\mathbb{Q}$.]

#8. Suppose $G\subseteq \mathrm{SL}(\ell,\mathbb{R})$, and G is defined over an algebraic number field $F\subset\mathbb{R}$. Show $\prod_{\sigma\in S^\infty}G^\sigma$ is defined over $\mathbb{Q}$, with respect to the $\mathbb{Q}$-form on $\mathrm{End}_{\mathbb{R}}(\bigoplus_{\sigma\in S^\infty}(F_\sigma)^\ell)$ induced by the $\mathbb{Q}$-form $\Delta(F^\ell)$ on $\bigoplus_{\sigma\in S^\infty}(F_\sigma)^\ell$.

[*Hint:* This is a generalization of Exercise 4. See the hint to Exercise 7.]

#9. Show, for all $m,n\geq 1$, with $m+n\geq 3$, that there exist a lattice Γ in $\mathrm{SO}(m,n)$, and a homomorphism $\rho\colon\Gamma\to\mathrm{SO}(m+n)$, such that $\rho(\Gamma)$ is dense in $\mathrm{SO}(m+n)$.

§5.6. Only isotypic groups have irreducible lattices

Intuitively, the **complexification** $G_{\mathbb{C}}$ of G is the complex Lie group that is obtained from G by replacing real numbers with complex numbers. For example, $\mathrm{SL}(n,\mathbb{R})_{\mathbb{C}}=\mathrm{SL}(n,\mathbb{C})$, and $\mathrm{SO}(n)_{\mathbb{C}}=\mathrm{SO}(n,\mathbb{C})$. (See Section 18.1 for more discussion of this.)

(5.6.1) **Definition.** G is **isotypic** if all of the simple factors of $G_{\mathbb{C}}$ are isogenous to each other.

    For example, $\mathrm{SL}(2,\mathbb{R})\times\mathrm{SL}(3,\mathbb{R})$ is not isotypic, because $\mathrm{SL}(2,\mathbb{C})$ is not isogenous to $\mathrm{SL}(3,\mathbb{C})$. Similarly, $\mathrm{SL}(5,\mathbb{R})\times\mathrm{SO}(2,3)$ is not isotypic, because the complexification of $\mathrm{SL}(5,\mathbb{R})$ is $\mathrm{SL}(5,\mathbb{C})$, but the complexification of $\mathrm{SO}(2,3)$ is (isomorphic to) $\mathrm{SO}(5,\mathbb{C})$. Therefore, the following consequence of the arithmeticity theorem implies that neither $\mathrm{SL}(2,\mathbb{R})\times\mathrm{SL}(3,\mathbb{R})$ nor $\mathrm{SL}(5,\mathbb{R})\times\mathrm{SO}(2,3)$ has an irreducible lattice.

(5.6.2) **Theorem** (Margulis). *Assume that G has no compact factors. If G has an irreducible lattice, then G is isotypic.*

Proof. Suppose Γ is an irreducible lattice in G. We may assume that G is not simple (otherwise, the desired conclusion is trivially true), so G is neither $\mathrm{SO}(1,n)$ nor $\mathrm{SU}(1,n)$. Therefore, from the Margulis Arithmeticity

Theorem (5.2.1), we know that Γ is arithmetic. Then, since Γ is irreducible, Corollary 5.5.15 implies there is a simple subgroup G' of some $SL(\ell, \mathbb{R})$, and a compact group K, such that

- G' is defined over a number field F, and
- $G \times K$ is isogenous to $\prod_{\sigma \in S^\infty} (G')^\sigma$.

So the simple factors of $G \times K$ are all in $\{ (G')^\sigma \mid \sigma \in S^\infty \}$ (up to isogeny). It then follows from Lemma 5.6.5 below that G is isotypic. □

(5.6.3) *Remarks.*

1) We will prove the converse of Theorem 5.6.2 in Proposition 18.7.5 (without the assumption that G has no compact factors).

2) By arguing just a bit more carefully, it can be shown that Theorem 5.6.2 remains valid when the assumption that G has no compact factors is replaced with the weaker hypothesis that G is not isogenous to $SO(1, n) \times K$ or $SU(1, n) \times K$, for any nontrivial, connected compact group K (see Exercise 2).

The following example shows that a nonisotypic group can have irreducible lattices, so some restriction on G is necessary in Theorem 5.6.2.

(5.6.4) **Example.** $SL(2, \mathbb{R}) \times K$ has an irreducible lattice, for any connected, compact Lie group K (cf. Exercise 4.9#7).

We now complete the proof of Theorem 5.6.2:

(5.6.5) **Lemma.** *Assume G is defined over an algebraic number field F. If σ is a place of F, and G is simple, then the complexification of G is isogenous to the complexification of G^σ.*

Proof. Extend σ to an automorphism $\hat{\sigma}$ of $\mathbb{C}$. Then $\hat{\sigma}(G_\mathbb{C}) = (G^\sigma)_\mathbb{C}$, so it is clear that $G_\mathbb{C}$ is isomorphic to $(G^\sigma)_\mathbb{C}$. Unfortunately, however, the automorphism $\hat{\sigma}$ is not continuous (not even measurable) unless it happens to be the usual complex conjugation, so we have only an isomorphism of abstract groups, not an isomorphism of Lie groups. Hence, this observation is not a proof, although it is suggestive. To give a rigorous proof, it is easier to work at the Lie algebra level.

First, let us make an observation that will also be pointed out in Remark 18.2.2. If $G = SL(n, \mathbb{C})$, or, more generally, if G is isogenous to a complex group $G'_\mathbb{C}$, then $G_\mathbb{C} = G \times G$ (because $\mathbb{C} \otimes_\mathbb{R} \mathbb{C} \cong \mathbb{C} \oplus \mathbb{C}$). So $G_\mathbb{C}$ is not simple. However, it can be shown that this is the only situation in which the complexification of a simple group fails to be simple: if G is simple, but $G_\mathbb{C}$ is not simple, then G is isogenous to a complex simple group $G'_\mathbb{C}$. Therefore, although the complexification of a simple group is not always simple, it is always isotypic.

Now assume, for definiteness, that $F \subset \mathbb{R}$ (see Exercise 6). Since G is defined over F, its Lie algebra $\mathfrak{g}$ is also defined over F. This means there is a basis $\{v_1, \ldots, v_n\}$ of $\mathfrak{g}$, such that the corresponding structure constants $\{c_{j,k}^{\ell}\}_{j,k,\ell=1}^{n}$ all belong to F; recall that the structure constants are defined by the formula

$$[v_j, v_k] = \sum_{\ell=1}^{n} c_{j,k}^{\ell} v_{\ell}.$$

Because G is isogenous to a group that is defined over $\mathbb{Q}$ (see Proposition 5.1.5), there is also a basis $\{u_1, \ldots, u_n\}$ of $\mathfrak{g}$ whose structure constants are in $\mathbb{Q}$. Write $v_k = \sum_{\ell=1}^{n} \alpha_k^{\ell} u_{\ell}$ with each $\alpha_k^{\ell} \in \mathbb{R}$, and define

$$v_k^{\sigma} = \sum_{\ell=1}^{n} \hat{\sigma}(\alpha_k^{\ell}) u_{\ell}.$$

Then $v_1^{\sigma}, \ldots, v_n^{\sigma}$ is a basis of $\mathfrak{g} \otimes_{\mathbb{R}} \mathbb{C}$ whose structure constants are $\{\sigma(c_{j,k}^{\ell})\}_{j,k,\ell=1}^{n}$. These are obviously the structure constants of the Lie algebra $\mathfrak{g}^{\sigma}$ of G^{σ}.

If $\sigma(F) \subset \mathbb{R}$, then the $\mathbb{R}$-span of $\{v_1^{\sigma}, \ldots, v_n^{\sigma}\}$ is (isomorphic to) $\mathfrak{g}^{\sigma}$, so its $\mathbb{C}$-span is $\mathfrak{g}^{\sigma} \otimes_{\mathbb{R}} \mathbb{C}$. Since $v_1^{\sigma}, \ldots, v_n^{\sigma}$ is also a basis of $\mathfrak{g} \otimes_{\mathbb{R}} \mathbb{C}$, we conclude that $(G^{\sigma})_{\mathbb{C}}$ is isogenous to $G_{\mathbb{C}}$.

Finally, if $\sigma(F) \not\subset \mathbb{R}$, then the $\mathbb{C}$-span of $\{v_1^{\sigma}, \ldots, v_n^{\sigma}\}$ is (isomorphic to) $\mathfrak{g}^{\sigma}$, so $\mathfrak{g} \otimes_{\mathbb{R}} \mathbb{C} = \mathfrak{g}^{\sigma}$. This implies that $(G^{\sigma})_{\mathbb{C}}$ is isogenous to $G_{\mathbb{C}}$. $\square$

(5.6.6) *Remark.* The proof of Lemma 5.6.5 used our standing assumption that G is semisimple only to show that G is isogenous to a group that is defined over $\mathbb{Q}$. See Exercise 3 for an example of a Lie group H, defined over an algebraic number field $F \subset \mathbb{R}$, and an embedding σ of F in $\mathbb{R}$, such that $H \times H^{\sigma}$ is not isotypic.

Exercises for §5.6.

#1. Show, for $m, n \geq 2$, that $\mathrm{SL}(m, \mathbb{R}) \times \mathrm{SL}(n, \mathbb{R})$ has an irreducible lattice if and only if $m = n$.

#2. Suppose G is not isogenous to $\mathrm{SO}(1, n) \times K$ or $\mathrm{SU}(1, n) \times K$, for any nontrivial, connected compact group K. Show that if G has an irreducible lattice, then G is isotypic.

[*Hint:* Use Remark 5.2.3 to modify the proof of Theorem 5.6.2.]

#3. (optional) For $\alpha \in \mathbb{C} \setminus \{0, -1\}$, let $\mathfrak{h}_{\alpha}$ be the 7-dimensional, nilpotent Lie algebra over $\mathbb{C}$, generated by $\{x_1, x_2, x_3\}$, such that
  - $[\mathfrak{h}_{\alpha}, x_1, x_1] = [\mathfrak{h}_{\alpha}, x_2, x_2] = [\mathfrak{h}_{\alpha}, x_3, x_3] = 0$, and
  - $[x_2, x_3, x_1] = \alpha[x_1, x_2, x_3]$.
  a) Show that $[x_3, x_1, x_2] = -(1 + \alpha)[x_1, x_2, x_3]$.
  b) For $h \in \mathfrak{h}_{\alpha}$, show that $[\mathfrak{h}_{\alpha}, h, h] = 0$ if and only if there exists $x \in \{x_1, x_2, x_3\}$ and $t \in \mathbb{C}$, such that $h \in tx + [\mathfrak{h}_{\alpha}, \mathfrak{h}_{\alpha}]$.
  c) Show $\mathfrak{h}_{\alpha} \cong \mathfrak{h}_{\beta}$ iff $\beta \in \left\{\alpha, \frac{1}{\alpha}, -(1 + \alpha), -\frac{1}{1+\alpha}, -\frac{\alpha}{1+\alpha}, -\frac{1+\alpha}{\alpha}\right\}$.

   d) Show that if the degree of $\mathbb{Q}(\alpha)$ over $\mathbb{Q}$ is at least 7, then there is a place σ of $\mathbb{Q}(\alpha)$, such that $\mathfrak{h}_\alpha$ is not isomorphic to $(\mathfrak{h}_\alpha)^\sigma$.

#4. (optional) In the notation of Exercise 3, show that if the degree of $\mathbb{Q}(\alpha)$ over $\mathbb{Q}$ is at least 7, then $\mathfrak{h}_\alpha$ is not isomorphic to any Lie algebra that is defined over $\mathbb{Q}$.

#5. (optional) In the notation of Exercise 3, show, for $\alpha = \sqrt{2} - (1/2)$, that $\mathfrak{h}_\alpha$ is isomorphic to a Lie algebra that is defined over $\mathbb{Q}$.
   [*Hint:* Let $y_1 = x_1 + x_2$ and $y_2 = (x_1 - x_2)/\sqrt{2}$. Show that the $\mathbb{Q}$-subalgebra of $\mathfrak{h}_\alpha$ generated by $\{y_1, y_2, x_3\}$ is a $\mathbb{Q}$-form of $\mathfrak{h}_\alpha$.]

#6. Carry out the proof of Lemma 5.6.5 for the case where $F \not\subset \mathbb{R}$.
   [*Hint:* Write $\mathfrak{g} = \mathfrak{g}' \otimes_\mathbb{R} \mathbb{C}$ and let $\{u_1, \ldots, u_n\}$ be a basis of $\mathfrak{g}'$ with rational structure constants. Show that G is isogenous to either G^σ or $(G^\sigma)_\mathbb{C}$.]

Notes

The fact that G is unirational (used in Exercise 5.1#3) is proved in [4, Thm. 18.2, p. 218].

The Margulis Arithmeticity Theorem (5.2.1) was proved by Margulis [9, 11] under the assumption that $\operatorname{rank}_\mathbb{R} G \geq 2$. (Proofs also appear in [12, Thm. A, p. 298] and [16].) Much later, the superrigidity theorems of Corlette [6] and Gromov-Schoen [7] extended this to all groups except $SO(1, n)$ and $SU(1, n)$.

Proposition 5.1.5 is a weak version of a theorem of Borel [2]. (A proof also appears in [14, Chap. 14].)

The Commensurability Criterion (5.2.5(1)) is due to Margulis [10]. We will see it again in Theorem 16.3.3, and it is proved in [1], [12], and [16].

The fact that all noncocompact lattices have unipotent elements (that is, the generalization of Theorem 5.3.3 to the nonarithmetic case) is due to D. Kazhdan and G. A. Margulis [8] (or see [3] or [14, Cor. 11.13, p. 180]).

The standard reference on restriction of scalars is [15, §1.3, pp. 4–9]. (A discussion can also be found in [13, §2.1.2, pp. 49–50].)

Proposition 5.5.12 (and Remark 5.5.13) is due to A. Borel and J. Tits [5, 6.21(ii), p. 113].

See [12, Cor. IX.4.5, p. 315] for a proof of Theorem 5.6.2.

References

[1] N. A'Campo and M. Burger: Réseaux arithmétiques et commensurateur d'après G. A. Margulis, *Invent. Math.* 116 (1994) 1–25. MR 1253187, http://eudml.org/doc/144182

[2] A. Borel: Compact Clifford-Klein forms of symmetric spaces, *Topology* 2 (1963) 111–122. MR 0146301, http://dx.doi.org/10.1016/0040-9383(63)90026-0

[3] A. Borel: Sous-groupes discrets de groups semi-simples (d'après
    D. A. Kajdan et G. A. Margoulis), *Séminaire Bourbaki* 1968/1969,
    no. 358. *Springer Lecture Notes in Math.* 175 (1971) 199–215.
    MR 3077127, http://eudml.org/doc/109759

[4] A. Borel: *Linear Algebraic Groups, 2nd ed.*. Springer, New York,
    1991. ISBN 0-387-97370-2, MR 1102012

[5] A. Borel and J. Tits: Groupes réductifs, *Inst. Hautes Études Sci. Publ.
    Math.* 27 (1965) 55–150. MR 207712,
    http://www.numdam.org/item?id=PMIHES_1965__27__55_0

[6] K. Corlette: Archimedean superrigidity and hyperbolic geometry,
    Ann. Math. 135 (1992) 165–182. MR 1147961,
    http://dx.doi.org/10.2307/2946567

[7] M. Gromov and R. Schoen: Harmonic maps into singular spaces and
    p-adic superrigidity for lattices in groups of rank one, *Publ. Math.
    Inst. Hautes Études Sci.* 76 (1992) 165–246. MR 1215595,
    http://www.numdam.org/item?id=PMIHES_1992__76__165_0

[8] D. A. Každan and G. A. Margulis: A proof of Selberg's Conjecture,
    Math. USSR–Sbornik 4 (1968), no. 1, 147–152. MR 0223487,
    http://dx.doi.org/10.1070/SM1968v004n01ABEH002782

[9] G. A. Margulis: Arithmetic properties of discrete subgroups, *Russian
    Math. Surveys* 29:1 (1974) 107–156. Translated from Uspekhi Mat.
    Nauk 29:1 (1974) 49–98. MR 0463353,
    http://dx.doi.org/10.1070/RM1974v029n01ABEH001281

[10] G. A. Margulis: Discrete groups of motions of manifolds of
     non-positive curvature, *Amer. Math. Soc. Translations* 109 (1977)
     33–45. MR 0492072

[11] G. A. Margulis: Arithmeticity of the irreducible lattices in the
     semi-simple groups of rank greater than 1. (Appendix to Russian
     translation of [14], 1977.) English translation in: *Invent. Math.* 76
     (1984) 93–120. MR 0739627, http://eudml.org/doc/143118

[12] G. A. Margulis: *Discrete Subgroups of Semisimple Lie Groups.*
     Springer, New York, 1991. ISBN 3-540-12179-X, MR 1090825

[13] V. Platonov and A. Rapinchuk: *Algebraic Groups and Number
     Theory.* Academic Press, Boston, 1994. ISBN 0-12-558180-7,
     MR 1278263

[14] M. S. Raghunathan: *Discrete Subgroups of Lie Groups.* Springer, New
     York, 1972. ISBN 0-387-05749-8, MR 0507234

[15] A. Weil: *Adeles and Algebraic Groups.* Birkhäuser, Boston, 1982.
     ISBN 3-7643-3092-9, MR 0670072

[16] R. J. Zimmer: *Ergodic Theory and Semisimple Groups.* Birkhäuser,
     Boston, 1984. ISBN 3-7643-3184-4, MR 0776417

Chapter 6

Examples of
Arithmetic Groups

§6.1. Arithmetic subgroups of SL(2, ℝ) via orthogonal groups

$\mathrm{SL}(2, \mathbb{Z})$ is the obvious example of an arithmetic subgroup of $\mathrm{SL}(2, \mathbb{R})$. Later in this section, we will show that (up to commensurability and conjugates) it is the only one that is not cocompact (see Proposition 6.1.5). In contrast, there are infinitely many cocompact, arithmetic subgroups. They can be constructed by several different methods. Perhaps the easiest way is to note that $\mathrm{SL}(2, \mathbb{R})$ is isogenous to the special orthogonal group $\mathrm{SO}(2, 1)$.

(6.1.1) **Notation.** In this chapter (and others), we will see many different special orthogonal groups over a field F. They can be specified in (at least) three different, but equivalent ways:

    1) (Gram matrix) For a symmetric, invertible matrix $A \in \mathrm{Mat}_{\ell \times \ell}(F)$, we define
$$\mathrm{SO}(A; F) = \{\, g \in \mathrm{SL}(n, F) \mid g^T A g = A \,\}.$$

Recall: The Standing Assumptions (4.0.0 on page 41) are in effect, so, as always, Γ is a lattice in the semisimple Lie group $G \subseteq \mathrm{SL}(\ell, \mathbb{R})$.

Main prerequisites for this chapter: definition of arithmetic subgroup (Section 5.1), Godement Criterion (Proposition 5.3.1), and restriction of scalars (Section 5.5).

This is the approach taken to the definition of $\mathrm{SO}(m,n)$ in Example A2.3.

2) (Bilinear form) A symmetric, bilinear form B on F^ℓ is **nondegenerate** if, for all nonzero $v \in F^\ell$, there exists $w \in F^\ell$, such that $B(v,w) \neq 0$. We define
$$\mathrm{SO}(B;F) = \{\, g \in \mathrm{SL}(\ell,F) \mid B(gv,gw) = B(v,w), \ \forall v, w \in F^\ell \,\}.$$

3) (Quadratic form) A **quadratic form** on F^ℓ is a homogeneous polynomial $Q(x_1,\ldots,x_\ell)$ of degree 2. It is **nondegenerate** if the corresponding bilinear form B_Q is nondegenerate, where
$$B_Q(v,w) = \tfrac{1}{4}\big(Q(v+w) - Q(v-w)\big).$$
We define
$$\mathrm{SO}(Q;F) = \{\, g \in \mathrm{SL}(\ell,F) \mid Q(gv) = Q(v), \ \forall v \in F^\ell \,\}.$$

The three approaches give rise to exactly the same groups (see Exercise 1), and it is straightforward to translate between them, so we will use whichever notation is most convenient in a particular context.

(6.1.2) **Examples.**

1) Fix positive integers a and b, and let
$$G = \mathrm{SO}(ax^2 + by^2 - z^2; \mathbb{R}) \cong \mathrm{SO}(2,1).$$
If $(0,0,0)$ is the only integer solution of the Diophantine equation $ax^2 + by^2 = z^2$, then $G_{\mathbb{Z}}$ is a cocompact, arithmetic subgroup of G (see Proposition 5.3.4). See Exercise 2 for some examples of a and b satisfying the hypotheses.

2) Restriction of scalars (see Section 5.5) allows us to use algebraic number fields other than $\mathbb{Q}$. Let
   - $F \neq \mathbb{Q}$ be a **totally real** algebraic number field (that is, an algebraic number field with no complex places),
   - $a,b \in F^+$, such that $\sigma(a)$ and $\sigma(b)$ are negative, for every place $\sigma \neq \mathrm{Id}$,
   - $\mathcal{O}$ be the ring of integers of F, and
   - $G = \mathrm{SO}(ax^2 + by^2 - z^2; \mathbb{R}) \cong \mathrm{SO}(2,1)$.

   Then the group $G_{\mathcal{O}}$ is a cocompact, arithmetic subgroup of G (cf. Example 5.5.4, or see Proposition 5.5.8 and Corollary 5.5.10). See Exercise 3 for an example of F, a, and b satisfying the hypotheses.

3) In both (1) and (2), the group G is conjugate to $\mathrm{SO}(2,1)$, via the diagonal matrix
$$g = \mathrm{diag}(\sqrt{a}, \sqrt{b}, 1).$$
Therefore, $g^{-1}(G_{\mathbb{Z}})g$ or $g^{-1}(G_{\mathcal{O}})g$ is a cocompact, arithmetic subgroup of $\mathrm{SO}(2,1)$.

(6.1.3) *Remark.* For a and b as in Example 6.1.2(2), $(0,0,0)$ is the only solution in $\mathcal{O}^3$ of the equation $ax^2 + by^2 = z^2$ (see Exercise 4). Therefore,

Example 6.1.2(1) and Example 6.1.2(2) could fairly easily be combined into a single construction, but we separated them to keep them a bit less complicated.

(6.1.4) **Proposition.** *The only cocompact, arithmetic subgroups of* $\mathrm{SO}(2,1)$ *are the arithmetic subgroups constructed in Example 6.1.2 (up to commensurability and conjugates).*

More precisely, any cocompact, arithmetic subgroup of $\mathrm{SO}(2,1)$ *has a conjugate that is commensurable to an arithmetic subgroup constructed in Example 6.1.2.*

Proof. Let Γ be a cocompact, arithmetic subgroup of $\mathrm{SO}(2,1)$. Ignoring the minor technical issue that not all automorphisms are inner (cf. Remark A6.4), it suffices to show that there is an automorphism α of $\mathrm{SO}(2,1)$, such that $\alpha(\Gamma)$ is commensurable to one of the arithmetic subgroups constructed in Example 6.1.2.

Step 1. There are

- *an algebraic number field $F \subset \mathbb{R}$, with ring of integers $\mathcal{O}$,*
- *a symmetric, bilinear form $B(x,y)$ on F^3, and*
- *an isomorphism $\phi\colon \mathrm{SO}(B;\mathbb{R}) \to \mathrm{SO}(2,1)$,*

such that $\phi(\mathrm{SO}(B;\mathcal{O}))$ is commensurable to Γ. We give two proofs.

First, we note that this follows from the classification results that will be proved in Chapter 18. Namely, a group of the form $\mathrm{SO}(m,n)$ does not appear in Proposition 18.5.6, and it arises as the right-hand side of two different parts of Proposition 18.5.7. However, $m + n = 1 + 2 = 3$ is odd in our situation, so only one of the listings is relevant: G_F must be $\mathrm{SO}(A;F)$, for some algebraic number field $F \subset \mathbb{R}$. This means that Γ is commensurable to $\mathrm{SO}(A;\mathcal{O})$, where $\mathcal{O}$ is the ring of integers of F.

Second, let us give a direct proof that does not rely on the results of Chapter 18. Because all (irreducible) arithmetic subgroups are obtained by restriction of scalars, and G is simple, Corollary 5.5.16 tells us there are

- an algebraic number field $F \subset \mathbb{R}$, with ring of integers $\mathcal{O}$,
- a simple Lie group $H \subseteq \mathrm{SL}(\ell, \mathbb{R})$ that is defined over F, and
- an isogeny $\phi\colon H \to \mathrm{SO}(2,1)$,

such that $\phi(H_{\mathcal{O}})$ is commensurable to Γ. All that remains is to show that we may identify H_F with $\mathrm{SO}(B;F)$, for some symmetric bilinear form B on F^3.

The Killing form

$$\kappa(u,v) = \mathrm{trace}((\mathrm{ad}_{\mathfrak{h}}\, u)(\mathrm{ad}_{\mathfrak{h}}\, v))$$

is a symmetric, bilinear form on the Lie algebra $\mathfrak{h}$. It is invariant under $\mathrm{Ad}\,H$, so Ad_H is an isogeny from H to $\mathrm{SO}(\kappa;\mathbb{R})$. Pretending that Ad_H is an isomorphism, not just an isogeny, we may identify H with $\mathrm{SO}(\kappa; \mathbb{R})$. Note that $\kappa(\mathfrak{h}_F, \mathfrak{h}_F) \subseteq F$, so, by identifying $\mathfrak{h}_F$ with F^3, we may think of κ as a bilinear form on F^3.

Step 2. We may assume that $B(x,x) = ax_1^2 + bx_2^2 - x_3^2$ for some $a, b \in F^+$. By choosing an orthogonal basis that diagonalizes the form, we may assume $B(x,x) = ax_1^2 + bx_2^2 + cx_3^2$. Since $\mathrm{SO}(B; \mathbb{R}) \approx \mathrm{SO}(2, 1)$, we know that $\pm B(x,x)$ has signature $(2, 1)$. So we may assume $a, b, -c \in F^+$. Dividing by c (which does not change the orthogonal group) yields the desired form.

Step 3. F is totally real, and both $\sigma(a)$ and $\sigma(b)$ are negative, for all places $\sigma \neq \mathrm{Id}$. Since $\Delta(G_{\mathcal{O}})$ is an irreducible lattice in $\prod_{\sigma \in S^\infty} G^\sigma$ (see Proposition 5.5.8), but the projection to the first factor, namely G, is Γ, which is discrete, we know that G^σ is compact, for all $\sigma \neq \mathrm{Id}$. This implies $G^\sigma \cong \mathrm{SO}(3)$, so $F_\sigma = \mathbb{R}$, and the three real numbers $\sigma(a)$, $\sigma(b)$, and $\sigma(-1)$ all have the same sign.

Step 4. B is anisotropic over F. Since $G_{\mathcal{O}}$ is cocompact, it has no nontrivial unipotent elements (see Corollary 4.4.4). Therefore $B(x,x) \neq 0$, for every nonzero $x \in F^3$ (see Exercise 5). □

(6.1.5) **Proposition.** $\mathrm{SL}(2, \mathbb{Z})$ *is the only noncocompact, arithmetic subgroup Γ of $\mathrm{SL}(2, \mathbb{R})$ (up to commensurability and conjugates).*

Proof. Let us consider the isogenous group $\mathrm{SO}(2, 1)$, instead of $\mathrm{SL}(2, \mathbb{R})$.

Step 1. There are

- *a symmetric, bilinear form $B(x,y)$ on $\mathbb{Q}^3$, and*
- *an isogeny $\phi \colon \mathrm{SO}(B; \mathbb{R}) \to \mathrm{SO}(2, 1)$,*

such that $\phi(\mathrm{SO}(B; \mathbb{Z}))$ is commensurable to Γ. Since Γ is not cocompact, there is an isogeny $\phi \colon G \to \mathrm{SO}(2, 1)$, such that G is defined over $\mathbb{Q}$ and $\phi(G_{\mathbb{Z}})$ is commensurable to Γ (see Corollary 5.3.2). The argument in Steps 1 and 3 of the proof of Proposition 6.1.4 shows that we may assume $G = \mathrm{SO}(B; \mathbb{R})$.

Step 2. We may assume $B(x,x) = x_1^2 + x_2^2 - x_3^2$. Because Γ is not cocompact, we know that B is isotropic over F (see Proposition 5.3.4). So there is some nonzero $u \in F^3$, such that $B(u, u) = 0$. Choose $v \in F^3$, such that $B(u, v) \neq 0$. By adding a scalar multiple of u to v, we may assume $B(v, v) = 0$. Now choose a nonzero $w \in F^3$ that is orthogonal to both u and v. After multiplying B and u by appropriate scalars, we may assume $B(w, w) = 2B(u, v) = 1$. Then B has the desired form with respect to the basis $w, u + v, u - v$. □

(6.1.6) *Remark.* As a source of counterexamples, it is useful to remember that $SL(2,\mathbb{Z})$ contains a free subgroup of finite index (see Exercise 4.9#5 or 4.9#6). This implies that (finitely generated) nonabelian free groups are lattices in $SL(2,\mathbb{R})$.

Exercises for §6.1.

#1. Show:
   a) If Q is nondegenerate quadratic form on F^ℓ, and B_Q is defined as in Notation 6.1.1, then B_Q is a bilinear form, and we have $SO(B_Q;F) = B(Q;F)$.
   b) If B is a nondegenerate bilinear form on F^ℓ, and we define $Q(x) = B(x,x)$, then $Q(x)$ is a quadratic form, and $B = B_Q$.
   c) If A is a symmetric, invertible matrix in $\mathrm{Mat}_{\ell\times\ell}(F)$, and we define $B(v,w) = v^T A w$, then B is a nondegenerate bilinear form, and $SO(B;F) = SO(A;F)$.
   d) If B is a nondegenerate bilinear form on F^ℓ, and $\{\varepsilon_1,\ldots,\varepsilon_\ell\}$ is the standard basis of F^ℓ, then the matrix $A = (B(\varepsilon_i,\varepsilon_j))$ is invertible and symmetric, and we have $SO(A;F) = SO(B;F)$.

#2. Suppose p is a prime, such that $x^2 + y^2 \equiv 0 \pmod p$ has only the trivial solution $x \equiv y \equiv 0 \pmod p$. (For example, p could be 3.) Show that $(0,0,0)$ is the only integer solution of the Diophantine equation $px^2 + py^2 = z^2$.

#3. Let $F = \mathbb{Q}[\sqrt{2}, \sqrt{3}]$, and $a = b = \sqrt{2} + \sqrt{3} - 3$. Show
   a) F is a totally real extension of $\mathbb{Q}$,
   b) a is positive, and
   c) $\sigma(a)$ is negative, for every place $\sigma \neq \mathrm{Id}$.

#4. If a and b are elements of an algebraic number field F, and there is a real place σ of F, such that $\sigma(a)$ and $\sigma(b)$ are negative, show $(0,0,0)$ is the only solution in F^3 of the equation $ax^2 + by^2 = z^2$.

#5. In Step 4 of the proof of Proposition 6.1.4, verify the assertion that $B(x,x) \neq 0$, for every nonzero $x \in F^3$.
   [*Hint:* If $B(x,x) = 0$ for some nonzero x, then, after a change of basis, $B(x,x)$ is a scalar multiple of the form $x_1 x_3 + x_2^2$, which is invariant under the unipotent transformation $x_1 \mapsto x_1 - 2x_2 - x_3, x_2 \mapsto x_2 + x_3, x_3 \mapsto x_3$.]

§6.2. Arithmetic subgroups of SL(2, ℝ) via quaternion algebras

In the preceding section, we constructed the cocompact, arithmetic subgroups of $SL(2,\mathbb{R})$ from orthogonal groups. As an alternative approach, we will now explain what quaternion algebras are, and how they can be used to construct those same arithmetic subgroups. In later sections (and later chapters), the use of quaternion algebras will sometimes be necessary, not an alternative approach.

(6.2.1) **Definitions.**

1) For any field F, and any nonzero $a, b \in F$, the corresponding *quaternion algebra* over F is the ring
$$\mathbb{H}_F^{a,b} = \{ p + qi + rj + sk \mid p, q, r, s \in F \},$$
where
   - addition is defined in the obvious way, and
   - multiplication is determined by the relations
$$i^2 = a, \quad j^2 = b, \quad ij = k = -ji,$$
   together with the requirement that every element of F is in the center of D. (Note that $k^2 = k \cdot k = (-ji)(ij) = -aj^2 = -ab$.)

2) The *reduced norm* of $x = p + qi + rj + sk \in \mathbb{H}_F^{a,b}$ is
$$N_{\text{red}}(x) = x\overline{x} = p^2 - aq^2 - br^2 + abs^2 \in F,$$
   where $\overline{x} = p - qi - rj - sk$ is the *conjugate* of x. (Note that $\overline{xy} = \overline{y}\,\overline{x}$.)

(6.2.2) **Example.**

1) We have $\mathbb{H}_\mathbb{R}^{-1,-1} = \mathbb{H}$.

2) We have $\mathbb{H}_F^{t^2 a, t^2 b} \cong \mathbb{H}_F^{a,b}$ for any nonzero $a, b, t \in F$ (see Exercise 1).

3) We have $\mathbb{H}_F^{a^2, b} \cong \text{Mat}_{2\times 2}(F)$, for any nonzero $a, b \in F$ (see Exercise 2).

4) We have $N_{\text{red}}(gh) = N_{\text{red}}(g) \cdot N_{\text{red}}(h)$ for $g, h \in \mathbb{H}_F^{a,b}$.

(6.2.3) **Lemma.** *We have* $\mathbb{H}_\mathbb{C}^{a,b} \cong \text{Mat}_{2\times 2}(\mathbb{C})$, *for all* $a, b \in \mathbb{C}$, *and*
$$\mathbb{H}_\mathbb{R}^{a,-1} \cong \begin{cases} \text{Mat}_{2\times 2}(\mathbb{R}) & \text{if } a > 0, \\ \mathbb{H} & \text{if } a < 0. \end{cases}$$

Proof. This follows from the observations in Example 6.2.2. $\square$

(6.2.4) **Proposition.** *Fix positive integers a and b, and let*
$$G = \text{SL}(1, \mathbb{H}_\mathbb{R}^{a,b}) = \{ g \in \mathbb{H}_\mathbb{R}^{a,b} \mid N_{\text{red}}(g) = 1 \}.$$
Then:

1) $G \cong \text{SL}(2, \mathbb{R})$,

2) $G_\mathbb{Z} = \text{SL}(1, \mathbb{H}_\mathbb{Z}^{a,b})$ *is an arithmetic subgroup of G, and*

3) *the following are equivalent:*
   (a) $G_\mathbb{Z}$ *is cocompact in G.*
   (b) $(0,0,0,0)$ *is the only integer solution (p, q, r, s) of the Diophantine equation*
$$w^2 - ax^2 - by^2 + abz^2 = 0.$$

(c) *Every nonzero element of $\mathbb{H}_{\mathbb{Q}}^{a,b}$ has a multiplicative inverse (so $\mathbb{H}_{\mathbb{Q}}^{a,b}$ is a "division algebra").*

Proof. (1) Define an ℝ-linear bijection $\phi \colon \mathbb{H}_{\mathbb{R}}^{a,b} \to \mathrm{Mat}_{2\times 2}(\mathbb{R})$ by $\phi(1) = \mathrm{Id}$,

$$\phi(i) = \begin{bmatrix} \sqrt{a} & 0 \\ 0 & -\sqrt{a} \end{bmatrix}, \quad \phi(j) = \begin{bmatrix} 0 & 1 \\ b & 0 \end{bmatrix}, \quad \phi(k) = \begin{bmatrix} 0 & \sqrt{a} \\ -b\sqrt{a} & 0 \end{bmatrix}.$$

It is straightforward to check that ϕ preserves multiplication, so ϕ is a ring isomorphism.

For $g = p + qi + rj + sk \in \mathbb{H}_{\mathbb{R}}^{a,b}$, we have

$$\begin{aligned} \det(\phi(g)) &= (p + q\sqrt{a})(p - q\sqrt{a}) - (r + s\sqrt{a})(br - bs\sqrt{a}) \\ &= p^2 - aq^2 - br^2 + abs^2 \\ &= \mathrm{N_{red}}(g). \end{aligned}$$

Therefore, $\phi(G) = \mathrm{SL}(2, \mathbb{R})$.

(2) For $g \in G$, define $T_g \colon \mathbb{H}_{\mathbb{R}}^{a,b} \to \mathbb{H}_{\mathbb{R}}^{a,b}$ by $T_g(v) = gv$. Then T_g is ℝ-linear. For $y \in \mathbb{H}_{\mathbb{R}}^{a,b}$, we have $T_y(\mathbb{H}_{\mathbb{Z}}^{a,b}) \subset \mathbb{H}_{\mathbb{Z}}^{a,b}$ if and only if $y \in \mathbb{H}_{\mathbb{Z}}^{a,b}$. So $G_{\mathbb{Z}} = G \cap \mathbb{H}_{\mathbb{Z}}^{a,b}$ is an arithmetic subgroup of G.

(3c ⇒ 3a) We prove the contrapositive. Suppose $G_{\mathbb{Z}}$ is not cocompact. Then the Godement Criterion (5.3.1) tells us that it has a nontrivial unipotent element y. So 1 is an eigenvalue of T_y; that is, there is some nonzero $v \in \mathbb{H}_{\mathbb{Z}}^{a,b}$, such that $T_y(v) = v$. By definition of T_y, this means $yv = v$. Hence $(y - 1)v = 0$. Since $y \neq 1$ and $v \neq 0$, this implies v is a zero divisor, so it certainly does not have a multiplicative inverse.

(3a ⇒ 3c) We prove the contrapositive. Suppose $\mathbb{H}_{\mathbb{Q}}^{a,b}$ is not a division algebra. Then $\mathbb{H}_{\mathbb{Q}}^{a,b} \cong \mathrm{Mat}_{2\times 2}(\mathbb{Q})$ (see Exercise 3). This implies $\mathrm{SL}(1, \mathbb{H}_{\mathbb{Z}}^{a,b}) \approx \mathrm{SL}(2, \mathbb{Z})$ is not cocompact. (It has nontrivial unipotent elements.)

(3b ⇔ 3c) See Exercise 4. □

The following can be proved similarly (see Exercise 5).

(6.2.5) **Proposition.** *Let*

- *F be a totally real algebraic number field (with $F \neq \mathbb{Q}$),*
- *$\mathcal{O}$ be the ring of integers of F,*
- *$a, b \in \mathcal{O}$, such that a and b are positive, but $\sigma(a)$ and $\sigma(b)$ are negative, for every place $\sigma \neq \mathrm{Id}$, and*
- *$G = \mathrm{SL}(1, \mathbb{H}_{\mathbb{R}}^{a,b})$.*

Then:

1) *$G \cong \mathrm{SL}(2, \mathbb{R})$, and*
2) *$G_{\mathcal{O}} = \mathrm{SL}(1, \mathbb{H}_{\mathcal{O}}^{a,b})$ is a cocompact, arithmetic subgroup of G.*

(6.2.6) **Proposition.** *Every cocompact, arithmetic subgroup of* $\mathrm{SL}(2, \mathbb{R})$ *appears in either Proposition 6.2.4 or 6.2.5 (up to commensurability and conjugates).*

Proof. This can be proved directly, but we will instead derive it as a corollary of Proposition 6.1.4. For each arithmetic subgroup Γ of $\mathrm{SO}(2,1)$, constructed in Example 6.1.2, we find an isogeny $\phi \colon \mathrm{SL}(2, \mathbb{R}) \to \mathrm{SO}(2,1)$, such that $\phi(\Gamma')$ is commensurable to an arithmetic subgroup constructed in Proposition 6.2.4 or 6.2.5.

(1) First, let us show that every arithmetic subgroup of type 6.1.2(1) appears in (6.2.4). Given positive integers a and b, such that $(0,0,0)$ is the only rational solution of the equation $ax^2 + by^2 = z^2$, let

$$G = \mathrm{SL}(1, \mathbb{H}_{\mathbb{R}}^{a,b}) \cong \mathrm{SL}(2, \mathbb{R}).$$

One can show that $(0,0,0,0)$ is the only rational solution of the equation $w^2 - ax^2 - by^2 + abz^2 = 0$ (see Exercise 6), so $G_{\mathbb{Z}}$ is a cocompact, arithmetic subgroup of G (see Proposition 6.2.4).

As a subspace of $\mathbb{H}_{\mathbb{R}}^{a,b}$, the Lie algebra $\mathfrak{g}$ of G is

$$\mathfrak{g} = \{\, v \in \mathbb{H}_{\mathbb{R}}^{a,b} \mid \operatorname{Re} v = 0 \,\}$$

(see Exercise 7). For $g \in G$ and $v \in \mathfrak{g}$, we have $(\mathrm{Ad}_G g)(v) = gvg^{-1}$, so $\mathrm{N}_{\mathrm{red}} \,|_{\mathfrak{g}}$ is a quadratic form on $\mathfrak{g}$ that is invariant under $\mathrm{Ad}\, G_F$. For $v = xi + yj + zk \in \mathfrak{g}$, we have

$$\mathrm{N}_{\mathrm{red}}(v) = -ax^2 - by^2 + abz^2.$$

After the change of variables $x \mapsto by$ and $y \mapsto ax$, this becomes $-ab(ax^2 + by^2 - z^2)$, which is a scalar multiple of the quadratic form in 6.1.2(1). Therefore, after identifying $\mathfrak{g}$ with $\mathbb{R}^3$ by an appropriate choice of basis, the arithmetic subgroup constructed in Example 6.1.2(1) (for the given values of a and b) is commensurable to $\mathrm{Ad}_G G_{\mathbb{Z}}$.

(2) Similarly, every arithmetic subgroup of type 6.1.2(2) appears in (6.2.5) (see Exercise 8). $\qquad\square$

Exercises for §6.2.

#1. Show $\mathbb{H}_F^{u^2 a, v^2 y} \cong \mathbb{H}_F^{a,b}$, for any nonzero $u, v \in F$.

[*Hint:* An isomorphism is given by $1 \mapsto 1$, $i \mapsto ui$, $j \mapsto vj$, $k \mapsto uvk$.]

#2. Show $\mathbb{H}_F^{a^2,b} \cong \mathrm{Mat}_{2\times 2}(F)$, for any field F, and any $a, b \in F$.

[*Hint:* See the proof of Proposition 6.2.4(1).]

#3. Show that if the ring $\mathbb{H}_{\mathbb{Q}}^{a,b}$ is not a division algebra, then it is isomorphic to $\mathrm{Mat}_{2\times 2}(\mathbb{Q})$.

[*Hint:* This follows from Wedderburn's Theorem (6.8.5), but can also be proved directly: if x is not invertible, then $xy = 0$ for some y, so the left ideal generated by x is a 2-dimensional subspace on which $\mathbb{H}_{\mathbb{Q}}^{a,b}$ acts faithfully.]

#4. Show that every nonzero element of $\mathbb{H}_F^{a,b}$ has a multiplicative inverse if and only if the reduced norm of every nonzero element is nonzero.

[*Hint:* If $N_{red}(x) \neq 0$, then multiply the conjugate of x by an element of F to obtain a multiplicative inverse of x. If $N_{red}(x) = 0$, then x is a zero divisor.]

#5. For G, F, $\mathcal{O}$, a, and b as in Proposition 6.2.5, show:
  a) $G \cong SL(2, \mathbb{R})$,
  b) $G_\mathcal{O}$ is an arithmetic subgroup of G,
  c) if $g \in \mathbb{H}_F^{a,b}$ with $N_{red}(g) = 0$, then $g = 0$, and
  d) $G_\mathcal{O}$ is cocompact in G.

#6. Let a and b be nonzero elements of a field F. Show that if there is a nonzero solution of the equation $w^2 - ax^2 - by^2 + abz^2 = 0$, then there is a nonzero solution of the equation $w^2 - ax^2 - by^2 = 0$.

[*Hint:* By assumption, there is a nonzero element g of $\mathbb{H}_F^{a,b}$, such that $N_{red}(g) = 0$. There is some nonzero $\alpha \in F + Fi$, such that the k-component of αg is zero.]

#7. For $a, b \in \mathbb{R}$, the set

$$G = \{ g \in \mathbb{H}_\mathbb{R}^{a,b} \mid N_{red}(g) = 1 \}$$

is a submanifold of $\mathbb{H}_\mathbb{R}^{a,b}$. Show that the tangent space $T_1 G$ is

$$\{ v \in \mathbb{H}_\mathbb{R}^{a,b} \mid \operatorname{Re} v = 0 \}.$$

[*Hint:* $T_1 G$ is the kernel of the derivative $d(N_{red})_1$.]

#8. Carry out Part (2) of the proof of Proposition 6.2.6.

§6.3. Arithmetic subgroups of SL(2, ℝ) via unitary groups

Unitary groups provide yet another construction of the cocompact, arithmetic subgroups of $SL(2, \mathbb{R})$. In later sections (and later chapters), they will join quaternion algebras as another essential tool, not an alternative approach.

In fact, unitary groups can be applied in two different ways. The simpler of the two approaches is based on the fact that $SL(2, \mathbb{R})$ is isomorphic to $SU(1, 1)$ (see Exercise 2). (This is very similar to the construction in Section 6.1 that is based on the fact that $SL(2, \mathbb{R})$ is isogenous to $SO(2, 1)$.) However, the required isogeny has no higher-dimensional analogue, so this method will not provide any lattices in $SL(n, \mathbb{R})$ when $n > 2$.

The following method is much more important, because it will be used in later sections to construct arithmetic subgroups of $SL(n, \mathbb{R})$ for all n, not just $n = 2$.

(6.3.1) **Example.** Let

- $a, b \in \mathbb{Q}^+$,
- $L = \mathbb{Q}[\sqrt{a}] \subset \mathbb{R}$,

- $\mathcal{O}$ be the ring of integers of L (so $\mathcal{O} \doteq \mathbb{Z}[\sqrt{a}]$),
- τ denote the nontrivial element of $\mathrm{Gal}(L/\mathbb{Q})$,
- $A = \mathrm{diag}(b, -1) = \left[\begin{smallmatrix} b & 0 \\ 0 & -1 \end{smallmatrix}\right]$, and
- $G_\mathcal{O} = \mathrm{SU}(A, \tau; \mathcal{O}) = \{ g \in \mathrm{SL}(2, \mathcal{O}) \mid \tau(g^T) A g = A \} \subset \mathrm{SL}(2, \mathbb{R})$.

If $x = (0, 0)$ is the only solution in L^2 of the equation $\tau(x^T) A x = 0$, then $G_\mathcal{O}$ is a cocompact, arithmetic subgroup of $\mathrm{SL}(2, \mathbb{R})$.

Proof. It is not at all difficult to verify that $G_\mathcal{O}$ is commensurable to an arithmetic group constructed from a quaternion algebra in Proposition 6.2.4 (see Exercise 1), but a direct proof is more instructive.

To see that $G_\mathcal{O}$ is an arithmetic subgroup, we apply restriction of scalars. The Galois automorphism $\tau \colon L \to L$ is $\mathbb{Q}$-linear. Therefore, if we think of L as a (2-dimensional) vector space over $\mathbb{Q}$, then τ is a polynomial with $\mathbb{Q}$-coefficients (with respect to any basis of L over $\mathbb{Q}$). Since matrix multiplication and transpose are also defined by polynomial functions, this implies that if we write $g = X + \sqrt{a}\, Y$, where $X, Y \in \mathrm{Mat}_{2 \times 2}(\mathbb{Q})$, then the equation $\tau(g^T) A g = A$ is a system of polynomial equations with $\mathbb{Q}$-coefficients, in terms of the matrix entries of X and Y. Therefore, it determines a group that is defined over $\mathbb{Q}$. More precisely, letting $G = \mathrm{SL}(2, \mathbb{R})$, define:

- $\Delta \colon L \to L^2$ by $\Delta(s) = (s, A\tau(s))$, so $\mathcal{L} = \Delta(\mathcal{O})$ is a $\mathbb{Z}$-lattice in $\mathbb{R}^2$, and
- $\phi \colon G \to G \times G$ by $\phi(g) = (g, (g^T)^{-1})$.

The import of the above argument is that $\phi(G)$ is defined over $\mathbb{Q}$, with respect to the $\mathbb{Q}$-form $\Delta(L)$ of $\mathbb{R}^2$. Since it is not difficult to verify that $G_\mathcal{O} = \rho^{-1}(\rho(G)_{\Delta(\mathcal{O})})$, we see that $G_\mathcal{O}$ is an arithmetic subgroup of G.

If $G_\mathcal{O}$ is not cocompact, then it has a nontrivial unipotent element u, so there exist nonzero $x, y \in L^2$, such that $ux = x$ and $uy = x + y$. Define $B \colon L^2 \times L^2 \to L$ by $B(x_1, x_2) = \tau(x_1^T) A x_2$. Since $u \in G_\mathcal{O}$, the definition of $G_\mathcal{O}$ implies

$$B(x, y) = B(ux, uy) = B(x, x + y) = B(x, x) + B(x, y).$$

Therefore $B(x, x) = 0$. By assumption, this contradicts the fact that $x \neq 0$. □

(6.3.2) **Example.** The preceding example can be modified, much as in Proposition 6.2.5, to obtain all of the other cocompact lattices in $\mathrm{SL}(2, \mathbb{R})$. Namely, replace $\mathbb{Q}$ with a totally real number field $F \neq \mathbb{Q}$, and let:

- $a, b \in F^+$, such that $\sigma(a) < 0$ and $\sigma(b) < 0$, for all nonidentity places of F,
- $L = F[\sqrt{a}] \subset \mathbb{R}$, and
- $\mathcal{O}, \tau, A$, and $G_\mathcal{O}$ be defined as in Example 6.3.1.

Then $G_{\mathcal{O}}$ is a cocompact, arithmetic subgroup of $\mathrm{SL}(2, \mathbb{R})$.

Proof. Let $\mathcal{O}_F$ be the ring of integers of F. From the second paragraph of the proof of Example 6.3.1 (with F in the place of $\mathbb{Q}$) we see that $G_{\mathcal{O}}$ is the $\mathcal{O}_F$-points of a certain F-form G_F of $\mathrm{SL}(2, \mathbb{R})$. Then restriction of scalars (5.5.8) implies that $\Delta(G_{\mathcal{O}})$ is an arithmetic subgroup of $\prod_{\sigma \in S^\infty} G^\sigma$.

For any nonidentity place σ of F, we have $\sigma(a) < 0$, so

$$L_\sigma = F_\sigma\left[\sqrt{\sigma(a)}\right] = \mathbb{C}.$$

Then, since $\sigma(b)$ and -1 are both negative, we have

$$G^\sigma = \mathrm{SU}(\sigma(A), \tau_{\mathbb{C}}; \mathbb{C}) = \mathrm{SU}(\mathrm{diag}(\sigma(b), -1), \tau_{\mathbb{C}}; \mathbb{C}) \cong \mathrm{SU}(2) \text{ is compact.}$$

Therefore, all factors of $\prod_{\sigma \in S^\infty} G^\sigma$ other than G are compact, so we can mod them out, to conclude that $G_{\mathcal{O}}$ is an arithmetic subgroup of the group $G = \mathrm{SL}(2, \mathbb{R})$ (cf. Definition 5.1.19). Furthermore, the existence of compact factors implies that the arithmetic subgroup is cocompact (see Corollary 5.5.10). ☐

Exercises for §6.3.

#1. Let $a, b \in \mathbb{Z}^+$, let $\phi \colon \mathbb{H}^{a,b}_{\mathbb{R}} \to \mathrm{Mat}_{2 \times 2}(\mathbb{R})$ be as in the proof of Proposition 6.2.4, let $\mathcal{O} = \mathbb{Z}[\sqrt{a}]$, and let $G_{\mathcal{O}}$ be as in Example 6.3.1. Show $\phi(\mathrm{SL}(1, \mathbb{H}^{a,b}_{\mathbb{Z}})) = G_{\mathcal{O}}$.

#2. Let
- $a, b \in \mathbb{Q}^+$,
- $L = \mathbb{Q}[\sqrt{-a}]$,
- $\mathcal{O}$ be the ring of integers of L (so $\mathcal{O} \doteq \mathbb{Z}[\sqrt{-a}]$),
- τ denote complex conjugation (the only nontrivial element of $\mathrm{Gal}(\mathbb{C}/\mathbb{R})$, and also of $\mathrm{Gal}(L/\mathbb{Q})$),
- $A = \mathrm{diag}(b, -1) = \begin{bmatrix} b & 0 \\ 0 & -1 \end{bmatrix}$, and
- $G = \mathrm{SU}(A, \tau; \mathbb{C}) = \{ g \in \mathrm{SL}(2, \mathbb{C}) \mid \tau(g^T) A g = A \} \cong \mathrm{SU}(1,1)$.

Show that if $x = (0,0)$ is the only solution in L^2 of the equation $\tau(x^T) A x = 0$, then $G_{\mathcal{O}}$ is a cocompact, arithmetic subgroup of G.

§6.4. Arithmetic subgroups of SO(1, n)

(6.4.1) Proposition. *Let*
- *$a_1, \ldots, a_n$ be positive integers, and*
- *$G = \mathrm{SO}(x_0^2 - a_1 x_1^2 - \cdots - a_n x_n^2; \mathbb{R}) \cong \mathrm{SO}(1, n)$.*

*If $n \geq 4$, then $G_{\mathbb{Z}}$ is an arithmetic subgroup of G that is **not** cocompact.*

Proof. Since $a_1, \ldots, a_n > 0$ it is obvious that $G \cong \mathrm{SO}(1, n)$. Also, since $a_1, \ldots, a_n \in \mathbb{Q}$, it is clear that G is defined over $\mathbb{Q}$, so $G_{\mathbb{Z}}$ is an arithmetic subgroup of G.

Since we are assuming $n \geq 4$, a theorem of Number Theory (called *Meyer's Theorem*) tells us that the equation $a_1 x_1^2 + \cdots + a_n x_n^2 = x_0^2$ has a nontrivial integral solution. (This is related to, but more difficult than, the fact that every integer is a sum of four squares.) Therefore, $G_{\mathbb{Z}}$ is noncocompact.   □

In most cases, the above construction is exhaustive:

(6.4.2) **Proposition** (see Corollary 18.6.3). *If $n \notin \{3,7\}$, then the arithmetic subgroups constructed in Proposition 6.4.1 are the only noncocompact, arithmetic subgroups of $\mathrm{SO}(1,n)$ (up to commensurability and conjugates).*

(6.4.3) *Remarks.*

1) The case $n = 7$ is genuinely exceptional: there exist some exotic arithmetic subgroups of $\mathrm{SO}(1,7)$ (see Remark 18.5.10).

2) The groups $\mathrm{SO}(1,2)$ and $\mathrm{SO}(1,3)$ are isogenous to $\mathrm{SL}(2,\mathbb{R})$ and $\mathrm{SL}(2,\mathbb{C})$, respectively. Therefore, Propositions 6.1.5 and 6.2.6 describe all of the arithmetic subgroups of $\mathrm{SO}(1,2)$. Similar constructions yield the arithmetic subgroups of $\mathrm{SL}(2,\mathbb{C}) \approx \mathrm{SO}(1,3)$.

Cocompact arithmetic subgroups of $\mathrm{SO}(1,n)$ can be constructed by using an algebraic extension of $\mathbb{Q}$, much as in Example 6.1.2:

(6.4.4) **Proposition.** *Let*

- *F be an algebraic number field that is totally real,*
- *$\mathcal{O}$ be the ring of integers of F,*
- *$a_1, \ldots, a_n \in \mathcal{O}$, such that*
  - *each a_j is positive, and*
  - *each $\sigma(a_j)$ is negative, for every place $\sigma \neq \mathrm{Id}$, and*
- *$G = \mathrm{SO}(x_0^2 - a_1 x_1^2 - \cdots - a_n x_n^2; \mathbb{R}) \cong \mathrm{SO}(1,n)$.*

Then $G_{\mathcal{O}}$ is a cocompact, arithmetic subgroup of G.

This construction is exhaustive when n is even:

(6.4.5) **Proposition** (see Corollary 18.6.1). *If n is even, then the arithmetic subgroups constructed in Proposition 6.4.4 are the only cocompact, arithmetic subgroups of $\mathrm{SO}(1,n)$ (up to commensurability and conjugates).*

(6.4.6) *Remark.* Theoretically, it is easy to tell whether two choices of $a_1, \ldots, a_n$ give essentially the same arithmetic subgroup (see Exercise 3).

When n is odd, we can construct additional arithmetic subgroups of $\mathrm{SO}(1,n)$ by using quaternion algebras. This requires a definition:

(6.4.7) **Definition.** Suppose $\mathbb{H}_F^{a,b}$ is a quaternion algebra over a field F.

1) Define $\tau_r \colon \mathbb{H}_F^{a,b} \to \mathbb{H}_F^{a,b}$ by

$$\tau_r(x_0 + x_1 i + x_2 j + x_3 k) = x_0 + x_1 i - x_2 j + x_3 k.$$

This is the ***reversion*** anti-involution of $\mathbb{H}_F^{a,b}$ (cf. Exercise A2#2).

2) For $A \in \mathrm{GL}(m, \mathbb{H}_F^{a,b})$, with $\tau_r(A^T) = A$, let

$$\mathrm{SU}(A, \tau_r; \mathbb{H}_F^{a,b}) = \{\, g \in \mathrm{SL}(m, \mathbb{H}_F^{a,b}) \mid \tau_r(g^T) A g = A \,\}.$$

Now, here is the main idea of the construction:

(6.4.8) **Proposition.** *Let*

- *$a, b \in \mathbb{Q} \smallsetminus \{0\}$, with $a > 0$,*
- *$a_1, \ldots, a_m$ be invertible elements of $\mathbb{H}^{a,b}$, such that $\tau_r(a_\ell) = a_\ell$, for each ℓ,*
- *$A = \mathrm{diag}(a_1, \ldots, a_m) \in \mathrm{GL}(m, \mathbb{H}_{\mathbb{Q}}^{a,b})$,*
- *$G = \mathrm{SU}(A, \tau_r; \mathbb{H}_{\mathbb{R}}^{a,b})$, and*
- *$\mathcal{O}$ be a $\mathbb{Z}$-lattice in $\mathbb{H}_F^{a,b}$, such that $\mathcal{O}$ is also a subring.*

Then:

1) *$G \cong \mathrm{SO}(p, q)$, for some p and q with $p + q = 2m$, and*
2) *$\mathrm{SU}(A, \tau_r; \mathcal{O})$ is an arithmetic subgroup of G.*

Proof. To make things a bit easier, let us assume $b < 0$ (see Exercise 8). Exercise 5 provides an isomorphism $\phi \colon \mathbb{H}_{\mathbb{R}}^{a,b} \to \mathrm{Mat}_{2 \times 2}(\mathbb{R})$, such that:

- $\phi(\tau_r(x)) = \phi(x)^T$, for all $x \in \mathbb{H}_{\mathbb{R}}^{a,b}$, and
- $\phi(x)$ is symmetric, for all $x \in \mathbb{H}_{\mathbb{R}}^{a,b}$, such that $\tau_r(x) = x$.

Then $\phi(A)$ is symmetric, and G is isomorphic to $\mathrm{SO}_{2m}(\phi(A))$ (see Exercise 6). This establishes (1).

As a vector space over $\mathbb{R}$, $(\mathbb{H}_{\mathbb{R}}^{a,b})^m$ is isomorphic to $\mathbb{R}^{4m}$. With this identification, and considering $(\mathbb{H}_{\mathbb{R}}^{a,b})^m$ as a vector space over $\mathbb{H}_{\mathbb{R}}^{a,b}$ via scalar multiplication on the right, we have

$$\mathrm{GL}(m, \mathbb{H}_{\mathbb{R}}^{a,b}) = \left\{ g \in \mathrm{GL}(4m, \mathbb{R}) \;\middle|\; \begin{array}{l} g(\vec{x} t) = (g\vec{x})t \text{ for all} \\ \vec{x} \in (\mathbb{H}_{\mathbb{R}}^{a,b})^m \text{ and } t \in \mathbb{H}_{\mathbb{R}}^{a,b} \end{array} \right\}.$$

Since $\mathbb{H}_{\mathbb{Q}}^{a,b}$ is dense in $\mathbb{H}_{\mathbb{R}}^{a,b}$, we may restrict t to belong to $\mathbb{H}_{\mathbb{Q}}^{a,b}$. This implies G is defined over $\mathbb{Q}$, with respect to the $\mathbb{Q}$-form $(\mathbb{H}_{\mathbb{Q}}^{a,b})^m$ of $(\mathbb{H}_{\mathbb{R}}^{a,b})^m$. For this $\mathbb{Q}$-form, we have $G_{\mathbb{Z}} = \mathrm{SU}(A, \tau_r; \mathcal{O})$. This establishes (2). □

(6.4.9) *Remark.* Since $p + q = 2m$ must be even, the preceding proposition cannot yield any arithmetic subgroups of $\mathrm{SO}(1, n)$ unless $1 + n$ is even, which means that n is odd.

Proposition 6.4.8 yields an arithmetic subgroup of some $SO(p,q)$, but not necessarily a subgroup of $SO(1,n)$. Obtaining a particular value of p requires us to prescribe the number of positive eigenvalues of the symmetric matrix $\phi(A)$ that appears in the proof. Since $\phi(A)$ is made from $\phi(a_1), \ldots, \phi(a_m)$, this is achieved by calculating the number $\varepsilon_{a,b}(a_\ell)$ of positive eigenvalues of each $\phi(a_\ell)$; the formula is in Notation 6.4.10 below.

However, as in Proposition 6.4.1, Meyer's Theorem implies that arithmetic subgroups obtained in this way are never cocompact (unless G is compact or $m \le 2$). To construct cocompact lattices, restriction of scalars is applied, as usual: choose an extension field F of $\mathbb{Q}$, and arrange for G^σ to be compact at all but one place. The outcome of these considerations is stated in Proposition 6.4.11 below.

(6.4.10) **Notation** (cf. Exercises 9 and 10). Suppose
- a and b are nonzero elements of $\mathbb{R}$, such that either a or b is positive, and
- x is an invertible element of $\mathbb{H}^{a,b}_{\mathbb{R}}$, such that $\tau_r(x) = x$.

Write $x = p + qi + sk$, for some $p, q, s \in \mathbb{R}$. For convenience, let
$$N_{a,b}(x) = x\overline{x} = p^2 - aq^2 + abs^2,$$
and note that $N_{a,b}(x) \ne 0$ (since x is invertible). Define
$$
\varepsilon_{a,b}(x) =
\begin{cases}
1 & \text{if } b\,N_{a,b}(x) > 0, \\
2 & \text{if } b\,N_{a,b}(x) < 0, \text{ and} \\
& \text{either } \begin{cases} b < 0 \text{ and } p > 0, \text{ or} \\ b > 0 \text{ and } (a+1)q + (a-1)s\sqrt{b} > 0, \end{cases} \\
0 & \text{otherwise.}
\end{cases}
$$

(6.4.11) **Proposition.** *Let*
- *F be a totally real algebraic number field (such that $F \ne \mathbb{Q}$),*
- *a and b be nonzero elements of F, such that, for each place σ of F, either $\sigma(a)$ or $\sigma(b)$ is positive,*
- *$a_1, \ldots, a_m \in \mathbb{H}^{a,b}_F$, such that*
  - *$\tau_r(a_\ell) = a_\ell$ for each ℓ,*
  - *$\sigma(a_\ell)$ is invertible, for each ℓ, and each place σ,*
  - *$\sum_{\ell=1}^{m} \varepsilon_{a,b}(a_\ell) = 1$, and*
  - *$\sum_{\ell=1}^{m} \varepsilon_{\sigma(a),\sigma(b)}(\sigma(a_\ell)) \in \{0, 2m\}$ for each place $\sigma \ne \mathrm{Id}$,*
- *$\mathcal{O}$ be a $\mathbb{Z}$-lattice in $\mathbb{H}^{a,b}_F$, such that $\mathcal{O}$ is also a subring, and*
- *$G = SU(\mathrm{diag}(a_1, \ldots, a_m), \tau_r; \mathbb{H}^{a,b}_{\mathbb{R}})^\circ$.*

Then:
1) *$G \cong SO(1, 2m-1)^\circ$, and*
2) *$G_\mathcal{O}$ is a cocompact, arithmetic subgroup of G.*

(6.4.12) Proposition. *If $n \notin \{3, 7\}$, then the arithmetic subgroups constructed in Propositions 6.4.4 and 6.4.11 are the only cocompact, arithmetic subgroups of* SO$(n, 1)$ *(up to commensurability and conjugates).*

Remark 18.5.10 briefly explains the need to assume $n \neq 7$.

Exercises for §6.4.

#1. Use restriction of scalars (see Section 5.5) to construct cocompact arithmetic subgroups SO(m, n) for all m and n.

#2. Suppose G is an irreducible subgroup of GL$(\ell, \mathbb{C})$. (This means there is no nonzero, proper, G-invariant subspace of $\mathbb{C}^\ell$.) Show that if B_1 and B_2 are (nonzero) G-invariant quadratic forms on $\mathbb{C}^\ell$, then there exists $\lambda \in \mathbb{C}$, such that $B_1 = \lambda B_2$.
   [*Hint:* Let A_1 and A_2 be the symmetric matrices that represent B_1 and B_2, and write $A_2 = A_1 L$. For any $g \in G$, we have $A_1 L = g^T A_1 L g = A_1 (g^{-1} L g)$.]

#3. Let
   - F, $\mathcal{O}$, $a_1, \ldots, a_n$, and G be as in Proposition 6.4.4,
   - $\Gamma = h^{-1} G_{\mathcal{O}} h$, where $n = \mathrm{diag}(1, \sqrt{a_1}, \ldots, \sqrt{a_n})$, and
   - F', $\mathcal{O}'$, $a'_1, \ldots, a'_n$, G', Γ', and h' be defined similarly.

   Show $g^{-1} \Gamma g$ is commensurable to Γ', for some $g \in$ O$(1, n)$, if and only if there exists $\lambda \in F^\times$ and $g' \in$ GL$(n + 1, F)$, such that
   $$(g')^T \, \mathrm{diag}(-1, a_1, \ldots, a_n) \, g' = \lambda \, \mathrm{diag}(-1, a'_1, \ldots, a'_n).$$
   [*Hint:* ($\Rightarrow$) For $g' = h' g h^{-1}$, we have $(g')^{-1}$ SO$(B; \mathcal{O}) g' \subseteq$ SO$(B'; \mathbb{R})$, so the Borel Density Theorem implies $(g')^{-1}$ SO$(B; \mathbb{R}) g' \subseteq$ SO$(B'; \mathbb{R})$. Apply Exercise 2 with $G = (g')^{-1}$ SO$(B; \mathbb{R}) g'$.]

#4. Let F, $\mathcal{O}$, $a_1, \ldots, a_n$, $a'_1, \ldots, a'_n$, Γ, and Γ' be as in Exercise 3.
   Show that if n is odd, and there exists $g \in$ O$(1, n)$, such that $g\Gamma g^{-1}$ is commensurable to Γ' then
   $$\frac{a_1 \cdots a_n}{a'_1 \cdots a'_n} \in (F^\times)^2.$$

   [*Hint:* The **discriminant** of a quadratic form $B(x)$ on F^{n+1} is defined to be the determinant of the Gram matrix of B, with respect to any basis $\mathcal{B}$ of F^{n+1}. This is not uniquely determined by B, but show that it is well-defined up to multiplication by a nonzero square in $F^\times$.]

#5. Suppose a and b are real numbers, such that $a > 0$ and $b < 0$. Show that there is an isomorphism $\phi \colon \mathbb{H}_\mathbb{R}^{a,b} \to \mathrm{Mat}_{2\times 2}(\mathbb{R})$, such that:
   a) $\phi(\tau_r(x)) = \phi(x)^T$, for all $x \in \mathbb{H}_\mathbb{R}^{a,b}$, and
   b) $\phi(x)$ is symmetric, for all $x \in \mathbb{H}_\mathbb{R}^{a,b}$, such that $\tau_r(x) = x$.
   [*Hint:* Let $\phi(i) = \begin{bmatrix} \sqrt{a} & 0 \\ 0 & -\sqrt{a} \end{bmatrix}$ and $\phi(j) = \begin{bmatrix} 0 & \sqrt{|b|} \\ -\sqrt{|b|} & 0 \end{bmatrix}$.]

#6. Assume the notation of the proof of Proposition 6.4.8. Show G is isomorphic to $SO_{2m}(\phi(A))$.

[*Hint:* Apply the isomorphism ϕ to both sides of the equation $\tau_r(g^T)Ag = A$.]

#7. Suppose a and b are nonzero real numbers, such that $b > 0$, and let $w = \begin{bmatrix} 0 & 1 \\ 1 & 0 \end{bmatrix}$. Show there is an isomorphism $\phi \colon \mathbb{H}_{\mathbb{R}}^{a,b} \to \mathrm{Mat}_{2\times 2}(\mathbb{R})$, such that:

a) $\phi(\tau_r(x)) = w\phi(x)^T w$, for all $x \in \mathbb{H}_{\mathbb{R}}^{a,b}$, and

b) $w\phi(x)$ is symmetric, for all $x \in \mathbb{H}_{\mathbb{R}}^{a,b}$, such that $\tau_r(x) = x$.

[*Hint:* Let $\phi(i) = \begin{bmatrix} 0 & 1 \\ a & 0 \end{bmatrix}$ and $\phi(j) = \begin{bmatrix} \sqrt{b} & 0 \\ 0 & -\sqrt{b} \end{bmatrix}$.]

#8. Prove Proposition 6.4.8 under the additional assumption that $b > 0$.

[*Hint:* Use Exercise 7 and show $G \cong SO_{2m}(\mathrm{diag}(w\phi(a_1),\ldots,w\phi(a_m)))$.]

#9. In the situation of Exercise 5, show that ϕ can be chosen so that if $x = p + qi + rj + sk$ is an invertible element of $\mathbb{H}_{\mathbb{R}}^{a,b}$, and $\tau_r(x) = x$ (so $r = 0$), then the number of positive eigenvalues of $\phi(x)$ is

$$\begin{cases} 1 & \text{if } N_{a,b}(x) < 0, \\ 2 & \text{if } N_{a,b}(x) > 0 \text{ and } p > 0, \\ 0 & \text{otherwise.} \end{cases}$$

[*Hint:* Since both eigenvalues of $\phi(x)$ are real (and nonzero), the number of positive eigenvalues is determined by the determinant and trace.]

#10. In the situation of Exercise 7, show that ϕ can be chosen so that if $x = p + qi + rj + sk$ is an invertible element of $\mathbb{H}_{\mathbb{R}}^{a,b}$, and $\tau_r(x) = x$ (so $r = 0$), then the number of positive eigenvalues of $w\phi(x)$ is

$$\begin{cases} 1 & \text{if } N_{a,b}(x) > 0, \\ 2 & \text{if } N_{a,b}(x) < 0 \text{ and } (a+1)q + (a-1)s\sqrt{b} > 0, \\ 0 & \text{otherwise.} \end{cases}$$

[*Hint:* See the hint to Exercise 9.]

§6.5. Some nonarithmetic lattices in $SO(1, n)$

Section 6.4 describes algebraic methods to construct all of the arithmetic lattices in $SO(1, n)$ (when $n \neq 7$). We now present a geometric method that is sometimes able to produce a new lattice by combining two known lattices. The result is often nonarithmetic. We assume some familiarity with hyperbolic geometry.

§6.5(i). Hyperbolic manifolds. For geometric purposes, it is more convenient to consider the locally symmetric space $\Gamma \backslash \mathfrak{H}^n$, instead of the lattice Γ.

(6.5.1) **Definition.** A connected, Riemannian n-manifold M is **hyperbolic** if

1) M is locally isometric to $\mathfrak{H}^n$ (that is, each point of M has a neighborhood that is isometric to an open set in $\mathfrak{H}^n$),
2) M is complete, and
3) M is orientable.

(6.5.2) **Other terminology.** Many authors do not require M to be complete or orientable. Our requirement (1) is equivalent to the assertion that M has constant sectional curvature -1; some authors relax this to require the sectional curvature to be a negative constant, but do not require it to be normalized to -1.

(6.5.3) **Notation.** Let $\mathrm{PO}(1, n) = \mathrm{O}(1, n) / \{\pm \mathrm{Id}\}$.
Note that:

- $\mathrm{PO}(1, n)$ is isogenous to $\mathrm{SO}(1, n)$,
- $\mathrm{PO}(1, n) \cong \mathrm{Isom}(\mathfrak{H}^n)$, and
- $\mathrm{PO}(1, n)$ has two connected components (one component consists of orientation-preserving isometries of $\mathfrak{H}^n$, and the other consists of orientation-reversing isometries).

The following observation is easy to prove (see Exercise 1).

(6.5.4) **Proposition.** *A connected Riemannian manifold M of finite volume is hyperbolic if and only if there is a torsion-free lattice Γ in $\mathrm{PO}(1, n)^{\circ}$, such that M is isometric to $\Gamma \backslash \mathfrak{H}^n$.*

§6.5(ii). Hybrid manifolds and totally geodesic hypersurfaces. We wish to combine two (arithmetic) hyperbolic manifolds M_1 and M_2 into a single hyperbolic manifold. The idea is that we will choose closed hypersurfaces C_1 and C_2 of M_1 and M_2, respectively, such that C_1 is isometric to C_2. Let M_j' be the manifold with boundary that results from cutting M_j open, by slicing along C_j (see Figure 6.5A and Exercise 3).

The boundary of M_1' (namely, two copies of C_1) is isometric to the boundary of M_2' (namely, two copies of C_2) (see Exercise 3). So we may glue M_1' and M_2' together, by identifying $\partial M_1'$ with $\partial M_2'$ (see Figure 6.5B), as described in the following well-known proposition.

(6.5.5) **Proposition.** *Suppose*

- *M_1' and M_2' are connected n-manifolds with boundary, and*
- *$f' : \partial M_1' \to \partial M_2'$ is any homeomorphism.*

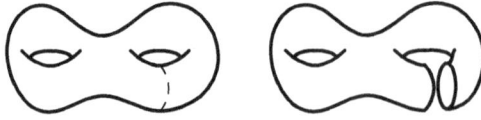

FIGURE 6.5A. Cutting open a manifold by slicing along a closed hypersurface (dashed) results in a manifold with boundary.

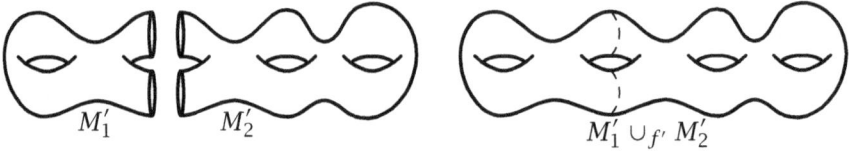

FIGURE 6.5B. Gluing M_1' to M_2' along their boundaries results in a manifold without boundary.

Define a topological space $M_1' \cup_{f'} M_2'$, by gluing M_1' to M_2' along their boundaries:

- *let $M_1' \sqcup M_2'$ be the disjoint union of M_1' and M_2',*
- *define an equivalence relation on $M_1' \sqcup M_2'$ by specifying that we have $m \sim f'(m)$, for every $m \in \partial M_1'$, and*
- *let $M_1' \cup_{f'} M_2' = (M_1' \sqcup M_2')/\sim$ be the quotient of $M_1' \sqcup M_2'$ by this equivalence relation.*

Then $M_1' \cup_{f'} M_2'$ is an n-manifold (without boundary).

(6.5.6) **Corollary.** *Suppose*

- M_1 *and* M_2 *are connected, orientable n-manifolds,*
- C_j *is a closed $(n-1)$-submanifold of M_j, and*
- $f: C_1 \to C_2$ *is any homeomorphism.*

Define $M_1 \#_f M_2 = M_1' \cup_{f'} M_2'$, where

- M_j' *is the manifold with boundary that is obtained by slicing M_j open along C_j, and*
- $f': \partial M_1' \to \partial M_2'$ *is defined by $f'(c,k) = (f(c),k)$, under a natural identification of $\partial M_j'$ with $C_j \times \{1,2\}$.*

Then $M_1 \#_f M_2$ is a (connected) n-manifold (without boundary). Furthermore,

1) $M_1 \#_f M_2$ *is compact if and only if both M_1 and M_2 are compact, and*
2) $M_1 \#_f M_2$ *is connected if and only if either $M_1 \smallsetminus C_1$ or $M_2 \smallsetminus C_2$ is connected.*

(6.5.7) **Other terminology.** Gromov and Piatetski-Shapiro [5] call the manifold $M_1 \#_f M_2$ a *hybrid* of M_1 and M_2, and they call this construction *interbreeding*.

Unfortunately, gluing two Riemannian manifolds together does not always result in a Riemannian manifold (in any natural way), even if the gluing map f is an isometry from $\partial M_1'$ to $\partial M_2'$.

(6.5.8) **Example.** Let $M_1' = M_2'$ be the closed unit disk in $\mathbb{R}^2$, and let $f: \partial M_1' \to \partial M_2'$ be the identity map. Then $M_1' \cup_f M_2'$ is homeomorphic to the 2-sphere S^2. The Riemannian metrics on M_1' and M_2' are flat, so the resulting Riemannian metric on S^2 would also be flat. However, there is no flat Riemannian metric on S^2. (This follows, for example, from the Gauss-Bonnet Theorem.)

We can eliminate this problem by putting a restriction on the hypersurface C_j.

(6.5.9) **Definition.** Let M be a hyperbolic n-manifold. A *totally geodesic hypersurface* in M is a (closed, nonempty) connected submanifold C of M, such that, for each point c of C, there are

- a neighborhood U of c in M,
- a point x in $\mathfrak{H}^{n-1} = \{ v \in \mathfrak{H}^n \mid v_1 = 0 \}$,
- a neighborhood V of x in $\mathfrak{H}^n$, and
- a Riemannian isometry $g: U \to V$, such that $g(U \cap C) = V \cap \mathfrak{H}^{n-1}$.

(6.5.10) *Remark.* If C is a totally geodesic hypersurface in a hyperbolic n-manifold of finite volume, then there are

- a lattice Γ in PO$(1, n)$, and
- an isometry $f: M \to \Gamma \backslash \mathfrak{H}^n$,

such that $f(C)$ is the image of $\mathfrak{H}^{n-1}$ in $\Gamma \backslash \mathfrak{H}^n$.

(6.5.11) **Proposition.** *If*

- *M_1 and M_2 are hyperbolic n-manifolds,*
- *C_j is a totally geodesic hypersurface in M_j,*
- *$f: C_1 \to C_2$ is a Riemannian isometry, and*
- *M_1 and M_2 have finite volume,*

then $M_1 \#_f M_2$ is a hyperbolic n-manifold of finite volume.

Proof. The main issue is to show that each point of $\partial M_1'$ has a neighborhood U in $M_1' \cup_{f'} M_2'$, such that U is isometric to an open subset of $\mathfrak{H}^n$. This is not difficult (see Exercise 4).

We have vol$(M_1 \#_f M_2) =$ vol$(M_1) +$ vol$(M_2) < \infty$.

If $M_1 \#_f M_2$ is compact, then it is obviously complete. More generally, since M_1' and M_2' are complete, and their union is all of $M_1' \cup_f M_2'$, it seems rather obvious that every Cauchy sequence in $M_1' \cup_f M_2'$ has a convergent subsequence. Hence, it seems to be more-or-less obvious that $M_1' \cup_f M_2'$ is complete.

Unfortunately, if $M_1 \#_f M_2$ is not compact, then there is a technical difficulty arising from the possibility that, theoretically, the Riemannian isometry f may not be an isometry with respect to the topological metrics that C_1 and C_2 inherit as submanifolds of M_1 and M_2, respectively. We will ignore this issue. $\qquad\qquad\square$

The following lemma describes how we will construct the totally geodesic hypersurface C_j.

(6.5.12) **Lemma.** *Suppose*

- Γ *is a torsion-free lattice in* $\mathrm{PO}(1,n)^\circ$,
- C *is the image of* $\mathfrak{H}^{n-1}$ *in* $\Gamma \backslash \mathfrak{H}^n$,
- $\tau \colon \mathfrak{H}^n \to \mathfrak{H}^n$ *is the reflection across* $\mathfrak{H}^{n-1}$, *so*

$$\tau(v_0, v_1, \ldots, v_n) = (v_0, v_1, \ldots, v_{n-1}, -v_n),$$

- $\Gamma \cap \mathrm{PO}(1, n-1)$ *is a lattice in* $\mathrm{PO}(1, n-1)$, *and*
- Γ *is contained in a torsion-free lattice* Γ' *of* $\mathrm{PO}(1,n)^\circ$, *such that* Γ' *is normalized by* τ.

Then C is a totally geodesic hypersurface in $\Gamma \backslash \mathfrak{H}^n$, and C has finite volume (as an $(n-1)$-manifold).

Proof. It is clear, from the definition of C, that we need only show C is a (closed, embedded) submanifold of $\Gamma \backslash \mathfrak{H}^n$.

Let $\Gamma_0 = \{ \gamma \in \Gamma \mid \gamma(\mathfrak{H}^{n-1}) = \mathfrak{H}^{n-1} \}$. (Then $\Gamma \cap \mathrm{PO}(1, n-1)$ is a subgroup of index at most two in Γ_0.) The natural map

$$\phi \colon \Gamma_0 \backslash \mathfrak{H}^{n-1} \to \Gamma \backslash \mathfrak{H}^n$$

is proper (cf. Exercise 4.4#3), so C, being the image of ϕ, is closed.

Because ϕ is obviously an immersion (and is a proper map), all that remains is to show that ϕ is injective. This follows from the assumption on Γ' (see Exercise 5). $\qquad\qquad\square$

§6.5(iii). Construction of nonarithmetic lattices. The following theorem is the key to the construction of nonarithmetic lattices. We postpone the proof until later in the section (see Subsection 6.5(iv) and Exercise 11).

(6.5.13) **Definition.** A hyperbolic n-manifold of finite volume is **arithmetic** if the corresponding lattice Γ in $\mathrm{PO}(1, n)$ (see Proposition 6.5.4) is arithmetic. (Note that Γ is well-defined, up to conjugacy (see Exercise 2), so this definition is independent of the choice of Γ.)

(6.5.14) **Theorem.** *Suppose*

- M_1 *and* M_2 *are hyperbolic n-manifolds,*
- C_j *is a totally geodesic hypersurface in* M_j,
- $f \colon C_1 \to C_2$ *is a Riemannian isometry,*
- M_1 *and* M_2 *have finite volume (as n-manifolds),*
- C_1 *and* C_2 *have finite volume (as* $(n-1)$*-manifolds), and*
- *each of* $M_1 \smallsetminus C_1$ *and* $M_2 \smallsetminus C_2$ *is connected.*

If the hyperbolic manifold $M_1 \#_f M_2$ *is arithmetic, then* $M_1 \#_f M_2$ *is commensurable to* M_1; *that is, there are*

1) *a finite cover* $\widetilde{M}$ *of* $M_1 \cup_f M_2$, *and*

2) *a finite cover* $\widetilde{M_1}$ *of* M_1,

such that $\widetilde{M}$ *is isometric to* $\widetilde{M_1}$.

(6.5.15) **Corollary.** *In the situation of Theorem 6.5.14, if the hyperbolic manifold* $M_1 \#_f M_2$ *is arithmetic, then* M_1 *is commensurable to* M_2.

Proof. From Theorem 6.5.14, we know that $M_1 \#_f M_2$ is commensurable to M_1. By interchanging M_1 and M_2, we see that $M_1 \#_f M_2$ is also commensurable to M_2. By transitivity, M_1 is commensurable to M_2. $\qquad\square$

(6.5.16) **Corollary.** *There exist nonarithmetic lattices* Γ_{cpct} *and* Γ_{non} *in* SO$(1,n)$, *such that* Γ_{cpct} *is cocompact, and* Γ_{non} *is not cocompact.*

Proof. We construct only Γ_{non}. (See Exercise 6 for the construction of Γ_{cpct}, which is similar.)

Define quadratic forms $B_1(x)$ and $B_2(x)$ on $\mathbb{Q}^{n+1}$ by

$$B_1(x) = x_0^2 - x_1^2 - x_2^2 - \cdots - x_{n-1}^2 - x_n^2$$

and

$$B_2(x) = x_0^2 - x_1^2 - x_2^2 - \cdots - x_{n-1}^2 - 2x_n^2.$$

Let

- $\Gamma_1 \approx \mathrm{SO}(B_1; \mathbb{Z})$,
- $\Gamma_2 \approx h^{-1}\,\mathrm{SO}(B_2; \mathbb{Z})\,h$, where
$$h = \mathrm{diag}(1,1,\ldots,1,\sqrt{2}) \in \mathrm{GL}(n+1, \mathbb{R}),$$
- $M_j = \Gamma_j \backslash \mathfrak{H}^n$,
- C_j be the image of $\mathfrak{H}^{n-1}$ in M_j, and
- $\hat{\Gamma}_j = \Gamma_j \cap \mathrm{SO}(1, n-1)$.

Then Proposition 6.4.4 tells us that Γ_1 and Γ_2 are noncocompact (arithmetic) lattices in SO$(1,n)$. By passing to finite-index subgroups, we may assume Γ_1 and Γ_2 are torsion free (see Theorem 4.8.2). Therefore,

M_1 and M_2 are hyperbolic n-manifolds of finite volume (see Proposition 6.5.4).

Because $\hat{\Gamma}_j \approx \mathrm{SO}(1, n-1; \mathbb{Z})$ is a lattice in $\mathrm{SO}(1, n-1)$, and $\mathrm{SO}(B_j; \mathbb{Z})$ is normalized by the involution τ of Lemma 6.5.12, we know C_j is a totally geodesic hypersurface in M_j that has finite volume (see Lemma 6.5.12).

Let us assume that $M_1 \smallsetminus C_1$ and $M_2 \smallsetminus C_2$ are connected. (See Exercise 8 for a way around this issue, or note that this hypothesis can be achieved by passing to finite covers of M_1 and M_2.)

We know that $\hat{\Gamma}_1 \approx \hat{\Gamma}_2$ (since both groups are commensurable to $\mathrm{SO}(1, n-1; \mathbb{Z})$). By taking a little bit of care in the choice of Γ_1 and Γ_2, we may arrange that $\hat{\Gamma}_1 = \hat{\Gamma}_2$ (see Exercise 9). Then

$$C_1 \cong \hat{\Gamma}_1 \backslash \mathfrak{H}^{n-1} = \hat{\Gamma}_2 \backslash \mathfrak{H}^{n-1} \cong C_2,$$

so there is an isometry $f \colon C_1 \to C_2$.

If n is odd, then M_1 is not commensurable to M_2 (see Exercise 6.4#4), so Corollary 6.5.15 implies that $M_1 \#_f M_2$ is not arithmetic; therefore, the corresponding lattice Γ_{non} is not arithmetic (see Definition 6.5.13). When n is even, an additional argument is needed; see Exercise 10.   □

§6.5(iv). **Proof of Theorem 6.5.14.** Let us recall the following lemma, which was proved in Exercise 5.1#7.

(6.5.17) **Lemma.** *If*

- *G has no compact factors,*
- *Γ_1 and Γ_2 are arithmetic lattices in G, and*
- *$\Gamma_1 \cap \Gamma_2$ is Zariski dense in G,*

then Γ_1 is commensurable to Γ_2.

(6.5.18) **Definition.** Let M' be a Riemmanian n-manifold with boundary. We say that M' is a **hyperbolic manifold with totally geodesic boundary** if

1) M' is complete,

2) each point of $M' \smallsetminus \partial M'$ has a neighborhood that is isometric to an open set in $\mathfrak{H}^n$, and

3) for each point p of $\partial M'$, there are
    - a neighborhood U of p in M',
    - a point x in $\mathfrak{H}^{n-1} = \{v \in \mathfrak{H}^n \mid v_1 = 0\}$,
    - a neighborhood V of x in $\mathfrak{H}^n$, and
    - an isometry $g \colon U \to V^+$, where
    $$V^+ = \{v \in V \mid v_1 \geq 0\}.$$
    (Note that $g(U \cap \partial M') = V \cap \mathfrak{H}^{n-1}$.)

The following is a generalization of Theorem 6.5.14 (see Exercise 11).

(6.5.19) **Theorem.** *Suppose*

- M_1 *and* M_2 *are hyperbolic n-manifolds,*
- M'_j *is a connected, n-dimensional submanifold of M_j with totally geodesic boundary,*
- $f' : \partial M'_1 \to \partial M'_2$ *is an isometry,*
- M_1 *and* M_2 *have finite volume (as n-manifolds),*
- $\partial M'_j$ *has only finitely many components, and*
- $\partial M'_1$ *and* $\partial M'_2$ *have finite volume (as $(n-1)$-manifolds).*

If the hyperbolic manifold $M'_1 \cup_{f'} M'_2$ is arithmetic, then $M'_1 \cup_{f'} M'_2$ is commensurable to M_1.

Proof.

- Let $M = M'_1 \cup_{f'} M'_2$.
- Write $M = \Gamma \backslash \mathfrak{H}^n$, for some torsion-free lattice Γ in PO$(1, n)$.
- Let $\phi \colon \mathfrak{H}^n \to M$ be the resulting covering map.
- Let $B = \phi^{-1}(\partial M'_1)$. Because M'_1 has totally geodesic boundary, we know that B is a union of disjoint hyperplanes. (That is, each component of B is of the form $g(\mathfrak{H}^{n-1})$, for some $g \in \mathrm{O}(1, n)$.)
- Let V be the closure of some connected component of $\mathfrak{H}^n \smallsetminus B$ that contains a point of $\phi^{-1}(M'_1)$.
- Let
$$\Gamma' = \{ \gamma \in \Gamma \mid \gamma V = V \} = \{ \gamma \in \Gamma \mid \mathrm{interior}(\gamma V \cap V) \neq \varnothing \}$$
(see Exercise 12), so $M'_1 = \phi(V) \cong \Gamma' \backslash V$.

By definition, V is an intersection of half-spaces, so it is (hyperbolically) convex; hence, it is simply connected. Therefore, V is the universal cover of M'_1, and Γ' can be identified with the fundamental group of M'_1.

Since $M'_1 \subseteq M_1$, we may define $\Gamma_1, \phi_1, B_1, V_1, \Gamma'_1$ as above, but with M_1 in the place of M. From the uniqueness of the universal cover of M'_1, we know that there is an isometry $\psi \colon V \to V_1$, and an isomorphism $\psi_* \colon \Gamma' \to \Gamma'_1$, such that $\psi(\gamma v) = \psi_*(\gamma)\psi(v)$, for all $\gamma \in \Gamma'$ and $v \in V$. Since ψ extends to an isometry of $\mathfrak{H}^n$, we may assume (after replacing Γ_1 with $\psi^{-1}\Gamma_1\psi$) that $V = V_1$ and $\psi_* = \mathrm{Id}$. Hence $\Gamma' = \Gamma'_1 \subset \Gamma \cap \Gamma_1$. It suffices to show (after replacing Γ by a conjugate subgroup) that the Zariski closure of Γ' contains PO$(1, n)^\circ$, for then Lemma 6.5.17 implies Γ is commensurable to Γ_1.

Claim. We may assume that the Zariski closure of Γ' contains PO$(1, n)^\circ$. We may assume $\mathfrak{H}^{n-1}$ is one of the connected components of ∂V. Since $\partial M'_1$ has finite volume, this means that

$$\Gamma' \cap \mathrm{SO}(1, n-1) \text{ is a lattice in PO}(1, n-1). \tag{6.5.20}$$

Let $\overline{\Gamma'}$ be the Zariski closure of Γ'. From (6.5.20) and the Borel Density Theorem (4.5.6), we know that $\overline{\Gamma'}$ contains $\mathrm{PO}(1, n-1)^\circ$. Then, since $\mathrm{PO}(1, n-1)^\circ$ is a maximal connected subgroup of $\mathrm{PO}(1, n)$ (see Exercise 13), we may assume that $\overline{\Gamma'}^\circ = \mathrm{PO}(1, n-1)^\circ$. (Otherwise, the claim holds.) Because $\overline{\Gamma'}^\circ$ has finite index in $\overline{\Gamma'}$ (see A4.6), this implies that $\mathrm{PO}(1, n-1)^\circ$ contains a finite-index subgroup of Γ'. In fact,

$$\{ \gamma \in \Gamma' \mid \gamma H = H \} \text{ has finite index in } \Gamma',$$
$$\text{for every connected component } H \text{ of } \partial V. \tag{6.5.21}$$

This will lead to a contradiction.

Case 1. Assume ∂V is connected. We may assume $\partial V = \mathfrak{H}^{n-1}$. Then, by passing to a finite-index subgroup, we may assume that $\Gamma' \subset \mathrm{PO}(1, n-1)$ (see 6.5.21). Define $g \in \mathrm{Isom}(\mathfrak{H}^n)$ by

$$g(v_1, v_2, \ldots, v_n) = (-v_1, v_2, \ldots, v_n).$$

Then

- g centralizes Γ', and
- $\mathfrak{H}^n = V \cup g(V)$.

Since $\Gamma' \backslash V \cong M_1'$ has finite volume, we know that $\Gamma' \backslash g(V)$ also has finite volume. Therefore

$$\Gamma' \backslash \mathfrak{H}^n = (\Gamma' \backslash V) \cup (\Gamma' \backslash g(V))$$

has finite volume, so Γ' is a lattice in $\mathrm{PO}(1, n)$. But this contradicts the Borel Density Theorem (4.5.6) (since $\Gamma' \subset \mathrm{PO}(1, n-1)$).

Case 2. Assume ∂V is not connected. Let H_1 and H_2 be two distinct connected components of ∂V. Replacing Γ' by a finite-index subgroup, let us assume that each of H_1 and H_2 is invariant under Γ' (see 6.5.21).

To simplify the argument, let us assume that $\partial M_1'$ is compact, rather than merely that it has finite volume. (See Exercise 14 for the general case.) Therefore, $\Gamma' \backslash H_1$ is compact, so there is a compact subset C of H_1, such that $\Gamma' C = H_1$. Let

$$\delta = \min\{ \mathrm{dist}(c, H_2) \mid c \in C \} > 0.$$

Because Γ' acts by isometries, we have $\delta = \mathrm{dist}(H_1, H_2)$. Now, since $\mathfrak{H}^n$ is negatively curved, there is a unique point p in H_1, such that $\mathrm{dist}(p, H_2) = \delta$. The uniqueness implies that p is fixed by every element of Γ'. Since Γ acts freely on $\mathfrak{H}^n$ (recall that it is a group of deck transformations), we conclude that Γ' is trivial. This contradicts the fact that $\Gamma' \backslash H_1$ is compact. (Note that $H_1 \cong \mathfrak{H}^{n-1}$ is not compact.)   □

Exercises for §6.5.

#1. Prove Proposition 6.5.4.

#2. Show that if Γ_1 and Γ_2 are torsion-free lattices in $\mathrm{PO}(1, n)$, such that $\Gamma_1 \backslash \mathfrak{H}^n$ is isometric to $\Gamma_2 \backslash \mathfrak{H}^n$, then Γ_1 is conjugate to Γ_2.

[*Hint:* Any isometry $\phi \colon \Gamma_1 \backslash \mathfrak{H}^n \to \Gamma_2 \backslash \mathfrak{H}^n$ lifts to an isometry of $\mathfrak{H}^n$.]

#3. Let C be a closed, connected hypersurface in an orientable Riemannian manifold M, and let M' be the manifold with boundary that results from cutting M open, by slicing along C. Show:
   a) If C is orientable, then the boundary of M is two copies of C.
   b) If C is not orientable, then the boundary is the orientable double cover of C.
   c) If C is isometric to a closed, connected hypersurface C_0 in an orientable Riemannian manifold M_0, and M'_0 is the manifold with boundary that results from cutting M_0 open, by slicing along C_0, then the boundary of M' is isometric to the boundary of M'_0.

#4. For M_1, M_2, and f as in Exercise 4, show that if $p \in \partial M'_1$, then p has a neighborhood U in $M'_1 \cup_f M'_2$, such that U is isometric to an open subset of $\mathfrak{H}^n$.

[*Hint:* Find a ball V around a point x in $\mathfrak{H}^{n-1}$, and isometries $g_1 \colon U_1 \to V^+$ and $g_2 \colon U_2 \to V^-$, where U_j is a neighborhood of p in M'_j, with $g_1|_{\partial M'_1} = (g_2 \circ f)|_{\partial M'_1}$.]

#5. For $\phi \colon \Gamma_0 \backslash \mathfrak{H}^{n-1} \to \Gamma \backslash \mathfrak{H}^n$, as defined in the proof of Lemma 6.5.12, show that ϕ is injective.

[*Hint:* Suppose $yx = y$, for some $y \in \Gamma$ and $x, y \in \mathfrak{H}^{n-1}$. Then $y^{-1}\tau y \tau$ is an element of Γ' that fixes x, so it is trivial. Hence, the fixed-point set of τ is y-invariant.]

#6. Assume n is odd, and construct a cocompact, nonarithmetic lattice Γ in SO$(1, n)$.

[*Hint:* Let $F = \mathbb{Q}[\sqrt{2}]$, define $B_1(x) = \sqrt{2}x_0^2 - x_1^2 - x_2^2 - \cdots - x_{n-1}^2 - x_n^2$ and $B_2(x) = \sqrt{2}x_0^2 - x_1^2 - x_2^2 - \cdots - x_{n-1}^2 - 3x_{n+1}^2$, and use the proof of Corollary 6.5.16.]

#7. In the notation of the proof of Corollary 6.5.16, assume that $M_1 \smallsetminus C_1$ and $M_2 \smallsetminus C_2$ are *not* connected; let M'_j be the closure of a component of $M_j \smallsetminus C_j$. Show that if $f' \colon C_1 \to C_j$ is any isometry (and n is odd), then $M'_1 \cup_{f'} M'_2$ is a *nonarithmetic* hyperbolic n-manifold of finite volume.

#8. Eliminate the assumption that $M_1 \smallsetminus C_1$ and $M_2 \smallsetminus C_2$ are connected from the proof of Corollary 6.5.16.

[*Hint:* Define $B_3(x) = x_0^2 - x_1^2 - x_2^2 - \cdots - x_{n-1}^2 - 3x_n^2$. If $M_j \smallsetminus C_j$ has the same number of components as $M_k \smallsetminus C_k$ (and $j \neq k$), then either Exercise 7 or the proof of Corollary 6.5.16 applies.]

#9. For $B_1(x)$ and $B_2(x)$ as in the proof of Corollary 6.5.16, show that there are finite-index subgroups Γ_1 and Γ_2 of SO$(B_1; \mathbb{Z})$ and SO$(B_2; \mathbb{Z})$, respectively, such that
   a) Γ_1 and Γ_2 are torsion free, and
   b) $\Gamma_1 \cap$ SO$(1, n-1) = \Gamma_2 \cap$ SO$(1, n-1)$.

[*Hint:* Let $\Gamma_j = \Lambda \cap SO(B_j; \mathbb{Z})$, where Λ is a torsion-free subgroup of finite index in $SL(n+1, \mathbb{Z})$.]

#10. In the notation of the proof of Corollary 6.5.16, show that if n is even (and $n \geq 4$), then Γ_{non} is not arithmetic.

[*Hint:* If Γ_{non} is arithmetic, then its intersection with $SO(1, n-1)$ is arithmetic in $SO(1, n-1)$, and $n-1$ is odd.]

#11. Derive Theorem 6.5.14 as a corollary of Theorem 6.5.19.

[*Hint:* Apply Theorem 6.5.19 to $\widetilde{M_j} = M_j \#_{f_j} M_j$, where $f_j : C_j \to C_j$ is the identity map. Note that $\widetilde{M_j}$ is a double cover of M_j, so $\widetilde{M_j}$ is commensurable to M_j.]

#12. For Γ and V as in the proof of Theorem 6.5.19, let $\overset{\circ}{V}$ be the interior of V, and show, for each $y \in \Gamma$, that if $y\overset{\circ}{V} \cap \overset{\circ}{V} \neq \varnothing$, then $y\overset{\circ}{V} = \overset{\circ}{V}$.

#13. Show that if H is a connected subgroup of $PO(1, n)$ that contains $PO(1, n-1)^\circ$, then $H = PO(1, n-1)^\circ$.

#14. Eliminate the assumption that $\partial M_1'$ is compact from Case 2 of the proof of Theorem 6.5.19.

[*Hint:* The original proof applies unless $\mathrm{dist}(H_1, H_2) = 0$, which would mean that H_1 and H_2 intersect at infinity. This intersection is a single point, and it is invariant under Γ', which contradicts the Zariski density of Γ'.]

§6.6. Noncocompact arithmetic subgroups of $SL(3, \mathbb{R})$

We saw in Proposition 6.1.5 that $SL(2, \mathbb{Z})$ is essentially the only noncocompact, arithmetic subgroup of $SL(2, \mathbb{R})$. So it may be surprising that $SL(3, \mathbb{Z})$ is *not* the only one in $SL(3, \mathbb{R})$.

(6.6.1) Proposition. *Let*

- *L be a real quadratic extension of $\mathbb{Q}$, so $L = \mathbb{Q}[\sqrt{r}]$, for some square-free positive integer $r \geq 2$,*

- *σ be the nontrivial Galois automorphism of L,*

- *$\tilde{\sigma}$ be the automorphism of $\mathrm{Mat}_{3 \times 3}(L)$ induced by applying σ to each entry of a matrix,*

- *$J_3 = \begin{bmatrix} 0 & 0 & 1 \\ 0 & 1 & 0 \\ 1 & 0 & 0 \end{bmatrix}$, and*

- *$\Gamma = SU(J_3, \sigma; \mathbb{Z}[\sqrt{r}]) = \{ g \in SL(3, \mathbb{Z}[\sqrt{r}]) \mid \tilde{\sigma}(g^T) J_3 \, g = J_3 \}$.*

Then:

1) *Γ is an arithmetic subgroup of $SL(3, \mathbb{R})$,*

2) *Γ is not cocompact, and*

3) *no conjugate of Γ is commensurable to $SL(3, \mathbb{Z})$.*

Proof. (1) This is a special case of Proposition 18.5.7(6), but we provide a concrete, explicit proof (using the methods of Sections 5.4 and 5.5).

Define

- $\Delta: L^3 \to \mathbb{R}^6$ by $\Delta(v) = (v, J_3\,\sigma(v))$,
- $V_{\mathbb{Q}} = \Delta(L^3)$,
- $\mathcal{L} = \Delta(\mathbb{Z}[\sqrt{r}]^3)$, and
- $\rho: \mathrm{SL}(3, \mathbb{R}) \to \mathrm{SL}(6, \mathbb{R})$ by

$$\rho(A)(v, w) = (Av, (A^T)^{-1}w) \text{ for } v, w \in \mathbb{R}^3.$$

Then

- $V_{\mathbb{Q}}$ is a $\mathbb{Q}$-form of $\mathbb{R}^6$ (cf. Exercise 5.5#5(b)),
- $\mathcal{L}$ is a $\mathbb{Z}$-lattice in $V_{\mathbb{Q}}$ (cf. Exercise 5.5#5(c)),
- ρ is a homomorphism,
- $\rho(\mathrm{SL}(3, \mathbb{R}))$ is defined over $\mathbb{Q}$ (with respect to the $\mathbb{Q}$-form $V_{\mathbb{Q}}$) (see (6.6.2) below), and
- $\Gamma = \{\, g \in \mathrm{SL}(3, \mathbb{R}) \mid \rho(g)\mathcal{L} = \mathcal{L} \,\}$ (cf. Exercise 5.5#1).

Hence, Proposition 5.4.5(1) (together with Theorem 5.1.11) implies that Γ is an arithmetic subgroup of $\mathrm{SL}(3, \mathbb{R})$.

Now let us show that

$$\rho(\mathrm{SL}(3, \mathbb{R})) \text{ is defined over } \mathbb{Q}. \tag{6.6.2}$$

This can be verified directly, by finding appropriate $\mathbb{Q}$-polynomials, but let us, instead, show that $\rho(\mathrm{SL}(3, \mathbb{R}))_{\mathbb{Q}}$ is dense in $\rho(\mathrm{SL}(3, \mathbb{R}))$.

Define U_1 as in (6.6.3) below, but allowing a, b, c to range over all of $\mathbb{Q}$, instead of only $2\mathbb{Z}$. Then $\rho(U_1)V_{\mathbb{Q}} \subset V_{\mathbb{Q}}$ (see Exercise 1), so we have $\rho(U_1) \subseteq \rho(\mathrm{SL}(3, \mathbb{R}))_{\mathbb{Q}}$. Furthermore, U_1 is dense in

$$U = \begin{bmatrix} 1 & * & * \\ 0 & 1 & * \\ 0 & 0 & 1 \end{bmatrix}.$$

Similarly, there is a dense subgroup U_2 of U^T, with $\rho(U_2) \subseteq \rho(\mathrm{SL}(3, \mathbb{R}))_{\mathbb{Q}}$ (see Exercise 2). Since $\langle U, U^T \rangle = \mathrm{SL}(3, \mathbb{R})$, we know that $\langle U_1, U_2 \rangle$ is dense in $\mathrm{SL}(3, \mathbb{R})$, so $\rho(\mathrm{SL}(3, \mathbb{R}))_{\mathbb{Q}}$ is dense in $\rho(\mathrm{SL}(3, \mathbb{R}))$. Therefore $\rho(\mathrm{SL}(3, \mathbb{R}))$ is defined over $\mathbb{Q}$ (see 5.1.8).

(2) By calculation, one may verify, directly from the definition of Γ, that the subgroup

$$U_{\Gamma} = \left\{ \begin{bmatrix} 1 & a + b\sqrt{r} & -(a^2 - rb^2)/2 + c\sqrt{r} \\ 0 & 1 & -a + b\sqrt{r} \\ 0 & 0 & 1 \end{bmatrix} \;\middle|\; a, b, c \in 2\mathbb{Z} \right\} \tag{6.6.3}$$

is contained in Γ. Then, since every element of U_{Γ} is unipotent, it is obvious that Γ has nontrivial unipotent elements. So the Godement Criterion (5.3.1) implies that G/Γ is not compact.

(3) We sketch a proof. Choose an element $\omega \in \mathbb{Z}[\sqrt{r}]$, such that $\sigma(\omega) = 1/\omega$. Then $\mathrm{diag}(\omega, 1, \omega^{-1})$ is a hyperbolic element of Γ that

normalizes the maximal unipotent subgroup U_Γ. On the other hand, it is easy to see that if $U'_{\mathbb{Z}}$ is any subgroup of $\mathrm{SL}(3, \mathbb{Z})$ that is commensurable to the maximal unipotent subgroup $U_{\mathbb{Z}}$, then $U'_{\mathbb{Z}}$ has finite index in $\mathcal{N}_{\mathrm{SL}(3,\mathbb{Z})}(U'_{\mathbb{Z}})$. Since all of the maximal unipotent subgroups of $\mathrm{SL}(3,\mathbb{Z})$ are conjugate (up to commensurability) under $\mathrm{SL}(3, \mathbb{Q})$, this implies that no conjugate of Γ is commensurable to $\mathrm{SL}(3, \mathbb{Z})$.

Here is a more complete argument that is based on the notion of $\mathbb{Q}$-rank, which will be explained in Chapter 9. Define a nondegenerate σ-Hermitian form $B(x, y)$ on L^3 by $B(x, y) = \sigma(x^T) J_3\, y$. Then $v = (1, 0, 0)$ is an isotropic vector for B (i.e., $B(v, v) = 0$). On the other hand, because B is nondegenerate, the dimension of the orthogonal complement of any subspace is equal to the codimension of the subspace. Since L^3 is 3-dimensional, this implies there is no 2-dimensional subspace that consists entirely of isotropic vectors. Therefore, $\mathrm{rank}_{\mathbb{Q}}\, \Gamma = 1$ (cf. Example 9.1.5(2)). However, we have $\mathrm{rank}_{\mathbb{Q}}\, \mathrm{SL}(3, \mathbb{Z}) = 2$ (cf. Example 9.1.5(1)). Two lattices with different $\mathbb{Q}$-ranks cannot be conjugate. (They cannot even be abstractly commensurable.) $\square$

(6.6.4) *Remarks.*

1) From Proposition 6.6.1(3), we know that none of the arithmetic subgroups in Proposition 6.6.1 are conjugate to a subgroup that is commensurable to $\mathrm{SL}(3, \mathbb{Z})$.

   Indeed, let $X = \mathrm{SL}(3, \mathbb{R}) / \mathrm{SO}(3)$ be the symmetric space associated to $\mathrm{SL}(3, \mathbb{R})$. Theorem 2.2.8 implies that if Γ is one of the arithmetic subgroups constructed in Proposition 6.6.1, then the geometry of the locally symmetric space $\Gamma \backslash X$ is very different from that of $\mathrm{SL}(3, \mathbb{Z}) \backslash X$. Namely, $\Gamma \backslash X$ is only mildly noncompact: it merely has cusps, which means that its asymptotic cone is a union of finitely many rays. In contrast, the asymptotic cone of $\mathrm{SL}(3, \mathbb{Z}) \backslash X$ is a 2-complex, not just a union of rays. Even from a distance, $\Gamma \backslash X$ and $\mathrm{SL}(3, \mathbb{Z}) \backslash X$ look completely different.

2) Different values of r always give essentially different arithmetic subgroups (see Exercise 4), but this is not so obvious.

The classification results in Chapter 18 imply that these are the only arithmetic subgroups of $\mathrm{SL}(3, \mathbb{R})$ that are not cocompact:

(6.6.5) **Proposition** (see Proposition 18.6.4). *$\mathrm{SL}(3, \mathbb{Z})$ and the arithmetic subgroups constructed in Proposition 6.6.1 are the only noncocompact arithmetic subgroups of $\mathrm{SL}(3, \mathbb{R})$ (up to commensurability and conjugates).*

Exercises for §6.6.

#1. For U_1, ρ, and $V_{\mathbb{Q}}$ as in the proof of Proposition 6.6.1(1), show that $\rho(U_1)V_{\mathbb{Q}} \subseteq V_{\mathbb{Q}}$.

#2. In the notation of the proof of Proposition 6.6.1, find a dense subgroup U_2 of

$$\begin{bmatrix} 1 & 0 & 0 \\ * & 1 & 0 \\ * & * & 1 \end{bmatrix},$$

such that $\rho(U_2) \subseteq \rho(\mathrm{SL}(3, \mathbb{R}))_{\mathbb{Q}}$.

#3. Assume the notation of the proof of Proposition 6.6.1, and let $G = \rho(\mathrm{SL}(3, \mathbb{R}))$.
  a) Show that G is **quasisplit**. That is, show that some Borel subgroup of G is defined over $\mathbb{Q}$.
  b) Show that every proper parabolic $\mathbb{Q}$-subgroup of G is a Borel subgroup of G.
  [*Hint:* Let B be the group of upper-triangular matrices in $\mathrm{SL}(3, \mathbb{R})$. Then B is a Borel subgroup of $\mathrm{SL}(3, \mathbb{R})$, and $\rho(B)$ is defined over $\mathbb{Q}$.]

#4. Let Γ_1 and Γ_2 be noncocompact arithmetic subgroups of $\mathrm{SL}(3, \mathbb{R})$ that correspond to two different values of r, say r_1 and r_2. Show that Γ_1 is not commensurable to any conjugate of Γ_2.
  [*Hint:* There is a diagonal matrix in Γ_1 whose trace is not in $\mathbb{Z}[\sqrt{r_2}]$.]

§6.7. Cocompact arithmetic subgroups of SL(3, ℝ)

Example 6.3.2 used unitary groups over a totally real extension to construct cocompact, arithmetic subgroups of $\mathrm{SL}(2, \mathbb{R})$. The same technique can be applied to $\mathrm{SL}(3, \mathbb{R})$:

(6.7.1) **Proposition.** *Let*
- F *be a totally real algebraic number field, such that* $F \neq \mathbb{Q}$,
- $t, a, b \in F$, *such that*
  - $t, a, b > 0$, *but*
  - $\sigma(t), \sigma(a), \sigma(b) < 0$ *for every place* $\sigma \neq \mathrm{Id}$,
- $L = F[\sqrt{t}]$,
- τ *be the Galois automorphism of* L *over* F,
- $\mathcal{O}$ *be the ring of integers of* L, *and*
- $\Gamma = \mathrm{SU}(\mathrm{diag}(a, b, -1), \tau; \mathcal{O})$.

Then Γ *is a cocompact, arithmetic subgroup of* $\mathrm{SL}(3, \mathbb{R})$.

Here is a specific example:

(6.7.2) **Corollary.** *Let*
- $t = \sqrt{2}$,

- $F = \mathbb{Q}[t] = \mathbb{Q}[\sqrt{2}]$,
- $L = F[\sqrt{t}] = \mathbb{Q}[\sqrt[4]{2}]$,
- τ be the Galois automorphism of L over F,
- $\mathcal{O} \doteq \mathbb{Z}[\sqrt[4]{2}]$ be the ring of integers of L, and
- $\Gamma = \mathrm{SU}(\mathrm{Id}_{3\times3}, \tau; \mathcal{O})$.

Then Γ is a cocompact, arithmetic subgroup of $\mathrm{SL}(3, \mathbb{R})$.

(6.7.3) *Remark.* It is necessary to assume $F \neq \mathbb{Q}$ in Proposition 6.7.1 (in other words, there is no analogue of Example 6.3.1 for $\mathrm{SL}(3, \mathbb{R})$), because unitary groups over $\mathbb{Q}$ yield only noncompact lattices in $\mathrm{SL}(3, \mathbb{R})$ (as in Proposition 6.6.1), not cocompact ones (see Exercise 1).

Here is a quite different construction (not using unitary groups) that yields additional examples of cocompact, arithmetic subgroups. See Example 6.7.6 for explicit examples of L and p that satisfy the hypotheses.

(6.7.4) **Proposition.** *Let*

- *L be a cubic, Galois extension of $\mathbb{Q}$ (that is, a Galois extension of $\mathbb{Q}$, such that $|L : \mathbb{Q}| = 3$),*
- *σ be a generator of $\mathrm{Gal}(L/\mathbb{Q})$ (note that $\mathrm{Gal}(L/\mathbb{Q})$, being of order 3, is cyclic),*
- *$\mathcal{O}$ be the ring of integers of L,*
- *$p \in \mathbb{Z}^+$,*
- *$\phi : L^3 \to \mathrm{Mat}_{3\times3}(L)$ be given by*

$$\phi(x, y, z) = \begin{bmatrix} x & y & z \\ p\,\sigma(z) & \sigma(x) & \sigma(y) \\ p\,\sigma^2(y) & p\,\sigma^2(z) & \sigma^2(x) \end{bmatrix}, \qquad (6.7.5)$$

and

- *$\Gamma = \{\, \gamma \in \phi(\mathcal{O}^3) \mid \det \gamma = 1 \,\}$.*

Then:

1) *Γ is an arithmetic subgroup of $\mathrm{SL}(3, \mathbb{R})$.*
2) *Γ is cocompact if and only if $p \neq t\,\sigma(t)\,\sigma^2(t)$, for all $t \in L$.*

Proof. (1) It is easy to see that:

- $L \subset \mathbb{R}$ (see Exercise 2).
- $\phi(L^3)$ and $\phi(\mathcal{O}^3)$ are subrings of $\mathrm{Mat}_{3\times3}(L)$ (even though ϕ is *not* a ring homomorphism).
- $\phi(L^3)$ is a $\mathbb{Q}$-form of $\mathrm{Mat}_{3\times3}(\mathbb{R})$.
- $\phi(\mathcal{O}^3)$ is a $\mathbb{Z}$-lattice in $\phi(L^3)$.
- If we define $\rho : \mathrm{Mat}_{3\times3}(\mathbb{R}) \to \mathrm{End}_{\mathbb{R}}(\mathrm{Mat}_{3\times3}(\mathbb{R}))$ by $\rho(g)(v) = gv$, then $\rho(\mathrm{SL}(3, \mathbb{R}))$ is defined over $\mathbb{Q}$ (with respect to the $\mathbb{Q}$-form $\phi(L^3)$) (see Exercise 3)).

- $\Gamma = \{ g \in \mathrm{SL}(3,\mathbb{R}) \mid g\,\phi(\mathcal{O}^3) = \phi(\mathcal{O}^3) \}$.

So Γ is an arithmetic subgroup of $\mathrm{SL}(3,\mathbb{R})$ (see Proposition 5.4.5(1)).

($2 \Leftarrow$) If $\mathrm{SL}(3,\mathbb{R})/\Gamma$ is not compact, then the Godement Criterion (5.3.1) tells us there is a nontrivial unipotent element u in Γ. This means 1 is an eigenvalue of u (indeed, it is the only eigenvalue of u), so there is some nonzero $v \in \mathbb{R}^3$ with $uv = v$. Hence $(u - 1)v = 0$. Since $u \neq \mathrm{Id}$ and $v \neq 0$, we conclude that $\phi(L^3)$ has a nonzero element that is not invertible.

Hence, letting $D = \phi(L^3)$, it suffices to show that every nonzero element of D is invertible. (That is, D is a "division algebra.") For convenience, define $N \colon L \to \mathbb{Q}$ by $N(t) = t\,\sigma(t)\,\sigma^2(t)$. (In Algebraic Number Theory, N is called the "norm" from L to $\mathbb{Q}$.) We know that $p \neq N(t)$, for all $t \in L$. It is easy to see that $N(t_1 t_2) = N(t_1)\,N(t_2)$.

Note that if $xyz = 0$, but $(x,y,z) \neq (0,0,0)$, then $\phi(x,y,z)$ is invertible. For example, if $z = 0$, then $\det\phi(x,y,z) = N(x) + p\,N(y)$. Since $p \neq N(-x/y) = -N(x)/N(y)$ (assuming $y \neq 0$), we see that $\det\phi(x,y,z) \neq 0$, as desired. The other cases are similar.

For any $x, y, z \in L$, with $z \neq 0$, we have

$$\phi\left(1, -\frac{x}{p\,\sigma(z)}, 0\right)\phi(x,y,z) = \phi(0, *, *)$$

is invertible, so $\phi(x,y,z)$ is invertible.

($2 \Rightarrow$) If $p = t\,\sigma(t)\,\sigma^2(t)$, for some $t \in L$, then $\phi(L^3) \cong \mathrm{Mat}_{3\times 3}(\mathbb{Q})$ (see Exercise 5). From this, it is easy clear that $\phi(\mathcal{O}^3)$ contains nonidentity unipotent matrices. Since the determinant of any unipotent matrix is 1, these unipotents belong to Γ. Therefore Γ is not cocompact.   □

(6.7.6) **Example.** Let

- $\zeta = 2\cos(2\pi/7)$,
- $L = \mathbb{Q}[\zeta]$, and
- p be any prime that is congruent to either 3 or 5, modulo 7.

Then

1) L is a cubic, Galois extension of $\mathbb{Q}$, and
2) $p \neq t\,\sigma(t)\,\sigma^2(t)$, for all $t \in L$, and any generator σ of $\mathrm{Gal}(L/\mathbb{Q})$.

To see this, let $\omega = e^{2\pi i/7}$ be a primitive 7th root of unity, so $\zeta = \omega + \omega^6$. Now it is well known that the Galois group of $\mathbb{Q}[\omega]$ is cyclic of order 6, generated by $\tau(\omega) = \omega^3$ (see Proposition B3.4). So the fixed field L of τ^3 is a cyclic extension of degree $6/2 = 3$.

Now suppose $t\,\sigma(t)\,\sigma^2(t) = p$, for some $t \in L^\times$. Clearing denominators, we have $s\,\sigma(s)\,\sigma^2(s) = pm$, where

- $m \in \mathbb{Z}^+$,
- $s = a + b(\omega + \omega^6) + c(\omega + \omega^6)^2$, with $a, b, c \in \mathbb{Z}$ and $p \nmid \gcd(a,b,c)$.

Replacing ω with the variable x, we obtain integral polynomials $s_1(x)$, $s_2(x)$, and $s_3(x)$, such that

$$s_1(x)s_2(x)s_3(x) = pm = 0 \text{ in } \frac{\mathbb{Z}_p[x]}{\langle x^6 + x^5 + \cdots + 1 \rangle}.$$

This implies that $x^6 + x^5 + \cdots + 1$ is *not* irreducible in $\mathbb{Z}_p[x]$. This contradicts the choice of p (see Exercise 6).

(6.7.7) *Remark.* The famous Kronecker-Weber Theorem tells us that if L is a Galois extension of $\mathbb{Q}$, with abelian Galois group, then L is contained in an extension obtained by adjoining an nth root of unity to $\mathbb{Q}$ (for some n). (*Warning:* this does not hold for abelian extensions of algebraic number fields other than $\mathbb{Q}$.) As a very special case, this implies that all of the cubic, Galois extension fields L of $\mathbb{Q}$ can be constructed quite explicitly, in the manner of Example 6.7.6:

- Choose $n \in \mathbb{Z}^+$, such that $\varphi(n)$ is divisible by 3 (where
$$\varphi(n) = \#\{\, k \mid 1 \leq k \leq n, \ \gcd(k,n) = 1 \,\}$$
is the Euler φ-function).
- Let $\omega = e^{2\pi i/n}$ be a primitive n^{th} root of unity.
- Let H be any subgroup of index 3 in the multiplicative group $(\mathbb{Z}_n)^\times$ of units modulo n.
- Let $\zeta = \sum_{k \in H} \omega^k = \sum_{k \in H} \cos(2\pi k/n)$.
- Let $L = \mathbb{Q}[\zeta]$.

We have now seen that cocompact arithmetic subgroups of $\mathrm{SL}(3, \mathbb{R})$ can be constructed by two different methods: some are constructed by using unitary groups (as in Proposition 6.7.1) and others are constructed by using "division algebras" $D = \phi(L^3)$ (as in Proposition 6.8.8). We will see in the following section that these two methods can be combined: some cocompact arithmetic subgroups are constructed by using *both* unitary groups *and* division algebras. The classification results in Section 18.4 show that all cocompact arithmetic subgroups of $\mathrm{SL}(3, \mathbb{R})$ can be obtained from these methods, using either unitary groups, division algebras, or a combination of the two (perhaps also combined with restriction of scalars). The same is true for the cocompact arithmetic subgroups of any $\mathrm{SL}(n, \mathbb{R})$, with $n \geq 3$.

Exercises for §6.7.

#1. Assume the situation of Proposition 6.7.1, except that $F = \mathbb{Q}$. More precisely, let
- $a, b, c \in \mathbb{Q}$ (all nonzero),
- L be a real quadratic extension of $\mathbb{Q}$,
- τ be the Galois automorphism of L over $\mathbb{Q}$,

- $\mathcal{O}$ be the ring of integers of L,
- $A = \mathrm{diag}(a, b, c)$, and
- $\Gamma = \mathrm{SU}(A, \tau; \mathcal{O})$, so Γ is an arithmetic subgroup of $\mathrm{SL}(3, \mathbb{R})$.

Show that Γ is <u>not</u> cocompact.

[*Hint:* The equation $\tau(x^T)Ax = 0$ for $x \in L^3$ can be considered as an equation in 6 variables over $\mathbb{Q}$, so the Number Theory fact mentioned in the proof of Proposition 6.4.1 implies it has a nontrivial solution.]

#2. Let L be a Galois extension of $\mathbb{Q}$, with $|L : \mathbb{Q}|$ odd. Show $L \subset \mathbb{R}$.

#3. Assume the notation of the proof of Proposition 6.7.4. For $h \in L^3$, define $T_h \in \mathrm{End}_{\mathbb{R}}(\mathrm{Mat}_{3\times3}(\mathbb{R}))$ by $T_h(v) = \phi(h) v$.
   a) Show that $\phi(h) \in \mathrm{End}_{\mathbb{R}}(\mathrm{Mat}_{3\times3}(\mathbb{R}))_{\mathbb{Q}}$, where the $\mathbb{Q}$-form is induced by the $\mathbb{Q}$-form $\phi(L^3)$ of $\mathrm{Mat}_{3\times3}(\mathbb{R})$.
   [*Hint:* Show $\phi(h)\,\phi(L^3) \subseteq \phi(L^3)$.]
   b) Show that $\rho(\mathrm{Mat}_{3\times3}(\mathbb{R}))$ is the centralizer of $\{\, T_h \mid h \in L^3 \,\}$.
   c) Show that $\rho(\mathrm{SL}(3, \mathbb{R}))$ is defined over $\mathbb{Q}$.

#4. In the notation of Proposition 6.7.4, show that if $p = t\,\sigma(t)\,\sigma^2(t)$, then the element $\phi\big(1, 1/t, 1/(t\,\sigma(t))\big)$ of $\phi(L^3)$ is not invertible.

#5. In the notation of Proposition 6.7.4, show that if $p = t\,\sigma(t)\,\sigma^2(t)$, for some $t \in L$, then $\phi(L^3) \cong \mathrm{Mat}_{3\times3}(\mathbb{Q})$.

[*Hint:* $\{(a, t\sigma(a), t\sigma(t)\sigma^2(a))\}$ is a 3-dimensional, $\phi(L)$-invariant $\mathbb{Q}$-subspace of L^3.]

#6. Let p and q be distinct primes, and
$$f(x) = x^{q-1} + \cdots + x + 1.$$
Show that $f(x)$ is reducible over $\mathbb{Z}_p$ if and only if there exists $r \in \{1, 2, \ldots, q - 2\}$, such that $p^r \equiv 1 \pmod{q}$.

[*Hint:* Let $g(x)$ be an irreducible factor of $f(x)$, and let $r = \deg g(x) < q - 1$. Then $f(x)$ has a root α in a finite field F of order p^r. Since α is an element of order q in $F^{\times}$, we must have $q \mid \#F^{\times}$.]

§6.8. Arithmetic subgroups of SL(n, ℝ)

We will briefly explain how the previous results on $\mathrm{SL}(3, \mathbb{R})$ can be generalized to higher dimensions. (The group $\mathrm{SL}(2, \mathbb{R})$ is a special case that does not fit into this pattern.) The proofs are similar to those for $\mathrm{SL}(3, \mathbb{R})$.

In Section 6.2, and in the proof of Proposition 6.7.4, we have seen that cocompact arithmetic subgroups of $\mathrm{SL}(n, \mathbb{R})$ can sometimes be constructed by using rings in which every nonzero element has a multiplicative inverse. More such "division algebras" will be needed in order to construct all the arithmetic subgroups of $\mathrm{SL}(n, \mathbb{R})$ with $n > 3$.

§6.8(i). Division algebras.

(6.8.1) **Definition.** An associative ring D is a ***division algebra*** over a field F if

1) D contains F in its center (that is, $xf = fx$ for all $x \in D$ and $f \in F$),
2) the element $1 \in F$ is the identity element of D,
3) D is finite-dimensional as a vector space over F, and
4) every nonzero element of D has a multiplicative inverse.

Furthermore, it is ***central*** over F if the entire center of D is precisely F.

(6.8.2) *Remarks.*

1) D is an *algebra* over F if (1) and (2) hold.
2) The word *division* requires (4). We also assume (3), although not all authors require this.

(6.8.3) **Other terminology.** Division algebras are also called ***skew fields***.

(6.8.4) **Examples.**

1) Any extension field of F is a division algebra over F (but is not usually central).
2) $\mathbb{H} = \mathbb{H}_{\mathbb{R}}^{-1,-1}$ is a central division algebra over $\mathbb{R}$.
3) More generally, a quaternion algebra $\mathbb{H}_F^{a,b}$ is a central division algebra over F if and only if $N_{red}(x) \neq 0$, for every nonzero $x \in \mathbb{H}_F^{a,b}$ (see Exercise 6.2#4). Note that this is consistent with (2).

The following famous theorem shows that division algebras are the building blocks of simple algebras:

(6.8.5) **Theorem** (Wedderburn's Theorem). *Let A be a finite-dimensional algebra over a field F. If A is **simple** (that is, if A has no nonzero, proper, two-sided ideals), then $A \cong \mathrm{Mat}_{n \times n}(D)$, for some n and some division algebra D over F.*

Proof. Since A is finite-dimensional, we may let I be a minimal left ideal. Then I is a left A-module that is ***simple*** (that is, has no nonzero, proper submodules). So $\mathrm{End}_A(I)$ is a division algebra (see Exercise 2); call it D.

We have $IA = A$, since IA is a 2-sided ideal and A is simple. Hence, the minimality of I implies $A = Ia_1 \oplus \cdots \oplus Ia_n$, for some $a_1, \ldots, a_n \in A$ (see Exercise 3).

For A considered as a left A-module, it is easy to see that each element of $\mathrm{End}_A(A)$ is multiplication on the right by an element of A (see Exercise 4); therefore $\mathrm{End}_A(A) \cong A$. On the other hand, it is easy to see that Ia_i is isomorphic to I as a left A-module (see Exercise 5), so we have

$$\mathrm{End}_A(A) = \mathrm{End}_A(Ia_1 \oplus \cdots \oplus Ia_n) \cong \mathrm{End}_A(I^n)$$
$$\cong \mathrm{Mat}_{n \times n}(\mathrm{End}_A(I)) = \mathrm{Mat}_{n \times n}(D).$$

Therefore $A \cong \mathrm{Mat}_{n \times n}(D)$. $\square$

(6.8.6) Corollary. *If D is a central division algebra over F, then we have* $\dim_F D = d^2$, *for some* $d \in \mathbb{Z}^+$. *(This integer d is called the degree of D over F.)*

Proof. Let $\overline{F}$ be the algebraic closure of F. Then (from Wedderburn's Theorem), we see that $D \otimes_F \overline{F} \cong \mathrm{Mat}_{d \times d}(D')$, for some d and some central division algebra D' over $\overline{F}$. Since $\overline{F}$ is algebraically closed, we must have $D' = \overline{F}$ (see Exercise 6), so
$$\dim_F D = \dim_{\overline{F}}(D \otimes_F \overline{F}) = \dim_{\overline{F}} \mathrm{Mat}_{d \times d}(\overline{F}) = d^2.$$ $\square$

In order to produce arithmetic groups from division algebras, the following lemma provides an analogue of the ring of integers in an algebraic number field F.

(6.8.7) Lemma. *If D is a division algebra over an algebraic number field F, then there is a subring $\mathcal{O}_D$ of D, such that $\mathcal{O}_D$ is a $\mathbb{Z}$-lattice in D. Any such subring is called an order in D.*

Proof. Let $\{v_0, v_1, \ldots, v_r\}$ be a basis of D over $\mathbb{Q}$, with $v_0 = 1$. Let $\{c_{j,k}^{\ell}\}_{j,k,\ell=0}^r$ be the structure constants of D with respect to this basis. That is, for $j, k \in \{0, \ldots, r\}$, we have $v_j v_k = \sum_{\ell=0}^r c_{j,k}^{\ell} v_{\ell}$. There is some nonzero $m \in \mathbb{Z}$, such that $m c_{j,k}^{\ell} \in \mathbb{Z}$, for all j, k, ℓ. Let $\mathcal{O}_D$ be the $\mathbb{Z}$-span of $\{1, m v_1, \ldots, m v_r\}$. $\square$

In the proof of Proposition 6.7.4, we showed that $\phi(L^3)$ is a division algebra if $p \neq t\,\sigma(t)\,\sigma^2(t)$. Conversely, it is known that every division algebra of degree 3 arises from the above construction. (In the terminology of ring theory, this means that every central division algebra of degree 3 is "cyclic.") Therefore, we can restate the proposition in the following more abstract form.

(6.8.8) Proposition. *Let*
- *L be a cubic, Galois extension of $\mathbb{Q}$,*
- *D be a central division algebra of degree 3 over $\mathbb{Q}$, such that D contains L as a subfield, and*
- *$\mathcal{O}_D$ be an order in D (see Lemma 6.8.7).*

Then there is an embedding $\phi: D \to \mathrm{Mat}_{3 \times 3}(\mathbb{R})$, such that
1) *$\phi(\mathrm{SL}(1, D))$ is a $\mathbb{Q}$-form of $\mathrm{Mat}_{3 \times 3}(\mathbb{R})$, and*
2) *$\phi(\mathrm{SL}(1, \mathcal{O}_D))$ is a cocompact, arithmetic subgroup of $\mathrm{SL}(3, \mathbb{R})$.*

Furthermore, $\phi(\mathrm{SL}(1, \mathcal{O}_D))$ is essentially independent of the choice of $\mathcal{O}_D$ or of the embedding ϕ. Namely, if $\mathcal{O}_D'$ and ϕ' are some other choices, *then there is an automorphism α of* $\mathrm{SL}(3, \mathbb{R})$, *such that* $\alpha\phi'(\mathrm{SL}(1, \mathcal{O}_D'))$ *is commensurable to* $\phi(\mathrm{SL}(1, \mathcal{O}_D))$.

This generalizes in an obvious way to provide cocompact, arithmetic subgroups of $\mathrm{SL}(n, \mathbb{R})$. By replacing $\mathrm{SL}(1, \mathcal{O}_D)$ with the more general $\mathrm{SL}(m, \mathcal{O}_D)$, we can also obtain arithmetic subgroups that are not cocompact (if n is not prime).

(6.8.9) **Proposition.** *Let*
- *D be a central division algebra of degree d over $\mathbb{Q}$, such that D splits over $\mathbb{R}$,*
- *$m \in \mathbb{Z}^+$, and*
- *$\mathcal{O}_D$ be $\mathbb{Z}$-lattice in D that is also a subring of D.*

Then $\phi(\mathrm{SL}(m, \mathcal{O}_D))$ is an arithmetic subgroup of $\mathrm{SL}(dm, \mathbb{R})$, for any embedding $\phi: D \to \mathrm{Mat}_{d \times d}(\mathbb{R})$, such that $\phi(D)$ is a $\mathbb{Q}$-form of $\mathrm{Mat}_{d \times d}(\mathbb{R})$.
It is cocompact if and only if $m = 1$.

§6.8(ii). Unitary groups over division algebras.
The definition of a unitary group is based on the Galois automorphism of a quadratic extension. This is a field automorphism of order 2. The following analogue makes it possible to define unitary groups over division algebras that are not required to be fields.

(6.8.10) **Definition.** Let D be a central division algebra. A map $\tau: D \to D$ is an ***anti-involution*** if $\tau^2 = \mathrm{Id}$ and τ is an **anti**-automorphism; that is, $\tau(x + y) = \tau(x) + \tau(y)$ and $\tau(xy) = \tau(y)\tau(x)$. (Note that τ reverses the order of the factors in a product.)

(6.8.11) **Other terminology.** Some authors call τ an involution, rather than an anti-involution, but, to avoid confusion, our terminology emphasizes the fact that τ is *not* an automorphism (unless D is commutative).

(6.8.12) **Examples.** Let D be a quaternion division algebra. Then:
1) The map $\tau_c: D \to D$ defined by
$$\tau_c(a + bi + cj + dk) = a - bi - cj - dk$$
   is an anti-involution. It is called the ***standard anti-involution*** of D, or the ***conjugation*** on D, so $\tau_c(x)$ can also be denoted $\overline{x}$.
2) The map $\tau_r: D \to D$ defined by
$$\tau_r(a + bi + cj + dk) = a + bi - cj + dk$$
   is an anti-involution. It is called the ***reversion*** on D.

(6.8.13) **Definitions.** Let τ be an anti-involution of a division algebra D over F.
1) A matrix $A \in \mathrm{Mat}_{n \times n}(D)$ is said to be **Hermitian** (or, more precisely, τ-*Hermitian*) if $(A^\tau)^T = A$.

2) Given a Hermitian matrix A, we let
$$\mathrm{SU}(A, \tau; D) = \{ g \in \mathrm{SL}(n, D) \mid (A^\tau)^T A g = A \}.$$

This notation makes it possible to state a version of Proposition 6.7.1 that replaces the quadratic extension L with a larger division algebra.

(6.8.14) **Proposition.** *Let*
- *L be a real quadratic extension of $\mathbb{Q}$,*
- *D be a central simple division algebra of degree d over L,*
- *τ be an anti-involution of D, such that $\tau|_L$ is the Galois automorphism of L over $\mathbb{Q}$,*
- *$b_1, \ldots, b_m \in D^\times$, such that $\tau(b_j) = b_j$ for each j,*
- *$\mathcal{O}_D$ be an order in D, and*
- *$\Gamma = \mathrm{SU}(\mathrm{diag}(b_1, b_2, \ldots, b_m), \tau; \mathcal{O}_D)$.*

Then:
1) *Γ is an arithmetic subgroup of $\mathrm{SL}(md, \mathbb{R})$.*
2) *Γ is cocompact if and only if, for all nonzero $x \in D^m$, we have*
$$\tau(x^T) \, \mathrm{diag}(b_1, b_2, \ldots, b_m) \, x \neq 0.$$

Additional examples of cocompact arithmetic subgroups can be obtained by generalizing Proposition 6.8.14 to allow L to be a totally real quadratic extension of a totally real algebraic number field F (as in Proposition 6.7.1). However, in this situation, one must require $b_1, \ldots, b_m$ to be chosen in such a way that $\mathrm{SU}(\mathrm{diag}(b_1, b_2, \ldots, b_m), \tau; \mathcal{O}_D)^\sigma$ is compact, for every place σ of F, such that $\sigma \neq \mathrm{Id}$. For $n \geq 3$, every arithmetic subgroup of $\mathrm{SL}(n, \mathbb{R})$ is obtained either from this unitary construction or from Proposition 6.8.9 (see Theorem 18.4.1).

Exercises for §6.8.

#1. Show $\mathrm{N}_{\mathrm{red}}(xy) = \mathrm{N}_{\mathrm{red}}(x) \, \mathrm{N}_{\mathrm{red}}(y)$ for all elements x and y of a quaternion algebra $\mathbb{H}_F^{a,b}$.

#2. (*Schur's Lemma*) Suppose A is a (finite-dimensional) F-algebra, and M is a simple A-module (that is finite-dimensional as a vector space over F). Show $\mathrm{End}_A(M)$ is a division algebra, where
$$\mathrm{End}_A(M) = \{ \varphi \colon M \to M \mid \varphi(am) = a\,\varphi(m), \ \forall a \in A, \ m \in M \}.$$
[*Hint:* If φ is not invertible, then it has a nontrivial kernel, which is a nontrivial A-submodule of M.]

#3. Show that if I is any minimal left-ideal of a finite-dimensional algebra A, then there exist $a_1, \ldots, a_n \in A$, such that $A = Ia_1 \oplus \cdots \oplus Ia_n$.
[*Hint:* By finite-dimensionality, $A = Ia_1 + \cdots + Ia_n$ for some $a_1, \ldots, a_n \in A$. If n is minimal, then $Ia_n \not\subset a_1 + \cdots + Ia_{n-1}$, so the minimality of I implies $Ia_n \cap (a_1 + \cdots + Ia_{n-1}) = \{0\}$.]

#4. Show that if A is a ring with identity, and we consider A to be a left A-module, then, for every $\varphi \in \operatorname{End}_A(A)$, there exists $a \in A$, such that $\varphi(x) = xa$ for all $x \in A$.
[*Hint:* Let $a = \varphi(1)$.]

#5. For any minimal left ideal I of a ring A, and any $a \in A$, such that $Ia \neq \{0\}$, show $I \cong Ia$ as left A-modules.
[*Hint:* $i \mapsto ia$ is a homomorphism of modules that is obviously surjective. The minimality of I implies it is also injective.]

#6. Show that if D is a division algebra over an algebraically closed field $\overline{F}$, then $D = \overline{F}$.
[*Hint:* Multiplication on the left by any $x \in D$ is a linear transformation, which must have an eigenvalue $\lambda \in \overline{F}$. Then $x - \lambda$ is not invertible.]

#7. Suppose $\mathbb{H}_F^{a,b}$ is a quaternion algebra over some field F, and let $L = F + Fi \subseteq \mathbb{H}_F^{a,b}$.
  a) Show that if a is not a square in F, then L is a subfield of $\mathbb{H}_F^{a,b}$.
  b) Show that $\mathbb{H}_F^{a,b}$ is a two-dimensional (left) vector space over L.
  c) For each $x \in \mathbb{H}_F^{a,b}$, define $R_x \colon \mathbb{H}_F^{a,b} \to \mathbb{H}_F^{a,b}$ by $R_x(v) = vx$, and show that R_x is an L-linear transformation.
  d) For each $x \in \mathbb{H}_F^{a,b}$, show $\det(R_x) = \mathrm{N}_{\mathrm{red}}(x)$.

#8. Let τ be an anti-involution on a division algebra D.
  a) For any $J \in \operatorname{Mat}_{n \times n}(D)$, define $B_J \colon D^n \times D^n \to D$ by
$$B_J(x, y) = \tau(x^T) J y$$
  for all $x, y \in D^n = \operatorname{Mat}_{n \times 1}(D)$. Show that B_J is a Hermitian form if and only if $\tau(J^T) = J$.
  b) Conversely, show that if B is a Hermitian form on D^n, then $B = B_J$, for some $J \in \operatorname{Mat}_{n \times n}(D)$.

#9. Let D be a finite-dimensional algebra over a field F. Show that D is a division algebra if and only if D has no proper, nonzero left ideals. (We remark that, by definition, D is **simple** if and only if it has no proper, nonzero *two-sided* ideals.)

Notes

Generalizing the examples considered here, see Chapter 18 for the construction of all arithmetic subgroups of classical groups (except some strange arithmetic subgroups of groups, such as $\mathrm{SO}(1,7)$, whose complexification has $\mathrm{SO}(8, \mathbb{C})$ as a simple factor).

The construction of all arithmetic subgroups of $\mathrm{SL}(2, \mathbb{R})$ is discussed (from the point of view of quaternion algebras) in [7, Chap. 5].

See [14, Cor. 2 of §4.3.2, p. 43] for a proof of the fact (used in Proposition 6.4.1 and Exercise 6.7#1) that if $a_1, \ldots, a_{n+1} \in \mathbb{Q}$ are not all of the same sign, and $n \geq 4$, then the equation $a_1 x_1^2 + \cdots a_{n+1} x_{n+1}^2 = 0$ has a nontrivial integer solution. It is called Meyer's Theorem, and will be used again in Corollary 18.6.2.

The original paper of Gromov and Piatetski-Shapiro [5] on the construction of nonarithmetic lattices in $SO(1, n)$ (§6.5) is highly recommended. The exposition there is very understandable, especially for a reader with some knowledge of arithmetic groups and hyperbolic manifolds. A brief treatment also appears in [9, App. C.2, pp. 362–364].

It was known quite classically that there are nonarithmetic lattices in $SO(1, 2)$ (or, in other words, in $SL(2, \mathbb{R})$). This was extended to $SO(1, n)$, for $n \leq 5$, by Makarov [8] and Vinberg [15]. The nonarithmetic lattices of Gromov and Piatetski-Shapiro [5] came later. Nonarithmetic lattices in $SU(1, n)$ were constructed by Mostow [10] for $n = 2$, and by Deligne and Mostow [3] for $n = 3$. These results on $SO(1, n)$ and $SU(1, n)$ are presented briefly in [9, App. C, pp. 353–368].

The Kronecker-Weber Theorem can be found in books on Class Field Theory, such as [11, Thm. 5.1.10, p. 324] (or see [4]).

Wedderburn's Theorem (6.8.5) is proved in [12, Thm. 3.5, p. 49], and other introductory texts on noncommutative rings (often in the more general setting of semisimple Artinian rings).

The fact that division algebras of degree 3 are cyclic (mentioned on page 145) is due to Wedderburn [16], and a proof can be found in [6, Thm. 2.9.17, p. 69]. Much more generally, the famous (and much more difficult) Albert-Brauer-Hasse-Noether Theorem states that any division algebra (of any degree) over a finite extension of $\mathbb{Q}$ is cyclic. It was first proved in [1, 2]. See [13, proof of Thm. 32.20, p. 280] for references to more modern expositions of Class Field Theory that provide proofs.

References

[1] A. A. Albert and H. Hasse: A determination of all normal division algebras over an algebraic number field, *Trans. Amer. Math. Soc.* 34 (1932), no. 3, 722–726. MR 1501659, http://dx.doi.org/10.1090/S0002-9947-1932-1501659-X

[2] R. Brauer, H. Hasse, and E. Noether: Beweis eines Hauptsatzes in der Theorie der Algebren, *J. Reine Angew. Math.* 167 (1932) 399–404. Zbl 0003.24404, https://eudml.org/doc/149820

[3] P. Deligne and G. D. Mostow: Monodromy of hypergeometric functions and non-lattice integral monodromy, *Publ. Math. Inst. Hautes Études Sci.* 63 (1986) 5–89. MR 0849651, http://www.numdam.org/item?id=PMIHES_1986__63__5_0

[4] M. J. Greenberg: An elementary proof of the Kronecker-Weber theorem, *Amer. Math. Monthly* 81 (1974), 601–607. (Correction 82

(1975), no. 8, 803.) MR 0340214 (MR 0376605),
http://www.jstor.org/stable/2319208, (2319794)

[5] M. Gromov and I. Piatetski-Shapiro: Nonarithmetic groups in
Lobachevsky spaces, *Publ. Math. Inst. Hautes Études Sci.* 66 (1987)
93–103. MR 0932135, http://eudml.org/doc/104029

[6] N. Jacobson: *Finite-dimensional Division Algebras over Fields.*
Springer, Berlin, 1996. ISBN 3-540-57029-2, MR 1439248

[7] S. Katok: *Fuchsian Groups.* Univ. of Chicago, Chicago, 1992.
MR 1177168

[8] V. S. Makarov: On a certain class of discrete groups of Lobachevsky
space having an infinite fundamental region of finite measure,
Soviet Math. Dokl. 7 (1966) 328–331. MR 0200348

[9] G. A. Margulis: *Discrete Subgroups of Semisimple Lie Groups.*
Springer, New York, 1991. ISBN 3-540-12179-X, MR 1090825

[10] G. D. Mostow: Existence of a non-arithmetic lattice in $SU(2,1)$, *Proc.
Nat. Acad. Sci. USA* 75 (1978) 3029–3033. MR 0499658,
http://www.pnas.org/content/75/7/3029

[11] J. Neukirch: *Algebraic Number Theory.* Springer, Berlin, 1999. ISBN
3-540-65399-6, MR 1697859

[12] R. S. Pierce: *Associative Algebras.* Springer, New York, 1982. ISBN
0-387-90693-2, MR 0674652

[13] I. Reiner: *Maximal Orders.* Oxford University Press, Oxford, 2003.
ISBN 0-19-852673-3, MR 1972204

[14] J.-P. Serre: *A Course in Arithmetic,* Springer, New York 1973.
MR 0344216

[15] E. B. Vinberg: Discrete groups generated by reflections in
Lobachevsky spaces, *Math. USSR Sbornik* 1 (1967) 429–444.
MR 0207853,
http://dx.doi.org/10.1070/SM1967v001n03ABEH001992

[16] J. H. M. Wedderburn: On division algebras, *Trans. Amer. Math. Soc.*
22 (1921), no. 2, 129–135. MR 1501164,
http://dx.doi.org/10.1090/S0002-9947-1921-1501164-3

SL(n, $\mathbb{Z}$) is a lattice in SL(n, $\mathbb{R}$)

In this chapter, we describe two different proofs of the following crucial fact, which is the basic case of the fundamental fact that if G is defined over $\mathbb{Q}$, then $G_\mathbb{Z}$ is a lattice in G (see Theorem 5.1.11). This special case was specifically mentioned (without proof) in Example 5.1.12(2).

(7.0.1) Theorem. SL(n, $\mathbb{Z}$) *is a lattice in* SL(n, $\mathbb{R}$).

The case $n = 2$ of Theorem 7.0.1 was established in Example 1.3.7, by constructing a subset $\mathcal{F}$ of SL(2, $\mathbb{R}$), such that

1) SL(2, $\mathbb{Z}$) $\cdot$ $\mathcal{F}$ = SL(2, $\mathbb{R}$), and

2) $\mathcal{F}$ has finite measure.

Our first proof of Theorem 7.0.1 shows how to generalize this approach to other values of n, by choosing $\mathcal{F}$ to be an appropriate "Siegel set" (see Sections 7.2 and 7.3).

(7.0.2) Remarks.

1) As was mentioned on page 87, the *statement* that $G_\mathbb{Z}$ is a lattice in G is more important than the *proof*. The same is true of the special case in Theorem 7.0.1, but it is advisable to understand at least the statements of the three main ingredients of our first proof:

Main prerequisites for this chapter: definition of $\mathbb{Z}$-lattices (Definition 5.4.1) and Moore Ergodicity Theorem (Section 4.10).

(a) the definition of a Siegel set (see Section 7.2),
(b) the fact that every Siegel set has finite measure (see Proposition 7.2.5), and
(c) the fact that some Siegel set is a coarse fundamental domain for $SL(n, \mathbb{Z})$ in $SL(n, \mathbb{R})$ (cf. Theorem 7.3.1).

2) This subject is often called **Reduction Theory**. The idea is that, given an element g of G, we would like to multiply g by an element γ of Γ to make the matrix γg as simple as possible. That is, we would like to "reduce" g to a simpler form by multiplying it by an element of Γ. This is a generalization of the classical reduction theory of quadratic forms, which goes back to Gauss and others.

Unfortunately, serious complications arise when using Siegel sets to establish in general that $G_\mathbb{Z}$ is a lattice in G Theorem 5.1.11 (see the proof in Section 19.4). Therefore, we will give a second proof with the virtue that it can easily be extended to establish that all arithmetic subgroups are lattices (see Section 7.4 for this proof of Theorem 7.0.1, and see Exercise 7.4#20 for the generalization to a proof of Theorem 5.1.11). However, this argument relies on a fact about $SL(n, \mathbb{R}) / SL(n, \mathbb{Z})$ that we will not prove in general (see Theorem 7.4.7).

(7.0.3) **Warning. The Standing Assumptions (4.0.0 on page 41) are *not* in effect in this chapter,** because we are *proving* that $\Gamma = SL(n, \mathbb{Z})$ is a lattice, instead of *assuming* that it is a lattice.

§7.1. Iwasawa decomposition: $SL(n, \mathbb{R}) = KAN$

The definition of a "Siegel set" is based on the following fundamental structure theorem:

(7.1.1) **Theorem** (Iwasawa Decomposition of $SL(n, \mathbb{R})$). *In $G = SL(n, \mathbb{R})$, let*

$$K = SO(n), \qquad N = \left\{ \begin{bmatrix} 1 & & * \\ & 1 & \\ 0 & & \ddots \\ & & & 1 \end{bmatrix} \right\}, \qquad A = \left\{ \begin{bmatrix} a_1 & & 0 \\ & a_2 & \\ 0 & & \ddots \\ & & & a_n \end{bmatrix} \right\}^\circ.$$

*Then $G = KAN$. In fact, every $g \in G$ has a **unique** representation of the form $g = kau$ with $k \in K$, $a \in A$, and $u \in N$.*

Proof. It is important to note that, because of the superscript "$\circ$" in its definition, A is only the *identity component* of the group of diagonal matrices; the entire group of diagonal matrices has a nontrivial intersection with K. With this in mind, the uniqueness of the decomposition is easy (see Exercise 1).

We now prove the existence of k, a, and u. To get started, let $\varepsilon_1, \ldots, \varepsilon_n$ be the standard basis of $\mathbb{R}^n$. Then, for any $g \in G$, the set $\{g\varepsilon_1, \ldots, g\varepsilon_n\}$ is a basis of $\mathbb{R}^n$.

The Gram-Schmidt Orthogonalization process constructs a corresponding orthonormal basis $w_1, \ldots, w_n$. We briefly recall how this is done: for $1 \le i \le n$, inductively define

$$w_i^* = v_i - \sum_{j=1}^{i-1} \langle v_i \mid w_j \rangle w_j \quad \text{and} \quad w_i = \frac{1}{\|w_i^*\|} w_i^*, \quad \text{where } v_i = g\varepsilon_i.$$

It is easy to verify that $w_1, \ldots, w_n$ is an orthonormal basis of $\mathbb{R}^n$ (see Exercise 2).

Since $\{w_1, \ldots, w_n\}$ and $\{\varepsilon_1, \ldots, \varepsilon_n\}$ are orthonormal, there is an orthogonal matrix $k \in O(n)$, such that $kw_i = \varepsilon_i$ for all i. Then

$$kw_i^* = k \cdot \|w_i^*\| w_i = \|w_i^*\|(k\, w_i) = \|w_i^*\| \varepsilon_i,$$

so there is a diagonal matrix a (with positive entries on the diagonal), such that

$$kw_i^* = a\varepsilon_i \quad \text{for all } i.$$

Also, it is easy to see (by induction) that $w_i \in \langle v_1, \ldots, v_i \rangle$ for every i. With this in mind, we have

$$g^{-1}w_i^* = g^{-1}v_i - g^{-1}\sum_{j=1}^{i-1}\langle v_i \mid w_j \rangle w_j$$

$$\in g^{-1}v_i + g^{-1}\langle v_1, \ldots, v_{i-1}\rangle$$

$$= \varepsilon_i + \langle \varepsilon_1, \ldots, \varepsilon_{i-1}\rangle,$$

so there exists $u \in N$, such that

$$g^{-1}w_i^* = u\varepsilon_i \quad \text{for all } i.$$

Therefore

$$u^{-1}g^{-1}w_i^* = \varepsilon_i = a^{-1}kw_i^*,$$

so $u^{-1}g^{-1} = a^{-1}k$. Hence, $g = k^{-1}au^{-1} \in KAN$ (see Exercise 4).   □

Exercises for §7.1.

#1. Show that if $k_1a_1u_1 = k_2a_2u_2$, with $k_i \in K$, $a_i \in A$, and $u_i \in N$, then $k_1 = k_2$, $a_1 = a_2$, and $u_1 = u_2$.

[*Hint:* Show $k_1^{-1}k_2 = a_1u_1u_2^{-1}a_2^{-1} \in K \cap AN = \{e\}$, so $k_1 = k_2$. This implies $a_1^{-1}a_2 = u_1u_2^{-1} \in A \cap N = \{e\}$, so $a_1 = a_2$ and $u_1 = u_2$.]

#2. In the notation of the proof of Theorem 7.1.1, show $\{w_1, \ldots, w_n\}$ is an orthonormal basis of $\mathbb{R}^n$.

[*Hint:* Calculating an inner product shows that $w_i^* \perp w_k$ whenever $i > k$.]

#3. Show that the components k, a, and u in the Iwasawa decomposition $g = kau$ are real analytic functions of g.

[*Hint:* The matrix entries of a and k^{-1} can be written explicitly in terms of the vectors w_i^* and w_i, which are real-analytic functions of g. Then $u = a^{-1}k^{-1}g$ is also real analytic.]

#4. In the proof of Theorem 7.1.1, note that:
   - a is a diagonal matrix, but we do not know the determinant of a, so it is not obvious that $a \in$ SL($n, \mathbb{R}$), and
   - $k \in$ O(n), but $K =$ SO(n), so it is not obvious that $k \in K$.

   From the fact that $g = k^{-1}au^{-1}$, show $a \in A$ and $k \in K$.
   [*Hint:* We know $\det k \in \{\pm 1\}$, $\det a > 0$, $\det u = 1$, and the determinant of a product is the product of determinants.]

#5. Let $G =$ SL($n, \mathbb{R}$).
   a) Show $G = KNA = ANK = NAK$.
      [*Hint:* We have $AN = NA$ and $G = G^{-1}$.]
   b) (optional) (*harder*) Show $G \neq NKA$ (if $n \geq 2$).
      [*Hint:* For $n = 2$, the action of G by isometries on $\mathfrak{H}^2$ yields a simply transitive action on the set of unit tangent vectors. Let v be a vertical tangent vector at the point i, and let w be a horizontal tangent vector at the point $2i$. The N-orbit of w consists of horizontal vectors at points on the line $\mathbb{R} + 2i$, but vectors in the KA-orbit of v are horizontal only on the line $\mathbb{R} + i$.]

#6. Show that every compact subgroup of SL($n, \mathbb{R}$) is conjugate to a subgroup of SO(n).
   [*Hint:* For every compact subgroup C of SL($n, \mathbb{R}$), there is a C-invariant inner product on $\mathbb{R}^n$, defined by $\langle v \mid w \rangle = \int_C (cv \cdot cw) \, dc$. Since $\langle v \mid w \rangle = gv \cdot gw$ for some $g \in$ SL($n, \mathbb{R}$), the usual dot product is invariant under some conjugate of C. This conjugate is contained in SO(n).]

§7.2. Siegel sets for SL($n, \mathbb{Z}$)

(7.2.1) **Example.** Let $\Gamma =$ SL($2, \mathbb{Z}$) and $G =$ SL($2, \mathbb{R}$). Figure 7.2A(a) (on page 155) depicts a well-known fundamental domain for the action of Γ on the upper half plane $\mathfrak{H}$. (We have already seen this in Figure 1.3A.) For convenience, let us give it a name, say $\overline{\mathcal{F}_0}$. There is a corresponding fundamental domain $\mathcal{F}_0$ for Γ in G, namely

$$\mathcal{F}_0 = \{\, g \in G \mid g(i) \in \overline{\mathcal{F}_0} \,\}$$

(cf. Exercise 1).

   Unfortunately, the shape of $\overline{\mathcal{F}_0}$ is not entirely trivial, because the bottom edge is curved. Furthermore, the shape of a fundamental domain gets much more complicated when G is larger than just SL($2, \mathbb{R}$). Therefore, we will content ourselves with finding a set that is easier to describe, and is close to being a fundamental domain.

(7.2.2) **Example.** To construct a region that is simpler than $\overline{\mathcal{F}_0}$, we can replace the curved edge with an edge that is straight. Also, because we do not need to find precisely a fundamental domain, we can be a bit sloppy about exactly where to place the edges, so we can enlarge the

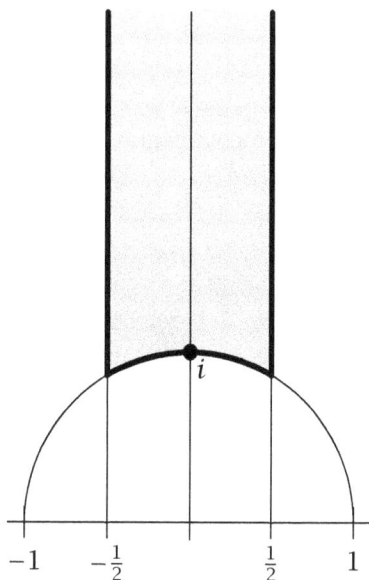

FIGURE 7.2A(a). A fundamental domain $\overline{\mathcal{F}_0}$.

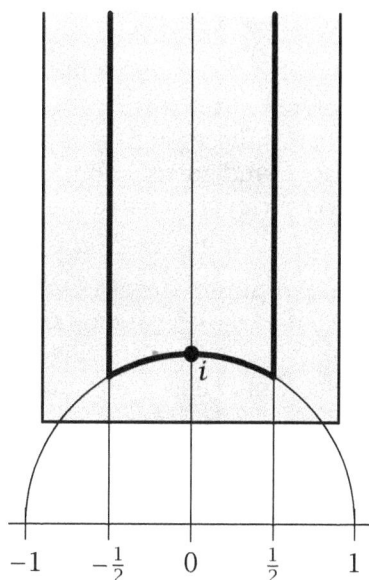

FIGURE 7.2A(b). A coarse fundamental domain $\overline{\mathcal{F}}$.

region slightly by moving the edges out a bit. The result is depicted in Figure 7.2A(b). This new region $\overline{\mathcal{F}}$ is slightly larger than a fundamental domain, but it is within bounded distance of a fundamental domain, and it suffices for many purposes. In particular, it is a coarse fundamental domain, in the sense of Definition 4.7.2 (see Exercise 6).

An important virtue of this particular coarse fundamental domain is that it can be specified quite easily:

$$\overline{\mathcal{F}} = \left\{ x + yi \,\middle|\, \begin{array}{c} c_1 \leq x \leq c_2, \\ y \geq c_3 \end{array} \right\}$$

for appropriate $c_1, c_2, c_3 \in \mathbb{R}$.

By using the Iwasawa decomposition $G = NAK$ (see Exercise 7.1#5), we can give a fairly simple description of the corresponding coarse fundamental domain $\mathcal{F}$ in SL$(2, \mathbb{R})$:

(7.2.3) Example. Let

- $\mathcal{F} = \{ g \in G \mid g(i) \in \overline{\mathcal{F}} \}$,

- $N_{c_1,c_2} = \left\{ \begin{bmatrix} 1 & t \\ 0 & 1 \end{bmatrix} \,\middle|\, c_1 \leq t \leq c_2 \right\}$,

- $A_{c_3} = \left\{ \begin{bmatrix} e^t & \\ & e^{-t} \end{bmatrix} \,\middle|\, e^{2t} \geq c_3 \right\}$, and

- $K = \mathrm{SO}(2)$.

Then (see Exercise 10)
$$\mathcal{F} = N_{c_1,c_2} A_{c_3} K.$$

Any set of the form $N_{c_1,c_2} A_{c_3} K$ is called a "Siegel set," so we can summarize this discussion by saying that Siegel sets provide good examples of coarse fundamental domains for SL($2, \mathbb{Z}$) in SL($2, \mathbb{R}$).

To construct a coarse fundamental domain for SL($n, \mathbb{Z}$) (with $n > 2$), we generalize the notion of Siegel set to SL($n, \mathbb{R}$).

(7.2.4) **Definition** (Siegel sets for SL($n, \mathbb{Z}$)). Let $G = $ SL($n, \mathbb{R}$), and consider the Iwasawa decomposition $G = NAK$ (see Exercise 7.1#5). To generalize Example 7.2.3, we construct a "Siegel set" $\mathfrak{S}$ by choosing appropriate subsets $\overline{N}$ of N and $\overline{A}$ of A, and letting $\mathfrak{S} = \overline{N}\,\overline{A}\,K$.

- The set $\overline{N}$ can be any (nonempty) compact subset of N. For example, we could let
$$\overline{N} = N_{c_1,c_2} = \{\, u \in N \mid c_1 \le u_{i,j} \le c_2 \text{ for } i < j \,\}.$$

- Note that the set A_{c_3} of Example 7.2.3 has the following alternate description:
$$A_{c_3} = \{\, a \in A \mid a_{1,1} \ge c_3\, a_{2,2} \,\}.$$

Therefore, we can generalize to SL($n, \mathbb{R}$) by defining
$$A_c = \{\, a \in A \mid a_{i,i} \ge c\, a_{i+i,i+1} \text{ for } i = 1, \dots, n-1 \,\}.$$

Thus, for $c_1, c_2 \in \mathbb{R}$ and $c_3 \in \mathbb{R}^+$, we have a Siegel set
$$\mathfrak{S}_{c_1,c_2,c_3} = N_{c_1,c_2} A_{c_3} K.$$

By calculating an appropriate multiple integral, it is not difficult to see that Siegel sets have finite measure:

(7.2.5) **Proposition** (see Exercise 14). $\mathfrak{S}_{c_1,c_2,c_3}$ *has finite measure (with respect to the Haar measure on* SL($n, \mathbb{R}$)).

Exercises for §7.2.

#1. Suppose H is a closed subgroup of G, and $\overline{\mathcal{F}}$ is a strict fundamental domain for the action of Γ on G/H. For every $x \in G/H$, show that
$$\mathcal{F} = \{\, g \in G \mid gx \in \overline{\mathcal{F}} \,\}$$
is a strict fundamental domain for Γ in G.

#2. Suppose $\mathcal{F}_1$ and $\mathcal{F}_2$ are coarse fundamental domains for Γ in G. Show that if $\mathcal{F}_1 \subseteq \mathcal{F} \subseteq \mathcal{F}_2$, then $\mathcal{F}$ is also a coarse fundamental domain for Γ in G.

#3. Suppose
- $\mathcal{F}$ is a coarse fundamental domain for the action of Γ on X, and
- F is a nonempty, finite subset of Γ.

Show that $F\mathcal{F} = \bigcup_{f \in F} f\mathcal{F}$ is also a coarse fundamental domain.

#4. Suppose H is a closed subgroup of G, and $\overline{\mathcal{F}}$ is a coarse fundamental domain for the action of Γ on G/H. For every $x \in G/H$, show that
$$\mathcal{F} = \{\, g \in G \mid gx \in \overline{\mathcal{F}} \,\}$$
is a coarse fundamental domain for Γ in G.

#5. In the notation of Figure 7.2A, show that the coarse fundamental domain $\overline{\mathcal{F}}$ is contained in the union of finitely many Γ-translates of the fundamental domain $\overline{\mathcal{F}_0}$.

#6. Show that the set $\overline{\mathcal{F}}$ depicted in Figure 7.2A(b) is indeed a coarse fundamental domain for the action of Γ on $\mathfrak{H}$.

[*Hint:* Exercises 3 and 5. You may assume (without proof) that $\overline{\mathcal{F}_0}$ is a fundamental domain.]

#7. Suppose
- $\mathcal{F}$ is a coarse fundamental domain for Γ in G, and
- F_1 is a nonempty, finite subset of Γ, and
- Γ_1 is a finite-index subgroup of Γ.

Show:
a) that $F_1 \mathcal{F}$ is also a coarse fundamental domain for Γ in G.
b) If $\Gamma_1 F_1 \mathcal{F} = G$, then $F_1 \mathcal{F}$ is a coarse fundamental domain for both Γ and Γ_1 in G.

#8. Assume Γ is infinite (or, equivalently, that G is not compact), and Γ_1 is a finite-index, proper subgroup of Γ. Show there exists a (strict) fundamental domain for Γ_1 in G that is *not* contained in any coarse fundamental domain for Γ in G.

[*Hint:* Construct a strict fundamental domain for Γ_1 that contains a strict fundamental domain $\mathcal{F}_0$ for Γ, but is not covered by finitely many Γ-translates of $\mathcal{F}_0$.]

#9. Suppose
- $\mathcal{F}$ is a coarse fundamental domain for Γ in G, and
- $g \in \mathcal{N}_G(\Gamma)$.

Show that $\mathcal{F}^g = g^{-1} \mathcal{F} g$ is also a coarse fundamental domain.

#10. Let $\mathcal{F}$ be the coarse fundamental domain for SL($2, \mathbb{Z}$) in SL($2, \mathbb{R}$) that is defined in Example 7.2.3. Verify that $\mathcal{F} = N_{c_1, c_2} A_{c_3} K$.

#11. Let $G = \mathrm{SL}(n, \mathbb{R})$. Given $c > 0$, show there exists $a \in A$, such that $A_c = a A^+$.

#12. *This exercise provides a description of the Haar measure on G.*

Let dg, dk, da, and du be the Haar measures on the unimodular groups G, K, A, and N, respectively, where $G = KAN$ is an Iwasawa decomposition. Also, for $a \in A$, let $\rho(a)$ be the modulus (or Jacobian) of the action of a on N by conjugation, so
$$\int_N f(a^{-1} u a)\, du = \int_N f(u)\, \rho(a)\, du \text{ for } f \in C_c(N).$$

Show, for $f \in C_c(G)$, that

$$\int_G f \, dg = \int_K \int_N \int_A f(kua) \, da \, du \, dk$$
$$= \int_K \int_A \int_N f(kau) \, \rho(a) \, du \, da \, dk.$$

[*Hint:* Since G is unimodular, dg is invariant under left translation by elements of K and right translation by elements of AN.]

#13. Let $G = \mathrm{SL}(n, \mathbb{R})$, choose N, A, and K as in Definition 7.2.4, and define ρ as in Exercise 12. Show

$$\rho\left(\begin{bmatrix} a_{1,1} & & & \\ & a_{2,2} & & \mathbf{0} \\ & & \ddots & \\ \mathbf{0} & & & a_{n,n} \end{bmatrix}\right) = \prod_{i<j} \frac{a_{j,j}}{a_{i,i}}.$$

#14. Let $c_1, c_2, c_3 \in \mathbb{R}$, with $c_1 < c_2$ and $c_3 > 0$. Show that the Siegel set $\mathfrak{S}_{c_1,c_2,c_3}$ in $\mathrm{SL}(n, \mathbb{R})$ has finite measure.

[*Hint:* See Exercises 12 and 13 for a description of the Haar measure on $\mathrm{SL}(n, \mathbb{R})$.]

§7.3. Constructive proof using Siegel sets

In this section, we prove the following result:

(7.3.1) Theorem. *Let*

- $G = \mathrm{SL}(n, \mathbb{R})$,
- $\Gamma = \mathrm{SL}(n, \mathbb{Z})$, *and*
- $\mathfrak{S}_{0,1,\frac{1}{2}} = N_{0,1} A_{1/2} K$ *be the Siegel set defined in Definition 7.2.4.*

Then $G = \Gamma \mathfrak{S}_{0,1,\frac{1}{2}}$.

This establishes Theorem 7.0.1:

Proof of Theorem 7.0.1. Combine the conclusion of Theorem 7.3.1 with Propositions 4.1.11 and 7.2.5. □

(7.3.2) Remarks.

1) Γ is written on the left in the conclusion of Theorem 7.3.1, because our definition of Siegel sets is motivated by a fundamental domain for the action of $\mathrm{SL}(2, \mathbb{Z})$ on $\mathfrak{H}^2$, and Γ acts on the left there. However, taking the transpose of both sides of the conclusion of Theorem 7.3.1 yields $G = \mathfrak{S}_{0,1,\frac{1}{2}}^T \Gamma$, where $\mathfrak{S}_{0,1,\frac{1}{2}}^T = K A_{1/2} N_{c_1,c_2}^T$. Thus, Γ can be written on the right, if the definition of Siegel set is modified appropriately.

2) Our definition of Siegel sets uses the upper-triangular group N, and Theorem 7.3.1 puts Γ on the left. Then (1) uses the lower-triangular group N^T (also called N^-), and puts Γ on the right. Some authors

reverse this, using N^- when Γ is on the left and using N when the action on the right. However, to accomplish this, the inequality in the definition of A_c needs to be reversed. (See Exercise 1 and the proof of Theorem 7.3.1.)

The following elementary observation is the crux of the proof of Theorem 7.3.1:

(7.3.3) **Lemma.** *If $\mathcal{L}$ is any $\mathbb{Z}$-lattice in $\mathbb{R}^n$, then there is an ordered basis $v_1, \ldots, v_n$ of $\mathbb{R}^n$, such that*

1) *$\{v_1, \ldots, v_n\}$ generates $\mathcal{L}$ as an abelian group, and*
2) *$\| \operatorname{proj}_i^\perp v_{i+1} \| \geq \frac{1}{2} \| \operatorname{proj}_{i-1}^\perp v_i \|$ for $1 \leq i < n$, where $\operatorname{proj}_i^\perp \colon \mathbb{R}^n \to V_i^\perp$ is the orthogonal projection onto the orthogonal complement of the subspace V_i spanned by $\{v_1, v_2, \ldots, v_i\}$.*

Proof. Choose v_1 to be a nonzero vector of minimal length in $\mathcal{L}$. Then define the remaining vectors $v_2, v_3, \ldots, v_n$ by induction, as follows:

Given $v_1, v_2, \ldots, v_i$, choose $v_{i+1} \in \mathcal{L}$ to make $\operatorname{proj}_i^\perp v_{i+1}$ as short as possible, subject to the constraint that v_{i+1} is linearly independent from $\{v_1, v_2, \ldots, v_i\}$ (so $\operatorname{proj}_i^\perp v_{i+1}$ is nonzero).

We now verify (1) and (2).

(1) For each i, let $\mathcal{L}_i$ be the abelian group generated by $v_1, v_2, \ldots, v_i$. If $\mathcal{L}_n \neq \mathcal{L}$, we may let i be minimal with $\mathcal{L}_{i+1} \neq \mathcal{L} \cap V_{i+1}$. Then we must have $\operatorname{proj}_i^\perp \mathcal{L}_{i+1} \subsetneq \operatorname{proj}_i^\perp (\mathcal{L} \cap V_{i+1})$ (see Exercise 2), so there is some $v \in \mathcal{L} \cap V_{i+1}$ with $\operatorname{proj}_i^\perp v_{i+1} = k \cdot \operatorname{proj}_i^\perp v$ for some $k \geq 2$ (see Exercise 3). This contradicts the minimality of $\| \operatorname{proj}_i^\perp v_{i+1} \|$.

(2) For simplicity, assume $i = 1$ (see Exercise 4), and let $v_2^* = \operatorname{proj}_1^\perp v_2$, so $v_2 = v_2^* + \alpha v_1$, with $\alpha \in \mathbb{R}$. Obviously, there exists $k \in \mathbb{Z}$, such that $|\alpha - k| \leq 1/2$. If $\| \operatorname{proj}_1^\perp v_2 \| < \frac{1}{2} \|v_1\|$, then

$$\|v_2 - kv_1\| = \|v_2^* + (\alpha - k)v_1\| \leq \|v_2^*\| + |\alpha - k| \cdot \|v_1\|$$

$$< \frac{1}{2}\|v_1\| + \frac{1}{2}\|v_1\| = \|v_1\|.$$

This contradicts the minimality of $\|v_1\|$. $\qquad\square$

Proof of Theorem 7.3.1. We wish to show $G = \Gamma \mathfrak{S}_{0,1,\frac{1}{2}} = \Gamma N_{0,1} A_{1/2} K$. However, since the proof uses an action of G, and most readers prefer to have this action on the left, we will instead prove an analogous result with Γ on the right: $G = \mathfrak{S}_{0,1,\frac{1}{2}}^- \Gamma$. Namely, given $g \in G$,

we will show $g \in K A_{1/2}^- N_{0,1} \Gamma$,

where $A_c^- = \{ a^{-1} \mid a \in A_c \} = \{ a \in A \mid a_{i,i} \leq a_{i+1,i+1}/c \text{ for all } i \}$.

For convenience, let $\mathcal{L} = g\mathbb{Z}^n$, and let $\{\varepsilon_1, \ldots, \varepsilon_n\}$ be the standard basis of $\mathbb{R}^n$. Lemma 7.3.3 provides us with a sequence $v_1, \ldots, v_n$ of elements of $\mathcal{L}$. From 7.3.3(1), we see that, after multiplying g on the right

by an element of Γ, we may assume
$$g\varepsilon_i = v_i \text{ for } i = 1, \ldots, n$$
(see Exercise 6).

From the Iwasawa decomposition $G = KAN$ (see Theorem 7.1.1), we may write $g = kau$ with $k \in K$, $a \in A$, and $u \in N$. For simplicity, let us assume k is trivial (see Exercise 7), so
$$g = au \text{ with } a \in A \text{ and } u \in N.$$
Since $g \in AN$, we know g is upper triangular (and its diagonal entries are exactly the same as the diagonal entries of a), so
$$\langle \varepsilon_1, \varepsilon_2, \ldots, \varepsilon_i \rangle = \langle g\varepsilon_1, g\varepsilon_2, \ldots, g\varepsilon_i \rangle = \langle v_1, v_2, \ldots, v_i \rangle \text{ for all } i.$$
This implies that the diagonal entry $a_{i,i}$ of a is given by
$$a_{i,i} = g_{i,i} = \| \text{proj}_{i-1}^{\perp} g\varepsilon_i \| = \| \text{proj}_{i-1}^{\perp} v_i \|$$
$$\leq 2\| \text{proj}_i^{\perp} v_{i+1} \| = 2\| \text{proj}_i^{\perp} g\varepsilon_{i+1} \| = 2g_{i+1,i+1} = 2a_{i+1,i+1}.$$
Therefore $a \in A_{1/2}^{-}$.

Also, there exists $y \in \Gamma \cap N$, such that $u \in N_{0,1} y$ (see Exercise 8). Therefore $g = au \in A_{1/2}^{-} N_{0,1} y \subseteq K A_{1/2}^{-} N_{0,1} \Gamma$, as desired. $\qquad \square$

(7.3.4) *Remark.* It can be shown that that the Siegel set $\mathfrak{S}_{0,1,\frac{1}{2}}$ is a coarse fundamental domain for SL$(n, \mathbb{Z})$ in SL$(n, \mathbb{R})$ (cf. Subsection 19.4(ii)), but this fact is not needed in the proof that SL$(n, \mathbb{Z})$ is a lattice in SL$(n, \mathbb{R})$.

Exercises for §7.3.

#1. Let
- $N_{c_1, c_2}^{-} = \left\{ \begin{bmatrix} 1 & 0 \\ t & 1 \end{bmatrix} \,\middle|\, c_1 \leq t \leq c_2 \right\},$
- $A_{c_3}^{-} = \left\{ \begin{bmatrix} e^t & \\ & e^{-t} \end{bmatrix} \,\middle|\, e^{2t} \leq c_3 \right\},$
- $K = \text{SO}(2)$, and
- $\mathcal{F}' = N_{c_1, c_2}^{-} A_{c_3}^{-} K.$

Show that $\mathcal{F}'$ is a coarse fundamental domain for SL$(2, \mathbb{Z})$ in SL$(2, \mathbb{R})$ if and only if the set $\mathcal{F} = N_{c_1, c_2} A_{c_3} K$ of Example 7.2.3 is a coarse fundamental domain.
[*Hint:* Conjugate by $\begin{bmatrix} 0 & 1 \\ 1 & 0 \end{bmatrix}$.]

#2. In the notation of Lemma 7.3.3, show that if X and Y are two subgroups of V_{i+1}, such that
$$X \subseteq Y, \quad X \cap V_i = Y \cap V_i, \quad \text{and} \quad \text{proj}_i^{\perp} X = \text{proj}_i^{\perp} Y,$$
then $X = Y$.

#3. In the notation of Lemma 7.3.3, show that the group $\text{proj}_i^{\perp}(\mathcal{L} \cap V_{i+1})$ is cyclic.
[*Hint:* Since $\dim \text{proj}_i^{\perp} V_{i+1} = 1$, it suffices to show $\text{proj}_i^{\perp}(\mathcal{L} \cap V_{i+1})$ is discrete.]

#4. Prove Lemma 7.3.3(2) without assuming $i = 1$.

[*Hint:* Mod out V_{i-1}, which is in the kernel of both $\text{proj}_{i-1}^{\perp}$ and $\text{proj}_i^{\perp}$.]

#5. For $g \in \text{GL}(n, \mathbb{R})$, show $g \in \text{GL}(n, \mathbb{Z})$ if and only if $g\mathbb{Z}^n \subseteq \mathbb{Z}^n$ and $g^{-1}\mathbb{Z}^n \subseteq \mathbb{Z}^n$.

#6. For every n-element generating set $\{v_1, \dots, v_n\}$ of the group $\mathbb{Z}^n$, show there exists $y \in \text{SL}(n, \mathbb{Z})$, such that $g\varepsilon_i = \pm v_i$ for every i.

[*Hint:* Show there exists $y \in \text{GL}(n, \mathbb{Z})$, such that $g\varepsilon_i = v_i$ for every i.]

#7. Complete the proof of Theorem 7.3.1 (without assuming the element k is trivial).

[*Hint:* The group K acts by isometries on $\mathbb{R}^n$, so replacing $\{v_1, \dots, v_n\}$ with its image under an element of K does not affect the validity of 7.3.3(2).]

#8. For all $c \in \mathbb{R}$, show $N = N_{c,c+1}N_{\mathbb{Z}}$.

§7.4. Elegant proof using nondivergence of unipotent orbits

We now present a very nice proof of Theorem 7.0.1 that relies on two key facts: the Moore Ergodicity Theorem (4.10.3), and an important observation about orbits of unipotent elements (Theorem 7.4.7). The statement of this observation will be more enlightening after some introductory remarks.

(7.4.1) **Example.** Let $a = \begin{bmatrix} 2 & 0 \\ 0 & 1/2 \end{bmatrix}$, or, more generally, let a be any element of $\text{SL}(2, \mathbb{R})$ that is diagonalizable over $\mathbb{R}$ (and is not $\pm \text{Id}$). Then a has one eigenvalue that is greater than 1, and one eigenvalue that is less than 1 (in absolute value), so it is obvious that there exist linearly independent vectors v_+ and v_- in $\mathbb{R}^2$, such that

$$a^k v_+ \to 0 \text{ and } a^{-k}v_- \to 0 \text{ as } k \to +\infty.$$

By the Mahler Compactness Criterion (4.4.7), this implies that some of the orbits of a on $\text{SL}(2, \mathbb{R}) / \text{SL}(2, \mathbb{Z})$ are "divergent" or "go off to infinity" or "leave compact all sets." That is, there exists $x \in \text{SL}(2, \mathbb{R}) / \text{SL}(2, \mathbb{Z})$, such that, for every compact subset C of $\text{SL}(2, \mathbb{R}) / \text{SL}(2, \mathbb{Z})$,

$$\{ k \in \mathbb{Z} \mid a^k x \in C \} \text{ is finite}$$

(see Exercise 1).

In contrast, if $u = \begin{bmatrix} 1 & 1 \\ 0 & 1 \end{bmatrix}$, then it is clear that there does *not* exist a nonzero vector $v \in \mathbb{R}^2$, such that $u^k v \to 0$ as $k \to \infty$. In fact, if v is not fixed by u (i.e., if $uv \neq v$), then

$$\|u^k v\| \to \infty \text{ as } k \to \pm\infty \qquad (7.4.2)$$

(see Exercise 2). Therefore, it is not very difficult to show that *none* of the orbits of u on $\text{SL}(2, \mathbb{R}) / \text{SL}(2, \mathbb{Z})$ go off to infinity:

(7.4.3) **Proposition.** *If u is any unipotent element of* SL($2,\mathbb{R}$), *then, for all* $x \in$ SL($2,\mathbb{R}$)/SL($2,\mathbb{Z}$), *there is a compact subset C of* SL($2,\mathbb{R}$)/SL($2,\mathbb{Z}$), *such that*

$$\{ k \in \mathbb{Z}^+ \mid u^k x \in C \} \text{ is infinite.}$$

Proof. We may assume u is nontrivial. Then, by passing to a conjugate (and perhaps taking the inverse), we may assume $u = \left[\begin{smallmatrix} 1 & 1 \\ 0 & 1 \end{smallmatrix}\right]$.

Choose a small neighborhood $\mathcal{O}$ of 0 in $\mathbb{R}^2$ so that, for all $g \in$ SL($2,\mathbb{R}$), there do not exist two linearly independent vectors in $\mathcal{O} \cap g\mathbb{Z}^2$ (see Exercise 4). Since $x\mathbb{Z}^2$ is discrete, we may assume $\mathcal{O}$ is small enough that

$$\mathcal{O} \cap x\mathbb{Z}^2 = \{0\}. \tag{7.4.4}$$

Since $\mathcal{O}$ is open and 0 is a fixed point of u (and the action of u^{-1} is continuous), there exists $r > 0$, such that

$$B_r(0) \cup u^{-1}B_r(0) \subseteq \mathcal{O}, \tag{7.4.5}$$

where $B_r(0)$ is the open ball of radius r around 0. Let

$$C = \left\{ c \in \text{SL}(2,\mathbb{R})/\text{SL}(2,\mathbb{Z}) \;\middle|\; c\mathbb{Z}^2 \cap B_r(0) = \{0\} \right\}.$$

The Mahler Compactness Criterion (4.4.7) tells us that C is compact.

Given $N \in \mathbb{Z}^+$, it suffices to show there exists $k \geq 0$, such that $u^{N+k}x \in C$. That is,

we wish to show there exists $k \geq 0$, such that $u^{N+k}x\mathbb{Z}^2 \cap B_r(0) = \{0\}$.

Let v be a nonzero vector of smallest length in $u^N x\mathbb{Z}^2$. We may assume $\|v\| < r$ (for otherwise we may let $k = 0$). Hence, (7.4.4) implies that v is not fixed by u. Then, from (7.4.2), we know there is some $k > 0$, such that $\|u^k v\| \geq r$, and we may assume k is minimal with this property. Therefore $\|u^{k-1}v\| < r$, so $u^{k-1}v \in B_r(0) \subseteq \mathcal{O}$ by (7.4.5).

From the choice of $\mathcal{O}$, we know that $\mathcal{O} \cap u^{N+k-1}x\mathbb{Z}$ does not contain any vector that is linearly independent from $u^{k-1}v$. Therefore $u^{N+k}x\mathbb{Z}^2$ does not contain any nonzero vectors of length less than r (see Exercise 5), as desired. $\qquad\square$

This result has a natural generalization to SL($n,\mathbb{R}$) (but the proof is more difficult; see Section 7.5):

(7.4.6) **Theorem** (Margulis). *Suppose*
- *u is a unipotent element of* SL($n,\mathbb{R}$), *and*
- *$x \in$ SL($n,\mathbb{R}$)/SL($n,\mathbb{Z}$).*

Then there exists a compact subset C of SL($n,\mathbb{R}$)/SL($n,\mathbb{Z}$), *such that*

$$\{ k \in \mathbb{Z}^+ \mid u^k x \in C \} \text{ is infinite.}$$

In other words, every unipotent orbit visits some compact set infinitely many times. In fact, it can be shown that the orbit visits the compact set quite often — it spends a nonzero fraction of its life in the set:

(7.4.7) **Theorem** (Dani-Margulis). *Suppose*

- *u is a unipotent element of* $SL(n, \mathbb{R})$, *and*
- *$x \in SL(n, \mathbb{R}) / SL(n, \mathbb{Z})$.*

Then there exists a compact subset C of $SL(n, \mathbb{R}) / SL(n, \mathbb{Z})$, *such that*

$$\liminf_{m \to \infty} \frac{\#\left\{ k \in \{1, 2, \ldots, m\} \mid u^k x \in C \right\}}{m} > 0.$$

Before saying anything about the proof of this important fact, let us see how it implies the main result of this chapter:

Proof of Theorem 7.0.1. Let

- $X = SL(n, \mathbb{R}) / SL(n, \mathbb{Z})$, and
- μ be an $SL(n, \mathbb{R})$-invariant measure on X (see Proposition 4.1.3).

We wish to show $\mu(X) < \infty$.

Fix a nontrivial unipotent element u of $SL(n, \mathbb{R})$. For each $x \in X$ and compact $C \subseteq X$, let

$$\rho_C(x) = \liminf_{m \to \infty} \frac{\#\left\{ k \in \{1, 2, \ldots, m\} \mid u^k x \in C \right\}}{m}.$$

Since X can be covered by countably many compact sets, Theorem 7.4.7 implies there is a compact set C, such that

$$\rho_C > 0 \text{ on a set of positive measure} \qquad (7.4.8)$$

(see Exercise 6). Letting χ_C be the characteristic function of C, we have

$$\int_X \rho_C \, d\mu = \int_X \liminf_{m \to \infty} \frac{\#\left\{ k \in \{1, 2, \ldots, m\} \mid u^k x \in C \right\}}{m} \, d\mu(x)$$

$$\leq \liminf_{m \to \infty} \int_X \frac{\#\left\{ k \in \{1, 2, \ldots, m\} \mid u^k x \in C \right\}}{m} \, d\mu(x) \quad \begin{pmatrix} \text{Fatou's} \\ \text{Lemma} \\ \text{(B6.4)} \end{pmatrix}$$

$$= \liminf_{m \to \infty} \frac{1}{m} \int_X (\chi_{u^{-1}C} + \chi_{u^{-2}C} + \cdots + \chi_{u^{-m}C}) \, d\mu$$

$$= \liminf_{m \to \infty} \frac{1}{m} \left(\int_X \chi_{u^{-1}C} \, d\mu + \int_X \chi_{u^{-2}C} \, d\mu + \cdots + \int_X \chi_{u^{-m}C} \, d\mu \right)$$

$$= \liminf_{m \to \infty} \frac{1}{m} \left(\mu(u^{-1}C) + \mu(u^{-2}C) + \cdots + \mu(u^{-m}C) \right)$$

$$= \liminf_{m \to \infty} \frac{1}{m} \left(\mu(C) + \mu(C) + \cdots + \mu(C) \right)$$

$$= \mu(C)$$

$$< \infty,$$

so $\rho_C \in \mathcal{L}^1(X, \mu)$.

It is easy to see that ρ_C is u-invariant (see Exercise 7), so the Moore Ergodicity Theorem (4.10.3) implies that ρ_C is constant (a.e.). Also, from

(7.4.8), we know that the constant is not 0. Therefore, we have a nonzero constant function that is in $\mathcal{L}^1(X, \mu)$, which tells us that $\mu(X)$ is finite. $\square$

Now, to begin our discussion of the proof of Theorem 7.4.7, we introduce a bit of terminology and notation, and make some simple observations. First of all, let us restate the result by using the Mahler Compactness Criterion (4.4.7), and also replace the discrete times $\{1, 2, 3, \ldots, m\}$ with a continuous interval $[0, T]$. Exercise 10 shows that this new version implies the original.

(7.4.9) **Definition.** For any $\mathbb{Z}$-lattice $\mathcal{L}$ in $\mathbb{R}^n$, there is some $g \in \text{GL}(n, \mathbb{R})$, such that $\mathcal{L} = g\mathbb{Z}^n$. We say $\mathcal{L}$ is **unimodular** if $\det g = \pm 1$.

(7.4.10) **Theorem** (restatement of Theorem 7.4.7). *Suppose*
- $\{u^t\}$ *is a one-parameter unipotent subgroup of* SL$(n, \mathbb{R})$,
- $\mathcal{L}$ *is a unimodular* $\mathbb{Z}$-*lattice in* $\mathbb{R}^n$, *and*
- l *is the usual Lebesgue measure (i.e., length) on* $\mathbb{R}$.

Then there exists a neighborhood $\mathcal{O}$ of 0 in $\mathbb{R}^n$, such that

$$\liminf_{T \to \infty} \frac{l\big(\{t \in [0, T] \mid u^t \mathcal{L} \cap \mathcal{O} = \{0\}\}\big)}{T} > 0.$$

(7.4.11) **Notation.** Suppose W is a discrete subgroup of $\mathbb{R}^n$.
- A vector $w \in W$ is **primitive** in W if $\lambda w \notin W$, for $0 < \lambda < 1$.
- Let $\widehat{W}$ be the set of primitive vectors in W.
- Let $\widehat{W}^+ \subseteq \widehat{W}$ be a set of representatives that contains either w or $-w$, but not both, for every $w \in \widehat{W}$. (Note that $\widehat{W} = -\widehat{W}$; see Exercise 11.)

For simplicity, let us assume now that $n = 2$ (see Section 7.5 for a discussion of the general case).

(7.4.12) **Lemma.**
1) *There is a neighborhood $\mathcal{O}_1$ of 0 in $\mathbb{R}^2$, such that if W is any unimodular $\mathbb{Z}$-lattice in $\mathbb{R}^2$, then $\#\big(\widehat{W}^+ \cap \mathcal{O}_1\big) \leq 1$.*

2) *Given any neighborhood $\mathcal{O}_1$ of 0 in $\mathbb{R}^2$, and any $\epsilon > 0$, there exists a neighborhood $\mathcal{O}_2$ of 0 in $\mathbb{R}^2$, such that if $x \in \mathbb{R}^2$, and $[a, b]$ is an interval in $\mathbb{R}$, such that there exists $t \in [a, b]$ with $u^t x \notin \mathcal{O}_1$, then*

$$l\big(\{t \in [a, b] \mid u^t x \in \mathcal{O}_2\}\big) \leq \epsilon\, l\big(\{t \in [a, b] \mid u^t x \in \mathcal{O}_1\}\big).$$

Proof. (1) A unimodular $\mathbb{Z}$-lattice in $\mathbb{R}^2$ cannot contain two linearly independent vectors of norm less than 1 (see Exercise 12).

(2) Note that $u^t x$ moves at constant velocity along a straight line (see Exercise 13). So we simply wish to choose $\mathcal{O}_2$ small enough that

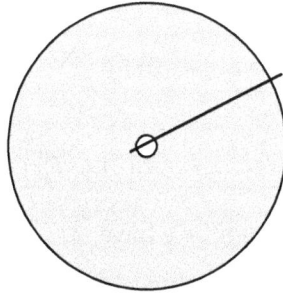

FIGURE 7.4A. Part (2) of Lemma 7.4.12: Any line segment that reaches the boundary of the large disk has only a small fraction of its length inside the tiny disk.

every line segment that reaches the boundary of $\mathcal{O}_1$ has only a small fraction of its length inside $\mathcal{O}_2$ (cf. Figure 7.4A).

By making $\mathcal{O}_1$ smaller, there is no harm in assuming it is a disk centered at 0. Let R be the radius of $\mathcal{O}_1$, and let $\mathcal{O}_2$ be a disk of radius r centered at 0, with r small enough that

$$\frac{2r}{R - r} < \epsilon.$$

Then, for any line segment L that reaches both $\mathcal{O}_2$ and the boundary of $\mathcal{O}_1$, we have:

- the length of $L \cap \mathcal{O}_2$ is $\leq$ the diameter $2r$ of $\mathcal{O}_2$, and

- the length of $L \cap \mathcal{O}_1$ is $\geq$ the distance $R - r$ from $\partial \mathcal{O}_1$ to $\partial \mathcal{O}_2$.

Therefore, the segment of L that is in $\mathcal{O}_2$ has length less than ϵ times the length of the segment that is $\mathcal{O}_1$ (cf. Figure 7.4A). $\qquad \square$

Proof of Theorem 7.4.10 when $n = 2$. Let $\mathcal{O}_1$ and $\mathcal{O}_2$ be as described in Lemma 7.4.12, with $\epsilon = 1/2$. We may assume $\mathcal{O}_1$ and $\mathcal{O}_2$ are convex, that they are small enough that they contain no nonzero elements of $\mathcal{L}$, and that $\mathcal{O}_2 \subseteq \mathcal{O}_1$.

Fix $T \in \mathbb{R}^+$. For each $x \in \hat{\mathcal{L}}^+$, and $k = 1, 2$, let

$$I_x^k = \{ t \in [0, T] \mid x u^t \in \mathcal{O}_k \}.$$

Since $\mathcal{O}_k$ is convex, and $u^t x$ traces out a line (see Exercise 13), we know that I_x^k is an interval (possibly empty). Note that:

1) from Lemma 7.4.12(2) (and the fact that $\epsilon = 1/2$), we see that $l(I_x^2) \leq \frac{1}{2} l(I_x^1)$, and

2) from Lemma 7.4.12(1), we see that $I_{x_1}^1$ is disjoint from $I_{x_2}^1$ whenever $x_1 \neq x_2$.

Therefore

$$l(\{t \in [0,T] \mid u^t\mathcal{L} \cap \mathcal{O}_2 \neq \{0\}) = \sum_{x \in \hat{L}^+} l(I_x^2) \leq \sum_{x \in \hat{L}^+} \frac{l(I_x^1)}{2}$$

$$= \frac{1}{2}l\left(\bigcup_{x \in \hat{L}^+} I_x^1\right) \leq \frac{1}{2}l([0,T]) = \frac{T}{2}.$$

So, passing to the complement, we have

$$l(\{t \in [0,T] \mid u^t\mathcal{L} \cap \mathcal{O}_2 = \{0\}) \geq \frac{T}{2}. \qquad \square$$

Unfortunately, Theorem 7.4.10 is not nearly as easy to prove when $n > 2$, because two basic complications arise.

1) The first difficulty is that the u^t-orbit of a vector is usually not a straight line (contrary to Exercise 13 for $n = 2$). However, the coordinates of $u^t x$ are always polynomials of bounded degree (see Exercise 14), so, for any fixed vector x,

the function $\|u^t x\|^2$ is a polynomial in t

and the degree of this polynomial is bounded (independent of x). Therefore, it is easy to prove that the appropriate analogue of Lemma 7.4.12(2) holds even if $n > 2$ (see Exercise 15), so the non-linearity is not a major problem.

2) A much more serious difficulty is the failure of 7.4.12(1): if $n > 2$, then a unimodular lattice in $\mathbb{R}^n$ may have two linearly independent primitive vectors that are very small (see Exercise 16). This means that the sets I_x^2 in the above proof may not be disjoint, which is a major problem. It is solved by looking at not only single vectors, but at larger sets of linearly independent vectors. More precisely, we look at the subgroups generated by sets of small vectors in $u^t\mathcal{L}$. These subgroups can intersect in rather complicated ways, and sorting this out requires a study of chains of these subgroups (ordered by inclusion) and a rather delicate proof by induction. Although the proof is completely elementary, using only some observations about polynomial functions, it is very clever and intricate. The main idea is presented in Section 7.5.

Exercises for §7.4.

#1. Suppose $a \in$ SL(2, $\mathbb{R}$), and there exist linearly independent vectors v_+ and v_- in $\mathbb{R}^2$, such that

$$a^n v_+ \to 0 \text{ as } n \to +\infty \quad \text{and} \quad a^n v_- \to 0 \text{ as } n \to -\infty.$$

Show $\exists x \in$ SL(2, $\mathbb{R}$)/ SL(2, $\mathbb{Z}$), such that $\{n \in \mathbb{Z} \mid a^n x \in C\}$ is finite, for every compact subset C of SL(2, $\mathbb{R}$)/ SL(2, $\mathbb{Z}$).

[*Hint:* There exists $g \in \mathrm{SL}(2, \mathbb{R})$ that takes the two standard basis vectors of $\mathbb{R}^2$ to vectors that are scalar multiples of v_+ and v_-.]

#2. Let $u = \left[\begin{smallmatrix} 1 & 1 \\ 0 & 1 \end{smallmatrix}\right]$. For every $v \in \mathbb{R}^2$, show that either
   - $u^n v = v$ for all $n \in \mathbb{Z}$, or
   - $\|u^n v\| \to \infty$ as $n \to \infty$.

#3. Generalize the preceding exercise to $\mathrm{SL}(n, \mathbb{R})$:
   Let u be any unipotent element of $\mathrm{SL}(n, \mathbb{R})$. For every $v \in \mathbb{R}^n$, show that either
   - $u^n v = v$ for all $n \in \mathbb{Z}$, or
   - $\|u^n v\| \to \infty$ as $n \to \infty$.

   [*Hint:* Each coordinate of $u^n v$ is a polynomial function of n, and non-constant polynomials cannot be bounded.]

#4. Suppose v_1 and v_2 are linearly independent vectors in $\mathbb{Z}^2$, and we have $g \in \mathrm{SL}(2, \mathbb{R})$. Show that if $\|g v_1\| < 1$, then $\|g v_2\| > 1$.

   [*Hint:* Since $g \in \mathrm{SL}(2, \mathbb{R})$, the area of the parallelogram spanned by the vectors $g v_1$ and $g v_2$ is the same as the area of the parallelogram spanned by v_1 and v_2, which is an integer.]

#5. Near the end of the proof of Proposition 7.4.3, verify the assertion that $u^{n+N} \times \mathbb{Z}^2$ does not contain any nonzero vectors of length less than r.

   [*Hint:* If $\|w\| < r$, then $u^{n-1} v$ and $u^{-1} w$ are linearly independent vectors in $\mathcal{O} \cap u^{n+N-1} \times \mathbb{Z}$.]

#6. Prove (7.4.8).

   [*Hint:* X cannot be the union of countably many sets of measure 0.]

#7. In the proof of Theorem 7.0.1, verify (directly from the definition) that ρ_C is u-invariant.

#8. Show Definition 7.4.9 is well-defined. More precisely, given any $g_1, g_2 \in \mathrm{GL}(n, \mathbb{R})$, such that $g_1 \mathbb{Z}^n = g_2 \mathbb{Z}^n$, show
$$\det g_1 \in \{\pm 1\} \iff \det g_2 \in \{\pm 1\}.$$

#9. Assume
   - u^t is a one-parameter unipotent subgroup of G,
   - $x \in G/\Gamma$, and
   - C^* is a compact subset of G/Γ.

   Show that if
$$\liminf_{T \to \infty} \frac{l\left(\{t \in [0, T] \mid u^t x \in C^*\}\right)}{T} > 0,$$
   then there is a compact subset C of G/Γ, such that
$$\liminf_{m \to \infty} \frac{\#\left\{k \in \{1, 2, \ldots, m\} \mid u^k x \in C\right\}}{m} > 0.$$

   [*Hint:* Let $C = \bigcup_{t \in [0,1]} u^t C^*$.]

#10. Show Theorem 7.4.10 implies Theorem 7.4.7.

[*Hint:* Mahler Compactness Criterion (4.4.7) and Exercise 9.]

#11. Suppose w is a nonzero element of a discrete subgroup W of $\mathbb{R}^n$. Show the following are equivalent:

    a) w is primitive in W.

    b) $\mathbb{R}w \cap W = \{\mathbb{Z}w\}$.

    c) If $kw' = w$, for some $k \in \mathbb{Z}$ and $w' \in W$, then $k \in \{\pm 1\}$.

    d) $-w$ is primitive in W.

#12. Suppose v and w are linearly independent vectors in a unimodular $\mathbb{Z}$-lattice in $\mathbb{R}^2$. Show $\|v\| \cdot \|w\| \geq 1$.

#13. Show that if $x \in \mathbb{R}^2$, and $\{u^t\}$ is any nontrivial one-parameter unipotent subgroup of SL$(2, \mathbb{R})$, then $u^t x$ moves at constant velocity along a straight line.

[*Hint:* Calculate the coordinates of $u^t x$ after choosing a basis so that $u^t = \left[\begin{smallmatrix} 1 & t \\ 0 & 1 \end{smallmatrix}\right]$.]

#14. Given $n \in \mathbb{Z}^+$, show there is a constant D, such that if $x \in \mathbb{R}^n$, and $\{u^t\}$ is any one-parameter unipotent subgroup of SL$(n, \mathbb{R})$, then the coordinates of $u^t x$ are polynomial functions of t, and the degrees of these polynomials are $\leq D$.

[*Hint:* We have $u^t = \exp(tv)$ for some $v \in \mathrm{Mat}_{n \times n}(\mathbb{R})$. Furthermore, v is nilpotent, because u^t is unipotent, so the power series $\exp(tv)$ is just a polynomial.]

#15. Given $R, D, \epsilon > 0$, show there exists $r > 0$, such that if

    • $f(x)$ is a (real) polynomial of degree $\leq D$, and

    • $[a, b]$ is an interval in $\mathbb{R}$, with $|f(t)| \geq R$ for some $t \in [a, b]$,

then

$$l(\{t \in [a, b] \mid |f(t)| < r\}) \leq \epsilon\, l(\{t \in [a, b] \mid |f(t)| < R\}).$$

[*Hint:* If not, then taking a limit yields a polynomial of degree D that vanishes on a set of positive measure, but is $\geq R$ at some point.]

#16. For every $\epsilon > 0$, find a unimodular $\mathbb{Z}$-lattice $\mathcal{L}$ in $\mathbb{R}^n$ with $n - 1$ linearly independent primitive vectors of norm $\leq \epsilon$.

#17. Assume G is defined over $\mathbb{Q}$ (and connected). Show there exist

    • a finite-dimensional real vector space V,

    • a vector v in V, and

    • a homomorphism $\rho\colon \mathrm{SL}(\ell, \mathbb{R}) \to \mathrm{SL}(V)$,

such that

    a) $G = \mathrm{Stab}_{\mathrm{SL}(\ell, \mathbb{R})}(v)^\circ$, and

    b) $\rho(\mathrm{SL}(\ell, \mathbb{Z}))v$ is discrete.

[*Hint:* See the hint to Exercise A4#8, and choose v to be the exterior product of polynomials with integer coefficients.]

#18. Show that if G is defined over $\mathbb{Q}$, then the natural embedding $G/G_{\mathbb{Z}} \hookrightarrow \mathrm{SL}(\ell, \mathbb{R})/\mathrm{SL}(\ell, \mathbb{Z})$ is a proper map.

[*Hint:* Use Exercise 17.]

#19. Prove Theorem 5.1.11 under the additional assumption that G is simple.

[*Hint:* The natural embedding $G/G_{\mathbb{Z}} \hookrightarrow \mathrm{SL}(\ell, \mathbb{R})/\mathrm{SL}(\ell, \mathbb{Z})$ is a proper map (see Exercise 18), so the G-invariant measure on $G/G_{\mathbb{Z}}$ provides a G-invariant measure μ on $\mathrm{SL}(\ell, \mathbb{R})/\mathrm{SL}(\ell, \mathbb{Z})$, such that all compact sets have finite measure. The proof of Theorem 7.0.1 (with $u \in G$) implies that μ is finite.]

#20. Prove Theorem 5.1.11 (without assuming that G is simple).

[*Hint:* You may assume Exercise 11.2#10 (without proof). This provides a version of the Moore Ergodicity Theorem for groups that are not simple.]

§7.5. Proof that unipotent orbits return to a compact set

The proof of Theorem 7.4.10 is rather complicated. To provide the gist of the argument, while eliminating some of the estimates that obscure the main ideas, we prove only Theorem 7.4.6, which is a qualitative version of the result. (The quantitative conclusion in Theorem 7.4.10 makes additional use of observations similar to Lemma 7.4.12(2) and Exercise 7.4#15.) This section is optional, because none of the material is needed elsewhere in the book.

By the Mahler Compactness Criterion (and an appropriate modification of Exercise 7.4#9), it suffices to prove the following statement:

(7.5.1) **Theorem** (restatement of Theorem 7.4.6). *Suppose*

- $\{u^t\}$ *is a one-parameter unipotent subgroup of* $\mathrm{SL}(n, \mathbb{R})$, *and*
- $\mathcal{L}$ *is a unimodular* $\mathbb{Z}$-*lattice in* $\mathbb{R}^n$.

Then there exists a neighborhood $\mathcal{O}$ *of* 0 *in* $\mathbb{R}^n$, *such that*
$$\left\{ t \in \mathbb{R}^+ \ \middle| \ u^t \mathcal{L} \cap \mathcal{O} = \{0\} \right\} \text{ is unbounded.}$$

(7.5.2) **Definition.** Suppose

- W is a discrete subgroup of $\mathbb{R}^n$, and
- k is the dimension of the linear span $\langle W \rangle$ of W.

We make the following definitions:

1) We define an inner product on the exterior power $\bigwedge^k \mathbb{R}^n$ by declaring $\{\varepsilon_{i_1} \wedge \varepsilon_{i_2} \wedge \cdots \wedge \varepsilon_{i_k}\}$ to be an orthonormal basis, where $\{\varepsilon_1, \ldots, \varepsilon_n\}$ is the standard basis of $\mathbb{R}^n$.

2) Since $\bigwedge^k W$ is cyclic (see Exercise 3), it has a generator $w_1 \wedge \cdots \wedge w_k$ that is unique up to sign, and we define
$$d(W) = \|w_1 \wedge \cdots \wedge w_k\|.$$
(However, by convention, we let $d(\{0\}) = 1$.)

(7.5.3) *Remark.* If W is the cyclic group generated by a nonzero vector $w \in \mathbb{R}^n$, then it is obvious that $d(W) = \|w\|$. Therefore, Definition 7.5.2(2) presents a notion that generalizes the norm of a vector.

The following generalization of Exercise 7.4#14 is straightforward to prove (see Exercise 5).

(7.5.4) Lemma. *Suppose*

- *$\{u^t\}$ is a one-parameter unipotent subgroup of* SL($n, \mathbb{R}$)*, and*
- *W is a discrete subgroup of $\mathbb{R}^n$.*

Then $d(u^t W)^2$ is a polynomial function of t, and the degree of this polynomial is bounded by a constant D that depends only on n.

Lemma 7.5.4 allows us to make good use of the following two basic properties of polynomials of bounded degree. (See Exercises 6 and 7 for the proofs.) The first follows from the observation that polynomials of bounded degree form a finite-dimensional real vector space, so any closed, bounded subset is compact. The second uses the fact that nonzero polynomials of degree D cannot have more than D zeroes.

(7.5.5) Lemma. *Suppose $D \in \mathbb{Z}^+$, $\epsilon > 0$, and f is any real polynomial of degree $\leq D$. Then there exists $C > 1$, depending only on D and ϵ, such that, for all $T, \tau > 0$:*

1) *If $f(s) \geq \tau$ for some $s \in [0, T]$, and $|f(T)| \leq \tau/C$, then there exists $t \in [0, \epsilon T]$, such that $|f(T + t)| = \tau/C$.*

2) *If $|f(s)| \leq \tau$ for all $s \in [0, T]$, and $f(T) = \tau$, then there exists $T_1 \in [T, 4^D T]$, such that*
$$\tau/C \leq |f(t)| \leq \tau C \quad \text{for all } t \in [T_1, 2T_1].$$

(7.5.6) Notation. Suppose $\mathcal{L}$ is a $\mathbb{Z}$-lattice in $\mathbb{R}^n$.

- A subgroup W of $\mathcal{L}$ is **full** if it is the intersection of $\mathcal{L}$ with a vector subspace of $\mathbb{R}^n$. (This is equivalent to requiring $\mathcal{L}/W$ to be torsion-free.)

- Let $\mathcal{S}(\mathcal{L})$ be the collection of all full, nontrivial subgroups of $\mathcal{L}$, partially ordered by inclusion.

- For $W \subseteq \mathcal{L}$, we let $\langle W \rangle_{\mathcal{L}}$ be the (unique) smallest full subgroup of $\mathcal{L}$ that contains W. In other words, $\langle W \rangle_{\mathcal{L}} = \langle W \rangle \cap \mathcal{L}$.

The following simple observation uses full subgroups of $\mathcal{L}$ to provide a crucial lower bound on the norms of vectors (see Exercise 8):

(7.5.7) Lemma. *If $W \in \mathcal{S}(\mathcal{L})$ and $v \in \mathcal{L} \smallsetminus W$, then $\|v\| \geq \dfrac{d(\langle W, v \rangle_{\mathcal{L}})}{d(\langle W \rangle_{\mathcal{L}})}$.*

We can now prove Theorem 7.5.1. However, to avoid the need for a proof by induction, we assume $n = 3$.

Proof of Theorem 7.5.1 when $n = 3$. It is easy to see that
$$\{ W \in \mathcal{S}(\mathcal{L}) \mid d(W) < 1 \} \text{ is finite}$$

(see Exercise 10). Hence, there exists $\tau > 0$, such that
$$d(W) > \tau, \text{ for all } W \in \mathcal{S}(\mathcal{L}).$$
Let:

- D be the constant provided by Lemma 7.5.4,
- $\epsilon = 4^{-(D+1)}$, and
- C be the constant provided by Lemma 7.5.5.

Given $T > 0$, it suffices to find $R \geq 0$, such that $\|u^{T+R}v\| \geq \tau/C^2$ for all nonzero $v \in \mathcal{L}$.

Let
$$\mathcal{D} = \{ W \in \mathcal{S}(\mathcal{L}) \mid d(u^T W) < \tau/C \}.$$
We assume $\mathcal{D} \neq \varnothing$ (otherwise, we could let $R = 0$). For each $W \in \mathcal{D}$, Lemma 7.5.5(1) implies

there exists $t_W \in [0, \epsilon T]$, such that $d(u^{T+t_W}W) = \tau/C$.

By choosing t_W minimal, we may assume
$$d(u^{T+t}W) < \tau/C \text{ for all } t \in [0, t_W).$$
Since $\mathcal{D}$ is finite (see Exercise 10), we may

fix some $W^+ \in \mathcal{D}$ that maximizes t_W.

From Lemma 7.5.5(2), we see that there exists
$$T_1 \in [t_{W^+}, 4^D t_{W^+}] \subseteq \left[t_{W^+}, \frac{T}{2} \right],$$
such that
$$\tau/C^2 \leq d(u^{T+t}W^+) \leq \tau C^2 \text{ for all } t \in [T_1, 2T_1].$$

Since $\dim\langle \mathcal{L} \rangle = n = 3$, we know $\dim\langle W^+ \rangle$ is either 1 or 2. To be concrete, let us assume it is 2. (See Exercise 11 for the other case.) Then, for any $v \in \mathcal{L} \setminus W^+$, we have $\langle W^+, v \rangle_{\mathcal{L}} = \mathcal{L}$, so Lemma 7.5.7 implies $\|u^{T+t}v\| \geq 1/\tau$ for all $t \in [T_1, 2T_1]$. Hence, it is only the vectors in W^+ that can be small anywhere in this interval.

Therefore, we may assume there is some nonzero $v_0 \in W^+$, such that $\|u^{T+T_1}v_0\| < \tau/C^2$. There is no harm in assuming that $\mathbb{Z}v_0$ is a full subgroup of $\mathcal{L}$. Then, since $T_1 \geq t_{W^+}$, the maximality of t_{W^+} implies $\|u^{T+s}v_0\| \geq \tau/C$ for some $s \in [0, T_1]$. Therefore, Lemma 7.5.5(1) provides some $t \in [T_1, 2T_1]$, such that $\|u^{T+t}v_0\| = \tau/C^2$. Now, for any nonzero $v \in \mathcal{L}$,

either $\langle v \rangle_{\mathcal{L}} = \langle v_0 \rangle_{\mathcal{L}}$, or $\langle v_0, v \rangle_{\mathcal{L}} = W^+$, or $\langle v, W^+ \rangle_{\mathcal{L}} = \mathcal{L}$.

In each case, we see (by using Lemma 7.5.7 in the latter two cases) that $\|u^{T+t}v\| \geq \tau/C^2$ (if $\tau \leq 1$). $\qquad \square$

Exercises for §7.5.

#1. Show that Theorem 7.5.1 is a corollary of Theorem 7.4.10.

#2. Use Theorem 7.5.1 (and *not* Theorem 7.4.7 or Theorem 7.4.10) to show that if $\{u^t\}$, X, and x are as in Theorem 7.4.7, then there exists a compact subset K of X, such that

$$\left\{ t \in \mathbb{R}^+ \mid u^t x \in K \right\} \text{ is unbounded.}$$

#3. In the notation of Definition 7.5.2, show $\bigwedge^k W$ is cyclic.

[*Hint:* If $\{w_1, \ldots, w_k\}$ generates W, then $\bigwedge^k W$ is generated by $w_1 \wedge w_2 \wedge \cdots \wedge w_k$.]

#4. Suppose
  - W is (nontrivial) discrete subgroup of $\mathbb{R}^n$, and
  - $M \in \mathrm{SO}(n)$.

Show $d(MW) = d(W)$.

#5. Prove Lemma 7.5.4.

#6. Prove Lemma 7.5.5(1).

[*Hint:* Since rescaling does not change the degree of a polynomial, we may assume $T = \tau = 1$. If C does not exist, then taking a limit results in a polynomial of degree $\leq D$ that is 1 at some point of $[0,1]$, but vanishes on all of $[1, 1+\epsilon]$.]

#7. Prove Lemma 7.5.5(2).

[*Hint:* Assume, without loss of generality, that $T = \tau = 1$. The polynomials of degree $\leq D$ that are ≤ 1 on $[0,1]$ form a compact set, so they are uniformly bounded by some constant on $[1, 4^{D+1}]$. For $T_1 \in \{1, 4, \ldots, 4^D\}$, the intervals $[T_1, 2T_2]$ are pairwise disjoint. If f is not bounded away from 0 on any of these intervals, then taking a limit results in a nonzero polynomial of degree $\leq D$ that vanishes at $D + 1$ distinct points.]

#8. Prove Lemma 7.5.7.

[*Hint:* This is easy if W is generated by scalar multiples of the standard basis vectors of $\mathbb{R}^k$, and $v \in \mathbb{R}^{k+1}$.]

#9. Show that if $\mathcal{L}$ is a discrete subgroup of $\mathbb{R}^n$, and $1 \leq k \leq n$, then $\bigwedge^k \mathcal{L}$ is a discrete subset of $\bigwedge^k \mathbb{R}^n$.

[*Hint:* By choosing an appropriate basis, you can assume $\mathcal{L} \subseteq \mathbb{Z}^n$.]

#10. Assume
  - $\mathcal{L}$ is a $\mathbb{Z}$-lattice in $\mathbb{R}^n$, and
  - $\delta > 0$.

Show there are only finitely many full subgroups of $\mathcal{L}$, such that $d(W) < \delta$.

[*Hint:* Exercise 9. (If W_1 and W_2 are two different k-dimensional subspaces of $\mathbb{R}^n$, then $\bigwedge^k W_1 \neq \bigwedge^k W_2$.)]

#11. Complete the proof of Theorem 7.5.1 in the special case where $\dim\langle W^+ \rangle = 1$ (and $n = 3$).

[*Hint:* If there exist $v \in \mathcal{L} \smallsetminus W^+$ and $t \in [T_1, 2T_1]$, such that $\|u^{T+t}v\| < 1/C$, then $d(u^{T+R}\langle W^+, v\rangle_{\mathcal{L}}) = \tau/C$ for some $R \in [T_1, 2T_1]$.]

Notes

See [1, §1] or [7, §4.2] for more information on Siegel sets in SL(n, $\mathbb{R}$), and the proof of Theorem 7.0.1 that appears in Sections 7.2 and 7.3.

A brief discussion of the connection with the reduction theory of positive-definite quadratic forms can be found in [1, §2, pp. 20–24].

See [7, Prop. 3.12, p. 129] for a proof of Theorem 7.1.1. A generalization to other semisimple groups will be stated in Theorem 8.4.9.

The clever proof in Section 7.4 is by G. A. Margulis [6, Rem. 3.12(II)].

Theorem 7.4.6 is due to G. A. Margulis [5]. (Section 7.5 is adapted from the nice exposition in [3, Appendix, pp. 162–173], where all details can be found.) The result had been announced previously (without proof), and J. Tits [8, p. 59] commented that:

> "For a couple of years, Margulis' proof remained unpublished and every attempt by other specialists to supply it failed. When it finally appeared ..., the proof came as a great surprise, both for being rather short and using no sophisticated technique: it can be read without any special knowledge and gives a good idea of the extraordinary inventiveness shown by Margulis throughout his work."

The quantitative version stated in Theorem 7.4.7 is due to S. G. Dani [2]. See [4] for a recent generalization, and applications to number theory.

References

[1] A. Borel: *Introduction aux Groupes Arithmétiques.* Hermann, Paris, 1969. MR 0244260

[2] S. G. Dani: On invariant measures, minimal sets and a lemma of Margulis, *Invent. Math.* 51 (1979), no. 3, 239–260. MR 0530631, http://eudml.org/doc/142633

[3] S. G. Dani and G. A. Margulis: Values of quadratic forms at integral points: an elementary approach, *Enseign. Math.* (2) 36 (1990), no. 1-2, 143–174. MR 1071418, http://dx.doi.org/10.5169/seals-57906

[4] D. Kleinbock: An extension of quantitative nondivergence and applications to Diophantine exponents, *Trans. Amer. Math. Soc.* 360 (2008), no. 12, 6497–6523. MR 2434296, http://dx.doi.org/10.1090/S0002-9947-08-04592-3

[5] G. A. Margulis: On the action of unipotent groups in the space of lattices, in I. M. Gel'fand, ed.: *Lie Groups and Their Representations (Budapest, 1971).* Wiley (Halsted Press), New York, 1975, pp. 365–370. ISBN 0-470-29600-3, MR 0470140

[6] G. A. Margulis: Lie groups and ergodic theory, in L. L. Avramov and K. B. Tchakerian, eds.: *Algebra—Some Current Trends (Varna, 1986)*. Springer, New York, 1988, pp. 130–146. ISBN 3-540-50371-4, MR 0981823

[7] V. Platonov and A. Rapinchuk: *Algebraic Groups and Number Theory*. Academic Press, Boston, 1994. ISBN 0-12-558180-7, MR 1278263

[8] J. Tits: The work of Gregori Aleksandrovitch Margulis, in *Proceedings of the International Congress of Mathematicians (Helsinki, 1978)*. Acad. Sci. Fennica, Helsinki, 1980, vol. 1, pp. 57–63. MR 0562596, http://www.mathunion.org/ICM/ICM1978.1

Part III

Important Concepts

Chapter 8

Real Rank

§8.1. ℝ-split tori and ℝ-rank

(8.1.1) **Definition.** A closed, connected subgroup T of G is a ***torus*** if it is diagonalizable over $\mathbb{C}$; that is, if there exists $g \in \mathrm{GL}(n, \mathbb{C})$, such that gTg^{-1} consists entirely of diagonal matrices. A torus is ***ℝ-split*** if it is diagonalizable over $\mathbb{R}$; that is, if g may be chosen to be in $\mathrm{GL}(n, \mathbb{R})$.

(8.1.2) **Examples.**

1) Let A be the identity component of the group of diagonal matrices in $\mathrm{SL}(n, \mathbb{R})$. Then A is obviously an ℝ-split torus.

2) $\mathrm{SO}(1, 1)^\circ$ is an ℝ-split torus in $\mathrm{SL}(2, \mathbb{R})$ (see Exercise 1).

3) $\mathrm{SO}(2)$ is a torus in $\mathrm{SL}(2, \mathbb{R})$ that is *not* ℝ-split. It is diagonalizable over $\mathbb{C}$ (see Exercise 2), but not over $\mathbb{R}$ (see Exercise 3).

(8.1.3) **Warning.** An ℝ-split torus is *never* homeomorphic to the topologist's torus $\mathbb{T}^n$ (except in the trivial case $n = 0$).

(8.1.4) *Remarks.*

1) If T is an ℝ-split torus, then every element of T is hyperbolic (see Definition A5.1). In particular, no nonidentity element of T is elliptic or unipotent.

Recall: The Standing Assumptions (4.0.0 on page 41) are in effect, so, as always, Γ is a lattice in the semisimple Lie group $G \subseteq \mathrm{SL}(\ell, \mathbb{R})$.

Main prerequisites for this chapter: none.

2) When G is compact, every torus in G is isomorphic to $SO(2)^n$, for some n. This is homeomorphic to $\mathbb{T}^n$, which is the reason for the terminology "torus."

It is a key fact in the theory of semisimple Lie groups that *maximal* $\mathbb{R}$-split tori are conjugate:

(8.1.5) Theorem. *If A_1 and A_2 are maximal $\mathbb{R}$-split tori in G, then there exists $g \in G$, such that $A_1 = gA_2g^{-1}$.*

This implies that all maximal $\mathbb{R}$-split tori have the same dimension, which is called the "real rank" (or "$\mathbb{R}$-rank") of G, and is denoted $\operatorname{rank}_{\mathbb{R}} G$:

(8.1.6) Definition. $\operatorname{rank}_{\mathbb{R}} G$ is the dimension of a maximal $\mathbb{R}$-split torus A in G. This is independent of both the choice of A and the choice of the embedding of G in $SL(\ell, \mathbb{R})$.

(8.1.7) Examples.

1) $\operatorname{rank}_{\mathbb{R}}(SL(n, \mathbb{R})) = n - 1$. (Let A be the identity component of the group of all diagonal matrices in $SL(n, \mathbb{R})$.)

2) We have $\operatorname{rank}_{\mathbb{R}}(SL(n, \mathbb{C})) = \operatorname{rank}_{\mathbb{R}}(SL(n, \mathbb{H})) = n - 1$. This is because only the *real* diagonal matrices remain diagonal when $SL(n, \mathbb{C})$ or $SL(n, \mathbb{H})$ is embedded in $SL(2n, \mathbb{R})$ or $SL(4n, \mathbb{R})$, respectively.

3) $\operatorname{rank}_{\mathbb{R}} G = 0$ if and only if G is compact (see Exercise 9).

(8.1.8) Proposition. $\operatorname{rank}_{\mathbb{R}} SO(m, n) = \min\{m, n\}$.

Proof. Since $SO(m, n)$ contains a copy of $SO(1, 1)^{\min\{m,n\}}$ (see Exercise 7), and the identity component of this subgroup is an $\mathbb{R}$-split torus (cf. Exercise 1), we have

$$\operatorname{rank}_{\mathbb{R}} SO(m, n) \geq \dim(SO(1, 1)^{\min\{m,n\}})^\circ = \min\{m, n\}.$$

We now establish the reverse inequality. Let A be a maximal $\mathbb{R}$-split torus. We may assume A is nontrivial. (Otherwise $\operatorname{rank}_{\mathbb{R}} SO(m, n) = 0$, so the desired inequality is obvious.) Therefore, there is some nontrivial $a \in A$. Since a is diagonalizable over $\mathbb{R}$, and nontrivial, there is an eigenvector v of a, such that $av \neq v$; hence, $av = \lambda v$ for some $\lambda \neq 1$. Now, if we let $\langle \cdot \mid \cdot \rangle_{m,n}$ be an $SO(m, n)$-invariant bilinear form on $\mathbb{R}^{m,n}$, we have

$$\langle v \mid v \rangle_{m,n} = \langle av \mid av \rangle_{m,n} = \langle \lambda v \mid \lambda v \rangle_{m,n} = \lambda^2 \langle v \mid v \rangle_{m,n}.$$

By choosing a to be near e, we may assume $\lambda \approx 1$, so $\lambda \neq -1$. Since, by assumption, we know $\lambda \neq 1$, this implies $\lambda^2 \neq 1$. So we must have $\langle v \mid v \rangle_{m,n} = 0$; that is, v is an **isotropic** vector. Hence, we have shown that if the real rank is ≥ 1, then there is an isotropic vector in $\mathbb{R}^{m+n}$.

By arguing more carefully, it is not difficult to see that if the real rank is at least k, then there is a k-dimensional subspace of $\mathbb{R}^{m+n}$ that consists entirely of isotropic vectors (see Exercise 10). Such a subspace is said to be **totally isotropic**. The maximum dimension of a totally isotropic subspace is $\min\{m, n\}$ (see Exercise 11), so we conclude that $\min\{m, n\} \geq \text{rank}_\mathbb{R} \text{SO}(m, n)$, as desired. □

(8.1.9) *Remarks.*

1) Other classical groups, not just $\text{SO}(m, n)$, have the property that their real rank is the maximal dimension of a totally isotropic subspace. More concretely, we have

$$\text{rank}_\mathbb{R} \text{SU}(m, n) = \text{rank}_\mathbb{R} \text{Sp}(m, n) = \min\{m, n\}.$$

2) The Mostow Rigidity Theorem (15.1.1) will tell us that if Γ is (isomorphic to) a lattice in both G and G_1, then G° is isomorphic to G_1', modulo compact groups. Modding out a compact subgroup does not affect the real rank (cf. Exercise 9), so this implies that the real rank of G is uniquely determined by the algebraic structure of Γ.

3) Although it is not usually very useful in practice, we now state an explicit relationship between Γ and $\text{rank}_\mathbb{R} G$. Let S_r be the set of all elements y of Γ, such that the centralizer $C_\Gamma(y)$ is commensurable to a subgroup of the free abelian group $\mathbb{Z}^r$ of rank r. Then it can be shown that

$$\text{rank}_\mathbb{R} G = \min \left\{ r \geq 0 \;\middle|\; \begin{array}{l} \Gamma \text{ is covered by finitely} \\ \text{many translates of } S_r \end{array} \right\}.$$

We omit the proof, which is based on the very useful (and nontrivial) fact that if T is any maximal torus of G, then there exists $g \in G$, such that $gTg^{-1}/(\Gamma \cap gTg^{-1})$ is compact.

Exercises for §8.1.

#1. Show that the identity component of $\text{SO}(1, 1)$ is an ℝ-split torus.

[*Hint:* Let $g = \begin{bmatrix} 1 & 1 \\ 1 & -1 \end{bmatrix}$. Alternatively, note that each element of $\text{SO}(1, 1)$ is a symmetric matrix (hence, diagonalizable via an orthogonal matrix), and use the fact that any set of commuting diagonalizable matrices is simultaneously diagonalizable.]

#2. For $g = \begin{bmatrix} 1 & -i \\ 1 & i \end{bmatrix}$, show every element of $g \, \text{SO}(2) g^{-1}$ is diagonal.

#3. Show that $\text{SO}(2)$ is not diagonalizable over ℝ.

[*Hint:* If T is diagonalizable over ℝ, then eigenvalues of the elements of T are real.]

#4. Show that every ℝ-split torus is abelian.

#5. Suppose
- T is an ℝ-split torus in G, and
- A is a maximal ℝ-split torus in G.

Show that T is conjugate to a subgroup of A.

[*Hint:* By considering dimension, it is obvious that T is contained in some maximal ℝ-split torus of G.]

#6. Show that every maximal $\mathbb{R}$-split torus in G is almost Zariski closed.

#7. Assume $m \geq n$. Then $m + n \geq 2n$, so there is a natural embedding of $SO(1,1)^n$ in $SL(m + n, \mathbb{R})$. Show that $SO(m,n)$ contains a conjugate of this copy of $SO(1,1)^n$.

   [*Hint:* Permute the basis vectors.]

#8. Prove, directly from Definition 8.1.1, that if G_1 is conjugate to G_2 in $GL(\ell, \mathbb{R})$, then $\text{rank}_{\mathbb{R}}(G_1) = \text{rank}_{\mathbb{R}}(G_2)$.

#9. Show $\text{rank}_{\mathbb{R}} G = 0$ if and only if G is compact.

   [*Hint:* Remarks A5.2 and A2.6(2).]

#10. Show that if $\text{rank}_{\mathbb{R}} SO(m,n) = r$, then there is an r-dimensional subspace V of $\mathbb{R}^{m+n}$, such that $\langle v \mid w \rangle_{m,n} = 0$ for all $v, w \in V$.

   [*Hint:* Because A is diagonalizable over $\mathbb{R}$, there is a basis $\{v_1, \ldots, v_{m+n}\}$ of $\mathbb{R}^{m+n}$ whose elements are eigenvectors for every element of A. Since $\dim A = r$, we may assume, after renumbering, that for all $\lambda_1, \ldots, \lambda_r \in \mathbb{R}^+$, there exists $a \in A$, such that $av_i = \lambda_i v_i$, for $1 \leq i \leq r$. This implies $\langle v_1, \ldots, v_r \rangle$ is totally isotropic.]

#11. Show that if V is a subspace of $\mathbb{R}^{m+n}$ that is totally isotropic for $\langle \cdot \mid \cdot \rangle_{m,n}$, then $\dim V \leq \min\{m, n\}$.

   [*Hint:* If $v \neq 0$ and the last n coordinates of v are 0, then $\langle v \mid v \rangle_{m,n} > 0$.]

#12. Show $\text{rank}_{\mathbb{R}}(G_1 \times G_2) = \text{rank}_{\mathbb{R}} G_1 + \text{rank}_{\mathbb{R}} G_2$.

#13. Show $\text{rank}_{\mathbb{R}} G \geq 1$ if and only if G contains a subgroup that is isogenous to $SL(2, \mathbb{R})$.

   [*Hint:* Remark A2.6.]

#14. Show that Γ contains a subgroup that is isomorphic to $\mathbb{Z}^r$, where $r = \text{rank}_{\mathbb{R}} G$.

   [*Hint:* You may assume the fact stated in the last sentence of Remark 8.1.9(3).]

§8.2. Groups of higher real rank

In some situations, there is a certain subset S of G, such that the centralizer of each element of S is well-behaved, and it would be helpful to know that these centralizers generate G. The results in this section illustrate that an assumption on the real rank of G may be exactly what is needed. (However, we will often only prove the special case where $G = SL(3, \mathbb{R})$. A reader familiar with the theory of "real roots" should have no difficulty generalizing the arguments.)

(8.2.1) **Proposition.** *Let A be a maximal $\mathbb{R}$-split torus in G. Then we have $\text{rank}_{\mathbb{R}} G \geq 2$ if and only if there exist nontrivial elements a_1 and a_2 of A, such that $G = \langle C_G(a_1), C_G(a_2) \rangle$.*

Proof. ($\Rightarrow$) Assume, for simplicity, that $G = \mathrm{SL}(3, \mathbb{R})$. (See Exercise 1(a) for another special case.) Then we may assume A is the group of diagonal matrices (after replacing it by a conjugate). Let

$$a_1 = \begin{bmatrix} 2 & 0 & 0 \\ 0 & 2 & 0 \\ 0 & 0 & 1/4 \end{bmatrix} \quad \text{and} \quad a_2 = \begin{bmatrix} 1/4 & 0 & 0 \\ 0 & 2 & 0 \\ 0 & 0 & 2 \end{bmatrix}.$$

Then

$$C_G(a_1) = \begin{bmatrix} * & * & 0 \\ * & * & 0 \\ 0 & 0 & * \end{bmatrix} \quad \text{and} \quad C_G(a_2) = \begin{bmatrix} * & 0 & 0 \\ 0 & * & * \\ 0 & * & * \end{bmatrix}.$$

These generate G.

($\Leftarrow$) Suppose $\mathrm{rank}_{\mathbb{R}}\, G = 1$, so $\dim A = 1$. Then, since A is almost Zariski closed (and contains $\langle a_1 \rangle$), we have $C_G(a_1) = C_G(A) = C_G(a_2)$, so

$$\langle C_G(a_1), C_G(a_2) \rangle = C_G(A).$$

It is obvious that $C_G(A) \neq G$ (because the center of G is finite, and therefore cannot contain the infinite group A). □

The following explicit description of $C_G(A)$ will be used in some of the proofs.

(8.2.2) Lemma. *If A is any maximal $\mathbb{R}$-split torus in G, then $C_G(A) = A \times C$, where C is compact.*

Proof. (optional) A subgroup of $\mathrm{SL}(\ell, \mathbb{R})$ is said to be **reductive** if it is isogenous to $M \times T$, where M is semisimple and T is a torus. It is known that the centralizer of any torus is reductive (see Exercise 2), so, if we assume, for simplicity, that $C_G(A)$ is connected, then we may write $C_G(A) = M \times A$, where M is reductive (see Exercise 3). The maximality of A implies that M does not contain any $\mathbb{R}$-split tori, so M is compact (see Exercise 8.1#9). □

(8.2.3) Proposition (see Exercise 4). $\mathrm{rank}_{\mathbb{R}}\, G \geq 2$ *if and only if there exist a nontrivial hyperbolic element a and a nontrivial unipotent element u, such that $au = ua$.*

For use in the proof of the proposition that follows it, we mention a very useful characterization of a somewhat different flavor:

(8.2.4) Lemma (see Exercise 5). $\mathrm{rank}_{\mathbb{R}}\, G \leq 1$ *if and only if every nontrivial unipotent subgroup of G is contained in a **unique** maximal unipotent subgroup.*

(8.2.5) Proposition. $\mathrm{rank}_{\mathbb{R}}\, G \geq 2$ *if and only if there exist nontrivial unipotent subgroups $U_1, \ldots, U_k$, such that*

- $\langle U_1, \ldots, U_k \rangle = G$, *and*
- U_i *centralizes U_{i+1} for each i.*

Proof. ($\Rightarrow$) Assume, for simplicity, that $G = \mathrm{SL}(3, \mathbb{R})$. Then we take the sequence

$$\begin{bmatrix} 1 & * & 0 \\ 0 & 1 & 0 \\ 0 & 0 & 1 \end{bmatrix}, \begin{bmatrix} 1 & 0 & * \\ 0 & 1 & 0 \\ 0 & 0 & 1 \end{bmatrix}, \begin{bmatrix} 1 & 0 & 0 \\ 0 & 1 & * \\ 0 & 0 & 1 \end{bmatrix}, \begin{bmatrix} 1 & 0 & 0 \\ * & 1 & 0 \\ 0 & 0 & 1 \end{bmatrix}, \begin{bmatrix} 1 & 0 & 0 \\ 0 & 1 & 0 \\ * & 0 & 1 \end{bmatrix}, \begin{bmatrix} 1 & 0 & 0 \\ 0 & 1 & 0 \\ 0 & * & 1 \end{bmatrix}.$$

($\Leftarrow$) Since U_i commutes with U_{i+1}, we know that $\langle U_i, U_{i+1} \rangle$ is unipotent, so, if $\mathrm{rank}_{\mathbb{R}}\, G = 1$, then it is contained in a *unique* maximal unipotent subgroup $\overline{U}_i$ of G. Since $\overline{U}_i$ and $\overline{U}_{i+1}$ both contain U_{i+1}, we conclude that $\overline{U}_i = \overline{U}_{i+1}$ for all i. Hence, $\langle U_1, \ldots, U_k \rangle$ is contained in the unipotent group $\overline{U}_1$, and is therefore not all of G.   □

(8.2.6) *Remark.* See Lemma 16.5.7 for yet another result of the same type, which will be used in the proof of the Margulis Superrigidity Theorem in Section 16.5. A quite different characterization, based on the existence of subgroups of the form $\mathrm{SL}(2, \mathbb{R}) \ltimes \mathbb{R}^n$, appears in Exercise 13.2#2, and is used in proving Kazhdan's Property (T) in Chapter 13.

We know that $\mathrm{SL}(2, \mathbb{R})$ is the smallest group of real rank one (see Exercise 8.1#13). However, the smallest group of real rank two is not unique:

(8.2.7) **Proposition.** *Assume G is simple. Then $\mathrm{rank}_{\mathbb{R}}\, G \geq 2$ if and only if G contains a subgroup that is isogenous to either $\mathrm{SL}(3, \mathbb{R})$ or $\mathrm{SO}(2, 3)$.*

Exercises for §8.2.

#1. Prove the following results in the special case where $G = G_1 \times G_2$, and $\mathrm{rank}_{\mathbb{R}}\, G_i \geq 1$ for each i.
   a) Proposition 8.2.1($\Rightarrow$)
   b) Proposition 8.2.3($\Rightarrow$)
   c) Lemma 8.2.4($\Leftarrow$)
   d) Proposition 8.2.5($\Rightarrow$)

#2. (optional) It is known that if M is a subgroup that is almost Zariski closed, and $M^T = M$, then M is reductive (cf. Corollary A7.8). Assuming this, show that if T is a subgroup of the group of diagonal matrices, and $G^T = G$, then $C_G(T)$ is reductive.

#3. (optional) Suppose M is reductive, and A is an $\mathbb{R}$-split torus in the center of M. Show there exists a reductive subgroup L of M°, such that $M^\circ = L \times A$.
   [*Hint:* Up to isogeny, write $M = M_0 \times T$, with $A \subseteq T$. Then it suffices to show $T = E \times A$ for some E. You may assume, without proof, that, since T is a connected, abelian Lie group, it is isomorphic to $\mathbb{R}^m \times \mathbb{T}^n$ for some m and n.]

#4. a) Prove Proposition 8.2.3($\Rightarrow$) under the additional assumption that $G = \mathrm{SL}(3, \mathbb{R})$.
   b) Prove Proposition 8.2.3($\Leftarrow$).

#5. Find a nontrivial unipotent subgroup of $SL(3, \mathbb{R})$ that is contained in two different maximal unipotent subgroups.

#6. (*Assumes the theory of real roots*) Prove the general case of the following results.
   a) Lemma 8.2.2
   b) Proposition 8.2.1($\Rightarrow$)
   c) Proposition 8.2.3($\Rightarrow$)
   d) Lemma 8.2.4
   e) Proposition 8.2.5($\Rightarrow$)

#7. Show (without assuming G is simple): $\mathrm{rank}_\mathbb{R} G \geq 2$ if and only if G contains a subgroup that is isogenous to either $SL(3, \mathbb{R})$ or $SL(2, \mathbb{R}) \times SL(2, \mathbb{R})$.
   [*Hint:* Proposition 8.2.7. You may assume, without proof, that $SO(2, 2)$ is isogenous to $SL(2, \mathbb{R}) \times SL(2, \mathbb{R})$.]

§8.3. Groups of real rank one

As a complement to Section 8.2, here is an explicit list of the simple groups of real rank one.

(8.3.1) **Theorem.** *If G is simple, and $\mathrm{rank}_\mathbb{R} G = 1$, then G is isogenous to either*

- $SO(1, n)$ *for some $n \geq 2$,*
- $SU(1, n)$ *for some $n \geq 2$,*
- $Sp(1, n)$ *for some $n \geq 2$, or*
- F_4^{-20} *(also known as $F_{4,1}$), a certain exceptional group.*

(8.3.2) *Remark.* The special linear groups $SL(2, \mathbb{R})$, $SL(2, \mathbb{C})$ and $SL(2, \mathbb{H})$ have real rank one, but they are already on the list under different names, because

1) $SL(2, \mathbb{R})$ is isogenous to $SO(1, 2)$ and $SU(1, 1)$,
2) $SL(2, \mathbb{C})$ is isogenous to $SO(1, 3)$ and $Sp(1, 1)$, and
3) $SL(2, \mathbb{H})$ is isogenous to $SO(1, 4)$.

(8.3.3) *Remark.* Each of the simple groups of real rank one has a very important geometric realization. Namely, $SO(1, n)$, $SU(1, n)$, $Sp(1, n)$, and $F_{4,1}$ (respectively) are isogenous to the isometry groups of:

1) (real) hyperbolic n-space $\mathfrak{H}^n$,
2) *complex hyperbolic n-space* $\mathbb{C}\mathfrak{H}^n$,
3) *quaternionic hyperbolic n-space* $\mathbb{H}\mathfrak{H}^n$, and
4) the *Cayley plane*, which can be thought of as the hyperbolic plane over the (nonassociative) ring $\mathbb{O}$ of Cayley octonions.

§8.4. Minimal parabolic subgroups

The group of upper-triangular matrices plays a very important role in the study of $SL(n, \mathbb{R})$. In this section, we introduce subgroups that play the same role in other semisimple Lie groups:

(8.4.1) **Definition.** Let A be a maximal $\mathbb{R}$-split torus of G, and let a be a **generic** element of A, by which we mean that $C_G(a) = C_G(A)$. Then the corresponding **minimal parabolic subgroup** of G is

$$P = \left\{ g \in G \;\middle|\; \limsup_{n \to \infty} \|a^{-n} g a^n\| < \infty \right\}. \tag{8.4.2}$$

This is a Zariski closed subgroup of G.

(8.4.3) **Theorem.** *All minimal parabolic subgroups of G are conjugate.*

(8.4.4) **Examples.**

1) The group of upper triangular matrices is a minimal parabolic subgroup of $SL(n, \mathbb{R})$. To see this, let A be the group of diagonal matrices, and choose $a \in A$ with $a_{1,1} > a_{2,2} > \cdots > a_{n,n} > 0$ (see Exercise 1).

2) It is easier to describe a minimal parabolic subgroup of $SO(1, n)$ if we replace $\mathrm{Id}_{m,n}$ with a different symmetric matrix of the same signature: let $G = SO(A; \mathbb{R})$, for

$$A = \begin{pmatrix} 0 & 0 & 1 \\ 0 & \mathrm{Id}_{(n-1) \times (n-1)} & 0 \\ 1 & 0 & 0 \end{pmatrix}.$$

Then G is conjugate to $SO(1, n)$ (see Exercise 4), the (1-dimensional) group of diagonal matrices in G form a maximal $\mathbb{R}$-split torus, and a minimal parabolic subgroup in G is

$$\left\{ \begin{pmatrix} t & * & * \\ 0 & SO(n-1) & * \\ 0 & 0 & 1/t \end{pmatrix} \right\}$$

(see Exercise 2).

The following result explains that a minimal parabolic subgroup of a classical group is simply the stabilizer of a (certain kind of) flag. Recall that a subspace W of a vector space V, equipped with a bilinear (or Hermitian) form $\langle \cdot \mid \cdot \rangle$, is said to be **totally isotropic** if $\langle W \mid W \rangle = 0$.

(8.4.5) **Theorem** (see Exercise 3).

1) *A subgroup P of $SL(n, \mathbb{R})$ is a minimal parabolic if and only if there is a chain $V_0 \subsetneq V_1 \subsetneq \cdots \subsetneq V_n$ of subspaces of $\mathbb{R}^n$ (with $\dim V_i = i$), such that*

$$P = \{ g \in SL(n, \mathbb{R}) \mid \forall i,\ gV_i = V_i \}.$$

Similarly for SL$(n, \mathbb{C})$ *and* SL$(n, \mathbb{H})$, *taking chains of subspaces in* $\mathbb{C}^n$ *or* $\mathbb{H}^n$, *respectively.*

2) *A subgroup P of* SO(m, n) *is a minimal parabolic if and only if there is a chain* $V_0 \subsetneq V_1 \subsetneq \cdots \subsetneq V_r$ *of **totally isotropic** subspaces of* $\mathbb{R}^{m+n}$ *(with* dim $V_i = i$ *and* $r = \min\{m, n\}$), *such that*

$$P = \{ g \in SO(m, n) \mid \forall i, \ gV_i = V_i \}.$$

Similarly for SO$(n, \mathbb{C})$, SO$(n, \mathbb{H})$, Sp$(2m, \mathbb{R})$, Sp$(2m, \mathbb{C})$, SU(m, n) *and* Sp(m, n).

Note that any upper triangular matrix in SL$(n, \mathbb{R})$ can be written uniquely in the form mau, where

- a belongs to the $\mathbb{R}$-split torus A of diagonal matrices whose nonzero entries are positive,
- m is in the finite group M consisting of diagonal matrices whose nonzero entries are ± 1, and
- u belongs to the unipotent group N of upper triangular matrices with 1's on the diagonal.

The elements of every minimal parabolic subgroup have a decomposition of this form, except that the subgroup M may need to be compact, instead of only finite:

(8.4.6) **Theorem** (Langlands decomposition). *If P is a minimal parabolic subgroup of G, then we may write it in the form* $P = C_G(A) N = MAN$, *where*

- A *is a maximal* $\mathbb{R}$-*split torus,*
- M *is a compact subgroup of* $C_G(A)$, *and*
- N *is the unique maximal unipotent subgroup of P.*

Furthermore, N is a maximal unipotent subgroup of G, and, for some generic $a \in A$, *we have*

$$N = \left\{ u \in G \ \middle| \ \lim_{n \to \infty} a^{-n} u a^n = e \right\}. \tag{8.4.7}$$

Before discussing the proof (which is not so important for our purposes), let us consider a few examples:

(8.4.8) **Example.**
1) If $G = $ SL$(n, \mathbb{C})$, then, for the Langlands decomposition of the group P of upper-triangular matrices, we may let:
    - A be the group of diagonal matrices in G whose nonzero entries are positive real numbers (just as for SL$(n, \mathbb{R})$),
    - M be the group of diagonal matrices in G whose nonzero entries have absolute value 1, and
    - N be the group of upper triangular matrices with 1's on the diagonal.

The same description applies to $G = \mathrm{SL}(n, \mathbb{H})$ (and, actually, also to $\mathrm{SL}(n, \mathbb{R})$).

2) Assume $m \le m$, and let $G = \mathrm{SO}(A; \mathbb{R})$, where

$$A = \begin{bmatrix} 0 & 0 & J_m \\ 0 & \mathrm{Id}_{(n-m)\times(n-m)} & 0 \\ J_m & 0 & 0 \end{bmatrix} \quad \text{and} \quad J_m = \begin{bmatrix} & & & 1 \\ 0 & & 1 & \\ & \cdot^{\cdot^{\cdot}} & & \\ 1 & & 1 & 0 \end{bmatrix}$$

(and the size of the matrix J_m is $m \times m$). Then G is conjugate to $\mathrm{SO}(m, n)$ (see Exercise 4), and a minimal parabolic P of G is:

$$\left\{ \begin{pmatrix} b & * & * \\ 0 & k & * \\ 0 & 0 & b^\dagger \end{pmatrix} \;\middle|\; \begin{array}{c} b \in \mathrm{GL}(m, \mathbb{R}) \text{ is upper triangular,} \\ k \in \mathrm{SO}(n - m) \end{array} \right\}$$

where $x^\dagger = J_m (x^{-1})^T J_m$, so, for example,

$$\mathrm{diag}(b_1, \ldots, b_m)^\dagger = \mathrm{diag}(1/b_m, \ldots, 1/b_1).$$

Hence, we may let
- $A = \{ \mathrm{diag}(a_1, \ldots, a_m, 0, \ldots, 0, 1/a_m, \ldots, 1/a_1) \mid a_i > 0 \}$,
- $M \cong \mathrm{SO}(n - m) \times \{\pm 1\}^m$, and
- N be the group of upper triangular matrices with 1's on the diagonal that are in G.

Proof of Theorem 8.4.6 (optional). Choose a generic element a of A satisfying (8.4.2), and define N as in (8.4.7). Then, since a is diagonalizable over $\mathbb{R}$, it is not difficult to see that $P = C_G(a) N$ (see Exercise 5). Since a is a generic element of A, this means $P = C_G(A) N$.

It is easy to verify that N is normal in P (see Exercise 6); then, since $P/N \cong C_G(A) = A \times (\text{compact})$ (see Lemma 8.2.2), and therefore has no nontrivial unipotent elements, it is clear that N contains every unipotent element of P. Conversely, the definition of N implies that it is unipotent (see Exercise 7). Therefore, N is the unique maximal unipotent subgroup of P.

Suppose U is a unipotent subgroup of G that properly contains N. Since unipotent subgroups are nilpotent (see Exercise 9), then $\mathcal{N}_U(N)$ properly contains N (see Exercise 10). However, it can be shown that $\mathcal{N}_G(N) = P$ (see Exercise 8), so this implies $\mathcal{N}_U(N)$ is a unipotent subgroup of P that properly contains N, which contradicts the conclusion of the preceding paragraph. $\qquad\square$

The subgroups A and N that appear in the Langlands decomposition of P are two components of the Iwasawa decomposition of G:

(8.4.9) Theorem (Iwasawa decomposition). *Let*
- K *be a maximal compact subgroup of* G,
- A *be a maximal* $\mathbb{R}$*-split torus, and*

- *N be a maximal unipotent subgroup that is normalized by A.*

Then G = KAN.

*In fact, every $g \in G$ has a **unique** representation of the form $g = kau$ with $k \in K$, $a \in A$, and $u \in N$.*

(8.4.10) *Remark.* A stronger statement is true: if we define a function $\varphi \colon K \times A \times N \to G$ by $\varphi(k, a, u) = kau$, then φ is a (real analytic) diffeomorphism. Indeed, Theorem 8.4.9 tells us that φ is a bijection, and it is obviously real analytic. It is not so obvious that the inverse of φ is also real analytic, but this is proved in Exercise 7.1#3 when $G = \mathrm{SL}(n, \mathbb{R})$, and the general case can be obtained by choosing an embedding of G in $\mathrm{SL}(n, \mathbb{R})$ for which the subgroups K, A, and N of G are equal to the intersection of G with the corresponding subgroups of $\mathrm{SL}(n, \mathbb{R})$.

The Iwasawa decomposition implies $KP = G$ (since $AN \subseteq P$), so it has the following important consequence:

(8.4.11) **Corollary.** *If P is any minimal parabolic subgroup of G, then G/P is compact.*

(8.4.12) *Remark.* A subgroup of G is called **parabolic** if it contains a minimal parabolic subgroup.

1) Corollary 8.4.11 implies that if Q is any parabolic subgroup, then G/Q is compact. The converse does not hold. (For example, if $P = MAN$ is a minimal parabolic, then $G/(AN)$ is compact, but AN is not parabolic unless M is trivial.) However, passing to the "complexification" does yield the converse: Q is parabolic if and only if $G_{\mathbb{C}}/Q_{\mathbb{C}}$ is compact. Furthermore, Q is parabolic if and only if $Q_{\mathbb{C}}$ contains a maximal solvable subgroup ("**Borel subgroup**") of $G_{\mathbb{C}}$.

2) All parabolic subgroups can be described fairly completely (there are only finitely many that contain any given minimal parabolic), but we do not need the more general theory.

Exercises for §8.4.

#1. Let a be a diagonal matrix as described in Example 8.4.4(1), and show that the corresponding minimal parabolic subgroup is precisely the group of upper triangular matrices.

#2. Show that the subgroup at the end of Example 8.4.4(2) is indeed a minimal parabolic subgroup of $\mathrm{SO}(A; \mathbb{R})$.

#3. Show the minimal parabolic subgroups of each of the following groups are as described in Theorem 8.4.5:
   a) $\mathrm{SL}(n, \mathbb{R})$.
   b) $\mathrm{SO}(m, n)$.

[*Hint:* It suffices to find one minimal parabolic subgroup in order to understand all of them (see Theorem 8.4.3).]

#4. For A as in Example 8.4.8(2), show that $SO(A; \mathbb{R})$ is conjugate to $SO(m, n)$.

[*Hint:* Let $\alpha = 1/\sqrt{2}$, and define v_i to be: $\alpha(e_i + e_{n+1-i})$ for $i \le m$, e_i for $m < i \le n$, and $\alpha(e_i - e_{n+1-i})$ for $i > n$. Then $v_i^T A v_i$ is 1 for $i \le n$, and is -1 for $i > n$.]

#5. (optional) For P, a, and N as in the proof of Theorem 8.4.6, show $P = C_G(a) N$.

[*Hint:* Given $g \in P$, show that $a^{-n} g a^n$ converges to some element c of $C_G(a)$. Also show $c^{-1} g \in N$. You may assume a is diagonal, with $a_{11} \ge a_{22} \ge \cdots \ge a_{\ell\ell}$ (*why?*).]

#6. For P, a, and N as in the proof of Theorem 8.4.6, show N is a normal subgroup of P.

#7. Show that a subgroup N satisfying (8.4.7) must be unipotent.

[*Hint:* u has the same characteristic polynomial as $a^{-n} u a^n$.]

#8. For P and N as in Theorem 8.4.5(2), show $P = \mathcal{N}_G(N)$.

[*Hint:* P is the stabilizer of a certain flag, and the subgroup N also uniquely determines this same flag.]

#9. Show that every unipotent subgroup of $SL(\ell, \mathbb{R})$ is nilpotent. (Recall that a group N is **nilpotent** if there is a series
$$\{e\} = N_0 \lhd \cdots \lhd N_r = N$$
of subgroups of N, such that $[N, N_k] \subseteq N_{k-1}$ for each k.)

[*Hint:* Engel's Theorem (A5.7).]

#10. Show that if N is a proper subgroup of a nilpotent group U, then $\mathcal{N}_U(N) \nsubseteq N$.

[*Hint:* If $[N, U_k] \subseteq N$, then U_k normalizes N.]

#11. Assume K is a maximal compact subgroup of G. Show:

    a) G is diffeomorphic to the cartesian product $K \times \mathbb{R}^n$, for some n,

    b) G/K is diffeomorphic to $\mathbb{R}^n$, for some n,

    c) G is connected if and only if K is connected, and

    d) G is simply connected if and only if K is simply connected.

[*Hint:* Remark 8.4.10.]

Notes

The comprehensive treatise of Borel and Tits [1] is the standard reference on rank, parabolic subgroups, and other fundamental properties of reductive groups over any field. See [5, §7.7, pp. 474–487] for a discussion of parabolic subgroups of Lie groups (which is the special case in which the field is $\mathbb{R}$).

Remark 8.1.9(3) is due to Prasad-Raghunathan [7, Thms. 2.8 and 3.9].

Proofs of the Iwasawa decomposition for both SL$(n, \mathbb{R})$ (7.1.1) and the general case (8.4.9) can be found in [6, Prop. 3.12, p. 129, and Thm. 3.9, p. 131]. (Iwasawa's original proof is in [4, §3].) The decomposition also appears in many textbooks on Lie groups. In particular, Remark 8.4.10 is proved in [3, Thm. 6.5.1, pp. 270–271].

Regarding Remark 8.4.12(1), the obvious cocompact subgroups of G are parabolic subgroups and (cocompact) lattices. See [8] for a short proof that every cocompact subgroup is a combination of these two types. (A similar result had been proved previously in [2, (5.1a)].)

References

[1] A. Borel and J. Tits: Groupes réductifs, *Inst. Hautes Études Sci. Publ. Math.* 27 (1965) 55–150. MR 207712,
http://www.numdam.org/item?id=PMIHES_1965__27__55_0

[2] M. Goto and H.-C. Wang: Non-discrete uniform subgroups of semisimple Lie groups, *Math. Ann.* 198 (1972) 259–286. MR 0354934,
http://resolver.sub.uni-goettingen.de/purl?GDZPPN002306697

[3] S. Helgason: *Differential Geometry, Lie Groups, and Symmetric Spaces.* Academic Press, New York, 1978. ISBN 0-12-338460-5, MR 0514561

[4] K. Iwasawa: On some types of topological groups, *Ann. of Math.* (2) 50 (1949) 507–558. MR 0029911,
http://dx.doi.org/10.2307/1969548

[5] A. W. Knapp: *Lie Groups Beyond an Introduction, 2nd ed..* Birkhäuser, Boston, 2002. ISBN 0-8176-4259-5, MR 1920389

[6] V. Platonov and A. Rapinchuk: *Algebraic Groups and Number Theory.* Academic Press, Boston, 1994. ISBN 0-12-558180-7, MR 1278263

[7] G. Prasad and M. S. Raghunathan: Cartan subgroups and lattices in semi-simple groups, *Ann. of Math.* (2) 96 (1972), 296–317. MR 0302822, http://dx.doi.org/10.2307/1970790

[8] D. Witte: Cocompact subgroups of semisimple Lie groups, in G. Benkart and J. M. Osborn, eds.: *Lie Algebra and Related Topics (Madison, WI, 1988).* Amer. Math. Soc., Providence, RI, 1990, pp. 309–313. ISBN 0-8218-5119-5, MR 1079114

Chapter 9

ℚ-Rank

Algebraically, the definition of real rank extends in a straightforward way to a notion of rank over any field: if G is defined over F, then we can talk about $\mathrm{rank}_F\, G$. In the study of arithmetic groups, we assume G is defined over $\mathbb{Q}$, and the corresponding $\mathbb{Q}$-rank is an important invariant of the associated arithmetic group $\Gamma = G_{\mathbb{Z}}$.

Disclaimer. The reading of this chapter may be postponed without severe consequences (and can even be skipped entirely), because the material here will not arise elsewhere in this book (except marginally) other than in Chapter 19, where a coarse fundamental domain for Γ will be constructed. Furthermore, unlike the other chapters in this part of the book, the topic is of importance only for arithmetic groups and closely related subjects, not a broad range of areas of mathematics.

§9.1. ℚ-rank

(9.1.1) **Definition.** Assume G is defined over $\mathbb{Q}$. A closed, connected subgroup T of G is a $\mathbb{Q}$-*split torus* if

- T is defined over $\mathbb{Q}$, and
- T is diagonalizable over $\mathbb{Q}$. (That is, there exists $g \in \mathrm{GL}(\ell, \mathbb{Q})$, such that gTg^{-1} consists entirely of diagonal matrices.)

Recall: The Standing Assumptions (4.0.0 on page 41) are in effect, so, as always, Γ is a lattice in the semisimple Lie group $G \subseteq \mathrm{SL}(\ell, \mathbb{R})$.

Main prerequisites for this chapter: Real rank and minimal parabolic subgroups (Chapter 8), and groups defined over $\mathbb{Q}$ (Definition 5.1.2).

191

(9.1.2) **Example.**

1) $SO(1,1)^\circ$ is a $\mathbb{Q}$-split torus, because $g\,SO(1,1)g^{-1}$ consists of diagonal matrices if $g = \begin{bmatrix} 1 & 1 \\ 1 & -1 \end{bmatrix}$.

2) Although it is obvious that every $\mathbb{Q}$-split torus is an $\mathbb{R}$-split torus, the converse is not true (even if the torus is defined over $\mathbb{Q}$). For example, let $T = SO(x^2 - 2y^2; \mathbb{R})^\circ$. Then T is defined over $\mathbb{Q}$, and it is $\mathbb{R}$-split (because it is conjugate to $SO(1,1)^\circ$). However, it is **not** $\mathbb{Q}$-split. To see this, note that $\begin{bmatrix} 3 & 4 \\ 2 & 3 \end{bmatrix} \in T_\mathbb{Q}$, but the eigenvalues of this matrix are irrational (namely, $3 \pm 2\sqrt{2}$), so this rational matrix is not diagonalizable over $\mathbb{Q}$.

The following key fact implies that the maximal $\mathbb{Q}$-split tori of G all have the same dimension (which is called the "$\mathbb{Q}$-rank"):

(9.1.3) **Theorem.** *Assume G is defined over $\mathbb{Q}$. If S_1 and S_2 are maximal $\mathbb{Q}$-split tori in G, then $S_1 = gS_2g^{-1}$ for some $g \in G_\mathbb{Q}$.*

(9.1.4) **Definition** (for arithmetic lattices). Assume

- G is defined over $\mathbb{Q}$, and
- Γ is commensurable to $G_\mathbb{Z}$.

Then **rank$_\mathbb{Q}$** Γ is the dimension of any maximal $\mathbb{Q}$-split torus in G.

(More generally, if $\phi \colon G/K \overset{\cong}{\to} G'/K'$, where K and K' are compact, and $\phi(\overline{\Gamma})$ is commensurable to $\overline{G'_\mathbb{Z}}$ (see Definition 5.1.19), then rank$_\mathbb{Q}$ Γ is the dimension of any maximal $\mathbb{Q}$-split torus in G'.)

(9.1.5) **Examples.**

1) $\text{rank}_\mathbb{Q}(SL(n,\mathbb{Z})) = n - 1$. (Let S be the identity component of the group of all diagonal matrices in $SL(n,\mathbb{R})$.)

2) Let $G = SO(Q; \mathbb{R})$, where $Q(x_1, \ldots, x_\ell)$ is some quadratic form on $\mathbb{R}^\ell$, such that Q is defined over $\mathbb{Q}$. (That is, all of the coefficients of Q are rational.) Then G is defined over $\mathbb{Q}$, and the discussion of Example 8.1.7, with $\mathbb{Q}$ in place of $\mathbb{R}$, shows that rank$_\mathbb{Q}$ $G_\mathbb{Z}$ is the maximum dimension of a totally isotropic $\mathbb{Q}$-subspace of $\mathbb{Q}^\ell$.

   (a) For example, $\text{rank}_\mathbb{Q} SO(m,n)_\mathbb{Z} = \min\{m,n\}$. Similarly,
   $$\text{rank}_\mathbb{Q} SU(m,n)_\mathbb{Z} = \text{rank}_\mathbb{Q} Sp(m,n)_\mathbb{Z} = \min\{m,n\}.$$
   So rank$_\mathbb{Q}$ $G_\mathbb{Z} = $ rank$_\mathbb{R}$ G for these groups.

   (b) Let $G = SO(x_1^2 + x_2^2 + x_3^2 - 7x_4^2; \mathbb{R})$. Then, because the congruence $a^2 + b^2 + c^2 + d^2 \equiv 0 \pmod 8$ implies that all the variables are even, it is not difficult to see that this quadratic form has no nonzero isotropic vectors in $\mathbb{Q}^4$ (see Exercise 4). This means rank$_\mathbb{Q}$ $G_\mathbb{Z} = 0$.

   Note, however, that G is isomorphic to $SO(3,1)$, so its real rank is 1. Therefore, rank$_\mathbb{Q}$ $G_\mathbb{Z} \neq$ rank$_\mathbb{R}$ G.

3) $\operatorname{rank}_{\mathbb{Q}} \Gamma = 0$ if and only if G/Γ is compact (see Exercise 5).

4) $\operatorname{rank}_{\mathbb{Q}} \operatorname{SU}(B, \tau; \mathcal{O}_D)$ is the dimension (over D) of a maximal totally isotropic subspace of D^n, if B is a τ-Hermitian form on D^n, and D is a division algebra over F.

(9.1.6) **Warning.** In analogy with Exercise 8.1#13 and Exercise 5.3#7(e), one might suppose that $\operatorname{rank}_{\mathbb{Q}} \Gamma \neq 0$ if and only if Γ contains a subgroup that is isomorphic to $\operatorname{SL}(2, \mathbb{Z})$ (modulo finite groups). However, this is **false**: *every* lattice in G contains a subgroup that is abstractly commensurable to $\operatorname{SL}(2, \mathbb{Z})$ (unless G is compact). Namely, the Tits Alternative tells us that Γ contains a nonabelian free subgroup (see Corollary 4.9.2), and it is well known that $\operatorname{SL}(2, \mathbb{Z})$ has a finite-index subgroup that is free (see Exercise 4.9#5).

(9.1.7) *Remarks.*

1) The definition of $\operatorname{rank}_{\mathbb{Q}} \Gamma$ is somewhat indirect, because the $\mathbb{Q}$-split tori of G are not subgroups of Γ. Therefore, it would be more correct to say that we have defined $\operatorname{rank}_{\mathbb{Q}} G_{\mathbb{Q}}$.

2) Although different embeddings of G in $\operatorname{SL}(\ell, \mathbb{R})$ can yield maximal $\mathbb{Q}$-split tori of different dimensions, the theory of algebraic groups shows that the $\mathbb{Q}$-rank is the same for all of the embeddings in which Γ is commensurable to $G_{\mathbb{Z}}$ (see Corollary 9.4.7); therefore, $\operatorname{rank}_{\mathbb{Q}} \Gamma$ is well defined as a function of Γ.

3) We have $0 \leq \operatorname{rank}_{\mathbb{Q}} \Gamma \leq \operatorname{rank}_{\mathbb{R}} G$, since every $\mathbb{Q}$-split torus is $\mathbb{R}$-split. It can be shown that:

   (a) The extreme values are always realized: there exist lattices Γ_0 and Γ_r in G, such that $\operatorname{rank}_{\mathbb{Q}} \Gamma_0 = 0$ and $\operatorname{rank}_{\mathbb{Q}} \Gamma_r = \operatorname{rank}_{\mathbb{R}} G$ (see Theorem 18.7.1 and Exercise 7).

   (b) In some cases, there are intermediate values that are not realized. For example, the $\mathbb{Q}$-rank of every lattice in $\operatorname{SO}(2, 5)$ is either 0 or 2 (see Corollary 18.6.2).

4) Suppose Γ is defined by restriction of scalars (5.5.8), so Γ is commensurable to $G'_{\mathcal{O}}$, where G' is defined over a finite extension F of $\mathbb{Q}$, and $\mathcal{O}$ is the ring of integers of F. Then $\operatorname{rank}_{\mathbb{Q}} \Gamma$ is equal to the "F-rank" of G', or, in other words, the maximal dimension (over F_∞) of a subgroup of G' that is diagonalizable over F. For example, the $\mathbb{Q}$-rank of $\operatorname{SO}(B; \mathcal{O})$ is the dimension of a maximal totally isotropic F-subspace of F^n.

Definition 9.1.4 applies only to arithmetic lattices, but the Margulis Arithmeticity Theorem (5.2.1) allows the definition to be extended to all lattices:

(9.1.8) **Definition** (see Exercise 6). Up to isogeny, and modulo the maximal compact factor of G, we may write $G = G_1 \times \cdots \times G_s$, so that $\Gamma_i = \Gamma \cap G_i$

is an irreducible lattice in G_i for $i = 1, \ldots, r$ (see Proposition 4.3.3). We let

$$\operatorname{rank}_{\mathbb{Q}}(\Gamma) = \operatorname{rank}_{\mathbb{Q}}(\Gamma_1) + \cdots + \operatorname{rank}_{\mathbb{Q}}(\Gamma_s),$$

where:

1) If G/Γ_i is compact, then $\operatorname{rank}_{\mathbb{Q}} \Gamma_i = 0$.

2) If G/Γ_i is not compact, and $\operatorname{rank}_{\mathbb{R}} G = 1$, then $\operatorname{rank}_{\mathbb{Q}} \Gamma_i = 1$.

3) If G/Γ_i is not compact, and $\operatorname{rank}_{\mathbb{R}} G \geq 2$, then the Margulis Arithmeticity Theorem (5.2.1) implies that Γ_i is arithmetic, so Definition 9.1.4 applies.

Exercises for §9.1.

#1. Show that if T is a $\mathbb{Q}$-split torus, then $T_{\mathbb{Z}}$ is finite.

#2. Give an example of a torus T (that is defined over $\mathbb{Q}$), such that $T_{\mathbb{Z}}$ is infinite.

#3. Verify the claim in Example 9.1.5(2) that $\operatorname{rank}_{\mathbb{Q}} SO(Q;\mathbb{Z})$ is the dimension of a maximal totally isotropic subspace of $\mathbb{Q}^\ell$.

#4. Verify the claim in Example 9.1.5(2b) that $(0,0,0,0)$ is the only solution in $\mathbb{Q}^4$ of the equation $x_1^2 + x_2^2 + x_3^2 - 7x_4^2 = 0$.

#5. Prove Example 9.1.5(3).
[*Hint:* ($\Rightarrow$) See Exercise 5.3#7. ($\Leftarrow$) If a is diagonalizable over $\mathbb{Q}$, then there exists $v \in \mathbb{Z}^\ell$, such that $a^n v \to 0$ as $n \to +\infty$, so the Mahler Compactness Criterion (4.4.7) implies $G/G_{\mathbb{Z}}$ is not compact.]

#6. Show that Definition 9.1.8 is consistent with Definition 9.1.4. More precisely, assume Γ is arithmetic, and prove:
   a) G/Γ is compact if and only if $\operatorname{rank}_{\mathbb{Q}} \Gamma = 0$.
   b) If G/Γ is not compact, and $\operatorname{rank}_{\mathbb{R}} G = 1$, then $\operatorname{rank}_{\mathbb{Q}} \Gamma = 1$.
   c) If $\Gamma = \Gamma_1 \times \Gamma_2$ is reducible, then $\operatorname{rank}_{\mathbb{Q}} \Gamma = \operatorname{rank}_{\mathbb{Q}} \Gamma_1 + \operatorname{rank}_{\mathbb{Q}} \Gamma_2$.

#7. Suppose G is classical. Show that, for the natural embeddings described in Examples A2.3 and A2.4, we have $\operatorname{rank}_{\mathbb{Q}} G_{\mathbb{Z}} = \operatorname{rank}_{\mathbb{R}} G$.
[*Hint:* Example 9.1.5(1,2)).]

§9.2. Lattices of higher $\mathbb{Q}$-rank

This section closely parallels Section 8.2, because the results there on semisimple groups of higher real rank can be extended in a natural way to lattices of higher $\mathbb{Q}$-rank.

(9.2.1) **Assumption.** Throughout this section, if the statement of a result mentions $G_{\mathbb{Q}}$, $G_{\mathbb{Z}}$, or a $\mathbb{Q}$-split torus in G, then G is assumed to be defined over $\mathbb{Q}$.

(9.2.2) Proposition (see Exercise 1). *Let S be any maximal $\mathbb{Q}$-split torus in G. Then we have $\operatorname{rank}_{\mathbb{Q}} G_{\mathbb{Z}} \geq 2$ if and only if there exist nontrivial elements s_1 and s_2 of $S_{\mathbb{Q}}$, such that $G = \langle C_G(s_1), C_G(s_2) \rangle$.*

(9.2.3) Lemma. *If S is any maximal $\mathbb{Q}$-split torus in G, then we have $C_G(S) = S \times M = S \times CL$, where*

- *M, C, and L are defined over $\mathbb{Q}$,*
- *$\operatorname{rank}_{\mathbb{Q}} M = 0$,*
- *L is semisimple, and*
- *C is a torus that is the identity component of the center of M.*

(9.2.4) Proposition (see Exercise 3). *$\operatorname{rank}_{\mathbb{Q}} G_{\mathbb{Z}} \geq 2$ if and only if there exist nontrivial elements a and u of $G_{\mathbb{Q}}$, such that a belongs to a $\mathbb{Q}$-split torus of G, u is unipotent, and $au = ua$.*

(9.2.5) Lemma (see Exercise 4). *Assume Γ is commensurable to $G_{\mathbb{Z}}$. The following are equivalent:*

1) *$\operatorname{rank}_{\mathbb{Q}} \Gamma \leq 1$.*
2) *Every nontrivial unipotent subgroup of Γ is contained in a **unique** maximal unipotent subgroup of Γ.*
3) *Every nontrivial unipotent $\mathbb{Q}$-subgroup of G is contained in a **unique** maximal unipotent $\mathbb{Q}$-subgroup of G.*

(9.2.6) Proposition. *$\operatorname{rank}_{\mathbb{Q}} \Gamma \geq 2$ if and only if Γ contains nontrivial unipotent subgroups $U_1, \ldots, U_k$, such that*

- *$\langle U_1, \ldots, U_k \rangle$ is a finite-index subgroup of Γ, and*
- *U_i centralizes U_{i+1} for each i.*

(9.2.7) Proposition. *Assume Γ is irreducible. Then $\operatorname{rank}_{\mathbb{Q}} \Gamma \geq 2$ if and only if Γ contains a subgroup that is commensurable to either $\mathrm{SL}(3, \mathbb{Z})$ or $\mathrm{SO}(2,3)_{\mathbb{Z}}$.*

(9.2.8) Remarks.

1) Unfortunately, the list of lattices of $\mathbb{Q}$-rank one is longer and much more complicated than the list of simple groups of real rank one in Theorem 8.3.1. The classical arithmetic groups (of any $\mathbb{Q}$-rank) are described in Chapter 18 (see the table on 380), but there are also infinitely many different lattices of $\mathbb{Q}$-rank one in exceptional groups of type E_6 and F_4, and the nonarithmetic lattices of $\mathbb{Q}$-rank one in $\mathrm{SO}(1, n)$ and $\mathrm{SU}(1, n)$ have not yet been classified.

2) Suppose $\operatorname{rank}_{\mathbb{Q}} \Gamma \leq 1$. Proposition 9.2.6 shows that it is impossible to find a generating set $\{y_1, \ldots, y_r\}$ for Γ, such that each y_i is nontrivial and unipotent, and y_i commutes with y_{i+1}, for each i. However, it is possible, in some cases, to find a generating set $\{y_1, \ldots, y_r\}$ that has all of these properties *except* the requirement

that y_i is unipotent. For example, this is easy (up to finite index) if Γ is reducible (see Exercise 7).

Exercises for §9.2.

#1.   a) Prove Proposition 9.2.2($\Rightarrow$) for the special case where we have
$G_\mathbb{Q} = \mathrm{SL}(3, \mathbb{Q})$.
   b) Prove Proposition 9.2.2($\Leftarrow$).

#2. Prove the following results in the special case where $\Gamma = \Gamma_1 \times \Gamma_2$, and $\mathrm{rank}_\mathbb{Q} \Gamma_i \geq 1$ for each i.
   a) Proposition 9.2.2($\Rightarrow$)
   b) Proposition 9.2.4($\Rightarrow$)
   c) Lemma 9.2.5($\Leftarrow$)
   d) Proposition 9.2.6($\Rightarrow$)

#3.   a) Prove Proposition 9.2.4($\Rightarrow$) in the special case where we have
$G_\mathbb{Q} = \mathrm{SL}(3, \mathbb{Q})$.
   b) Prove Proposition 9.2.4($\Leftarrow$).

#4. For each of these groups, find a nontrivial unipotent subgroup that is contained in two different maximal unipotent subgroups.
   a) $\mathrm{SL}(3, \mathbb{Q})$.
   b) $\mathrm{SL}(3, \mathbb{Z})$.

#5. Prove Proposition 9.2.6.

#6. (*Assumes the theory of* $\mathbb{Q}$-*roots*) Prove the general case of the following results.
   a) Proposition 9.2.2.
   b) Lemma 9.2.3.
   c) Proposition 9.2.4($\Rightarrow$).
   d) Lemma 9.2.5.
   e) Proposition 9.2.6.
   f) Proposition 9.2.7.

#7. Show that if Γ is reducible, and G has no compact factors, then there is a finite subset $\{y_1, \ldots, y_r\}$ of Γ, such that
   a) $\{y_1, \ldots, y_r\}$ generates a finite-index subgroup of Γ,
   b) each y_i is nontrivial, and
   c) y_i commutes with y_{i+1}, for each i.

#8. Let Γ be a torsion-free, cocompact lattice in $\mathrm{SL}(3, \mathbb{R})$, constructed as in Proposition 6.7.4. Show that if y_1 and y_2 are any nontrivial elements of Γ, such that y_1 commutes with y_2, then $C_\Gamma(y_1) = C_\Gamma(y_2)$. (Hence, it is impossible to find a sequence of nontrivial generators of Γ, such that each generator commutes with the next.)

[*Hint:* Let $D = \phi(L^3)$, so D is a division ring of degree 3 over ℚ. Then $C_D(\gamma_1)$ is subring of D that contains the field $\mathbb{Q}[\gamma_1]$ in its center. Because the degree of D is prime, we conclude that $C_D(\gamma_1) = \mathbb{Q}[\gamma_1] \subseteq C_D(\gamma_2)$.]

§9.3. Minimal parabolic ℚ-subgroups

Minimal parabolic subgroups of G play an important role in the study of arithmetic subgroups, even when they are not defined over ℚ. However, for some purposes (especially when we construct a coarse fundamental domain in Chapter 19), we want a subgroup that is both defined over ℚ and is similar to a minimal parabolic subgroup:

(9.3.1) **Definition** (cf. Definition 8.4.1). Let S be a maximal ℚ-split torus of G, and let a be a generic element of S. Then the corresponding *minimal parabolic ℚ-subgroup* of G is

$$P = \left\{ g \in G \; \middle| \; \limsup_{n \to \infty} \| a^{-n} g a^n \| < \infty \right\}.$$

This is a Zariski closed subgroup of G that is defined over ℚ.

(9.3.2) **Examples.**

1) Since the group of upper triangular matrices is a minimal parabolic ℚ-subgroup of $\mathrm{SL}(n, \mathbb{R})$, we see that, in this case, the minimal parabolic ℚ-subgroup is also a minimal parabolic subgroup.

2) This is a special case of the fact that if $\mathrm{rank}_{\mathbb{Q}} \Gamma = \mathrm{rank}_{\mathbb{R}} G$, then every minimal parabolic ℚ-subgroup is also a minimal parabolic subgroup (see Exercise 1).

3) (Cf. Theorem 8.4.5(2)) Suppose Q is a nondegenerate quadratic form on $\mathbb{Q}^\ell$ that is defined over ℚ. A subgroup P of $\mathrm{SO}(Q; \mathbb{R})$ is a minimal parabolic ℚ-subgroup if and only if there is a chain $V_0 \subsetneq V_1 \subsetneq \cdots \subsetneq V_r$ of **totally isotropic** subspaces of $\mathbb{Q}^\ell$, such that
   - $\dim V_i = i$, for each i,
   - V_r is a maximal totally isotropic subspace, and
   - $P = \{ g \in \mathrm{SO}(Q; \mathbb{R}) \mid \forall i, \, gV_i = V_i \}$.

We have a Langlands decomposition over ℚ. However, unlike in the real case, where the subgroup M is compact (i.e., $\mathrm{rank}_{\mathbb{R}} M = 0$), we now have a subgroup that may be noncompact (but whose ℚ-rank is 0):

(9.3.3) **Theorem** (Langlands decomposition). *If P is a minimal parabolic ℚ-subgroup of G, then we may write P in the form $P = MSN = LCSN$, where*

1) *$M, S, N, L,$ and C are defined over ℚ,*

2) *S is a maximal ℚ-split torus,*

3) *$\mathrm{rank}_{\mathbb{Q}} M = 0$,*

4) $MS = C_G(S)$,

5) $M = LC$, where L is semisimple and C is the identity component of the center of M, and

6) N is the **unipotent radical** of P; that is, the unique maximal unipotent **normal** subgroup of P.

Furthermore, for some $a \in S_{\mathbb{Q}}$, we have

$$P = \left\{ g \in G \ \middle| \ \limsup_{n \to \infty} \|a^{-n}ga^n\| < \infty \right\} \tag{9.3.4}$$

and

$$N = \left\{ g \in G \ \middle| \ \lim_{n \to \infty} a^{-n}ga^n = e \right\}. \tag{9.3.5}$$

Proof. The examples and proof are essentially the same as for the real Langlands decomposition (8.4.6), but with $\mathbb{Q}$ in the place of $\mathbb{R}$, and groups of $\mathbb{Q}$-rank 0 in place of compact groups.   □

(9.3.6) **Proposition.** *Assume G is defined over $\mathbb{Q}$, and P is a minimal parabolic $\mathbb{Q}$-subgroup, with Langlands decomposition $P = MSN$. Then:*

1) *Every minimal parabolic $\mathbb{Q}$-subgroup of G is $G_{\mathbb{Q}}$-conjugate to P.*

2) *Every unipotent $\mathbb{Q}$-subgroup of G is $G_{\mathbb{Q}}$-conjugate to a subgroup of N.*

3) $P = \mathcal{N}_G(N) = \mathcal{N}_G(P)$.

(9.3.7) **Definition.** For P, M, S, N, L, and C as in Theorem 9.3.3, the **positive Weyl chamber** of S (with respect to P) is the set S^+ of all elements a of S, such that P is contained in the right-hand side of (9.3.4). (Equivalently, it is the closure of the set of elements a of S for which equality holds in (9.3.4).)

Exercises for §9.3.

#1. Show that if we have $\mathrm{rank}_{\mathbb{Q}} \Gamma = \mathrm{rank}_{\mathbb{R}} G$, then every minimal parabolic $\mathbb{Q}$-subgroup is also a minimal parabolic subgroup.

[*Hint:* Choose A to be both a maximal $\mathbb{Q}$-split torus and a maximal $\mathbb{R}$-split torus.]

#2. Show that the converse of Exercise 1 is not true.

[*Hint:* Proposition 6.6.1.]

#3. Show that every minimal parabolic $\mathbb{Q}$-subgroup of G contains a minimal parabolic subgroup.

[*Hint:* Choose a maximal $\mathbb{Q}$-split torus S. Then choose a maximal $\mathbb{R}$-split torus A that contains S. There is a generic element of A that is very close to a generic element of S.]

#4. If P is any minimal parabolic $\mathbb{Q}$-subgroup of G, and K is any maximal compact subgroup of G, show that $G = KP$.

[*Hint:* The Iwasawa decomposition (8.4.9) tells us $G = KAN$, and some conjugate of AN is contained in P.]

#5. Assume the notation of Theorem 9.3.3. Show that if $\mathrm{rank}_{\mathbb{Q}} G = 1$, then there is an isomorphism $\phi \colon S \xrightarrow{\cong} \mathbb{R}$, such that $\phi(S^+) = \mathbb{R}^+$.

#6. Show that if U is a unipotent $\mathbb{Q}$-subgroup of $\mathrm{SL}(\ell, \mathbb{R})$, then $U_{\mathbb{Z}}$ is a cocompact lattice in U.

[*Hint:* Induct on the nilpotence class of U (see Exercise 8.4#9). Note that the exponential map exp: $\mathfrak{u} \to U$ is a polynomial with rational coefficients, as is its inverse, so $U_{\mathbb{Z}}$ is Zariski dense in U.]

#7. Show that if U_1 and U_2 are maximal unipotent subgroups of Γ, and Γ is commensurable to $G_{\mathbb{Z}}$, then there exists $g \in G_{\mathbb{Q}}$, such that $g^{-1}U_1 g$ is commensurable to U_2.

§9.4. Isogenies over $\mathbb{Q}$

We have seen examples in which G is isogenous (or even isomorphic) to G', but the arithmetic subgroup $G_{\mathbb{Z}}$ is very different from $G'_{\mathbb{Z}}$. (For example, it may be the case that $G_{\mathbb{Z}}$ is cocompact, but $G'_{\mathbb{Z}}$ is not.) This does not happen if the isogeny is defined over $\mathbb{Q}$, in the following sense:

(9.4.1) **Definition.**

1) A homomorphism $\phi \colon G \to G'$ is **defined over** $\mathbb{Q}$ if $\phi(G_{\mathbb{Q}}) \subseteq G'_{\mathbb{Q}}$.

2) G_1 is **isogenous to** G_2 **over** $\mathbb{Q}$ (denoted $G_1 \approx_{\mathbb{Q}} G_2$) if there is a group G that is defined over $\mathbb{Q}$, and isogenies $\phi_i \colon G \to G_i$ that are defined over $\mathbb{Q}$.

The following result shows that any isogeny over $\mathbb{Q}$ can be thought of as a polynomial with rational coefficients.

(9.4.2) **Definition.** A function $\phi \colon G \to G'$ is a **polynomial with rational coefficients** if

- the matrix entries of $\phi(g)$ can be written as polynomial functions of the coefficients of g, and
- the polynomials can be chosen so that all of their coefficients are in $\mathbb{Q}$.

(9.4.3) **Proposition.** *If $G_1 \approx_{\mathbb{Q}} G_2$, then there is a group G that is defined over $\mathbb{Q}$, and isogenies $\phi_i \colon G \to G_i$ for $i = 1, 2$, that are polynomials with rational coefficients.*

Proof. Given isogenies $\phi_i \colon G \to G_i$ that are defined over $\mathbb{Q}$, let
$$G' = \{ (\phi_1(g), \phi_2(g)) \mid g \in G^{\circ} \}.$$

This is defined over $\mathbb{Q}$, since $G'_\mathbb{Q}$ is dense (see Proposition 5.1.8). The projection maps $\phi'_i \colon G' \to G_i$ defined by $\phi'_i(g_1, g_2) = g_i$ are polynomials with rational coefficients. $\square$

(9.4.4) **Warning.** There are examples where $\phi \colon G_1 \to G_2$ is an isomorphism, and ϕ is a polynomial, but ϕ^{-1} is not a polynomial. For example, the natural homomorphism $\phi \colon \mathrm{SL}(3, \mathbb{R}) \to \mathrm{PSL}(3, \mathbb{R})^\circ$ is an isomorphism (because $\mathrm{SL}(3, \mathbb{R})$ has no center). However, there is no isomorphism from $\mathrm{PSL}(3, \mathbb{C})$ to $\mathrm{SL}(3, \mathbb{C})$ (because one of these groups has a center and the other does not), so the inverse of ϕ cannot be a polynomial (because it does not extend to a well-defined map between the complexifications).

The following fundamental result implies that different embeddings of G with the same $\mathbb{Q}$-points have essentially the same $\mathbb{Z}$-points.

(9.4.5) **Proposition.** *Suppose $\phi \colon G \to G'$ is a surjective homomorphism that is defined over $\mathbb{Q}$. Then $\phi(G_\mathbb{Z})$ is commensurable to $G'_\mathbb{Z}$.*

Proof. From the proof of Proposition 9.4.3, we see that, after replacing G with an isogenous group, we may assume that ϕ is a polynomial with rational coefficients. Assume $G \subseteq \mathrm{SL}(\ell, \mathbb{R})$ and $G' \subseteq \mathrm{SL}(m, \mathbb{R})$.

Define $\widetilde{\phi} \colon G \to \mathrm{Mat}_{m \times m}(\mathbb{R})$ by $\widetilde{\phi}(x) = \phi(x - \mathrm{Id})$. Then $\widetilde{\phi}$ is a polynomial, so it is defined on all of $\mathrm{Mat}_{\ell \times \ell}(\mathbb{R})$. Since the coefficients are in $\mathbb{Q}$, there is some nonzero $n \in \mathbb{N}$, such that $\widetilde{\phi}(n \, \mathrm{Mat}_{\ell \times \ell}(\mathbb{R})) \subseteq \mathrm{Mat}_{m \times m}(\mathbb{Z})$. Therefore, letting Γ_n be the "principal congruence subgroup" of $G_\mathbb{Z}$ of level n (see page 66), we have $\phi(\Gamma_n) \subseteq G'_\mathbb{Z}$.

Because Γ_n is a lattice in G (and $\phi(\Gamma_n)$ is discrete), we know that $\phi(\Gamma_n)$ is a lattice in G'. Since $\phi(\Gamma_n) \subseteq G'_\mathbb{Z}$, this implies that $\phi(\Gamma_n)$ is commensurable to $G'_\mathbb{Z}$ (see Exercise 4.1#10). $\square$

The following fundamental fact is, unfortunately, not obvious from our definition of "$\mathbb{Q}$-split."

(9.4.6) **Proposition.** *Assume*
- *T and H are connected Lie groups that are defined over $\mathbb{Q}$, and*
- *$T \approx_\mathbb{Q} H$.*

Then T is a $\mathbb{Q}$-split torus if and only if H is a $\mathbb{Q}$-split torus.

(9.4.7) **Corollary.** *If $G \approx_\mathbb{Q} G'$, then $\mathrm{rank}_\mathbb{Q} G_\mathbb{Z} = \mathrm{rank}_\mathbb{Q} G'_\mathbb{Z}$.*

Proof. Suppose G is a $\mathbb{Q}$-group, and there is an isogeny $\varphi_i \colon G \to G_i$ that is defined over $\mathbb{Q}$ for $i = 1, 2$. If T_1 is a maximal $\mathbb{Q}$-split torus in G_1, then Proposition 9.4.6 implies that $\varphi_2(\varphi_1^{-1}(T_1)^\circ)$ is a $\mathbb{Q}$-split torus in G_2. Since isogeny preserves dimension, we conclude that $\mathrm{rank}_\mathbb{Q} G_1 \leq \mathrm{rank}_\mathbb{Q} G_2$. By symmetry, equality must hold. $\square$

Notes

As was mentioned in the notes of Chapter 8, the comprehensive treatise of Borel and Tits [3] is the standard reference on rank, parabolic subgroups, and other fundamental properties of reductive groups over any field (including $\mathbb{Q}$). Abbreviated accounts can be found in many texts, including [1, §10 and §11] and [2, Chap. 5].

See [1, Rem. 8.11, p. 60] for a proof of Proposition 9.4.5.

References

[1] A. Borel: *Introduction aux Groupes Arithmétiques.* Hermann, Paris, 1969. MR 0244260

[2] A. Borel: *Linear Algebraic Groups, 2nd ed..* Springer, New York, 1991. ISBN 0-387-97370-2, MR 1102012

[3] A. Borel and J. Tits: Groupes réductifs, *Inst. Hautes Études Sci. Publ. Math.* 27 (1965) 55–150. MR 207712, http://www.numdam.org/item?id=PMIHES_1965__27__55_0

Chapter 10

Quasi-Isometries

§10.1. Word metric and quasi-isometries

The field of Geometric Group Theory equips groups with a metric, which allows them to be studied as metric spaces:

(10.1.1) **Definition.** Fix a finite generating set S of Γ (see Theorem 4.7.10), and assume, for simplicity, that S is **symmetric**, which means $s^{-1} \in S$ for every $s \in S$.

1) For $g \in \Gamma$, the **word length** of g is the length ℓ of the shortest sequence $(s_1, s_2, \ldots, s_\ell)$ of elements of S, such that $s_1 s_2 \cdots s_\ell = g$. It is denoted $\ell(g)$. (By convention, $\ell(e) = 0$.)

2) For $g, h \in \Gamma$, we let $d(g, h) = \ell(g^{-1}h)$. This defines a metric on Γ, called the **word metric** (see Exercise 1).

The word metric has the important property that the action of Γ on itself by left-translations is an action by isometries (see Exercise 2).

(10.1.2) *Remark.* The word metric can be pictured geometrically, by constructing a **Cayley graph**. Namely, $\mathrm{Cay}(\Gamma; S)$ is a certain 1-dimensional simplicial complex (or "graph"):

Recall: The Standing Assumptions (4.0.0 on page 41) are in effect, so, as always, Γ is a lattice in the semisimple Lie group $G \subseteq \mathrm{SL}(\ell, \mathbb{R})$.

Main prerequisites for this chapter: none.

- its 0-skeleton is Γ, and
- it has a 1-simplex (or "edge") of length 1 joining v to vs, for every $v \in \Gamma$ and $s \in S$.

Define a metric on $\mathrm{Cay}(\Gamma; S)$ by letting $d(x, y)$ be the length of the shortest path from x to y. Then the restriction of this metric to the 0-skeleton is precisely the word metric on Γ.

Unfortunately, the word metric on Γ is not canonical, because it depends on the choice of the generating set S (see Exercise 3). However, it is "almost" well-defined, in that changing the generating set can only distort the distances by a bounded factor. This idea is formalized in the following notion:

(10.1.3) Definition. Let X_1 and X_2 be metric spaces, with metrics d_1 and d_2, respectively.

1) A function $f \colon X_1 \to X_2$ is a *quasi-isometry* if there is a constant $C > 0$, such that
   (a) for all $x, y \in X_1$ with $d_1(x, y) > C$, we have
   $$\frac{1}{C} < \frac{d_2(f(x), f(y))}{d_1(x, y)} < C,$$
   and
   (b) for all $x_2 \in X_2$, there exists $x_1 \in X_1$, such that
   $$d_2(f(x_1), x_2) < C.$$

   Note that f need not be continuous (and need not be one-to-one or onto, either).

2) We say X_1 is *quasi-isometric* to X_2 (and write $X_1 \overset{\mathrm{QI}}{\approx} X_2$) if there is a quasi-isometry from X_1 to X_2. This is an equivalence relation (see Exercise 4).

(10.1.4) Proposition (see Exercise 5). *Let*

- S_1 *and* S_2 *be two finite, symmetric generating sets for* Γ, *and*
- d_i *be the word metric on* Γ *corresponding to the generating set* S_i.

Then $(\Gamma, d_1) \overset{\mathrm{QI}}{\approx} (\Gamma, d_2)$.

Therefore, if Γ_1 and Γ_2 are quasi-isometric for some choice of the word metrics on the two groups, then they are quasi-isometric for all choices of the word metrics. So it makes sense to say that two groups are quasi-isometric, without any mention of generating sets (as long as both of the groups are finitely generated).

(10.1.5) Remark. A property is said to be *geometric* if is is invariant under quasi-isometry. For example, we will see (in Proposition 12.7.22 and Remark 13.4.10, respectively) that amenability is a geometric property, but Kazhdan's property (T) is not. In other words, if $\Lambda_1 \overset{\mathrm{QI}}{\approx} \Lambda_2$, and

Λ_1 is amenable, then Λ_2 is amenable, but the same cannot be said for Kazhdan's property (T). In general, quasi-isometric groups can be very different from each other, so most of the usual algebraic properties of groups are not geometric.

Quasi-isometries also arise from cocompact actions. Before stating the result, we introduce some terminology.

(10.1.6) **Definition.** Let (X, d) be a metric space, and let $C > 0$.

1) X is **proper** if the closed ball $B_r(x)$ is compact, for all $r > 0$ and all $x \in X$.

2) Let $x, y \in X$. A (C)-**coarse geodesic** from x to y is a finite sequence $x_0, x_1, \ldots, x_n$ in X, such that $x_0 = x$, $x_n = y$, and
$$\left| d(x_i, x_j) - |i - j| \right| < C \quad \text{for all } i, j.$$

3) X is $(C$-$)$**coarsely geodesic** if, for all $x, y \in X$, there is a C-coarse geodesic from x to y.

(10.1.7) **Proposition** (see Exercise 6). *Suppose*

- *(X, d) is a metric space that is proper and coarsely geodesic,*
- *Γ has a properly discontinuous action on X by isometries, such that $\Gamma \backslash X$ is compact, and*
- *d' is a word metric on Γ.*

Then $(\Gamma, d') \overset{\text{QI}}{\approx} (X, d)$.

More precisely, for any basepoint $x_0 \in X$, the map $\gamma \mapsto \gamma x_0$ is a quasi-isometry from Γ to X.

(10.1.8) **Corollary.** *If G/Γ is compact, then the inclusion $\Gamma \hookrightarrow G$ is a quasi-isometry, where we use any word metric on Γ, and we use any (coarsely geodesic, proper) metric on G that is invariant under left-translations.*

This implies that any two cocompact lattices in the same group are quasi-isometric:

(10.1.9) **Corollary.** *If Γ_1 and Γ_2 are cocompact lattices in G, then $\Gamma_1 \overset{\text{QI}}{\approx} \Gamma_2$.*

Proof. We have $\Gamma_1 \overset{\text{QI}}{\approx} G \overset{\text{QI}}{\approx} \Gamma_2$, so $\Gamma_1 \overset{\text{QI}}{\approx} \Gamma_2$ by transitivity. $\qquad \square$

We will see in Section 15.4 that the situation is usually very different for lattices that are not cocompact: in most cases, there are infinitely many different (noncocompact) lattices in G that are not quasi-isometric to each other.

Any (coarsely geodesic, proper) metric on G provides a metric on Γ, by restriction. In most cases, this restriction is the word metric (up to quasi-isometry):

(10.1.10) **Theorem** (Lubotzky-Mozes-Raghunathan). *If* $\mathrm{rank}_{\mathbb{R}}\, G \geq 2$, *and* Γ *is irreducible, then the inclusion* $\Gamma \hookrightarrow G$ *is a quasi-isometry onto its image.*

The assumption that $\mathrm{rank}_{\mathbb{R}}\, G \geq 2$ is essential:

(10.1.11) **Example.** Let
- $G = \mathrm{SL}(2, \mathbb{R})$,
- Γ be a free subgroup of finite index in $\mathrm{SL}(2, \mathbb{Z})$ (see Exercise 4.9#5),
- $u = \begin{bmatrix} 1 & k \\ 0 & 1 \end{bmatrix} \in \Gamma$, and
- $a^t = \begin{bmatrix} e^t & 0 \\ 0 & e^{-t} \end{bmatrix} \in G$.

Then:
1) For any word metric d_Γ on Γ, the function $d_\Gamma(u^n, e)$ grows linearly with n, because Γ is free.
2) For any left-invariant metric d_G on G, the function $d_G(u^n, e)$ grows only logarithmically, because $a^{\log n} u a^{-\log n} = u^{2n}$.

Therefore, the restriction of d_G to Γ is not quasi-isometric to d_Γ.

Exercises for §10.1.

#1. Show that the word metric is indeed a metric. More precisely, for $x, y, z \in \Gamma$, show
$$d(x, y) \geq 0, \qquad\qquad d(x, y) = 0 \iff x = y,$$
$$d(x, y) = d(y, x), \qquad\qquad d(x, y) \leq d(x, z) + d(z, y).$$

#2. Assume d is a word metric on Γ (with respect to a finite, symmetric generating set S). Show that $d(ax, ay) = d(x, y)$ for all $a, x, y \in \Gamma$. [*Warning:* $d(xa, ya)$ is usually *not* equal to $d(x, y)$.]

#3. Assume Γ is infinite. Show there exist two word metrics d_1 and d_2 on Γ (corresponding to finite, symmetric generating sets S_1 and S_2), such that the metric space (Γ_1, d_1) is not isometric to the metric space (Γ_2, d_2).
[*Hint:* A ball of radius r can have different cardinality for the two metrics.]

#4. Show that $\overset{QI}{\sim}$ is an equivalence relation.

#5. Prove Proposition 10.1.4.
[*Hint:* Show the identity map is a quasi-isometry from (Γ, d_1) to (Γ, d_2), by choosing C so that $d_1(e, s) \leq C$ for each $s \in S_2$.]

#6. Prove Proposition 10.1.7.
[*Hint:* Assume $S \supseteq \{ y \mid \exists x \in X,\ d(yx, x) \leq 3C \}$, and $\Gamma \cdot B_C(x) = X$ for all $x \in X$. Given $x_0, x_1, \ldots, x_n \in X$ with $d(x_i, x_{i+1}) \leq C$, there exists $y_i \in \Gamma$, such that $d(y_i x_0, x_i) \leq C$, so $\ell(y_n) \leq n$.]

§10.2. Hyperbolic groups

Manifolds of negative sectional curvature play an important role in differential geometry, both in applications and as a source of examples. The triangles in these manifolds have a very special "thinness" property that we will now explain. Groups whose triangles have this same property are said to be "negatively curved," or, in our terminology, "Gromov hyperbolic."

(10.2.1) **Definition** (Gromov). Let $\delta > 0$, and let X be a C-coarsely geodesic metric space.

1) A (C-coarse) ***triangle*** abc in X is a set $\{s_{ab}, s_{bc}, s_{ac}\}$, where s_{xy} is a C-coarse geodesic from x to y for $x, y \in \{a, b, c\}$.

2) A triangle abc is ***δ-thin*** if each of the three sides of the triangle is contained in the (closed) δ-neighborhood of the union of the other two sides. That is, each point in s_{ab} is at distance no more than δ from some point in $s_{ac} \cup s_{bc}$, and similarly for s_{ac} and s_{bc}.

3) X is ***Gromov hyperbolic*** if there exists some $\delta > 0$, such that every (C-coarse) triangle in X is δ-thin.

(10.2.2) **Theorem.** *The universal cover of any compact manifold of strictly negative sectional curvature is Gromov hyperbolic.*

Idea of proof. As an illustration, let us show that the hyperbolic plane $\mathfrak{H}^2$ is Gromov hyperbolic. We use the disk model.

Any three distinct points a, b, and c on $\partial\mathfrak{H}^2$ are the vertices of an ideal triangle (with geodesic sides). Choose a point p on $\overline{ab}$. Since the geodesic ray $\overrightarrow{pa}$ is asymptotic to $\overrightarrow{ca}$, there is some $\delta > 0$, such that every point of $\overrightarrow{pa}$ is in the δ-neighborhood of $\overrightarrow{ca}$. Similarly, every point of $\overrightarrow{pb}$ is in the δ-neighborhood of $\overrightarrow{cb}$ (after we enlarge δ). Therefore, all of $\overline{ab}$ is in the δ-neighborhood of the union of the other two sides. By applying the same argument to the sides $\overline{bc}$ and $\overline{ac}$, we see there is some δ, such that the triangle abc is δ-thin.

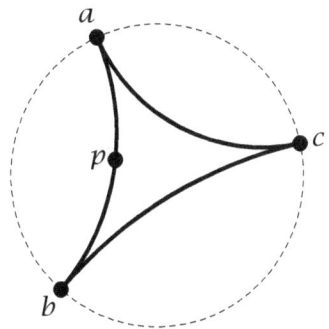

Since the isometry group $SL(2, \mathbb{R})$ acts transitively on the (unordered) triples of distinct points on the boundary, we conclude that every ideal triangle is δ-thin for this same value of δ. Having vertices on the boundary is the worst-case scenario, so this implies that all geodesic triangles in $\mathfrak{H}^2$ are δ-thin. $\square$

The following important "shadowing property" tells us that "quasi-geodesics" are always close to geodesics. Unfortunately, its proof is somewhat lengthy, so we omit it. (See Exercises 1 and 2 for some weaker results that are easier.)

(10.2.3) **Theorem.** *Suppose*
- *X is a C-coarsely geodesic δ-hyperbolic metric space, and*
- *$\{x_0, x_1, \ldots, x_n\}$ is finite sequence of points in X, such that, for all i, j we have:*
$$\frac{|i - j|}{C} - C \le d(x_i, x_j) \le C|i - j| + C.$$
Then there exists $C' > 0$ (depending only on C and δ), such that the set $\{x_0, x_1, \ldots, x_n\}$ is contained in the C'-neighborhood of every C-coarse geodesic from x_0 to x_1.

This implies that being Gromov hyperbolic is invariant under quasi-isometry:

(10.2.4) **Corollary** (see Exercise 4). *Assume*
- *X_1 and X_2 are coarsely geodesic metric spaces, and*
- *$X_1 \overset{QI}{\sim} X_2$.*

If X_1 is Gromov hyperbolic, then X_2 is Gromov hyperbolic.

(10.2.5) **Corollary.** *The fundamental group of any compact manifold M of strictly negative sectional curvature is Gromov hyperbolic.*

Proof. Since M is compact, the fundamental group $\pi_1(M)$ acts cocompactly on the universal cover $\widetilde{M}$ of M, so $\pi_1(M) \overset{QI}{\sim} \widetilde{M}$ (see Proposition 10.1.7). Now apply Theorem 10.2.2 and Corollary 10.2.4. $\square$

This observation allows us to determine precisely which lattices are Gromov hyperbolic:

(10.2.6) **Proposition.** *Γ is Gromov hyperbolic if and only if $\mathrm{rank}_{\mathbb{R}} G = 1$ and either G/Γ is compact, or the unique noncompact simple factor of G is isogenous to $\mathrm{SL}(2, \mathbb{R})$.*

Sketch of proof. With a bit more theory than has been presented here, it is not difficult to show that Gromov hyperbolic groups never contain a subgroup isomorphic to $\mathbb{Z} \times \mathbb{Z}$, so we may assume Γ is irreducible.

Case 1. Assume G/Γ is compact. From Proposition 10.1.7, we know that Γ is quasi-isometric to the symmetric space G/K associated to G.
- If $\mathrm{rank}_{\mathbb{R}} G = 1$, then G/K has negative sectional curvature, bounded away from 0, so it is Gromov hyperbolic.
- If $\mathrm{rank}_{\mathbb{R}} G \ge 2$, then G/K contains 2-dimensional flats, so it is not Gromov hyperbolic.

Case 2. Assume G/Γ is not compact. (⇒) We may assume that G has no compact factors. Let U be a maximal unipotent subgroup of $Γ$. We know that U does not contain a subgroup isomorphic to $\mathbb{Z} \times \mathbb{Z}$. Also, since $G/Γ$ is not compact, we know U is infinite (see Remark 4.4.5). Therefore, since U is nilpotent (and torsion-free), it is easy to see that U must be cyclic. It can be shown that this implies G is isogenous to $\mathrm{SL}(2, \mathbb{R})$.

(⇐) $Γ$ is virtually free, so it is Gromov hyperbolic (see Exercise 3). □

Exercises for §10.2.

#1. Show that the coarse geodesic between two points in a Gromov hyperbolic space is coarsely unique. More precisely, given C, show there is some $C' > 0$, such that if y and y' are two C-coarse geodesics with the same endpoints, then y is contained in the C'-neighborhood of y'.

[*Hint:* If a and b are the two endpoints, consider the (degenerate) triangle abb.]

#2. Show that if X is a C-coarsely geodesic δ-hyperbolic space, and $C' \geq C$, then there exists δ', such that every C'-coarse geodesic from a to b is in the δ'-neighborhood of every C-coarse geodesic from a to b. (This is a generalization of Exercise 1.)

[*Hint:* For any point c on a C'-coarse geodesic from a to b, there is a C-coarse triangle abc. If c is not in the δ-neighborhood of $\overline{ab}$, then there exist $a', b' \in \overline{ab}$ that are distance less than δ from points a'' and b'' on $\overline{ac}$ and $\overline{bc}$, respectively, such that $d(a', b') < C + 1$. Bound $d(a', c)$ by noting that c is on the C''-coarse geodesic $\overline{a''c} \cup \overline{cb''}$.]

#3. Show that free groups are Gromov hyperbolic.

[*Hint:* The word metric corresponding to a set of free generators is 0-hyperbolic.]

#4. Prove Corollary 10.2.4.

[*Hint:* Use Theorem 10.2.3 to show that coarse triangles in X_2 can be approximated by coarse triangles in X_1.]

Notes

Almost all of the material in this chapter can be found in any treatment of geometric group theory, such as [5], [7], or (more elementary) [1]. A detailed treatment of this and much more is in [2].

The Lubotzky-Mozes-Raghunathan Theorem (10.1.10) is proved in [9] (or see [8] for an exposition of the special case where $Γ = \mathrm{SL}(n, \mathbb{Z})$).

See [3] for an introduction to the theory of Gromov hyperbolic groups, or [4, 6] for much more information. The notion of δ-hyperbolic group is credited to E. Rips, who also proved some of the basic properties (such as that they are finitely presented), but much of the foundational work in the subject was done by M. Gromov [6].

Proposition 10.2.6 is well known.

References

[1] B. Bowditch: *A Course on Geometric Group Theory.* Mathematical Society of Japan, Tokyo, 2006. ISBN 4-931469-35-3, MR 2243589

[2] M. Bridson and A. Haefliger: *Metric Spaces of Non-Positive Curvature.* Springer, Berlin, 1999. ISBN 3-540-64324-9, MR 1744486

[3] É. Ghys: Les groupes hyperboliques, *Astérisque* 189-190 (1990) 203–238. MR 1099877,
http://www.numdam.org/item?id=SB_1989-1990__32__203_0

[4] É. Ghys and P. de la Harpe: Sur les Groupes Hyperboliques d'après Mikhael Gromov. Birkhäuser, Boston, 1990. ISBN 0-8176-3508-4, MR 1086648

[5] É. Ghys and P. de la Harpe: Infinite groups as geometric objects (after Gromov), in T. Bedford, et al., eds.: *Ergodic Theory, Symbolic Dynamics, and Hyperbolic Spaces (Trieste, 1989)*, Oxford Univ. Press, New York, 1991, pp. 299–314. ISBN 0-19-853390-X, MR 1130180

[6] M. Gromov: Hyperbolic groups, in S. M. Gersten, ed.: *Essays in Group Theory*, Springer, New York, 1987, pp. 75–263. ISBN 0-387-96618-8, MR 0919829

[7] P. de la Harpe: *Topics in Geometric Group Theory.* Univ. of Chicago Press, Chicago, 2000. ISBN 0-226-31719-6, MR 1786869

[8] A. Lubotzky, S. Mozes, and M. S. Raghunathan: Cyclic subgroups of exponential growth and metrics on discrete groups, *Comptes Rendus Acad. Sci. Paris*, Ser. A, 317 (1993), no. 8, 735–740. MR 1244421,
http://gallica.bnf.fr/ark:/12148/bpt6k5808224h/f739.image

[9] A. Lubotzky, S. Mozes, and M. S. Raghunathan: The word and Riemannian metrics on lattices of semisimple groups, *Publ. Math. IHES* 91 (2000) 5–53. MR 1828742,
http://www.numdam.org/item?id=PMIHES_2000__91__5_0

Chapter 11

Unitary Representations

Unitary representations are of the utmost importance in the study of Lie groups. For our purposes, one of the main applications is the proof of the Moore Ergodicity Theorem (4.10.3) in Section 11.2, but they are also the foundation of the definition (and study) of Kazhdan's Property (T) in Chapter 13.

§11.1. Definitions

(11.1.1) **Definition.** Assume $\mathcal{H}$ is a Hilbert space, with inner product $\langle \, | \, \rangle$, and H is a Lie group.

1) $\mathcal{U}(\mathcal{H})$ is the group of unitary operators on $\mathcal{H}$.

2) A **unitary representation** of the Lie group H on the Hilbert space $\mathcal{H}$ is a homomorphism $\pi \colon H \to \mathcal{U}(\mathcal{H})$, such that the map $h \mapsto \pi(h)\varphi$ is continuous, for each $\varphi \in \mathcal{H}$. (If we wish to spell out that a unitary representation is on a particular Hilbert space $\mathcal{H}$, we may refer it as $(\pi, \mathcal{H})$, rather than merely π.)

3) The **dimension** of a unitary representation $(\pi, \mathcal{H})$ is the dimension of the Hilbert space $\mathcal{H}$.

Recall: The Standing Assumptions (4.0.0 on page 41) are in effect, so, as always, Γ is a lattice in the semisimple Lie group $G \subseteq \mathrm{SL}(\ell, \mathbb{R})$.

Main prerequisites for this chapter: none.

4) Suppose $(\pi_1, \mathcal{H}_1)$ and $(\pi_2, \mathcal{H}_2)$ are unitary representations of H.

  (a) The **direct sum** of the representations π_1 and π_2 is the unitary representation $\pi_1 \oplus \pi_2$ of H on $\mathcal{H}_1 \oplus \mathcal{H}_2$ that is defined by

  $$(\pi_1 \oplus \pi_2)(h)(\varphi_1, \varphi_2) = (\pi_1(h)\varphi_1, \pi_2(h)\varphi_2),$$

  for $h \in H$ and $\varphi_i \in \mathcal{H}_i$.

  (b) π_1 and π_2 are **isomorphic** if there is a Hilbert-space isomorphism $T \colon \mathcal{H}_1 \overset{\cong}{\to} \mathcal{H}_2$ that **intertwines** the two representations. This means $T(\pi_1(h)\varphi) = \pi_2(h) T(\varphi)$, for all $h \in H$ and $\varphi \in \mathcal{H}_1$.

(11.1.2) **Example.** Every group H has a **trivial representation**, denoted by $\mathbb{1}$ (or $\mathbb{1}_H$, if it will avoid confusion). It is a unitary representation on the 1-dimensional Hilbert space $\mathbb{C}$, and is defined by $\mathbb{1}(h)\varphi = \varphi$ for all $h \in H$ and $\varphi \in \mathbb{C}$.

Here is a more interesting example:

(11.1.3) **Example.** Suppose

- H is a Lie group,

- H acts continuously on a locally compact, metrizable space X, and

- μ is an H-invariant Radon measure on X.

Then there is a unitary representation of H on $\mathcal{L}^2(X, \mu)$, defined by

$$(\pi(h)\varphi)(x) = \varphi(h^{-1}x)$$

(cf. Exercise 4.10#11). For the action of H on itself by translations (on the left), the resulting representation π_{reg} of H on $\mathcal{L}^2(H)$ is called the (left) **regular representation** of H.

(11.1.4) **Definition.** Suppose π is a unitary representation of H on $\mathcal{H}$, H' is a subgroup of H, and $\mathcal{K}$ is a closed subspace of $\mathcal{H}$.

  1) $\mathcal{K}$ is **H'-invariant** if $\pi(h')\mathcal{K} = \mathcal{K}$, for all $h' \in H'$. (If the representation π is not clear from the context, we may also say that $\mathcal{K}$ is $\pi(H')$-invariant.)

  2) For the special case where $H' = H$, an H-invariant subspace is simply said to be **invariant**, and the representation of H on any such subspace is called a **subrepresentation** of π. More precisely, if $\mathcal{K}$ is H-invariant (and closed), then the corresponding subrepresentation is the unitary representation $\pi_{\mathcal{K}}$ of H on $\mathcal{K}$, defined by $\pi_{\mathcal{K}}(h)\varphi = \pi(h)\varphi$, for all $h \in H$ and $\varphi \in \mathcal{K}$.

(11.1.5) **Lemma** (see Exercise 2). *If $(\pi, \mathcal{H})$ is a unitary representation of H, and $\mathcal{K}$ is a closed, H-invariant subspace of $\mathcal{H}$, then $\pi \cong \pi_{\mathcal{K}} \oplus \pi_{\mathcal{K}^{\perp}}$.*

The above lemma shows that any invariant subspace leads to a decomposition of the representation into a direct sum of subrepresentations. This suggests that the fundamental building blocks are the representations that do not have any (interesting) subrepresentations. Such representations are called "irreducible:"

(11.1.6) **Definitions.** Let H be a Lie group.
  1) A unitary representation $(\pi, \mathcal{H})$ of H is **irreducible** if it has no nontrivial, proper, closed, invariant subspaces. That is, the only closed, H-invariant subspaces of $\mathcal{H}$ are $\{0\}$ and $\mathcal{H}$.
  2) The set of all irreducible representations of H (up to isomorphism) is called the **unitary dual** of H, and is denoted $\hat{H}$.

(11.1.7) **Warnings.**
  1) Unfortunately, it is usually not the case that every unitary representation of H is a direct sum of irreducible representations. (This is a generalization of the fact that if U is a unitary operator on $\mathcal{H}$, then $\mathcal{H}$ may not be a direct sum of eigenspaces of U.) However, it will be explained in Section 11.6 that every unitary representation is a "direct integral" of irreducible representations. In the special case where $H = \mathbb{Z}$, this is a restatement of the Spectral Theorem for unitary operators (cf. Proposition B7.12).
  2) Although the unitary dual $\hat{H}$ has a fairly natural topology, it can be quite bad. In particular, the topology may not be Hausdorff. Indeed, in some cases, the topology is so bad that there does not exist an injective, Borel measurable function from $\hat{H}$ to $\mathbb{R}$. Fortunately, though, the worst problems do not arise for semisimple Lie groups: the unitary dual is always "tame" (measurably, at least) in this case.

Exercises for §11.1.

#1. Suppose π is a unitary representation of H on $\mathcal{H}$, and define a map $\xi \colon H \times \mathcal{H} \to \mathcal{H}$ by $\xi(h, v) = \pi(h)v$. Show that ξ is continuous.
  [*Hint:* Use the fact that $\pi(H)$ consists of unitary operators.]

#2. Prove Lemma 11.1.5.
  [*Hint:* If $\mathcal{K}$ is invariant, then $\mathcal{K}^\perp$ is also invariant. Define $T \colon \mathcal{K} \oplus \mathcal{K}^\perp \to \mathcal{H}$ by $T(\varphi, \psi) = \varphi + \psi$.]

#3. (Schur's Lemma) Suppose $(\pi, \mathcal{H})$ is an irreducible unitary representation of H, and T is a bounded operator on $\mathcal{H}$ that commutes with every element of $\pi(\mathcal{H})$. Show there exists $\lambda \in \mathbb{C}$, such that $T\varphi = \lambda \varphi$, for every $\varphi \in \mathcal{H}$.
  [*Hint:* Assume T is normal, by considering $T + T^*$ and $T - T^*$, and apply the Spectral Theorem (B7.12).]

#4. Suppose $(\pi, \mathcal{H})$ is an irreducible unitary representation of H, and $\langle \mid \rangle'$ is another H-invariant inner product on $\mathcal{H}$ that defines the same topology on $\mathcal{H}$ as the original inner product $\langle \mid \rangle$. Show there exists $c \in \mathbb{R}^+$, such that $\langle \mid \rangle' = c\langle \mid \rangle$.

[*Hint:* Each inner product provides an isomorphism of $\mathcal{H}$ with $\mathcal{H}^*$. Exercise 3 implies they are the same, up to a scalar multiple.]

#5. (optional) In the situation of Example 11.1.3, a weaker assumption on μ suffices to define a unitary representation on $\mathcal{L}^2(X, \mu)$. Namely, instead of assuming that μ is invariant, it suffices to assume only that the *class* of μ is invariant. This means, for every measurable subset A, and all $h \in H$, we have $\mu(A) = 0 \Leftrightarrow \mu(hA) = 0$. Then, for each $h \in H$, the Radon-Nikodym Theorem (B6.13) provides a function $D_h \colon X \to \mathbb{R}^+$, such that $h_*\mu = D_h\mu$. Show that the formula

$$(\pi(h)\varphi) \;=\; \sqrt{D_h(x)}\; \varphi(h^{-1}x)$$

defines a unitary representation of H on $\mathcal{L}^2(X, \mu)$.

§11.2. Proof of the Moore Ergodicity Theorem

Recall the following result that was proved only in a special case:

(11.2.1) **Theorem** (Moore Ergodicity Theorem (4.10.3)). *Suppose*

- *G is connected and simple,*
- *H is a closed, noncompact subgroup of G,*
- *Λ is a discrete subgroup of G, and*
- *ϕ is an H-invariant $\mathcal{L}^p$-function on G/Λ (with $1 \le p < \infty$).*

Then ϕ is constant (a.e.).

This is an easy consequence of the following result in representation theory (see Exercise 1).

(11.2.2) **Theorem** (Decay of matrix coefficients). *If*

- *G is simple,*
- *π is a unitary representation of G on a Hilbert space $\mathcal{H}$, such that no nonzero vector is fixed by $\pi(G)$, and*
- *$\{g_j\}$ is a sequence of elements of G, such that $\|g_j\| \to \infty$,*

then $\langle \pi(g_j)\phi \mid \psi \rangle \to 0$, for every $\phi, \psi \in \mathcal{H}$.

Proof. Assume, for simplicity, that

$$G = \mathrm{SL}(2, \mathbb{R}).$$

(A reader familiar with the theory of real roots and Weyl chambers should have little difficulty in extending this proof to the general case; cf. Exercise 6.) Let

$$A = \begin{bmatrix} * & \\ & * \end{bmatrix} \subset G.$$

Further assume, for simplicity, that
$$\{g_j\} \subseteq A.$$
(It is not difficult to eliminate this hypothesis; see Exercise 5.) By passing to a subsequence, we may assume $\pi(g_j)$ converges weakly, to some operator E; that is,
$$\langle \pi(g_j)\phi \mid \psi \rangle \to \langle E\phi \mid \psi \rangle \quad \text{for every } \phi, \psi \in \mathcal{H}$$
(see Exercise 2). Let
$$U = \{\, v \in G \mid g_j^{-1} v g_j \to e \,\} \tag{11.2.3}$$
and
$$U^- = \{\, u \in G \mid g_j u g_j^{-1} \to e \,\}. \tag{11.2.4}$$
For $u \in U^-$ and $\phi, \psi \in \mathcal{H}$, we have
$$
\begin{aligned}
\langle E\pi(u)\phi \mid \psi \rangle &= \lim \langle \pi(g_j u)\phi \mid \psi \rangle \\
&= \lim \langle \pi(g_j u g_j^{-1})\, \pi(g_j)\phi \mid \psi \rangle \\
&= \lim \langle \pi(g_j)\phi \mid \psi \rangle \qquad \text{(see Exercise 3)} \\
&= \langle E\phi \mid \psi \rangle,
\end{aligned}
$$
so $E\pi(u) = E$. Therefore, letting
$$\mathcal{H}^{U^-} = \{\, \phi \in \mathcal{H} \mid \pi(u)\phi = \phi \text{ for all } u \in U^- \,\}$$
be the space of U^--invariant vectors in $\mathcal{H}$, we have
$$(\mathcal{H}^{U^-})^\perp \subseteq \ker E \tag{11.2.5}$$
(see Exercise 4). Similarly, since
$$\langle E^*\phi \mid \psi \rangle = \langle \phi \mid E\psi \rangle = \lim \langle \phi \mid \pi(g_j)\psi \rangle = \lim \langle \pi(g_j^{-1})\phi \mid \psi \rangle,$$
the same argument, with E^* in the place of E and g_j^{-1} in the place of g_j, shows that
$$(\mathcal{H}^U)^\perp \subseteq \ker E^*.$$

We also have
$$
\begin{aligned}
\langle \pi(g_j)\phi \mid \pi(g_k)\phi \rangle &= \langle \pi(g_k^{-1} g_j)\phi \mid \phi \rangle && (\pi(g_k^{-1}) \text{ is unitary}) \\
&= \langle \pi(g_j g_k^{-1})\phi \mid \phi \rangle && (A \text{ is abelian}) \\
&= \langle \pi(g_k^{-1})\phi \mid \pi(g_j^{-1})\phi \rangle.
\end{aligned}
$$
Applying $\lim_{j \to \infty} \lim_{k \to \infty}$ to both sides yields $\|E\phi\|^2 = \|E^*\phi\|^2$, and this implies $\ker E = \ker E^*$. Hence,
$$
\begin{aligned}
\ker E = \ker E + \ker E^* &\supset (\mathcal{H}^{U^-})^\perp + (\mathcal{H}^U)^\perp \\
&= (\mathcal{H}^{U^-} \cap \mathcal{H}^U)^\perp = (\mathcal{H}^{\langle U, U^- \rangle})^\perp.
\end{aligned}
$$
Now, by passing to a subsequence of $\{g_j\}$, we may assume $\langle U, U^- \rangle = G$ (see Exercise 7). Then $\mathcal{H}^{\langle U,U^- \rangle} = \mathcal{H}^G = 0$, so $\ker E \supset 0^\perp = \mathcal{H}$. This

implies that, for all $\phi, \psi \in \mathcal{H}$, we have
$$\lim\langle \pi(g_j)\phi \mid \psi\rangle = \langle E\phi \mid \psi\rangle = \langle 0 \mid \psi\rangle = 0. \qquad \square$$

(11.2.6) *Remark.* If A is a bounded operator on a Hilbert space $\mathcal{H}$, and $\phi, \psi \in \mathcal{H}$, then the inner product $\langle A\phi \mid \psi\rangle$ is called a **matrix coefficient** of A. The motivation for this terminology is that if $A \in \mathrm{Mat}_{n\times n}(\mathbb{R})$, and $\varepsilon_1, \ldots, \varepsilon_n$ is the standard basis of $\mathcal{H} = \mathbb{R}^n$, then $\langle A\varepsilon_j \mid \varepsilon_i\rangle$ is the (i,j) matrix entry of A.

The above argument yields the following more general result.

(11.2.7) **Corollary** (of proof). *Assume*
- *G has no compact factors,*
- *π is a unitary representation of G on a Hilbert space $\mathcal{H}$, and*
- *$\{g_n\} \to \infty$ in G/N, for every proper, normal subgroup N of G.*

Then $\langle g_n\phi \mid \psi\rangle \to 0$, for every $\phi, \psi \in (\mathcal{H}^G)^\perp$.

This has the following consequence, which implies the Moore Ergodicity Theorem (see Exercise 9).

(11.2.8) **Corollary** (Mautner phenomenon (see Exercise 8)). *Assume*
- *G has no compact factors,*
- *π is a unitary representation of G on a Hilbert space $\mathcal{H}$, and*
- *H is a closed subgroup of G.*

Then there is a closed, normal subgroup N of G, containing a cocompact subgroup of H, such that every $\pi(H)$-invariant vector in $\mathcal{H}$ is $\pi(N)$-invariant.

(11.2.9) *Remark.* Theorem 11.2.2 does not give any information about the *rate* at which the function $\langle \phi(g)\phi \mid \psi\rangle$ tends to 0 as $\|g\| \to \infty$. For some applications, it is helpful to know that, for many choices of the vectors ϕ and ψ, the inner product decays exponentially fast:

If G, π, and $\mathcal{H}$ are as in Theorem 11.2.2, then there is a dense, linear subspace $\mathcal{H}_\infty$ of $\mathcal{H}$, such that, for all $\phi, \psi \in \mathcal{H}_\infty$, there exist $a, b > 1$, such that
$$|\langle \phi(g)\phi \mid \psi\rangle| < \frac{b}{a^{\|g\|}} \qquad \text{for all } g \in G.$$

Specifically, if K is a maximal compact subgroup of G, then we may take
$$\mathcal{H}_\infty = \{\phi \in \mathcal{H} \mid \text{the linear span of } K\phi \text{ is finite-dimensional}\}.$$
(So $\mathcal{H}_\infty$ is the space of "**K-finite**" vectors.)

Exercises for §11.2.

#1. Show that Theorem 11.2.1 is a corollary of Theorem 11.2.2.

[*Hint:* If ϕ is an H-invariant function in $\mathscr{L}^p(G/\Gamma)$, let $\phi' = |\phi|^{p/2} \in \mathscr{L}^2(G/\Gamma)$. Then $\langle \phi', g_j\phi' \rangle = \langle \phi', \phi' \rangle$ for every $g_j \in H$.]

#2. Let $\{T_j\}$ be a sequence of unitary operators on a Hilbert space $\mathcal{H}$. Show there is a subsequence $\{T_{j_i}\}$ of $\{T_j\}$ and a bounded operator E on $\mathcal{H}$, such that $\langle T_{j_i} v \mid w \rangle \overset{i \to \infty}{\longrightarrow} \langle Ev \mid w \rangle$ for all $v, w \in \mathcal{H}$.

[*Hint:* Choose an orthonormal basis $\{e_p\}$. For each p, q, the sequence $\{\langle T_j e_p \mid e_q \rangle$ is bounded, and therefore has a subsequence that converges to some $\alpha_{p,q}$. Cantor diagonalization implies that we may assume, after passing to a subsequence, that $\langle T_j e_p \mid e_q \rangle \to \alpha_{p,q}$ for all p and q.]

#3. Suppose
- π is a unitary representation of G on $\mathcal{H}$,
- $\{\phi_j\}$ is a sequence of unit vectors in $\mathcal{H}$, and
- $u_j \to e$ in G.

Show $\lim \langle \pi(u_j)\phi_j \mid \psi \rangle = \lim \langle \phi_j \mid \psi \rangle$, for all $\psi \in \mathcal{H}$.

[*Hint:* Move $\pi(u_j)$ to the other side of the inner product. Then use the continuity of π and the boundedness of $\{\phi_j\}$.]

#4. Prove (11.2.5).

[*Hint:* Let $\mathcal{K}$ be the closure of $\{\pi(u)\phi - \phi \mid u \in U^-, \phi \in (\mathcal{H}^{U^-})^\perp\}$, and note that $\mathcal{K} \subseteq \ker E$. If $\psi \in \mathcal{K}^\perp$, then $\pi(u)\psi - \psi = 0$ for all $u \in U^-$ (why?), so $\psi \in \mathcal{H}^{U^-}$.]

#5. Eliminate the assumption that $\{g_j\} \subseteq A$ from the proof of Theorem 11.2.2.

[*Hint:* You may assume the Cartan decomposition, which states that $G = KAK$, where K is compact. Hence, $g_j = c_j a_j c'_j$, with $c_j, c'_j \in K$ and $a_j \in A$. Assume, by passing to a subsequence, that $\{c_j\}$ and $\{c'_j\}$ converge. Then

$$\lim \langle \pi(g_j)\phi \mid \psi \rangle = \lim \langle \pi(a_j)(\pi(c')\phi) \mid \pi(c)^{-1}\psi \rangle = 0$$

if $c_j \to c$ and $c'_j \to c'$.]

#6. Prove Theorem 11.2.2 for the special case where $G = \mathrm{SL}(n, \mathbb{R})$.

#7. For $G, A, \{g_j\}, U$, and U^- as in the proof of Theorem 11.2.2 (with $\{g_j\} \subseteq A$), show that if $\{g_j\}$ is replaced by an appropriate subsequence, then $\langle U, U^- \rangle = G$.

[*Hint:* Arrange that U is $\begin{bmatrix} 1 & * \\ & 1 \end{bmatrix}$ and U^- is $\begin{bmatrix} 1 & \\ * & 1 \end{bmatrix}$, or vice versa.]

#8. Derive Corollary 11.2.8 from Corollary 11.2.7.

#9. Derive Theorem 14.2.4 from Corollary 11.2.8.

[*Hint:* If f is H-invariant, then $\langle \pi(h)f \mid f \rangle = \langle f \mid f \rangle$ for all $h \in H$.]

#10. Suppose
- G is connected, with no compact factors,
- Λ is a discrete subgroup of G,

- H is a subgroup of G whose projection to every simple factor of G is not precompact, and
- ϕ is an H-invariant $\mathscr{L}^p$ function on G/Λ, with $1 \le p < \infty$.

Show that ϕ is constant (a.e.).

#11. Suppose
- H is a noncompact, closed subgroup of G,
- Γ is irreducible, and
- $\phi\colon G/\Gamma \to \mathbb{R}$ is any H-invariant, measurable function.

Show that ϕ is constant (a.e.).

[*Hint:* There is no harm in assuming that ϕ is bounded (why?), so it is in $\mathscr{L}^2(G/\Gamma)$ (why?). Apply Corollary 11.2.8.]

§11.3. Induced representations

It is obvious that if H is a subgroup of G, then any unitary representation of G can be restricted to a unitary representation of H. (That is, if we define $\pi|_H$ by $\pi|_H(h) = \pi(h)$ for $h \in H$, then $\pi|_H$ is a unitary representation of H.) What is not so obvious is that, conversely, every unitary representation of H can be "induced" to a unitary representation of G. We will need only the special case where $H = \Gamma$ is a lattice in G (but see Exercise 1 for the definition in general). This construction will be a key ingredient of the proof in Section 13.4 that Γ often has Kazhdan's property (T).

(11.3.1) **Definition** (Induced representation). Suppose π is a unitary representation of Γ on $\mathcal{H}$.

1) A measurable function $\varphi\colon G \to \mathcal{H}$ is said to be (essentially) right **Γ-equivariant** if, for each $\gamma \in \Gamma$, we have
$$\varphi(g\gamma^{-1}) = \pi(\gamma)\,\varphi(g) \text{ for a.e. } g \in G.$$

2) We use $\mathscr{L}_\Gamma(G;\mathcal{H})$ to denote the space of right Γ-equivariant measurable functions from G to $\mathcal{H}$, where two functions are identified if they agree almost everywhere.

3) For $\varphi \in \mathscr{L}_\Gamma(G;\mathcal{H})$, we have $\|\varphi(g\gamma)\|_{\mathcal{H}} = \|\varphi(g)\|_{\mathcal{H}}$ for every $\gamma \in \Gamma$ and a.e. $g \in G$ (see Exercise 2). Hence, $\|\varphi(g)\|_{\mathcal{H}}$ is a well-defined function on G/Γ (a.e.), so we may define the $\mathscr{L}^2$-norm of φ by
$$\|\varphi\|_2 = \left(\int_{G/\Gamma} \|\varphi(g)\|_{\mathcal{H}}^2 \, dg \right)^{1/2}.$$

4) We use $\mathscr{L}_\Gamma^2(G;\mathcal{H})$ to denote the subspace of $\mathscr{L}_\Gamma(G;\mathcal{H})$ consisting of the functions with finite $\mathscr{L}^2$-norm. It is a Hilbert space (see Exercise 3).

5) Note that G acts by unitary operators on $\mathscr{L}_\Gamma^2(G;\mathcal{H})$, via
$$(g \cdot \varphi)(x) = \varphi(g^{-1}x) \quad \text{for } g \in G,\ \varphi \in \mathscr{L}_\Gamma^2(G;\mathcal{H}),\ \text{and } x \in G$$

(see Exercise 4). This unitary representation of G is called the representation **induced** from π, and it is denoted $\operatorname{Ind}_\Gamma^G(\pi)$.

Exercises for §11.3.

#1. (optional) Suppose $(\pi, \mathcal{H})$ is a unitary representation of a closed subgroup H of G. Define $\operatorname{Ind}_H^G(\pi)$, without assuming that there is a G-invariant measure on G/H.

[*Hint:* Since G/H is a C^∞ manifold (see Proposition A6.2), we may use a nowhere-vanishing differential form to choose a measure μ on G/H, such that $f_* \mu$ is in the class of μ, for every diffeomorphism f of G/H. A unitary representation of G on $\mathcal{L}^2(G/H, \mu)$ can be defined by using Radon-Nikodym derivatives, as in Exercise 11.1#5, and the same idea yields a unitary representation on a space of H-equivariant functions.]

#2. Let $\varphi \in \mathcal{L}_\Gamma(G; \mathcal{H})$ and $\gamma \in \Gamma$, where π is a unitary representation of Γ on $\mathcal{H}$. Show $\|\varphi(g\gamma)\|_\mathcal{H} = \|\varphi(g)\|_\mathcal{H}$, for a.e. $g \in G$.

#3. Show $\mathcal{L}_\Gamma^2(G; \mathcal{H})$ is a Hilbert space (with the given norm, and assuming that two functions represent the same element of the space if and only if they are equal a.e.).

#4. Show that the formula in Definition 11.3.1(5) defines a unitary representation of G on $\mathcal{L}_\Gamma^2(G; \mathcal{H})$.

#5. Show that $\operatorname{Ind}_\Gamma^G(\mathbb{1})$ is (isomorphic to) the usual representation of G on $\mathcal{L}^2(G/\Gamma)$ (by left translation).

#6. Show that if $\operatorname{Ind}_\Gamma^G(\pi)$ is irreducible, then π is irreducible.

#7. Show that the converse of Exercise 6 is false.

[*Hint:* Is the representation of G on $\mathcal{L}^2(G/\Gamma)$ irreducible?]

§11.4. Representations of compact groups

(11.4.1) **Example.** Consider the circle $\mathbb{R}/\mathbb{Z}$. For each $n \in \mathbb{Z}$, define
$$e_n \colon \mathbb{R}/\mathbb{Z} \to \mathbb{C} \text{ by } e_n(t) = e^{2\pi i n t}.$$
The theory of Fourier Series tells us that $\{e_n\}$ is an orthonormal basis of $\mathcal{L}^2(\mathbb{R}/\mathbb{Z})$, which means we have the direct-sum decomposition
$$\mathcal{L}^2(\mathbb{R}/\mathbb{Z}) = \bigoplus_{n \in \mathbb{Z}} \mathbb{C} e_n.$$
Furthermore, it is easy to verify that each subspace $\mathbb{C} e_n$ is an invariant subspace for the regular representation (see Exercise 1), and, being 1-dimensional, is obviously irreducible. Hence, we have a decomposition of the regular representation into a direct sum of irreducible representations. In addition, it is not difficult to see that every irreducible representation of $\mathbb{T}$ occurs exactly once in this representation.

More generally, it is not difficult to show that every unitary representation of $\mathbb{T}$ is a direct sum of 1-dimensional representations.

The following theorem generalizes this to any compact group. However, for nonabelian groups, the irreducible representations cannot all be 1-dimensional (see Exercise 11.6#1).

(11.4.2) Theorem (Peter-Weyl Theorem). *Assume H is **compact**. Then:*

1) *Every unitary representation of H is (isomorphic to) a direct sum of irreducible representations.*
2) *Every irreducible representation of H is finite-dimensional.*
3) *$\widehat{H}$ is countable.*
4) *For the particular case of the regular representation $(\pi_{reg}, \mathcal{L}^2(H))$, we have*

$$\pi_{reg} \cong \bigoplus_{(\pi, \mathcal{H}) \in \widehat{H}} (\dim \mathcal{H}) \cdot \pi,$$

where $k \cdot \pi$ denotes the direct sum $\pi \oplus \cdots \oplus \pi$ of k copies of π. That is, the "multiplicity" of each irreducible representation is equal to its dimension.

Proof. In order to establish both (1) and (2) simultaneously, it suffices to show that if $(\pi, \mathcal{H})$ is any unitary representation of H, then $\mathcal{H}$ is a direct sum of finite-dimensional, invariant subspaces. Zorn's Lemma (B5.3) provides a subspace $\mathcal{M}$ of $\mathcal{H}$ that is maximal among those that are a direct sum of finite-dimensional, invariant subspaces. By passing to the orthogonal complement of $\mathcal{M}$, we may assume that $\mathcal{H}$ has no nonzero, finite-dimensional, invariant subspaces.

Let

- P be the orthogonal projection onto some nonzero subspace of $\mathcal{H}$ that is finite-dimensional,
- μ be the Haar measure on H, and
- $\overline{P} = \int_H \pi(h) \, P \, \pi(h^{-1}) \, d\mu(h)$.

Note that, since $\overline{P}$ commutes with $\pi(H)$ (see Exercise 3), every eigenspace of $\overline{P}$ is H-invariant (see Exercise 6).

Since P is self-adjoint and each $\pi(h)$ is unitary (so $\pi(h^{-1}) = \pi(h)^*$), it is not difficult to see that $\overline{P}$ is self-adjoint. It is also compact (see Exercise 4) and nonzero (see Exercise 5). Therefore, the Spectral Theorem (B7.14) implies that $\overline{P}$ has at least one eigenspace E that is finite-dimensional. By contradicting the fact that $\mathcal{H}$ has no nonzero, finite-dimensional, invariant subspaces, this completes the proof of (1) and (2).

Note that (3) is an immediate consequence of (4), since Hilbert spaces are assumed to be separable (see Assumption B7.7), and therefore cannot have uncountably many terms in a direct sum.

We now give the main idea in the proof of (4). Given an irreducible representation $(\pi, \mathbb{C}^k)$, we will not calculate the exact multiplicity of π, but only indicate how to obtain the correct lower bound by using properties of matrix coefficients. Write $\pi(x) = [f_{i,j}(x)]$. Then

$$[(\pi_{\mathrm{reg}}(h)f_{i,j})(x)] = [f_{i,j}(h^{-1}x)] = \pi(h^{-1}x)$$
$$= \pi(h^{-1})\,\pi(x) = \pi(h^{-1})[f_{i,j}(x)]. \qquad (11.4.3)$$

Now, for $1 \le j \le k$, define $T_j \colon \mathbb{C}^k \to \mathcal{L}^2(H)$, by

$$T_j(a_1, \ldots, a_k) = a_1 f_{1,j} + a_2 f_{2,j} + \cdots + a_k f_{k,j}.$$

Equating the jth columns of the two ends of (11.4.3) tells us that

$$T_j(\pi(h)v) = \pi_{\mathrm{reg}}(h)\,T_j(v),$$

so $T_j(\mathbb{C}^k)$ is an invariant subspace, and the corresponding subrepresentation is isomorphic to π. Therefore, there are (at least) k different copies of π in π_{reg} (one for each value of j). Since, by definition, $k = \dim \pi$, this establishes the correct lower bound for the multiplicity of π. $\qquad\square$

As an illustrative, simple case of the main results in Sections 11.5 and 11.6, we present two different reformulations of the Peter-Weyl Theorem for the special case of abelian groups, after some preliminaries.

(11.4.4) **Definition.** Let A be an abelian Lie group.

  1) A ***character*** of A is a continuous homomorphism $\chi \colon A \to \mathbb{T}$, where $\mathbb{T} = \{\, z \in \mathbb{C} \mid |z| = 1 \,\}$.

  2) The set of all characters of A is called the ***Pontryagin dual*** of A, and is denoted A^*. It is an abelian group under the operation of pointwise multiplication. (That is, the product $\chi_1 \chi_2$ is defined by $(\chi_1 \chi_2)(a) = \chi_1(a)\,\chi_2(a)$, for $\chi_1, \chi_2 \in A^*$ and $a \in A$.) Furthermore, if A/A° is finitely generated, then A^* is a Lie group (with the topology of uniform convergence on compact sets).

(11.4.5) **Observation.** *If A is any abelian Lie group (compact or not), then every irreducible representation $(\pi, \mathcal{H})$ of A is 1-dimensional (see Exercise 8). Therefore, the unitary dual $\hat{A}$ can be identified with the Pontryagin dual A^* (see Exercise 9).*

Hence, for the special case where $H = A$ is abelian, we have the following reformulation of the Peter-Weyl Theorem:

(11.4.6) **Corollary** (see Exercise 10). *Assume $(\pi, \mathcal{H})$ is a unitary representation of a compact, abelian Lie group A. For each $\chi \in A^*$, let*

  • *$\mathcal{H}_\chi = \{\, \varphi \in \mathcal{H} \mid \phi(a)\varphi = \chi(a)\,\varphi, \text{ for all } a \in A \,\}$, and*

  • *$P_\chi \colon \mathcal{H} \to \mathcal{H}_\chi$ be the orthogonal projection.*

Then $\mathcal{H} = \bigoplus_{\chi \in A^*} \mathcal{H}_\chi$, so, for all $a \in A$, we have

$$\pi(a) = \sum_{\chi \in A^*} \chi(a)\,P_\chi.$$

Here is another way of saying the same thing:

(11.4.7) Corollary (see Exercise 11). *Assume $(\pi, \mathcal{H})$ is a unitary representation of a compact, abelian Lie group A. Then there exist*

- *a Radon measure μ on a locally compact metric space Y, and*
- *a Borel measurable function $\chi: Y \to A^*: y \mapsto \chi_y$ (where the countable set A^* is given the discrete topology)*

such that π is isomorphic to the the unitary representation ρ_χ of A on $\mathcal{L}^2(Y, \mu)$ that is defined by

$$(\rho_\chi(a)\varphi)(y) = \chi_y(a)\,\varphi(y) \quad \text{for } a \in A, \varphi \in \mathcal{L}^2(Y, \mu), \text{ and } y \in Y.$$

An analogue of this result for semisimple groups will be stated in Section 11.6, after we define the "direct integral" of a family of representations.

Exercises for §11.4.

#1. In the notation of Example 11.4.1, show $\pi_{\text{reg}}(h)e_n = e^{-2\pi i h}e_n$, for all $h \in \mathbb{R}/\mathbb{Z}$.

#2. Suppose $(\pi, \mathcal{H})$ is a unitary representation of a compact group H, let $\varphi, \psi \in \mathcal{H}$, and define $f: H \to \mathbb{C}$ by $f(h) = \langle \pi(h)\varphi \mid \psi \rangle$. Show $f \in \mathcal{L}^2(H)$.
   [*Hint:* It is a bounded function on a compact set.]

#3. Suppose $(\pi, \mathcal{H})$ is a unitary representation of a compact group H. Show that if T is any bounded operator on $\mathcal{H}$, then

$$\overline{T} = \int_H \pi(h)\, T\, \pi(h^{-1})\, d\mu(h)$$

   is an operator that commutes with every element of $\pi(H)$.
   [*Hint:* The invariance of Haar measure implies $\pi(g)\overline{T}\pi(g^{-1}) = \overline{T}$.]

#4. Show that the operator $\overline{P}$ in the proof of Theorem 11.4.2 is compact.
   [*Hint:* Apply Proposition B7.11, by noting that any integral can be approximated by a finite sum, and the finite sum is an operator whose range is finite-dimensional.]

#5. Show that the operator $\overline{P}$ in the proof of Theorem 11.4.2 is nonzero.
   [*Hint:* Choose some nonzero $\varphi \in \mathcal{H}$, such that $P\varphi = \varphi$. Then $\langle \overline{P}\varphi \mid \varphi \rangle > 0$, since $\langle P\psi \mid \psi \rangle \geq 0$ for all $\psi \in \mathcal{H}$.]

#6. Suppose $(\pi, \mathcal{H})$ is a unitary representation of a Lie group H, T is a bounded operator on $\mathcal{H}$, $\lambda \in \mathbb{C}$, and $\varphi \in \mathcal{H}$. Show that if T commutes with every element of $\pi(H)$, and $T(\varphi) = \lambda\varphi$, then $T(\pi(h)\varphi) = \lambda\,\pi(h)\varphi$, for every $h \in H$.

#7. Assume H is compact. Show that H is finite if and only if it has only finitely many different irreducible unitary representations (up to isomorphism).
   [*Hint:* You may assume Theorem 11.4.2.]

#8. Show that every irreducible representation of any abelian Lie group is 1-dimensional.

[*Hint:* If $\pi(a)$ is not a scalar, for some a, then the Spectral Theorem (B7.12) yields an invariant subspace.]

#9. Let H be a Lie group. Show there is a natural bijection between the set of 1-dimensional unitary representations (modulo isomorphism) and the set of continuous homomorphisms from H to $\mathbb{T}$.

[*Hint:* Any 1-dimensional unitary representation is isomorphic to a representation on $\mathbb{C}$.]

#10. Derive Corollary 11.4.6 from Theorem 11.4.2.

#11. Prove Corollary 11.4.7.

[*Hint:* If $\mathcal{H}_\chi \neq \{0\}$, then $\mathcal{H}_\chi$ is isomorphic to some $\mathcal{L}^2(Y_\chi, \mu_\chi)$. Let $Y = \bigcup_\chi Y_\chi$.]

§11.5. Unitary representations of $\mathbb{R}^n$

Any character χ of $\mathbb{R}^n$ is of the form
$$\chi(a) = e^{2\pi i (a \cdot y)} \qquad \text{for some (unique) } y \in \mathbb{R}^n$$
(see Exercise 1). Therefore, the Pontryagin dual $(\mathbb{R}^n)^*$ (or, equivalently, the unitary dual $\widehat{\mathbb{R}^n}$) can be identified with $\mathbb{R}^n$ (by matching χ with the corresponding vector y). In particular, unlike in Theorem 11.4.2, the unitary dual is uncountable.

Unfortunately, however, not every representation of $\mathbb{R}^n$ is a direct sum of irreducibles. For example, the regular representation π_{reg} of $\mathbb{R}^n$ on $\mathcal{L}^2(\mathbb{R}^n)$ has no 1-dimensional, invariant subspaces (see Exercise 2), so it does not even contain a single irreducible representation and is therefore not a sum of them. Indeed, Fourier Analysis tells us that a function in $\mathcal{L}^2(\mathbb{R}^n)$ is not a *sum* of exponentials, but an *integral*:

$$\varphi(a) = \int_{\mathbb{R}^n} \hat{\varphi}(y) \, e^{2\pi i (a \cdot y)} \, dy,$$

where $\hat{\varphi}$ is the Fourier transform of φ. Now, for each Borel subset E of $\mathbb{R}^n$, let

$$\mathcal{H}_E = \{ f \in \mathcal{L}^2(\mathbb{R}^n) \mid \hat{f}(y) = 0 \text{ for a.e. } y \notin E \}. \qquad (11.5.1)$$

Then it is not difficult to show that $\mathcal{H}_E$ is a closed, π_{reg}-invariant subspace (see Exercise 3).

Now, let $P(E): \mathcal{H} \to \mathcal{H}_E$ be the orthogonal projection. Then we can think of P as a projection-valued measure on $\mathbb{R}^n$ (or on $(\mathbb{R}^n)^*$), and, for all $a \in \mathbb{R}^n$, we have

$$\pi_{\text{reg}}(a) = \int_{\mathbb{R}^n} e^{i(a \cdot y)} \, dP(y) = \int_{(\mathbb{R}^n)^*} \chi(a) \, dE(\chi).$$

If we let $\pi = \pi_{\text{reg}}$, this is a perfect analogue of the conclusion of Corollary 11.4.6, but with the sum replaced by an integral.

A version of the Spectral Theorem tells us that this generalizes in a natural way to all unitary representations of $\mathbb{R}^n$, or, in fact, of any abelian Lie group:

(11.5.2) Proposition. *Suppose π is a unitary representation of an abelian Lie group A on $\mathcal{H}$. Then there is a (unique) projection-valued measure P on A^*, such that*

$$\pi(a) = \int_{A^*} \chi(a)\, dP(\chi) \quad \text{for all } a \in A.$$

This can be reformulated as a generalization of Corollary 11.4.7:

(11.5.3) Corollary. *Let $(\pi, \mathcal{H})$ be a unitary representation of an abelian Lie group A. Then there exist*

- *a probability measure μ on a locally compact metric space Y, and*
- *a Borel measurable function $\chi \colon Y \to A^* \colon y \mapsto \chi_y$,*

such that π is isomorphic to the unitary representation ρ_χ of A on $\mathcal{L}^2(Y, \mu)$ that is defined by

$$(\rho_\chi(a)\varphi)(y) = \chi_y(a)\, \varphi(y) \quad \text{for } a \in A,\, \varphi \in \mathcal{L}^2(Y, \mu),\, \text{and } y \in Y.$$

Proof. Let P be the projection-valued measure given by Proposition 11.5.2. A closed subspace $\mathcal{H}'$ of $\mathcal{H}$ is said to be *cyclic* for P if there exists $\psi \in \mathcal{H}'$, such that the span of $\{\, P(E)\,\psi \mid E \subset A^* \,\}$ is a dense subspace of $\mathcal{H}'$. It is not difficult to see that $\mathcal{H}$ is an orthogonal direct sum of countably many cyclic subspaces (see Exercise 4). Therefore, we may assume $\mathcal{H}$ is cyclic (see Exercise 5) (and nonzero). So we may fix some unit vector ψ that generates a dense subspace of $\mathcal{H}$.

Define a probability measure μ on A^* by

$$\mu(E) = \langle P(E)\psi \mid \psi \rangle = \langle P(E)\psi \mid P(E)\psi \rangle,$$

and let Id be the identity map on A^*.

For the characteristic function f_E of each Borel subset E of A^*, define $\Phi(f_E) = P(E)\psi$. Then $\langle \Phi(f_{E_1}) \mid \Phi(f_{E_2}) \rangle = \langle f_{E_1} \mid f_{E_2} \rangle$, by the definition of μ, so Φ extends to a norm-preserving linear map Φ' from $\mathcal{L}^2(A^*, \mu)$ to $\mathcal{H}$. Since ψ is a cyclic vector for $\mathcal{H}$, we see that Φ' is surjective, so it is an isomorphism of Hilbert spaces. Indeed, Φ' is an isomorphism from ρ_{Id} to π (see Exercise 6). $\qquad\square$

Exercises for §11.5.

#1. Show that every character χ of $\mathbb{R}^n$ is of the form $\chi(a) = e^{2\pi i\, (a \cdot t)}$ for some $t \in \mathbb{R}^n$.

   [*Hint:* Since $\mathbb{R}^n$ is simply connected, any continuous homomorphism from $\mathbb{R}^n$ to $\mathbb{T}$ can be lifted to a homomorphism into the universal cover, which is $\mathbb{R}$. Apply Exercise A6#1.]

#2. Let H be a noncompact Lie group. Show that the regular representation of H has no 1-dimensional, invariant subspaces.

[*Hint:* If φ is in a 1-dimensional, invariant subspace of $\mathscr{L}^2(H)$, then $|\varphi|$ is constant (a.e.).]

#3. For every measurable subset E of $\mathbb{R}^n$, show that the subspace $\mathcal{H}_E$ defined in (11.5.1) is closed, and is invariant under $\pi_{\text{reg}}(\mathbb{R}^n)$.

[*Hint:* It is clear that $\{\hat{f} \mid f \in \mathcal{H}_E\}$ is closed. For invariance, note that the Fourier transform of $\pi_{\text{reg}}(a)f$ is $e^{-2\pi i(a \cdot y)}\hat{f}(y)$.]

#4. Given a projection-valued measure P on a Hilbert space $\mathcal{H}$, show that $\mathcal{H}$ is the orthogonal direct sum of countably many cyclic subspaces.

[*Hint:* Every vector in $\mathcal{H}$ is contained in a cyclic subspace, the orthogonal complement of a $P(E)$-invariant subspace is $P(E)$-invariant, and all Hilbert spaces are assumed to be separable.]

#5. In the notation of Corollary 11.5.3, suppose ρ_{χ_i} is the representation on $\mathscr{L}^2(Y_i, \mu_i)$ corresponding to some $\chi_i \colon Y_i \to A^*$. Show $\bigoplus_{i=1}^{\infty} \rho_{\chi_i} \cong \rho_\chi$, for some Y, μ, and χ.

[*Hint:* Let (Y, μ) be the disjoint union of (copies of) (Y_i, μ_i).]

#6. In the notation of the proof of Corollary 11.5.3, show that Φ' is an isomorphism from ρ_{Id} to π.

[*Hint:* Given $a \in A$ and $E \subset A^*$, write E as the disjoint union of small sets $E_1, \ldots, E_n$ (so $\chi \mapsto \chi(a)$ is almost constant on each E_i). Then

$$\Phi'(\rho_{\text{Id}}(a)f_E) \approx \Phi'\left(\sum_i \chi_i(a) f_{E_i}\right) = \sum_i \chi_i(a) P(E_i)\,\psi \approx \pi(P(E)\psi) = \pi(\Phi'(f_E)),$$

for any $\chi_i \in E_i$.]

§11.6. Direct integrals of representations

Before we define the direct integral of a collection of unitary representations, we first discuss the simpler case of a direct sum of a sequence $\{(\pi_n, \mathcal{H}_n)\}_{n=1}^{\infty}$ of unitary representations.

(11.6.1) **Definition.** If $\{\mathcal{H}_n\}_{n=1}^{\infty}$ is a sequence of Hilbert spaces, then the *direct sum* $\bigoplus_{n=1}^{\infty} \mathcal{H}_n$ consists of all sequences $\{\varphi_n\}_{n=1}^{\infty}$, such that

- $\varphi_n \in \mathcal{H}_n$ for each n, and
- $\sum_{n=1}^{\infty} \|\varphi_n\|^2 < \infty$.

This is a Hilbert space, under the inner product

$$\langle \{\varphi_n\} \mid \{\psi_n\} \rangle = \sum_{n=1}^{\infty} \langle \varphi_n \mid \psi_n \rangle.$$

It contains a copy of $\mathcal{H}_n$, for each n, such that $\mathcal{H}_m \perp \mathcal{H}_n$, for $m \neq n$.

Suppose, now, that all of the Hilbert spaces in the sequence are the same; say, $\mathcal{H}_n = \mathcal{H}$, for all n. Then $\bigoplus_{n=1}^{\infty} \mathcal{H}_n$ is equal to the set of square-integrable functions from $\mathbb{Z}^+$ to $\mathcal{H}$, which can be denoted $\mathcal{L}^2(\mathbb{Z}^+; \mathcal{H})$. In this notation, the direct sum of unitary representations is easy to describe:

(11.6.2) **Definition.** If $\{\pi_n\}_{n=1}^{\infty}$ is a sequence of unitary representations of H on a fixed Hilbert space $\mathcal{H}$, then $\bigoplus_{n=1}^{\infty} \pi_n$ is the unitary representation π on $\mathcal{L}^2(\mathbb{Z}^+; \mathcal{H})$ that is defined by

$$(\pi(h)\varphi)(n) = \pi_n(h)\,\varphi(n) \quad \text{for } h \in H, \varphi \in \mathcal{L}^2(\mathbb{Z}^+; \mathcal{H}), \text{ and } n \in \mathbb{Z}^+.$$

This description of the direct sum naturally generalizes to a definition of the direct integral of representations:

(11.6.3) **Definition.** Suppose
- $\mathcal{H}$ is a Hilbert space,
- $\{\pi_x\}_{x \in X}$ is a measurable family of unitary representations of H on $\mathcal{H}$, which means:
  - X is a locally compact metric space,
  - π_x is a unitary representation of H on $\mathcal{H}$, for each $x \in X$, and
  - for each fixed $\varphi, \pi \in \mathcal{H}$, the expression $\langle \pi_x(h)\varphi \mid \psi \rangle$ is a Borel measurable function on $X \times H$,

  and
- μ is a Radon measure on X.

Then $\int_X \pi_x \, d\mu(x)$ is the unitary representation π of H on $\mathcal{L}^2(X, \mu; \mathcal{H})$ that is defined by

$$(\pi(h)\varphi)(x) = \pi_x(h)\,\varphi(x) \quad \text{for } h \in H, \varphi \in \mathcal{L}^2(X, \mu; \mathcal{H}), \text{ and } x \in X.$$

This is called the ***direct integral*** of the family of representations $\{\pi_x\}$.

The above definition is limited by requiring all of the representations to be on the same Hilbert space. The construction can be generalized to eliminate this assumption (see Remark 11.6.5), but there is often no need:

(11.6.4) **Theorem.** *Assume*
- *π is a unitary representation of G,*
- *G is connected, and has no compact factors, and*
- *no nonzero vector is fixed by every element of $\pi(G)$.*

Then there exist $\mathcal{H}$, $\{\pi_x\}_{x \in X}$, and μ, as in Definition 11.6.3, such that
1) *$\pi \cong \int_X \pi_x \, d\mu(x)$, and*
2) *π_x is irreducible for every $x \in X$.*

(11.6.5) *Remark.* Up to isomorphism, there are only countably many different Hilbert spaces (since any two Hilbert spaces of the same dimension are isomorphic). It is therefore not difficult to generalize Definition 11.6.3 to deal with a family of representations in which the Hilbert

space varies with x. Such a generalization allows every unitary representation of any Lie group to be written as a direct integral of representations
that are irreducible.

Here is one way. Let us say that $\{(\pi_x, \mathcal{H}_x\}_{x \in X}$ is a measurable family
of unitary representations of H if:

- $X = \bigcup_{n=1}^{\infty}$ is the union of countably many locally compact metric
  spaces X_n,
- for each n, there is Hilbert space $\mathcal{H}_n$, such that $\mathcal{H}_x = \mathcal{H}_n$ for all
  $x \in X_n$
- π_x is a unitary representation of H on $\mathcal{H}_x$, for each $x \in X$,
- for each n and each $\varphi, \pi \in \mathcal{H}_n$, the expression $\langle \pi_x(h)\varphi \mid \psi \rangle$ is a
  Borel measurable function on $X_n \times H$, and
- μ is a Radon measure on X.

Given such a family of representations, we define

$$\int_X \pi_x \, d\mu(x) = \bigoplus_{n=1}^{\infty} \int_{X_n} \pi_x \, d\mu(x).$$

With this, more general, notion of direct integral, it can be proved that
every unitary representation of any Lie group is isomorphic to a direct
integral of a measurable family of irreducible unitary representations.

Exercises for §11.6.

#1. Let H be a Lie group. Show that if the regular representation of H
    is a direct integral of 1-dimensional representations, then H is
    abelian.

Notes

There are many books on the theory of unitary representations, including the classics of Mackey [7, 8]. Several books, such as [6], specifically focus on the representations of semisimple Lie groups.

The Moore Ergodicity Theorem (11.2.1) is due to C. C. Moore [10].

Corollary 11.2.7 is due to R. Howe and C. C. Moore [5, Thm. 5.1] and
(independently) R. J. Zimmer [12, Thm. 5.2]. The elementary proof we
give here was found by R. Ellis and M. Nerurkar [2]. Other proofs are in
[9, §2.3, pp. 85-92] and [13, §2.4, pp. 28-31].

A more precise form of the quantitative estimate in Remark 11.2.9
can be found in [4, Cor. 7.2]. (As stated there, the result requires the
matrix coefficient $\langle \pi(g)\phi \mid \psi \rangle$ to be an $\mathcal{L}^p$ function of g, for ϕ, ψ in
a dense subspace of $\mathcal{H}$, and for some $p < \infty$, but it was proved in [1,
Thm. 2.4.2] that this integrability hypothesis always holds.)

Theorem 11.4.2 is proved in [11, Chap. 3]

See [3, Chap. 2] for a nice proof of Proposition 11.5.2. (Although most of the proof is written for $n = 1$, it is mentioned on p. 31 that the argument works in general.)

See [7, Thm. 2.9, p. 108] for a proof of Remark 11.6.5's statement that every unitary representation is a direct integral of irreducibles. (This is a generalization of Theorem 11.6.4.)

Regarding Warning 11.1.7(2), groups for which the set of irreducible unitary representations admits an injective Borel map to $[0, 1]$ are called "Type I" (and the others are "Type II"). See [7, §2.3, pp. 77–85] for some discussion of this.

References

[1] M. Cowling: Sur les coefficients des représentations unitaires des groupes de Lie simples, in: Pierre Eymard et al., eds., *Analyse Harmonique sur les Groupes de Lie II*. Springer, Berlin, 1979, pp. 132–178. ISBN 3-540-09536-5, MR 0560837

[2] R. Ellis and M. Nerurkar: Enveloping semigroup in ergodic theory and a proof of Moore's ergodicity theorem, in: J. C. Alexander., ed., *Dynamical Systems*. Springer, New York, 1988, pp. 172–179. ISBN 3-540-50174-6, MR 0970554

[3] H. Helson: *The Spectral Theorem*. Springer, Berlin, 1986. ISBN 3-540-17197-5, MR 0873504

[4] R. Howe: On a notion of rank for unitary representations of the classical groups, in A. Figa-Talamanca, ed.: *Harmonic Analysis and Group Representations*, Liguori, Naples, 1982, pp. 223–331. ISBN 978-3-642-11115-0, MR 0777342

[5] R. E. Howe and C. C. Moore: Asymptotic properties of unitary representations, *J. Funct. Anal.* 32 (1979), no. 1, 72–96. MR 0533220, http://dx.doi.org/10.1016/0022-1236(79)90078-8

[6] A. W. Knapp: *Representation Theory of Semisimple Groups*. Princeton University Press, Princeton, 2001. ISBN 0-691-09089-0, MR 1880691

[7] G. Mackey: *The Theory of Unitary Group Representations*. University of Chicago Press, Chicago, 1976. ISBN 0-226-50052-7, MR 0396826

[8] G. Mackey: *Unitary Group Representations in Physics, Probability, and Number Theory*. Benjamin, Reading, Mass., 1978. ISBN 0-8053-6702-0, MR 0515581

[9] G. A. Margulis: *Discrete Subgroups of Semisimple Lie Groups*. Springer, New York, 1991. ISBN 3-540-12179-X, MR 1090825

[10] C. C. Moore: Ergodicity of flows on homogeneous spaces, *Amer. J. Math.* 88 1966 154–178. MR 0193188, http://www.jstor.org/stable/2373052

[11] M. Sepanski: *Compact Lie Groups.* Springer, New York, 2007. ISBN 978-0-387-30263-8, MR 2279709

[12] R. J. Zimmer: Orbit spaces of unitary representations, ergodic theory, and simple Lie groups, *Ann. Math.* (2) 106 (1977), no. 3, 573–588. MR 0466406, http://dx.doi.org/10.2307/1971068

[13] R. J. Zimmer: *Ergodic Theory and Semisimple Groups.* Birkhäuser, Basel, 1984. ISBN 3-7643-3184-4, MR 0776417

Chapter 12

Amenable Groups

The classical Kakutani-Markov Fixed Point Theorem (12.2.3) implies that any abelian group of continuous linear operators has a fixed point in any compact, convex, invariant set. This theorem can be extended to some non-abelian groups; the groups that satisfy such a fixed-point property are said to be "amenable," and they have quite a number of interesting features. Many important subgroups of G are amenable, so the theory is directly relevant to the study of arithmetic groups, even though we will see that G and Γ are usually not amenable. In particular, the theory yields an important equivariant map that will be constructed in Section 12.6.

§12.1. Definition of amenability

(12.1.1) **Assumption.** Throughout this chapter, H denotes a Lie group. The ideas here are important even in the special case where H is discrete.

(12.1.2) **Definition.** Suppose H acts continuously (by linear maps) on a locally convex topological vector space $\mathcal{V}$. Every H-invariant, compact, convex subset of $\mathcal{V}$ is called a *compact, convex H-space*.

(12.1.3) **Definition.** H is *amenable* if and only if H has a fixed point in every nonempty, compact, convex H-space.

Recall: The Standing Assumptions (4.0.0 on page 41) are in effect, so, as always, Γ is a lattice in the semisimple Lie group $G \subseteq \mathrm{SL}(\ell, \mathbb{R})$.

You can copy, modify, and distribute this work, even for commercial purposes, all without asking permission.   http://creativecommons.org/publicdomain/zero/1.0/

Main prerequisites for this chapter: none.

This is just one of many different equivalent definitions of amenability. (A few others are discussed in Section 12.3.) The equivalence of these diverse definitions is a testament to the fact that this notion is very fundamental.

(12.1.4) *Remarks.*

1) All locally convex topological vector spaces are assumed to be Hausdorff.

2) In most applications, the locally convex space $\mathcal{V}$ is the dual of a separable Banach space, with the weak* topology (see Definition B7.3). In this situation, every compact, convex subset C is second countable, and is therefore metrizable (see Remark 12.3.4). With these thoughts in mind, we feel free to assume metrizability when it eliminates technical difficulties in our proofs. In fact, we could restrict to these spaces in our definition of amenability, because it turns out that this modified definition results in exactly the same class of groups (if we only consider groups that are second countable) (see Exercise 12.3#17).

3) The choice of the term "amenable" seems to have been motivated by two considerations:
   (a) The word "amenable" can be pronounced "a-MEAN-able," and we will see in Section 12.3 that a group is amenable if and only if it admits certain types of means.
   (b) One definition of "amenable" from the *Oxford American Dictionary* is "capable of being acted on a particular way." In other words, in colloquial English, something is "amenable" if it is easy to work with. Classical analysis has averaging theorems and other techniques that were developed for the study of functions on the group $\mathbb{R}^n$. Many of these methods can be generalized to all amenable groups, so amenable groups are easy to work with.

Exercises for §12.1.

#1. Show that every finite group is amenable.

   [*Hint:* For some $c_0 \in C$, let $c = \frac{1}{\#H} \sum_{h \in H} h c_0$. Then $c \in C$ and c is fixed by H.]

#2. Show that quotients of amenable groups are amenable. That is, if H is amenable, and N is any closed, normal subgroup of H, then H/N is amenable.

#3. Suppose H_1 is amenable, and there is a continuous homomorphism $\varphi \colon H_1 \to H$ with dense image. Show H is amenable.

§12.2. Examples of amenable groups

In this section, we will see that:

- abelian groups are amenable (see 12.2.3),
- compact groups are amenable (see 12.2.4),
- solvable groups are amenable (see 12.2.7), because the class of amenable groups is closed under extensions (see 12.2.6), and
- closed subgroups of amenable groups are amenable (see 12.2.8).

On the other hand, however, it is important to realize that not all groups are amenable. In particular, we will see in Section 12.4 that:

- nonabelian free groups are not amenable, and
- $SL(2, \mathbb{R})$ is not amenable.

We begin by showing that $\mathbb{Z}$ is amenable:

(12.2.1) **Proposition.** *Cyclic groups are amenable.*

Proof. Assume $H = \langle T \rangle$ is cyclic. Given a nonempty, compact, convex H-space C, choose some $c_0 \in C$. For $n \in \mathbb{N}$, let

$$c_n = \frac{1}{n+1} \sum_{k=0}^{n} T^k(c). \qquad (12.2.2)$$

Since C is compact, the sequence $\{c_n\}$ must have an accumulation point $c \in C$. It is not difficult to see that c is fixed by T (see Exercise 1). Since T generates H, this means that c is a fixed point for H. □

(12.2.3) **Corollary** (Kakutani-Markov Fixed Point Theorem). *Every abelian group is amenable.*

Proof. Let us assume $H = \langle g, h \rangle$ is a 2-generated abelian group. (See Exercise 5 for the general case.) If C is any nonempty, compact, convex H-space, then Proposition 12.2.1 implies that the set C^g of fixed points of g is nonempty. It is easy to see that C^g is compact and convex (see Exercise 2), and, because H is abelian, that C^g is invariant under h (see Exercise 3). Hence, C^g is a nonempty, compact, convex $\langle h \rangle$-space. Therefore, Proposition 12.2.1 implies that h has a fixed point c in C^g. Now c is fixed by g (because it belongs to C^g), and c is fixed by h (by definition), so c is fixed by $\langle g, h \rangle = H$. □

Compact groups are also easy to work with:

(12.2.4) **Proposition.** *Compact groups are amenable.*

Proof. Assume H is compact, and let μ be a Haar measure on H. Given a nonempty, compact, convex H-space C, choose some $c_0 \in C$. Since μ is a probability measure, we may let

$$c = \int_H h(c_0)\, d\mu(h) \in C. \tag{12.2.5}$$

(In other words, c is the center of mass of the H-orbit of c_0.) The H-invariance of μ implies that c is a fixed point for H (see Exercise 6). □

It is easy to show that amenable extensions of amenable groups are amenable (see Exercise 7):

(12.2.6) Proposition. *If H has a closed, normal subgroup N, such that N and H/N are amenable, then H is amenable.*

Combining the above results has the following consequences:

(12.2.7) Corollary.

1) *Every solvable group is amenable.*

2) *If H has a solvable, normal subgroup N, such that H/N is compact, then H is amenable.*

Proof. Exercises 9 and 10.   □

The converse of Corollary 12.2.7(2) is true for connected groups (see Proposition 12.4.7).

(12.2.8) Proposition. *Every closed subgroup of any amenable group is amenable.*

Proof. This proof employs a bit of machinery, so we postpone it to Section 12.5. (For discrete groups, the result follows easily from some other characterizations of amenability; see Remarks 12.3.13 and 12.3.24 below.)   □

Exercises for §12.2.

#1. Suppose T is a continuous linear map on a locally convex space $\mathcal{V}$. Show that if c is any accumulation point of the sequence $\{c_n\}$ defined by (12.2.2), then c is T-invariant.

   [*Hint:* If $\|c_n - c\|$ is small, then $\|T(c_n) - T(c)\|$ is small. Show that $\|T(c_n) - c_n\|$ is small whenever n is large. Conclude that $\|T(c) - c\|$ is smaller than every ϵ.]

#2. Suppose C is a compact, convex H-space. Show that the set C^H of fixed points of H is compact and convex.

   [*Hint:* Closed subsets of C are compact.]

#3. Suppose H acts on a space C, A is a subgroup of H, and h is an element of the centralizer of A. Show that the set C^A of fixed points of A is invariant under h.

#4. Establish Exercise 3 under the weaker assumption that h is an element of the *normalizer* of A, not the centralizer.

#5. Prove Corollary 12.2.3.

[*Hint:* For each $h \in H$, let C^h be the set of fixed points of h. The given argument implies (by induction) that $\{\, C^h \mid h \in H \,\}$ has the finite intersection property, so the intersection of these fixed-point sets is nonempty.]

#6. Show that if μ is the Haar measure on H, and H is compact, then the point c defined in (12.2.5) is fixed by H.

#7. Prove Proposition 12.2.6.

[*Hint:* Exercises 2 and 4.]

#8. Show that $H_1 \times H_2$ is amenable if and only if H_1 and H_2 are both amenable.

#9. Prove Corollary 12.2.7(1).

[*Hint:* Proposition 12.2.6.]

#10. Prove Corollary 12.2.7(2).

[*Hint:* Proposition 12.2.6.]

#11. Suppose H is discrete, and H_1 is a finite-index subgroup. Show H is amenable if and only if H_1 is amenable.

#12. Show that if Λ is a lattice in H, and Λ is amenable, then H is amenable.

[*Hint:* Let $\mu = \int_{H/\Lambda} hv\, dh$, where v is a fixed point for Λ.]

#13. Assume H is discrete. Show that if every finitely generated subgroup of H is amenable, then H is amenable.

[*Hint:* For each $h \in H$, let C^h be the set of fixed points of h. Then $\{\, C^h \mid h \in H \,\}$ has the finite intersection property, so $\bigcap_h C^h \neq \emptyset$.]

#14. Let $P = \begin{bmatrix} * & & \\ * & * & \\ * & * & * \end{bmatrix} \subset \mathrm{SL}(3, \mathbb{R})$. Show that P is amenable.

[*Hint:* P is solvable.]

#15. Assume there exists a discrete group that is not amenable. Show the free group F_2 on 2 generators is not amenable.

[*Hint:* F_n is a subgroup of F_2.]

#16. Assume there exists a Lie group that is not amenable.
  a) Show the free group F_2 on 2 generators is not amenable.
  b) Show $\mathrm{SL}(2, \mathbb{R})$ is not amenable.

§12.3. Other characterizations of amenability

Here are a few of the many conditions that are equivalent to amenability. The necessary definitions are provided in the discussions that follow.

(12.3.1) **Theorem.** *The following are equivalent:*

1) *H is amenable.*

2) *H has a fixed point in every nonempty, compact, convex H-space.*

3) *If H acts continuously on a compact, metrizable topological space X, then there is an H-invariant probability measure on X.*

4) *There is a left-invariant mean on the space $C_{\mathrm{bdd}}(H)$ of all real-valued, continuous, bounded functions on H.*

5) *There is a left-invariant finitely additive probability measure ρ defined on the collection of all Lebesgue measurable subsets of H, such that $\rho(E) = 0$ for every set E of Haar measure 0.*

6) *The left regular representation of H on $\mathcal{L}^2(H)$ has almost-invariant vectors.*

7) *There exists a Følner sequence in H.*

The equivalence (1 $\Leftrightarrow$ 2) is the definition of amenability (12.1.3). Equivalence of the other characterizations will be proved in the remainder of this section.

§12.3(i). Invariant probability measures.

(12.3.2) **Definitions.** Let X be a complete metric space.

1) A measure μ on X is a ***probability measure*** if $\mu(X) = 1$.

2) Prob(X) denotes the space of all probability measures on X.

Any measure on X is also a measure on the one-point compactification X^+ of X, so, if X is locally compact, then the Riesz Representation Theorem (B6.10) tells us that every finite measure on X can be thought of as a linear functional on the Banach space $C(X^+)$ of continuous functions on X^+. This implies that Prob(X) is a subset of the closed unit ball in the dual space $C(X^+)^*$, and therefore has a weak* topology. If X is compact (so there is no need to pass to X^+), then the Banach-Alaoglu Theorem (B7.4) tells us that Prob(X) is compact (see Exercise 1).

(12.3.3) **Example.** If a group H acts continuously on a compact, metrizable space X, then Prob(X) is a compact, convex H-space (see Exercise 2).

(12.3.4) *Remark* (Urysohn's Metrization Theorem). Recall that a compact, Hausdorff space is metrizable if and only if it is second countable, so requiring a compact, separable, Hausdorff space to be metrizable is not a strong restriction.

(12.3.5) **Proposition** (1 ⟺ 3). *H is amenable if and only if for every contin-uous action of H on a compact, metrizable space X, there is an H-invariant probability measure μ on X.*

Proof. (⟹) If H acts on X, and X is compact, then $\mathrm{Prob}(X)$ is a nonempty, compact, convex H-space (see Example 12.3.3). So H has a fixed point in $\mathrm{Prob}(X)$; this fixed point is the desired H-invariant measure.

(⟸) Suppose C is a nonempty, compact, convex H-space. By replacing C with the closure of the convex hull of a single H-orbit, we may assume C is separable; then C is metrizable (see Exercise 3). Since H is amenable, this implies there is an H-invariant probability measure μ on C. Since C is convex and compact, the center of mass

$$p = \int_C c \, d\mu(c)$$

belongs to C (see Exercise 4). Since μ is H-invariant (and the H-action is by linear maps), a simple calculation shows that p is H-invariant (see Exercise 6).   □

§12.3(ii). Invariant means.

(12.3.6) **Definition.** Suppose $\mathcal{V}$ is some linear subspace of $\mathcal{L}^\infty(H)$, and assume $\mathcal{V}$ contains the constant function 1_H that takes the value 1 at every point of H. A *mean* on $\mathcal{V}$ is a linear functional λ on $\mathcal{V}$, such that

- $\lambda(1_H) = 1$, and
- λ is positive, i.e., $\lambda(f) \geq 0$ whenever $f \geq 0$.

(12.3.7) *Remark.* Any mean is a continuous linear functional; indeed, $\|\lambda\| = 1$ (see Exercise 8).

It is easy to construct means:

(12.3.8) **Example.** If ϕ is any unit vector in $\mathcal{L}^1(H)$, and μ is the left Haar measure on H, then defining

$$\lambda(f) = \int_H f \, |\phi| \, d\mu$$

produces a mean (on any subspace of $\mathcal{L}^\infty(H)$ that contains 1_H). Means constructed in this way are (weakly) dense in the set of all means (see Exercise 12).

Compact groups are the only ones with invariant probability mea-sures, but invariant means exist more generally:

(12.3.9) **Proposition** (1 ⟹ 4). *If H is amenable, then there exists a left-invariant mean on the space $C_{\mathrm{bdd}}(H)$ of bounded, continuous functions on H.*

Proof. The set of means on $C_{bdd}(H)$ is obviously nonempty, convex and invariant under left translation (see Exercise 13). Furthermore, it is a weak* closed subset of the unit ball in $C_{bdd}(H)^*$ (see Exercise 14), so it is compact by the Banach-Alaoglu Theorem (Proposition B7.4). Therefore, the amenability of H implies that some mean is left-invariant. (Actually, there is a slight technical problem here if H is not discrete: the action of H on $C_{bdd}(H)$ may not be continuous in the sup-norm topology, because continuous functions do not need to be uniformly continuous.)   □

(12.3.10) *Remark.* With a bit more work, it can be shown that if H is amenable, then there is a left-invariant mean on $\mathscr{L}^\infty(H)$, not just on $C_{bdd}(H)$ (see Exercise 14). Therefore, $C_{bdd}(H)$ can be replaced with $\mathscr{L}^\infty(H)$ in Theorem 12.3.1(4). Furthermore, there exists a mean on $\mathscr{L}^\infty(H)$ that is **bi-invariant** (both left-invariant *and* right-invariant) (cf. Exercise 16).

(12.3.11) **Proposition** (4 ⇒ 3). *Suppose H acts continuously on a compact, metrizable space X. If there is a left-invariant mean on $C_{bdd}(H)$, then there is an H-invariant probability measure on X.*

Proof. Fix some $x \in X$. Then we have a continuous, H-equivariant linear map from $C(X)$ to $C_{bdd}(H)$, defined by

$$\overline{f}(h) = f(hx).$$

Therefore, any left-invariant mean on $C_{bdd}(H)$ induces an H-invariant mean λ on $C(X)$ (see Exercise 15). Since X is compact, the Riesz Representation Theorem (B6.10) tells us that any continuous, positive linear functional on $C(X)$ is a measure; thus, this H-invariant mean λ can be represented by an H-invariant measure μ on X. Since λ is a mean, we have $\lambda(1) = 1$, so $\mu(X) = 1$, which means that μ is a probability measure.   □

§12.3(iii). Invariant finitely additive probability measures. The following proposition is based on the observation that, just as probability measures on X correspond to elements of the dual of $C(X)$, finitely additive probability measures correspond to elements of the dual of $\mathscr{L}^\infty(X)$.

(12.3.12) **Proposition** (4 ⇔ 5). *There is a left-invariant mean on $\mathscr{L}^\infty(X)$ if and only if there is a left-invariant finitely additive probability measure ρ defined on the collection of all Lebesgue measurable subsets of H, such that $\rho(E) = 0$ for every set E of Haar measure 0.*

Proof. (⇒) Because H is amenable, there exists a left-invariant mean λ on $\mathscr{L}^\infty(H)$ (see Remark 12.3.10). For a measurable subset E of H, let $\rho(E) = \lambda(\chi_E)$, where χ_E is the characteristic function of E. It is easy to verify that ρ has the desired properties (see Exercise 18).

($\Leftarrow$) We define a mean λ via an approximation by step functions: for $f \in \mathcal{L}^\infty(H)$, let

$$\lambda(f) = \inf \left\{ \sum_{i=1}^n a_i \rho(E_i) \; \middle| \; f \le \sum_{i=1}^n a_i \chi_{E_i} \text{ a.e.} \right\}.$$

Since ρ is finitely additive, it is straightforward to verify that λ is a mean on $\mathcal{L}^\infty(H)$ (see Exercise 19). Since ρ is bi-invariant, we know that λ is also bi-invariant.   □

(12.3.13) *Remark.*
  1) Proposition 12.3.12 easily implies that every subgroup of a discrete amenable group is amenable (see Exercise 20), establishing Proposition 12.2.8 for the case of discrete groups. In fact, it is not very difficult to prove the general case of Proposition 12.2.8 similarly (see Exercise 21).
  2) Because any amenable group H has a bi-invariant mean on $\mathcal{L}^\infty(H)$ (see Remark 12.3.10), the proof of Proposition 12.3.12($\Rightarrow$) shows that the finitely additive probability measure ρ can be taken to be bi-invariant.

§12.3(iv). Almost-invariant vectors.

(12.3.14) **Definition.** An action of H on a normed vector space $\mathcal{B}$ has *almost-invariant* vectors if, for every compact subset C of H and every $\epsilon > 0$, there is a unit vector $v \in \mathcal{B}$, such that

$$\|cv - v\| < \epsilon \quad \text{for all } c \in C. \tag{12.3.15}$$

(A unit vector satisfying (12.3.15) is said to be (ϵ, C)-invariant.)

(12.3.16) **Example.** Consider the regular representation of H on $\mathcal{L}^2(H)$.
  1) If H is a compact Lie group, then the constant function 1_H belongs to $\mathcal{L}^2(H)$, so $\mathcal{L}^2(H)$ has an H-invariant unit vector.
  2) If $H = \mathbb{R}$, then $\mathcal{L}^2(H)$ does not have any (nonzero) H-invariant vectors (see Exercise 22), but it does have almost-invariant vectors: Given C and ϵ, choose $n \in \mathbb{N}$ so large that $C \subseteq [-n, n]$ and $2/\sqrt{n} < \epsilon$. Let $\phi = \frac{1}{n}\chi_{n^2}$, where χ_{n^2} is the characteristic function of $[0, n^2]$. Then ϕ is a unit vector and, for $c \in C$, we have

$$\|c\phi - \phi\|^2 \le \int_{-n}^n \frac{1}{n^2}\,dx + \int_{n^2-n}^{n^2+n} \frac{1}{n^2}\,dx = \frac{4}{n} < \epsilon^2.$$

(12.3.17) *Remark.* $\mathcal{L}^2(H)$ has almost-invariant vectors if and only if $\mathcal{L}^1(H)$ has almost-invariant vectors (see Exercise 23). Therefore, $\mathcal{L}^2(H)$ may be replaced with $\mathcal{L}^1(H)$ in Theorem 12.3.1(6). (In fact, $\mathcal{L}^2(H)$ may be replaced with $\mathcal{L}^p(H)$, for any $p \in [1, \infty)$ (see Exercise 24).)

(12.3.18) **Proposition** (4 ⇔ 6). *There is a left-invariant mean on $\mathcal{L}^\infty(H)$ if and only if $\mathcal{L}^2(H)$ has almost-invariant vectors.*

Proof. Because of Remark 12.3.17, we may replace $\mathcal{L}^2(H)$ with $\mathcal{L}^1(H)$.

(⇐) By applying the construction of means in Example 12.3.8 to almost-invariant vectors in $\mathcal{L}^1(H)$, we obtain almost-invariant means on $\mathcal{L}^\infty(H)$. A limit of almost-invariant means is invariant (see Exercise 25).

(⇒) Because the means constructed in Example 12.3.8 are dense in the space of all means, we can approximate a left-invariant mean by an $\mathcal{L}^1$ function. Vectors close to an invariant vector are almost-invariant, so $\mathcal{L}^1(H)$ has almost-invariant vectors. However, there are technical issues here; one problem is that the approximation is in the weak* topology, but we are looking for vectors that are almost-invariant in the norm topology. See Exercise 26 for a correct proof in the case of discrete groups (using the fact that a convex set has the same closure in both the norm topology and the weak* topology). □

§12.3(v). Følner sequences.

(12.3.19) **Definition.** Let $\{F_n\}$ be a sequence of measurable sets in H, such that $0 < \mu(F_n) < \infty$ for every n. We say $\{F_n\}$ is a ***Følner sequence*** if, for every compact subset C of H, we have

$$\lim_{n \to \infty} \max_{c \in C} \frac{\mu(F_n \bigtriangleup cF_n)}{\mu(F_n)} = 0, \qquad (12.3.20)$$

where μ is the Haar measure on H.

(12.3.21) **Example.**

1) If $F_n = B_n(0)$ is the ball of radius n in $\mathbb{R}^\ell$, then $\{F_n\}$ is a Følner sequence in $\mathbb{R}^\ell$ (see Exercise 29).

2) The free group F_2 on 2 generators does not have Følner sequences (see Exercise 12.4#2).

The reason that $\mathbb{R}^\ell$ has a Følner sequence, but the free group F_2 does not, is that $\mathbb{R}^\ell$ is amenable, but F_2 is not:

(12.3.22) **Proposition** (6 ⇔ 7). *There is an invariant mean on $\mathcal{L}^2(H)$ if and only if H has a Følner sequence.*

Proof. (⇐) Normalized characteristic functions of Følner sets are almost invariant vectors in $\mathcal{L}^1(H)$ (see Exercise 30).

(⇒) Let us assume H is discrete. Given $\epsilon > 0$, and a finite subset C of H, we wish to find a finite subset F of H, such that

$$\frac{\#(F \bigtriangleup c(F))}{\#(F_n)} < \epsilon \quad \text{for all } c \in C.$$

Since H is amenable, we know $\mathcal{L}^1(H)$ has almost-invariant vectors (see Remark 12.3.17); hence, there exists $f \in \mathcal{L}^1(H)$, such that

1) $f \geq 0$,

2) $\|f\|_1 = 1$, and

3) $\|cf - f\|_1 < \epsilon/\#C$, for every $c \in C$.

Note that if f were the normalized characteristic function of a set F, then this set F would be what we want; for the general case, we will approximate f by a sum of such characteristic functions.

Approximating f by a step function, we may assume f takes only finitely many values. Hence, there exist:

- finite subsets $A_1 \subseteq A_2 \subseteq \cdots \subseteq A_n$ of H, and

- real numbers $\alpha_1, \ldots, \alpha_n > 0$,

such that

1) $\alpha_1 + \alpha_2 + \cdots + \alpha_n = 1$ and

2) $f = \alpha_1 f_1 + \alpha_2 f_2 + \cdots \alpha_n f_n$,

where f_i is the normalized characteristic function of A_i (see Exercise 33). For all i and j, and any $c \in H$, we have

$$A_i \smallsetminus cA_i \text{ is disjoint from } cA_j \smallsetminus A_j \qquad (12.3.23)$$

(see Exercise 34), so, for any $x \in H$, we have

$$f_i(x) > (cf_i)(x) \implies f_j(x) \geq (cf_j)(x)$$

and

$$f_i(x) < (cf_i)(x) \implies f_j(x) \leq (cf_j)(x).$$

Therefore

$$|(cf - f)(x)| = \sum_i \alpha_i|(cf_i - f_i)(x)| \quad \text{for all } x \in H.$$

Summing over H yields

$$\sum_i \alpha_i\|cf_i - f_i\|_1 = \|cf - f\|_1 < \frac{\epsilon}{\#C}.$$

Summing over C, we conclude that

$$\sum_i \alpha_i \sum_{c \in C} \|cf_i - f_i\|_1 < \epsilon.$$

Since $\sum_i \alpha_i = 1$ (and all terms are positive), this implies there is some i, such that

$$\sum_{c \in C} \|cf_i - f_i\|_1 < \epsilon.$$

Hence, $\|cf_i - f_i\|_1 < \epsilon$, for every $c \in C$, so we may let $F = A_i$. $\square$

(12.3.24) *Remark.* Følner sets provide an easy proof that subgroups of discrete amenable groups are amenable.

Proof. Let

- A be a closed subgroup of a discrete, amenable group H,
- C be a finite subset of A, and
- $\epsilon > 0$.

Since H is amenable, there is a corresponding Følner set F in H.

It suffices to show there is some $h \in H$, such that $Fh \cap A$ is a Følner set in A. We have

$$\#F = \sum_{Ah \in A \backslash H} \#(F \cap Ah)$$

and, letting $\epsilon' = \epsilon \# C$, we have

$$(1 + \epsilon')\#F \geq \#(CF) = \sum_{Ah \in A \backslash H} \#(C(F \cap Ah)),$$

so there must be some $Ah \in A \backslash H$, such that

$$\#(C(F \cap Ah)) \leq (1 + \epsilon')\#(F \cap Ah)$$

(and $F \cap Ah \neq \varnothing$). Then, letting $F' = Fh^{-1} \cap A$, we have

$$\#(CF') = \#(C(F \cap Ah)) \leq (1 + \epsilon')\#(F \cap Ah) = (1 + \epsilon')\#F',$$

so F' is a Følner set in A. $\qquad\qquad\square$

Exercises for §12.3.

Invariant probability measures

#1. In the setting of Example 12.3.3, show that $\mathrm{Prob}(X)$ is a compact, convex subset of $C(X)^*$.

[*Hint:* You may assume the Banach-Alaoglu Theorem (Proposition B7.4).]

#2. Suppose H acts continuously on a compact, metrizable space X. There is an induced action of H on $\mathrm{Prob}(X)$ defined by

$$(h_*\mu)(A) = \mu(h^{-1}A) \quad \text{for } h \in H, \mu \in \mathrm{Prob}(X), \text{ and } A \subseteq X.$$

Show that this induced action of H on $\mathrm{Prob}(X)$ is continuous (with respect to the weak* topology on $\mathrm{Prob}(X)$).

#3. Let A be a separable subset of a Fréchet space $\mathcal{V}$. Show
   a) A is second countable.
   b) If A is compact, then A is metrizable.
   [*Hint:* (b) Remark 12.3.4.]

#4. Let μ be a probability measure on a compact, convex subset C of a Fréchet space $\mathcal{V}$. The **center of mass** of C is a point $c \in \mathcal{V}$, such that, for every continuous linear functional λ on $\mathcal{V}$, we have

$$\lambda(c) = \int_C \lambda(x)\, d\mu(x).$$

Show the center of mass of μ exists and is unique, and is an element of C.

#5. Give an example of a probability measure μ on a Fréchet space, such that the center of mass of μ does not exist.

[*Hint:* There are probability measures on $\mathbb{R}^+$, such that the center of mass is infinite.]

#6. Show that if p is the center of mass of a probability measure μ on a Fréchet space $\mathcal{V}$, then p is invariant under every continuous, linear transformation of $\mathcal{V}$ that preserves μ.

#7. Suppose H acts continuously on a compact, metrizable space X. Show that the map

$$H \times \mathrm{Prob}(X)\colon (h, \mu) \mapsto h_*\mu$$

defines a continuous action of H on $\mathrm{Prob}(X)$.

Left-invariant means

#8. Verify Remark 12.3.7.

[*Hint:* $\lambda(1_H) = 1$ implies $\|\lambda\| \geq 1$. For the other direction, note that if $\|f\|_\infty \leq 1$, then $1_H - f \geq 0$ a.e., so $\lambda(1_H - f) \geq 0$; similarly, $\lambda(f + 1_H) \geq 0$.]

#9. Show that the restriction of a mean is a mean. More precisely, let $\mathcal{V}_1$ and $\mathcal{V}_2$ be linear subspaces of $\mathscr{L}^\infty(H)$, with $1_H \in \mathcal{V}_1 \subseteq \mathcal{V}_2$. Show that if λ is a mean on $\mathcal{V}_2$, then the restriction of λ to $\mathcal{V}_1$ is a mean on $\mathcal{V}_1$.

#10. Suppose λ is a mean on $C_{\mathrm{bdd}}(H)$, the space of bounded, continuous functions on H. For $f \in C_{\mathrm{bdd}}(H)$, show

$$\min f \leq \lambda(f) \leq \max f.$$

#11. For $h \in H$, define $\delta_h\colon C_{\mathrm{bdd}}(H) \to \mathbb{R}$ by $\delta_h(f) = f(h)$. Show δ_h is a mean on $C_{\mathrm{bdd}}(H)$.

#12. Let $\mathcal{B}$ be any linear subspace of $\mathscr{L}^\infty(H)$, such that $\mathcal{B}$ contains 1_H and is closed in the $\mathscr{L}^\infty$-norm. Show that the means constructed in Example 12.3.8 are weak* dense in the set of all means on $\mathcal{B}$.

[*Hint:* If not, then the Hahn-Banach Theorem implies there exist $\epsilon > 0$, a mean λ, and some $f \in (\mathcal{B}^*)^* = \mathcal{B}$, such that

$$\lambda(f) > \epsilon + \int_H f\, |\phi|\, d\mu,$$

for every unit vector ϕ in $\mathscr{L}^1(H)$. This contradicts the fact that $\lambda(f) \leq \mathrm{ess.}\, \sup f$.]

#13. Let $\mathcal{M}$ be the set of means on $C_{\mathrm{bdd}}(H)$. Show:

a) $\mathcal{M} \neq \varnothing$.

b) $\mathcal{M}$ is convex.

    c) $\mathcal{M}$ is H-invariant.

[*Hint:* (a) Evaluation at any point is a mean.]

#14. Let $\mathcal{M}$ be the set of means on $C_{\mathrm{bdd}}(H)$. Show:
    a) $\mathcal{M}$ is contained in the closed unit ball of $C_{\mathrm{bdd}}(H)^*$. (That is, we
      have $|\lambda(f)| \le \|f\|_\infty$ for every $f \in C_{\mathrm{bdd}}(H)$.)
    b) (M) is weak* closed.
    c) $\mathcal{M}$ is compact in the weak* topology.

[*Hint:* You may assume the Banach-Alaoglu Theorem (Proposition B7.4).]

    Show that if H is amenable, then there is a left-invariant mean
on $\mathscr{L}^\infty(H)$.

[*Hint:* Define $\lambda(f) = \mu_0(f * \eta)$, where λ_0 is a left-invariant mean on $C_{\mathrm{bdd}}(H)$, and η is a nonnegative function of integral 1.]

#15. Suppose $\psi \colon Y \to X$ is continuous, and λ is a mean on $C_{\mathrm{bdd}}(Y)$. Show
that $\psi_*\lambda$ (defined by $(\psi_*\lambda)(f) = \lambda(f \circ \psi)$) is a mean on $C_{\mathrm{bdd}}(X)$.

#16. Assume H is amenable and discrete. Show there is a bi-invariant
mean on $\mathscr{L}^\infty(H)$.

[*Hint:* Since $\mathscr{L}^\infty(H) = C_{\mathrm{bdd}}(H)$, amenability implies there is a left-invariant mean on $\mathscr{L}^\infty(H)$ (see Theorem 12.3.1(4)). Now H acts by right translations on the set of all such means, so amenability implies that some left-invariant mean is right-invariant.]

#17. (*harder*) Assume H has a fixed point in every *metrizable*, nonempty,
compact, convex H-space (and H is second countable). Show H is
amenable.

[*Hint:* To find a fixed point in C, choose some $c_0 \in C$. For each mean λ on $C_{\mathrm{bdd}}(H)$ and each $\rho \in \mathcal{V}^*$, define $\phi_\lambda(\rho) = \lambda(h \mapsto \rho(hc_0))$, so $\phi_\lambda \in (\mathcal{V}^*)'$, the algebraic dual of $\mathcal{V}^*$. If λ is a convex combination of evaluations at points of H, it is obvious there exists $c_\lambda \in C$, such that $\phi_\lambda(\rho) = \rho(c_\lambda)$. Since the map $\lambda \mapsto \phi_\lambda$ is continuous (with respect to appropriate weak topologies), this implies c_λ exists for every λ. The proof of Proposition 12.3.9 shows that λ may be chosen to be left-invariant, and then c_λ is H-invariant.]

Invariant finitely additive probability measures

#18. Verify that ρ, as defined in the proof of Proposition 12.3.12($\Rightarrow$), has
the properties specified in the statement of the proposition.

#19. Let ρ and λ be as in the proof of Proposition 12.3.12($\Leftarrow$).
    a) If $\sum_{i=1}^{m} a_i \chi_{E_i} = \sum_{j=1}^{n} b_j \chi_{F_j}$ a.e., show $\sum_{i=1}^{m} a_i \rho(E_i) = \sum_{j=1}^{n} b_j \rho(F_j)$.
    b) If $\sum_{i=1}^{m} a_i \chi_{E_i} \le \sum_{j=1}^{n} b_j \chi_{F_j}$ a.e., show $\sum_{i=1}^{m} a_i \rho(E_i) \le \sum_{j=1}^{n} b_j \rho(F_j)$.
    c) Show that $\lambda(1_H) = 1$.
    d) Show that if $f \ge 0$, then $\lambda(f) \ge 0$.

e) Show that

$$\lambda(f) = \sup\left\{ \sum_{i=1}^{n} a_i \rho(E_i) \ \middle| \ f \ge \sum_{i=1}^{n} a_i \chi_{E_i} \text{ a.e.} \right\}.$$

f) Show that λ is a mean on $\mathcal{L}^\infty(H)$.

[*Hint:* (a,b) By passing to a refinement, arrange that $\{E_i\}$ are pairwise disjoint, $\{F_j\}$ are pairwise disjoint, and each E_i is contained in some F_j.]

#20. Use Proposition 12.3.12 to prove that every subgroup A of a discrete amenable group H is amenable.

[*Hint:* Let X be a set of representatives of the right cosets of A in H, and let λ be a left-invariant finitely additive probability measure on H. For $E \subseteq A$, define $\bar{\lambda}(E) = \lambda(EX)$.]

#21. Use Proposition 12.3.12 to prove that every closed subgroup A of an amenable group H is amenable.

[*Hint:* Let X be a Borel set of representatives of the right cosets of A in H, and define $\bar{\lambda}$ as in the solution of Exercise 20. Fubini's Theorem implies that if E has measure 0 in A, then XA has measure 0 in H. You may assume (without proof) the fact that if $f \colon M \to N$ is a continuous function between manifolds M and N, and E is a Borel subset of M, such that the restriction of f to E is one-to-one, then $f(E)$ is a Borel set in N.]

Almost-invariant vectors

#22. a) For $v \in \mathcal{L}^2(H)$, show that v is invariant under translations if and only if v is constant (a.e.).

b) Show that H is compact if and only if $\mathcal{L}^2(H)$ has a nonzero vector that is invariant under translation.

#23. Show that $\mathcal{L}^2(H)$ has almost-invariant vectors if and only if $\mathcal{L}^1(H)$ has almost-invariant vectors.

[*Hint:* Note that $f^2 - g^2 = (f-g)(f+g)$, so $\|f^2 - g^2\|_1 \le \|f - g\|_2 \|f + g\|_2$. Conversely, for $f, g \ge 0$, we have $(f-g)^2 \le |f^2 - g^2|$, so $\|f - g\|_2^2 \le \|f^2 - g^2\|_1$.]

#24. For $p \in [1, \infty)$, show that $\mathcal{L}^1(H)$ has almost-invariant vectors if and only if $\mathcal{L}^p(H)$ has almost-invariant vectors.

[*Hint:* If $p < q$, then almost-invariant vectors in $\mathcal{L}^p(H)$ yield almost-invariant vectors in $\mathcal{L}^q(H)$, because $|(f-g)|^{q/p} \le |f^{q/p} - g^{q/p}|$. And almost-invariant vectors in $\mathcal{L}^p(H)$ yield almost-invariant vectors in $\mathcal{L}^{p/2}(H)$, by the argument of the first hint in Exercise 23.]

#25. Let
  - $\{C_n\}$ be an increasing sequence of compact subsets of H, such that $\bigcup_n C_n = H$,
  - $\epsilon_n = 1/n$,
  - ϕ_n be an (ϵ_n, C_n)-invariant unit vector in $\mathcal{L}^1(H)$,
  - λ_n be the mean on $\mathcal{L}^\infty$ obtained from ϕ_n by the construction in Example 12.3.8, and
  - λ be an accumulation point of $\{\lambda_n\}$.

Show that λ is invariant.

#26. Assume H is discrete. Let
$$\mathcal{P} = \{ \phi \in \mathcal{L}^1(H) \mid \phi \geq 0, \|\phi\|_1 = 1 \}.$$
Suppose $\{\phi_i\}$ is a net in $\mathcal{P}$, such that the corresponding means λ_i converge weak* to an invariant mean λ on $\mathcal{L}^\infty(H)$.
   a) For each $h \in H$, show that the net $\{h^*\phi_i - \phi_i\}$ converges weakly to 0.
   b) Take a copy $\mathcal{L}^1(H)_h$ of $\mathcal{L}^1(H)$ for each $h \in H$, and let
   $$\mathcal{V} = \underset{h \in H}{\times} \mathcal{L}^1(H)_h$$
   with the product of the norm topologies. Show that $\mathcal{V}$ is a Fréchet space.
   c) Show that the weak topology on $\mathcal{V}$ is the product of the weak topologies on the factors.
   d) Define a linear map $T \colon \mathcal{L}^1(H) \to \mathcal{V}$ by $T(f)_h = h^*f - f$.
   e) Show that the net $\{T(\phi_i)\}$ converges to 0 weakly.
   f) Show that 0 is in the strong closure of $T(\mathcal{P})$.
   g) Show that $\mathcal{L}^1(H)$ has almost-invariant vectors.

#27. Show that if H is amenable, then H has the **Haagerup property**. By definition, this means there is a unitary representation of H on a Hilbert space $\mathcal{H}$, such that there are almost-invariant vectors, and all matrix coefficients decay to 0 at ∞ as in the conclusion of Theorem 11.2.2. (A group with the Haagerup property is also said to be **a-T-menable**.)

Følner sequences

#28. Show that $\{F_n\}$ is a Følner sequence if and only if, for every compact subset C of H, we have
$$\lim_{n \to \infty} \frac{\mu(F_n \cup cF_n)}{\mu(F_n)} = 1.$$

#29. Justify Example 12.3.21(1).
   [*Hint:* $C \subseteq B_r(0)$, for some r. We have $\mu(B_{r+\ell}(0))/\mu(B_\ell(0)) \to 1$.]

#30. Prove Proposition 12.3.22($\Leftarrow$).
   [*Hint:* Normalizing the characteristic function of F_n yields an almost-invariant unit vector.]

#31. Show (12.3.20) is equivalent to
$$\lim_{n \to \infty} \max_{c \in C} \frac{\mu(F_n \cup cF_n)}{\mu(F_n)} = 1.$$

#32. Assume H is discrete. Show that a sequence $\{F_n\}$ of finite subsets of H is a Følner sequence if and only if, for every finite subset C of H, we have
$$\lim_{n \to \infty} \frac{\#(CF_n)}{\#(F_n)} = 1.$$

#33. Given a step function f, as in the proof of Proposition 12.3.22($\Rightarrow$), let
- $a_1 > a_2 > \cdots > a_n$ be the finitely many positive values taken by f,
- $A_i = \{h \in H \mid f(h) \geq a_i\}$, and
- f_i be the normalized characteristic function of A_i.

Show
  a) $A_1 \subseteq A_2 \subseteq \cdots \subseteq A_n$,
  b) there exist real numbers $\alpha_1, \ldots, \alpha_n > 0$, such that
  $$f = \alpha_1 f_1 + \cdots + \alpha_n f_n,$$
  and
  c) $\alpha_1 + \cdots + \alpha_n = 1$.

#34. Prove (34).

[*Hint:* Note that either $A_i \subseteq A_j$ or $A_j \subseteq A_i$.]

#35. (*harder*) Use Følner sets to prove Remark 12.3.24 (without assuming H is discrete).

[*Hint:* Adapt the proof of the discrete case. There are technical difficulties, but begin by replacing the sum over $A\backslash H$ with an integral over $A\backslash H$.]

#36. A finitely generated (discrete) group Λ is said to have ***subexponential growth*** if there exists a generating set S for Λ, such that, for every $\epsilon > 0$,
$$\#(S \cup S^{-1})^n \leq e^{\epsilon n} \text{ for all large } n.$$

Show that every group of subexponential growth is amenable.

#37. Give an example of an finitely generated, amenable group that does *not* have subexponential growth.

§12.4. Some nonamenable groups

Other proofs of the following proposition appear in Exercises 1 and 2.

(12.4.1) Proposition. *Nonabelian free groups are not amenable.*

Proof. For convenience, we consider only the free group F_2 on two generators a and b. Suppose F_2 has a left-invariant finitely additive probability measure ρ. (This will lead to a contradiction.)

We may write $F_2 = A \cup A^- \cup B \cup B^- \cup \{e\}$, where A, A^-, B, and B^- consist of the reduced words whose first letter is a, a^{-1}, b, or b^{-1}, respectively. Assume, without loss of generality, that $\rho(A \cup A^-) \leq \rho(B \cup B^-)$ and $\rho(A) \leq \rho(A^-)$. Then
$$\rho(B \cup B^- \cup \{e\}) \geq \frac{1}{2} \quad \text{and} \quad \rho(A) \leq \frac{1}{4}.$$

Then, by left-invariance, we have

$$\rho\big(a(B \cup B^- \cup \{e\})\big) = \rho(B \cup B^- \cup \{e\}) \geq \frac{1}{2} > \rho(A).$$

This contradicts the fact that $a(B \cup B^- \cup \{e\}) \subseteq A$.   □

Combining this with the fact that subgroups of discrete amenable groups are amenable (see Proposition 12.2.8), we have the following consequence:

(12.4.2) **Corollary.** *Suppose H is a discrete group. If H contains a non-abelian, free subgroup, then H is not amenable.*

(12.4.3) *Remarks.*

1) The converse of Corollary 12.4.2 is known as "von Neumann's Conjecture," but it is false: a nonamenable group with no nonabelian free subgroups was found by Ol'shanskii in 1980. (The name is misleading: apparently, the conjecture is due to M. Day, and was never stated by Von Neumann.)

2) The assumption that H is discrete cannot be deleted from the statement of Corollary 12.4.2. For example, the orthogonal group SO(3) is amenable (because it is compact), but the Tits Alternative (4.9.1) implies that it contains nonabelian free subgroups.

3) The nonamenability of nonabelian free subgroups of SO(3) is the basis of the famous Banach-Tarski Paradox: A 3-dimensional ball B can be decomposed into finitely many subsets $X_1, \ldots, X_n$, such that these subsets can be reassembled to form the union of two disjoint balls of the same radius as B. (More precisely, the union $B_1 \cup B_2$ of two disjoint balls of the same radius as B can be decomposed into subsets $Y_1, \ldots, Y_n$, such that Y_i is congruent to X_i, for each i.)

4) If H contains a *closed*, nonabelian, free subgroup, then H is not amenable.

Here is an example of a nonamenable connected group:

(12.4.4) **Proposition.** SL(2, ℝ) *is not amenable.*

Proof. Let $G = \mathrm{SL}(2, \mathbb{R})$. The action of G on $\mathbb{R} \cup \{\infty\} \cong S^1$ by linear-fractional transformations is transitive, and the stabilizer of the point 0 is the subgroup $P = \begin{bmatrix} * & * \\ & * \end{bmatrix}$, so G/P is compact. However, the Borel Density Theorem implies there is no G-invariant probability measure on G/P (see Exercise 4.6#2). (See Exercise 4 for a direct proof that there is no G-invariant probability measure.) So G is not amenable.   □

More generally:

(12.4.5) **Proposition.** *If a connected, semisimple Lie group G is not compact, then G is not amenable.*

Proof. The Jacobson-Morosov Lemma (A5.8) tells us that G contains a closed subgroup isogenous to $SL(2, \mathbb{R})$. Alternatively, recall that any lattice Γ in G must contain a nonabelian free subgroup (see Corollary 4.9.2), and, being discrete, this is a closed subgroup of G. □

(12.4.6) *Remark.* Readers familiar with the structure of semisimple Lie groups will see that the proof of Proposition 12.4.4 generalizes to the situation of Proposition 12.4.5: Since G is not compact, it has a proper parabolic subgroup P. Then G/P is compact, but the Borel Density Theorem implies that G/P has no G-invariant probability measure.

Combining this result with the structure theory of connected Lie groups yields the following classification of connected, amenable Lie groups:

(12.4.7) **Proposition.** *A connected Lie group H is amenable if and only if H contains a connected, closed, solvable normal subgroup N, such that H/N is compact.*

Proof. ($\Leftarrow$) Corollary 12.2.7(2).

($\Rightarrow$) The structure theory of Lie groups tells us that there is a connected, closed, solvable, normal subgroup R of H, such that H/R is semisimple. (The subgroup R is called the **radical** of H.) Since quotients of amenable groups are amenable (see Exercise 12.1#2), we know that H/R is amenable. So H/R is compact (see Proposition 12.4.5). □

Exercises for §12.4.

#1. a) Find a homeomorphism ϕ of the circle S^1, such that the only ϕ-invariant probability measure is the delta mass at a single point p.

b) Find two homeomorphisms ϕ_1 and ϕ_2 of S^1, such that the subgroup $\langle \phi_1, \phi_2 \rangle$ they generate has no invariant probability measure.

c) Deduce that the free group F_2 on 2 generators is not amenable.

[*Hint:* (a) Identifying S^1 with $[0,1]$, let $\phi(x) = x^2$. For any $x \in (0,1)$, we have $\phi((0,x)) = (0, x^2)$, so $\mu((x^2, x)) = 0$. Since $(0,1)$ is the union of countably many such intervals, this implies that $\mu((0,1)) = 0$.]

#2. Show explicitly that free groups do not have Følner sequences. More precisely, let F_2 be the free group on two generators a and b, and show that if F is any nonempty, finite subset of F_2, then there exists $c \in \{a, b, a^{-1}, b^{-1}\}$, such that $\#(F \smallsetminus cF) \geq (1/4)\#F$. This shows that F_2 free groups is not amenable.

[*Hint:* Suppose $F = A \cup B \cup A^- \cup B^-$, where words in A, B, A^-, B^- start with a, b, a^{-1}, b^{-1}, respectively. If $\#A \leq \#A^-$ and $\#(A \cup A^-) \leq \#(B \cup B^-)$, then $\#(aF \smallsetminus F) \geq \#(B \cup B^-) - \#A$.]

#3. Assume that H is discrete, and that H is isomorphic to a (not necessarily discrete) subgroup of $\mathrm{SL}(\ell, \mathbb{R})$. Show:
   a) H is amenable if and only if H has no nonabelian, free subgroups.
   b) H is amenable if and only if H has a solvable subgroup of finite index.
   [*Hint:* Tits Alternative (4.9.1).]

#4. Let $G = \mathrm{SL}(2, \mathbb{R})$ act on $\mathbb{R} \cup \{\infty\}$ by linear-fractional transformations, as usual.
   a) For $u = \left[\begin{smallmatrix} 1 & 1 \\ 0 & 1 \end{smallmatrix}\right] \in G$, show that the only u-invariant probability measure on $\mathbb{R} \cup \{\infty\}$ is concentrated on the fixed point of u.
   b) Since the fixed point of u is not fixed by all of G, conclude that there is no G-invariant probability measure on $\mathbb{R} \cup \{\infty\}$.
   [*Hint:* (a) The action of u is conjugate to the homeomorphism ϕ in the hint to Exercise 1(a), so a similar argument applies.]

#5. Show that if a semisimple Lie group G is not compact, then every lattice Γ in G is not amenable.

#6. Give an example of a nonamenable Lie group that has a closed, cocompact, amenable subgroup. (By Proposition 12.4.7, the subgroup cannot be normal.)

§12.5. Closed subgroups of amenable groups

Before proving that closed subgroups of amenable groups are amenable (Proposition 12.2.8), we introduce some notation and establish a lemma. (Proofs for the case of discrete groups have already been given in Remarks 12.3.13 and 12.3.24.)

(12.5.1) **Notation.**
   1) We use $\mathcal{L}^\infty(H; C)$ to denote the space of all measurable functions from the Lie group H to the compact, convex set C, where two functions are identified if they are equal a.e. (with respect to the Haar measure on H).
   2) If Λ is a closed subgroup of H, and C is a Λ-space, then
$$\mathcal{L}^\infty_\Lambda(H; C) = \left\{ \psi \in \mathcal{L}^\infty(H; C) \,\middle|\, \begin{array}{c} \psi \text{ is essentially} \\ \Lambda\text{-equivariant} \end{array} \right\}.$$
   (To say ψ is **essentially Λ-equivariant** means, for each $\lambda \in \Lambda$, that $\psi(\lambda h) = \lambda \cdot \psi(h)$ for a.e. $h \in H$.)

(12.5.2) **Examples.**

1) Suppose H is discrete. Then every function on H is measurable, so $\mathcal{L}^\infty(H;C) = C^H$ is the cartesian product of countably many copies of C. Therefore, in this case, Tychonoff's Theorem (B5.2) implies that $\mathcal{L}^\infty(H;C)$ is compact.

2) If C is the closed unit disk in the complex plane (and H is arbitrary), then $\mathcal{L}^\infty(H;C)$ is the closed unit ball in the Banach space $\mathcal{L}^\infty(H)$, so the Banach-Alaoglu Theorem (Proposition B7.4) states that it is compact in the weak* topology.

More generally, if we put a technical restriction on C, then there is a weak topology on $\mathcal{L}^\infty(H;C)$ that makes it into a compact, convex H-space:

(12.5.3) **Lemma.** *Assume*

- Λ *is a closed subgroup of H,*
- C *is a nonempty, compact, convex H-space, and*
- C *is contained in the dual of some separable Banach space $\mathcal{B}$.*

Then $\mathcal{L}^\infty(H;C)$ and $\mathcal{L}^\infty_\Lambda(H;C)$ are nonempty, compact, convex H-spaces.

Proof. Let $\mathcal{L}^\infty(H;\mathcal{B}^*)$ be the space of all bounded measurable functions from H to $\mathcal{B}^*$ (where two functions are identified if they are equal a.e.). This is the dual of the (separable) Banach space $\mathcal{L}^1(H;\mathcal{B})$, so it has a natural weak* topology. Since $\mathcal{L}^\infty(H;C)$ is a closed, bounded, convex subset of $\mathcal{L}^\infty(H;\mathcal{B}^*)$, the Banach-Alaoglu Theorem (B7.4) tells us that it is weak* compact. In addition, the action of H by right-translation on $\mathcal{L}^\infty(H;C)$ is continuous (see Exercise 1).

It is not difficult to see that $\mathcal{L}^\infty_\Lambda(H;C)$ is a nonempty, closed, convex, H-invariant subset (see Exercise 3). $\qquad\square$

Proof of Proposition 12.2.8. Let Λ be a closed subgroup of an amenable Lie group H. Given any continuous action of Λ on a compact, metrizable space X, it suffices to show there is a Λ-invariant probability measure on X (see Theorem 12.3.1(3)). From Lemma 12.5.3, we know that $\mathcal{L}^\infty_\Lambda(H;\mathrm{Prob}(X))$ is a nonempty, compact, convex H-space. Therefore, the amenability of H implies that H has a fixed point ψ in $\mathcal{L}^\infty_\Lambda(H;C)$. So ψ is essentially H-invariant. If we fix any $\lambda \in \Lambda$, then, for a.e. $h \in H$, we have

$$\lambda \cdot \psi(h) = \psi(\lambda h) \qquad (\psi \text{ is essentially } \Lambda\text{-equivariant})$$
$$= \psi(h) \qquad (\psi \text{ is essentially } H\text{-invariant}).$$

If we assume, for simplicity, that Λ is countable (see Exercise 4), then the quantifiers can be reversed (because the union of countably many null sets is a null set), so we conclude that the probability measure $\psi(h)$ is Λ-invariant. $\qquad\square$

Exercises for §12.5.

#1. Show that the action of H on $\mathscr{L}^{\infty}(H; C)$ by right translations is continuous in the weak*-topology.

#2. Suppose Λ is a closed subgroup of H, and that Λ acts measurably on a measure space Ω. Show there is a Λ-equivariant, measurable map $\psi \colon H \to \Omega$.
   [*Hint:* ψ can be defined arbitrarily on a strict fundamental domain for Λ in H.]

#3. Show that $\mathscr{L}^{\infty}_{\Lambda}(H; C)$ is a nonempty, closed, convex, H-invariant subset of $\mathscr{L}^{\infty}(H; C)$.

#4. Prove Proposition 12.2.8 without assuming Λ is countable.
   [*Hint:* Consider λ in a countable dense subset of Λ.]

§12.6. Equivariant maps from G/P to $\mathrm{Prob}(X)$

We now use amenability to prove a basic result that has important consequences for the theory of arithmetic groups. In particular, it is an ingredient in two fundamental results of G. A. Margulis: his Superrigidity Theorem (16.1.6) and his Normal Subgroups Theorem (17.1.1).

(12.6.1) **Proposition** (Furstenberg's Lemma). *If*
   - *P is a closed, amenable subgroup of G, and*
   - *Γ acts continuously on a compact metric space X,*

then there is a Borel measurable map $\psi \colon G/P \to \mathrm{Prob}(X)$, such that ψ is essentially Γ-equivariant.

Proof. Lemma 12.5.3 tells us that $\mathscr{L}^{\infty}_{\Gamma}(G; \mathrm{Prob}(X))$ is a nonempty, compact, convex G-space. By restriction, it is also a nonempty, compact, convex P-space, so P has a fixed point ψ_0 (under the action by right-translation). Then ψ_0 factors through to an (essentially) well-defined map $\psi \colon G/P \to \mathrm{Prob}(X)$. Because ψ_0 is Γ-equivariant, it is immediate that ψ is Γ-equivariant. $\square$

In applications of Proposition 12.6.1, the subgroup P is usually taken to be a minimal parabolic subgroup. Here is an example of this:

(12.6.2) **Corollary.** *If*
   - *$G = \mathrm{SL}(3, \mathbb{R})$,*
   - *$P = \begin{bmatrix} * & & \\ * & * & \\ * & * & * \end{bmatrix} \subset G$, and*
   - *Γ acts continuously on a compact metric space X,*

then there is a Borel measurable map $\psi \colon G/P \to \mathrm{Prob}(X)$, such that ψ is essentially Γ-equivariant.

Proof. P is amenable (see Exercise 12.2#14). □

(12.6.3) *Remark.* The function ψ that is provided by Furstenberg's Lemma (12.6.1) (or Corollary 12.6.2) can be thought of as being a "random" map from G/P to X; for each $z \in G/P$, the value of $\psi(z)$ is a probability distribution that defines a random value for the function at the point z. However, we will see in Section 16.7 that the theory of proximality makes it possible to show, in certain cases, that $\psi(z)$ is actually a single well-defined point of X, not a random value that varies over some range.

Exercises for §12.6.

#1. Show that every minimal parabolic subgroup of G is amenable.
   [*Hint:* Langlands decomposition (8.4.6).]

§12.7. More properties of amenable groups (optional)

In this section, we mention (without proof, and without even defining all of the terminology) a variety of very interesting properties of amenable groups. For simplicity,

$$\text{we assume } \Lambda \text{ is a discrete group.}$$

§12.7(i). Bounded harmonic functions.

(12.7.1) **Definition.** Fix a probability measure μ on Λ.
   1) A function $f: \Lambda \to \mathbb{R}$ is $\boldsymbol{\mu}$**-harmonic** if $f = \mu * f$. This means, for every $\lambda \in \Lambda$,

$$f(\lambda) = \sum_{x \in \Lambda} \mu(x) f(x\lambda).$$

   2) μ is **symmetric** if $\mu(A^{-1}) = \mu(A)$ for every $A \subseteq \Lambda$.

(12.7.2) **Theorem.** Λ *is amenable if and only if there exists a symmetric probability measure μ on Λ, such that*
   1) *the support of μ generates Λ, and*
   2) *every bounded, μ-harmonic function on Λ is constant.*

   Because any harmonic function is the Poisson integral of a function on the Poisson boundary (and vice-versa), this result can be restated in the following equivalent form:

(12.7.3) **Corollary.** Λ *is amenable if and only if there exists a symmetric probability measure μ on Λ, such that*
   1) *the support of μ generates Λ, and*
   2) *the Poisson boundary of Λ (with respect to μ) consists of a single point.*

§12.7(ii). Norm of a convolution operator.

(12.7.4) **Definition.** For any probability measure μ on Λ, there is a corresponding convolution operator C_μ on $\mathcal{L}^2(\Lambda)$, defined by

$$(C_\mu f)(\lambda) = \sum_{x \in \Lambda} \mu(x) \, f(x^{-1}\lambda).$$

(12.7.5) **Theorem.** *Let μ be any probability measure on Λ, such that the support of μ generates Λ. Then $\|C_\mu\| = 1$ if and only if Λ is amenable.*

§12.7(iii). Spectral radius.

In geometric terms, the following famous result characterizes amenability in terms of the spectral radius of random walks on Cayley graphs.

(12.7.6) **Theorem** (Kesten). *Let μ be a finitely supported, symmetric probability measure on Λ, such that the support of μ generates Λ. Then μ is amenable if and only if*

$$\lim_{n \to \infty} \left(\sum_{\substack{g_1, \ldots, g_{2n} \,\in\, \mathrm{supp}\,\mu \\ g_1 g_2 \cdots g_{2n} = e}} \mu(g_1)\,\mu(g_2) \cdots \mu(g_{2n}) \right)^{1/2n} = 1.$$

§12.7(iv). Positive-definite functions.

(12.7.7) **Definition** (cf. Terminology 13.6.4). A $\mathbb{C}$-valued function φ on Λ is *positive-definite* if, for all $a_1, \ldots, a_n \in \Lambda$, the matrix

$$[a_{i,j}]_{i,j=1}^n = [\varphi(a_i^{-1} a_j)]$$

is Hermitian and has no negative eigenvalues.

(12.7.8) **Theorem.** *Λ is amenable if and only if $\sum_{g \in \Lambda} \varphi(g) \geq 0$ for every (finitely supported) positive-definite function φ on Λ.*

§12.7(v). Growth.

(12.7.9) **Definition.** Assume Λ is finitely generated, and fix a symmetric generating set S for Λ.

1) For each $r \in \mathbb{Z}^+$, let $B_r(\Lambda)$ be the ball of radius r centered at e, More precisely,

$$B_r(\Lambda; S) = \{ \lambda \in \Lambda \mid \exists s_1, s_2, \ldots, s_r \in S \cup \{e\}, \lambda = s_1 s_2 \cdots s_r \}.$$

2) We say Λ has *subexponential growth* if for every $\epsilon > 0$, we have $\#B_r(\Lambda; S) < e^{\epsilon r}$, for all sufficiently large $r \in \mathbb{Z}^+$.

(12.7.10) **Proposition** (see Exercise 1). *If Λ has subexponential growth, then Λ is amenable.*

(12.7.11) **Warning.** The implication in Proposition 12.7.10 goes only one direction: there are many groups (including many solvable groups) that are amenable, but do not have subexponential growth (see Exercise 2).

§12.7(vi). Cogrowth.

(12.7.12) **Definition.** Assume Λ is finitely generated. Let:

1) $S = \{s_1, s_2, \ldots, s_k\}$ be a finite generating set of Λ.
2) F_k be the free group on k generators $x_1, \ldots, x_k$.
3) $\phi_S \colon F_k \to \Lambda$ be the homomorphism defined by $\phi(x_i) = s_i$.

The **cogrowth** of Λ (with respect to S) is

$$\lim_{r \to \infty} \frac{1}{r} \log_{2k-1} \#((\ker \phi_S) \cap B_r(F_k; x_1^{\pm 1}, \ldots, x_k^{\pm 1}).$$

Note that $\#B_r(F_k; x_1^{\pm 1}, \ldots, x_k^{\pm 1})$ is equal to the number of reduced words of length r in the symbols $x_1^{\pm 1}, \ldots, x_k^{\pm 1}$, which is approximately $(2k - 1)^r$. Therefore, it is easy to see that that the cogrowth of Λ is between 0 and 1 (see Exercise 3). The maximum value is obtained if and only if Λ is amenable:

(12.7.13) **Theorem.** Λ *is amenable if and only if the cogrowth of Λ is 1, with respect to some (or, equivalently, every) finite generating set S.*

§12.7(vii). Unitarizable representations.

(12.7.14) **Definition.** Let $\rho \colon \Lambda \to \mathcal{B}(\mathcal{H})$ be a (not necessarily unitary) representation of Λ on a Hilbert space $\mathcal{H}$.

1) ρ is **uniformly bounded** if there exists $C > 0$, such that $\|\rho(\lambda)\| < C$, for all $\lambda \in \Lambda$.
2) ρ is **unitarizable** if it is conjugate to a unitary representation. This means there is an invertible operator T on $\mathcal{H}$, such that the representation $\lambda \mapsto T^{-1} \rho(\lambda) T$ is unitary.

It is fairly obvious that every unitarizable representation is uniformly bounded (see Exercise 4). The converse is not true, although it holds for amenable groups:

(12.7.15) **Theorem.** *If Λ is amenable, then every uniformly bounded representation of Λ is unitarizable.*

(12.7.16) *Remark.* The converse of Theorem 12.7.15 is an open question.

§12.7(viii). Almost representations are near representations.

(12.7.17) **Definition.** Fix $\epsilon > 0$, and let φ be a function from Λ to the group $\mathcal{U}(\mathcal{H})$ of unitary operators on a Hilbert space $\mathcal{H}$.

1) φ is ϵ-*almost* a unitary representation if
$$\|\varphi(\lambda_1\lambda_2) - \varphi(\lambda_1)\,\varphi(\lambda_2)\| < \epsilon \text{ for all } \lambda_1, \lambda_2 \in \Lambda.$$

2) φ is ϵ-*near* a unitary representation if there exists a unitary representation $\rho\colon \Lambda\colon \mathcal{U}(\mathcal{H})$, such that
$$\|\varphi(\lambda) - \rho(\lambda)\| < \epsilon \text{ for every } \lambda \in \Lambda.$$

For amenable groups, every almost representation is near a representation:

(12.7.18) **Theorem.** *Assume Λ is amenable. Given $\epsilon > 0$, there exists $\delta > 0$, such that if φ is δ-almost a unitary representation, then φ is ϵ-near a unitary representation.*

§12.7(ix). Bounded cohomology. The bounded cohomology groups of Λ are defined just like the ordinary cohomology groups, except that all cochains are assumed to be bounded functions.

(12.7.19) **Definition.** Assume $\mathcal{B}$ is a Banach space.

1) $\mathcal{B}$ is a **Banach Λ-module** if Λ acts continuously on $\mathcal{B}$, by linear isometries.

2) $\mathcal{L}^1_{\mathrm{bdd}}(\Lambda; \mathcal{B}) = \mathcal{L}^1(\Lambda; \mathcal{B}) \cap \mathcal{L}^\infty(\Lambda; \mathcal{B})$.

3) $\mathcal{H}^1_{\mathrm{bdd}}(\Lambda; \mathcal{B}) = \mathcal{L}^1_{\mathrm{bdd}}(\Lambda; \mathcal{B})/\mathcal{B}^1(\Lambda; \mathcal{B})$.

(12.7.20) **Theorem.** Λ *is amenable if and only if* $\mathcal{H}^1_{\mathrm{bdd}}(\Lambda; \mathcal{B}) = 0$ *for every Banach Λ-module $\mathcal{B}$.*

(12.7.21) *Remark.* In fact, if Λ is amenable, then $\mathcal{H}^n_{\mathrm{bdd}}(\Lambda; \mathcal{B}) = 0$ for all n.

§12.7(x). Invariance under quasi-isometry.

(12.7.22) **Proposition** (see Exercise 6). *Assume Λ_1 and Λ_2 are finitely generated groups, such that Λ_1 is quasi-isometric to Λ_2 (see Definition 10.1.3). Then Λ_1 is amenable if and only if Λ_2 is amenable.*

§12.7(xi). Ponzi schemes. Assume Λ is finitely generated, and let d be the word metric on Λ, with respect to some finite, symmetric generating set S (see Definition 10.1.1).

(12.7.23) Definition. A function $f : \Lambda \to \Lambda$ is a *Ponzi scheme on* Λ if there is some $C > 0$, such that, for all $\lambda \in \Lambda$, we have:

1) $\# f^{-1}(\lambda) \geq 2$, and

2) $d(f(\lambda), \lambda) < C$.

(12.7.24) Theorem. Λ *is amenable if and only if there does not exist a Ponzi scheme on* Λ.

Exercises for §12.7.

#1. Prove Proposition 12.7.10.

[*Hint:* If no balls are Følner sets, then the group has exponential growth.]

#2. Choose a prime number p, and let

$$\Lambda = \left\{ \begin{bmatrix} p^k & mp^n \\ 0 & p^{-k} \end{bmatrix} \;\middle|\; k, m, n \in \mathbb{Z} \right\} \subset \mathrm{SL}(2, \mathbb{Q}),$$

with the discrete topology. Show Λ is an amenable group that does not have subexponential growth.

#3. In the notation of Definition 12.7.12, show:
   a) $\# B_r(F_k; x_1^{\pm 1}, \dots, x_k^{\pm 1}) = 2k(2k-1)^{r-1}$.
   b) If $\cog \Lambda$ is the cogrowth of Λ, then $0 \leq \cog \Lambda \leq 1$.

#4. Show that every unitarizable representation is uniformly bounded.

#5. For every $\epsilon > 0$, show there exists $\delta > 0$, such that if φ is δ-near a unitary representation, then φ is ϵ-almost a unitary representation.

#6. Prove Proposition 12.7.22.

[*Hint:* Show that if Λ is not amenable, then, for every k, it has a finite subset S, such that $\#(SF) \geq k \cdot \#F$ for every finite subset F of Λ.]

#7. Explicitly construct a Ponzi scheme on the free group with two generators.

#8. Show (without using Theorem 12.7.24) that if Λ is amenable, then there does not exist a Ponzi scheme on Λ.

[*Hint:* Følner sets.]

Notes

The notion of amenability is attributed to J. von Neumann [20], but he used the German word "messbar" (which can be translated as "measurable"). The term "amenable" was apparently introduced into the literature by M. Day [4, #507, p. 1054] in the announcement of a talk.

The monographs [16, 17] are standard references on amenability. Briefer treatments are in [2, App. G], [6], and [22, §4.1]. Quite a different approach to amenability appears in [21, Chaps. 10–12] (for discrete groups only).

The fact that closed subgroups of amenable groups are amenable (Proposition 12.2.8) is proved in [6, Thm. 2.3.2, pp. 30–32], [17, Prop. 13.3, p. 118], and [22, Prop. 4.2.20, p. 74].

See [6, p. 67] for a proof of Proposition 12.3.22($\Rightarrow$) that does not require H to be discrete.

Remark 12.3.17 is proved in [6, pp. 46–47].

The solution of Exercise 12.3#17 can be found in [17, Thm. 5.4, p. 45].

For a proof of the fact (mentioned in the hint to Exercise 12.3#21) that the one-to-one continuous image of a Borel set is Borel, see [1, Thm. 3.3.2, p. 70].

Our proof of Proposition 12.3.22($\Rightarrow$) is taken from [6, pp. 66-67].

Remark 12.4.3(1), the existence of a nonamenable group with no nonabelian free subgroup, is due to Olshanskii [15]. (In this example, called an "Olshanskii Monster" or "Tarski Monster," every proper subgroup of the group is a cyclic group of prime order, so there is obviously no free subgroup.) A much more elementary example has recently been constructed by N. Monod [13].

The book of S. Wagon [21] is one of the many places to read about the Banach-Tarski Paradox (Remark 12.4.3(3)).

Furstenberg's Lemma (12.6.1) appears in [5, Thm. 15.1]. Another proof can be found in [22, Prop. 4.3.9, p. 81].

Theorem 12.7.2 is due to Kaimanovich and Vershik [10, Thms. 4.2 and 4.4] and (independently) Rosenblatt [19, Props. 1.2 and 1.9 and Thm. 1.10].

Theorem 12.7.5 is due to H. Kesten (if μ is symmetric). See [2, G.4.4] for a proof.

A proof of Proposition 12.7.10 can be found in [17, Props. 12.5 and 12.5].

Theorem 12.7.13 was proved by R. I. Grigorchuk [8] and J. M. Cohen [3] (independently).

Theorem 12.7.15 was proved by J. Dixmier and M. Day in 1950 (independently). See [18] for historical remarks and progress on the converse. (Another result on the converse is proved in [14].)

Theorem 12.7.18 is due to D. Kazhdan [11].

Theorem 12.7.20 and Remark 12.7.21 are due to B. E. Johnson [9]. See [12] (and its many references) for an introduction to bounded cohomology.

Theorem 12.7.24 appears in [7, 6.17 and $6.17\frac{1}{2}$, p. 328].

References

[1] W. Arveson: *An Invitation to C*-algebras.* Springer, New York, 1976. ISBN 0-387-90176-0, MR 0512360

[2] B. Bekka, P. de la Harpe, and A. Valette: *Kazhdan's Property (T),* Cambridge U. Press, Cambridge, 2008. ISBN 978-0-521-88720-5, MR 2415834, http://perso.univ-rennes1.fr/bachir.bekka/KazhdanTotal.pdf

[3] J. M. Cohen: Cogrowth and amenability of discrete groups, *J. Funct. Anal.* 48 (1982), no. 3, 301–309. MR 0678175, http://dx.doi.org/10.1016/0022-1236(82)90090-8

[4] J. W. Green: The summer meeting in Boulder, *Bulletin of the American Mathematical Society* 55 (1949), no. 11, 1035–1081. http://projecteuclid.org/euclid.bams/1183514222

[5] H. Furstenberg: Boundary theory and stochastic processes on homogeneous spaces, in C. C. Moore, ed.: *Harmonic Analysis on Homogeneous Spaces (Williamstown, Mass., 1972).* Amer. Math. Soc., Providence, R.I., 1973, pp. 193–229. ISBN 0-8218-1426-5, MR 0352328

[6] F. P. Greenleaf: *Invariant Means on Topological Groups and Their Applications.* Van Nostrand, New York, 1969. ISBN 0-442-02857-1, MR 0251549

[7] M. Gromov: *Metric Structures for Riemannian and Non-Riemannian Spaces.* Birkhäuser, Boston, 2007. ISBN 978-0-8176-4582-3, MR 2307192

[8] R. I. Grigorchuk: Symmetrical random walks on discrete groups, in: *Multicomponent Random Systems.* Dekker, New York, 1980, pp. 285–325. ISBN 0-8247-6831-0, MR 0599539

[9] B. E. Johnson: *Cohomology in Banach algebras.* American Mathematical Society, Providence, R.I., 1972. MR 0374934

[10] V. A. Kaimanovich and A. M. Vershik Random walks on discrete groups: boundary and entropy, *Ann. Probab.* 11 (1983), no. 3, 457–490. MR 0704539, http://www.jstor.org/stable/2243645

[11] D. Kazhdan: On ε-representations, *Israel J. Math.* 43 (1982), no. 4, 315–323. MR 0693352, http://dx.doi.org/10.1007/BF02761236

[12] N. Monod: An invitation to bounded cohomology, in: *International Congress of Mathematicians, Vol. II,* Eur. Math. Soc., Zürich, 2006, pp. 1183-1211. MR 2275641, http://www.mathunion.org/ICM/ICM2006.2/

[13] N. Monod: Groups of piecewise projective homeomorphisms *Proc. Natl. Acad. Sci. USA* 110 (2013), no. 12, 4524-4527. MR 3047655, http://dx.doi.org/10.1073/pnas.1218426110

[14] N. Monod and N. Ozawa: The Dixmier problem, lamplighters and Burnside groups, *J. Funct. Anal.* 258 (2010), no. 1, 255-259. MR 2557962, http://dx.doi.org/10.1016/j.jfa.2009.06.029

[15] A. Yu. Ol'shanskii: On the problem of the existence of an invariant mean on a group. *Russian Mathematical Surveys* 35 (1980), No. 4, 180-181. MR 0586204, http://dx.doi.org/10.1070/RM1980v035n04ABEH001876

[16] A. L. T. Paterson: *Amenability.* American Mathematical Society, Providence, RI, 1988. ISBN 0-8218-1529-6, MR 0961261

[17] J.-P. Pier: *Amenable Locally Compact Groups.* Wiley, New York, 1984. ISBN 0-471-89390-0, MR 0767264

[18] G. Pisier: Are unitarizable groups amenable?, in L. Bartholdi et al., eds.: *Infinite Groups: Geometric, Combinatorial and Dynamical Aspects.* Birkhäuser, Basel, 2005, pp. 323-362. ISBN 3-7643-7446-2, MR 2195457

[19] J. Rosenblatt: Ergodic and mixing random walks on locally compact groups, *Math. Ann.* 257 (1981), no. 1, 31-42. MR 0630645, http://eudml.org/doc/163540

[20] J. von Neumann: Zur allgemeinen Theorie des Masses, *Fund. Math.* 13, no. 1 (1929) 73-116. Zbl 55.0151.02, http://eudml.org/doc/211916

[21] S. Wagon: *The Banach-Tarski Paradox.* Cambridge U. Press, Cambridge, 1993. ISBN 0-521-45704-1, MR 1251963

[22] R. J. Zimmer: *Ergodic Theory and Semisimple Groups.* Monographs in Mathematics, 81. Birkhäuser, Basel, 1984. ISBN 3-7643-3184-4, MR 0776417

Chapter 13

Kazhdan's Property (T)

Recall that if a Lie group H is not amenable, then $\mathcal{L}^2(H)$ does not have almost-invariant vectors (see Theorem 12.3.1(6)). Kazhdan's property (T) is the much stronger condition that **no** unitary representation of H has almost-invariant vectors (unless it has a vector that is fixed by H). Thus, in a sense, Kazhdan's property is the antithesis of amenability.

We already know that Γ is not amenable (unless it is finite) (see Exercise 12.4#5). In this chapter, we will see that Γ usually has Kazhdan's Property (T), and we will look at some of the consequences of this.

§13.1. Definition and basic properties

Part (1) of the following definition is repeated from Definition 12.3.14, but the second half is new.

(13.1.1) **Definition.** Let H be a Lie group.

1) An action of H on a normed vector space $\mathcal{B}$ has **almost-invariant** vectors if, for every compact subset C of H and every $\epsilon > 0$, there is a unit vector $v \in \mathcal{B}$, such that

$$\|cv - v\| < \epsilon \quad \text{for all } c \in C. \tag{13.1.2}$$

Recall: The Standing Assumptions (4.0.0 on page 41) are in effect, so, as always, Γ is a lattice in the semisimple Lie group $G \subseteq \mathrm{SL}(\ell, \mathbb{R})$.

Main prerequisites for this chapter: Unitary representations (Sections 11.1, 11.3, and 11.5).

(A unit vector satisfying (13.1.2) is said to be (ϵ, C)-invariant.)

2) H has **Kazhdan's property (T)** if every unitary representation of H that has almost-invariant vectors also has (nonzero) invariant vectors.

We often abbreviate "Kazhdan's property (T)" to "Kazhdan's property." Also, a group that has Kazhdan's property is often said to be a **Kazhdan group**.

(13.1.3) **Warning.** By definition, unitary representations are actions on Hilbert spaces, so Kazhdan's property says nothing at all about actions on other types of topological vector spaces. In particular, there are actions of Kazhdan groups by norm-preserving linear transformations on some Banach spaces that have almost-invariant vectors, without having invariant vectors (see Exercise 1). On the other hand, it can be shown that there are no such examples on $\mathscr{L}^p$ spaces (with $1 \leq p < \infty$).

(13.1.4) **Proposition.** *A Lie group is compact if and only if it is amenable and has Kazhdan's property.*

Proof. Exercises 2 and 3.   □

(13.1.5) **Corollary.** *A discrete group Λ is finite if and only if it is amenable and has Kazhdan's property.*

(13.1.6) **Example.** $\mathbb{Z}^n$ does not have Kazhdan's property, because it is a discrete, amenable group that is not finite.

(13.1.7) **Proposition.** *If Λ is a discrete group with Kazhdan's property, then:*

1) *every quotient Λ/N of Λ has Kazhdan's property,*
2) *the abelianization $\Lambda/[\Lambda, \Lambda]$ of Λ is finite, and*
3) *Λ is finitely generated.*

Proof. For (1) and (2), see Exercises 4 and 6.

(3) Let $\{\Lambda_n\}$ be the collection of all finitely generated subgroups of Λ. We have a unitary representation of Λ on each $\mathscr{L}^2(\Lambda/\Lambda_n)$, given by $(\gamma f)(x\Lambda_n) = f(\gamma^{-1}x\Lambda_n)$. The direct sum of these is a unitary representation on

$$\mathcal{H} = \mathscr{L}^2(\Lambda/\Lambda_1) \oplus \mathscr{L}^2(\Lambda/\Lambda_2) \oplus \cdots.$$

Any compact set $C \subseteq \Lambda$ is finite, so we have $C \subseteq \Lambda_n$, for some n. Then C fixes the base point $p = \Lambda_n/\Lambda_n$ in Λ/Λ_n, so, letting $f = \delta_p$ be a nonzero function in $\mathscr{L}^2(\Lambda/\Lambda_n)$ that is supported on $\{p\}$, we have $\gamma f = f$ for all $\gamma \in C$. Therefore, $\mathcal{H}$ has almost-invariant vectors, so there must be an H-invariant vector in $\mathcal{H}$.

So some $\mathscr{L}^2(\Lambda/\Lambda_n)$ has an invariant vector. Since Λ is transitive on Λ/Λ_n, an invariant function must be constant. So a (nonzero) constant function is in $\mathscr{L}^2(\Lambda/\Lambda_n)$, which means Λ/Λ_n is finite. Because Λ_n is finitely generated, this implies that Λ is finitely generated. $\square$

Since the abelianization of any (nontrivial) free group is infinite, we have the following example:

(13.1.8) **Corollary.** *Free groups do not have Kazhdan's property.*

(13.1.9) *Remark.* Proposition 13.1.7 can be generalized to groups that are not required to be discrete, if we replace the word "finite" with "compact" (see Exercises 5, 7, and 15). This leads to the following definition:

(13.1.10) **Definition.** A Lie group H is ***compactly generated*** if there exists a compact subset that generates H.

(13.1.11) **Warning.** Although discrete Kazhdan groups are always finitely generated (see Proposition 13.1.7(3)), they need not be finitely presented. (In fact, there are uncountably many non-isomorphic discrete groups with Kazhdan's property (T), and only countably many of them can be finitely presented.) However, it can be shown that every discrete Kazhdan group is a quotient of a finitely presented Kazhdan group.

Exercises for §13.1.

#1. Let $C_0(H)$ be the Banach space of continuous functions on H that tend to 0 at infinity (with the supremum norm). Show:
   a) $C_0(H)$ has almost-invariant vectors of norm 1, but
   b) $C_0(H)$ does not have H-invariant vectors other than 0, unless H is compact.
   [*Hint:* Choose a uniformly continuous function $f(h)$ that tends to $+\infty$ as h leaves compact sets. For large n, the function $h \mapsto n/(n + f(h))$ is almost invariant.]

#2. Prove Proposition 13.1.4($\Rightarrow$).
   [*Hint:* If H is compact, then almost-invariant vectors are invariant.]

#3. Prove Proposition 13.1.4($\Leftarrow$).
   [*Hint:* Amenability plus Kazhdan's property implies $\mathscr{L}^2(H)$ has an invariant vector.]

#4. Prove Proposition 13.1.7(1)
   [*Hint:* Any representation of Λ/N is also a representation of Λ.]

#5. Show that if H has Kazhdan's property, and N is a closed, normal subgroup of H, then H/N has Kazhdan's property.

#6. Prove Proposition 13.1.7(2).
   [*Hint:* $\Lambda/[\Lambda, \Lambda]$ is amenable and has Kazhdan's property.]

#7. Show that if H has Kazhdan's property, then $H/[H, H]$ is compact.

#8. Show that if N is a closed, normal subgroup of H, such that N and H/N have Kazhdan's property, then H has Kazhdan's property.
   [*Hint:* The space of N-invariant vectors is H-invariant (why?).]

Warning. The converse is not true: there are examples in which a normal subgroup of a Kazhdan group is not Kazhdan (see Exercise 13.3#5).

#9. Show that $H_1 \times H_2$ has Kazhdan's property if and only if H_1 and H_2 both have Kazhdan's property.

#10. Let (π, V) be a unitary representation of a Kazhdan group H. Show that almost-invariant vectors in V are near invariant vectors. More precisely, given $\epsilon > 0$, find a compact subset C of H and $\delta > 0$, such that if v is any (δ, C)-invariant vector in V, then there is an invariant vector v_0 in V, such that $\|v - v_0\| < \epsilon$.

[*Hint:* There are no almost-invariant unit vectors in $(V^H)^\perp$, the orthogonal complement of the space of invariant vectors.]

#11. Suppose S is a generating set of a discrete group Λ, and Λ has Kazhdan's property. Show there exists $\epsilon > 0$, such that if π is any unitary representation of Λ that has an (ϵ, S)-invariant vector, then π has an invariant vector. (The point here is to reverse the quantifiers: the same ϵ works for every π.) Such an ϵ is called a **Kazhdan constant** for Λ.

#12. Recall that we say H has the **Haagerup property** if it has a unitary representation, such that there are almost-invariant vectors, and all matrix coefficients decay to 0 at ∞. Show that if H is a noncompact group with Kazhdan's property, then H does not have the Haagerup property.

#13. Assume:
- $\varphi \colon H_1 \to H_2$ is a homomorphism with dense image, and
- H_1 has Kazhdan's property.

Show H_2 has Kazhdan's property.

#14. Show that a Lie group H is compactly generated if and only if H/H° is finitely generated.

[*Hint:* ($\Leftarrow$) Since H° is connected, it is generated by any subset with nonempty interior.]

#15. Show that every Lie group with Kazhdan's property is compactly generated.

[*Hint:* Either adapt the proof of Proposition 13.1.7(3), or use Proposition 13.1.7(3) together with Exercises 5 and 14.]

#16. Assume Γ has Kazhdan's property (T), and S is a finite generating set for Γ. Show there exists $\epsilon > 0$, such that if N is any finite-index normal subgroup of Γ, and A is any subset of Γ/N, then

$$\#(SA \cup A) \geq \min\left\{ (1 + \epsilon) \cdot \#A, \; \tfrac{1}{2}|\Gamma/N| \right\}.$$

(In graph-theoretic terminology, this means the Cayley graphs $\text{Cay}(\Gamma/N_k; S)$ form a family of *expander graphs* if $N_1, N_2, \ldots$ are finite-index normal subgroups, such that $|\Gamma/N_k| \to \infty$.)

§13.2. Semisimple groups with Kazhdan's property

(13.2.1) **Theorem** (Kazhdan). $\text{SL}(3, \mathbb{R})$ *has Kazhdan's property.*

This theorem is an easy consequence of the following lemma, which will be proved in Section 13.3.

(13.2.2) **Lemma.** *Assume*

- π *is a unitary representation of the natural semidirect product*

$$\text{SL}(2, \mathbb{R}) \ltimes \mathbb{R}^2 = \begin{bmatrix} * & * & * \\ * & * & * \\ 0 & 0 & 1 \end{bmatrix} \subset \text{SL}(3, \mathbb{R}),$$

and

- π *has almost-invariant vectors.*

Then π has a nonzero vector that is invariant under the subgroup $\mathbb{R}^2$.

Other terminology. Suppose R is a subgroup of a topological group H. The pair (H, R) is said to have *relative property* (*T*) if every unitary representation of H that has almost-invariant vectors must also have an R-invariant vector. In this terminology, Lemma 13.2.2 states that the pair $(\text{SL}(2, \mathbb{R}) \ltimes \mathbb{R}^2, \mathbb{R}^2)$ has relative property (*T*).

Proof of Theorem 13.2.1. Let

$$G = \text{SL}(3, \mathbb{R}), \quad R = \begin{bmatrix} 1 & 0 & * \\ 0 & 1 & * \\ 0 & 0 & 1 \end{bmatrix} \cong \mathbb{R}^2, \quad \text{and} \quad H = \text{SL}(2, \mathbb{R}) \ltimes R,$$

and suppose π is a unitary representation of G that has almost-invariant vectors. Then it is obvious that the restriction of π to H also has almost-invariant vectors (see Exercise 1), so Lemma 13.2.2 implies there is a nonzero vector v that is fixed by R. Then the Moore Ergodicity Theorem (11.2.8) implies that v is fixed by all of G. So π has a fixed vector (namely, v). $\qquad\square$

If G is simple, and $\text{rank}_{\mathbb{R}} G \geq 2$, then G contains a subgroup isogenous to $\text{SL}(2, \mathbb{R}) \ltimes \mathbb{R}^n$, for some n (cf. Exercise 2), so a modification of the above argument shows that G has Kazhdan's property. On the other hand, it is important to know that not all simple Lie groups have the property:

(13.2.3) **Example.** $\text{SL}(2, \mathbb{R})$ does *not* have Kazhdan's property.

Proof. Choose a torsion-free lattice Γ in $\text{SL}(2, \mathbb{R})$. Then Γ is either a surface group or a nonabelian free group. In either case, $\Gamma/[\Gamma, \Gamma]$ is infinite, so Γ does not have Kazhdan's property. Therefore, we conclude from Proposition 13.4.1 below that $\text{SL}(2, \mathbb{R})$ does not have Kazhdan's property. $\quad\square$

Alternate proof. A reader familiar with the unitary representation theory of $SL(2, \mathbb{R})$ can easily construct a sequence of representations in the principal series whose limit is the trivial representation. The direct sum of this sequence of representations has almost-invariant vectors. $\square$

We omit the proof of the following precise characterization of the semisimple groups that have Kazhdan's property:

(13.2.4) **Theorem.** *G has Kazhdan's property if and only if no simple factor of G is isogenous to* $SO(1, n)$ *or* $SU(1, n)$.

Exercises for §13.2.

#1. Assume π is a unitary representation of H that has almost-invariant vectors, and L is a subgroup of H. Show that the restriction of π to L has almost-invariant vectors.

#2. (*Assumes familiarity with real roots*) Assume G is simple. Show $\text{rank}_{\mathbb{R}} G \geq 2$ if and only if some connected subgroup L of G is isogenous to $SL(2, \mathbb{R})$ and normalizes (but does not centralize) a nontrivial, unipotent subgroup U of G.

    [*Hint:* ($\Rightarrow$) An entire maximal parabolic subgroup of G normalizes a nontrivial unipotent subgroup. ($\Leftarrow$) Construct two unipotent subgroups of G that both contain U, but generate a subgroup that is not unipotent.]

#3. Suppose H is a closed, noncompact subgroup of G, and G is simple. Show that the pair (G, H) has relative property (T) if and only if G has Kazhdan's property.

#4. Suppose G has Kazhdan's property. Show there is a compact subset C of G and some $\epsilon > 0$, such that every unitary representation of G with (ϵ, C)-invariant vectors has invariant vectors.

§13.3. Proof of relative property (T)

In this section, we prove Lemma 13.2.2, thereby completing the proof that $SL(3, \mathbb{R})$ has Kazhdan's property (T). The argument relies on a decomposition theorem for representations of $\mathbb{R}^n$.

Proof of Lemma 13.2.2. For convenience, let $H = SL(2, \mathbb{R}) \ltimes \mathbb{R}^2$. Given a unitary representation $(\pi, \mathcal{H})$ of H that has almost-invariant vectors, we wish to show that some nonzero vector in $\mathcal{H}$ is fixed by the subgroup $\mathbb{R}^2$ of H. In other words, if we let E be the projection-valued measure provided by Proposition 11.5.2 (for the restriction of π to $\mathbb{R}^2$), then we wish to show $E(\{0\})$ is nontrivial.

Letting $\mathcal{B}(\mathcal{H})$ be the algebra of bounded linear operators on $\mathcal{H}$, and using the fact that π has almost-invariant vectors, Exercise 1 provides a continuous, linear functional $\lambda: \mathcal{B}(\mathcal{H}) \to \mathbb{C}$, such that

- $\lambda(\mathrm{Id}) = 1$,
- $\lambda(E) \geq 0$ for every orthogonal projection E, and
- λ is bi-invariant under H. (More precisely, for all $h_1, h_2 \in H$ and $T \in \mathcal{B}(\mathcal{H})$ we have $\lambda(\pi(h_1) \, T \, \pi(h_2)) = \lambda(T)$.)

Now, let μ be the composition of λ with E (that is, let $\mu(A) = \lambda(E(A))$ for $A \subseteq \mathbb{R}^n$), so μ is a finitely additive probability measure on $\mathbb{R}^n$ (see Exercise 2). Since $\mathbb{R}^2 \lhd H$, there is an action of H on $\mathbb{R}^2$ by conjugation. One can show that the probability measure μ is invariant under this action (see Exercise 3).

On the other hand, the only $\mathrm{SL}(2, \mathbb{R})$-invariant, finitely additive probability measure on $\mathbb{R}^2$ is the point-mass supported at the origin (see Exercise 4). Therefore, we must have $\mu(\{0\}) = 1 \neq 0$. Hence, $E(\{0\})$ is nonzero, as desired. □

Exercises for §13.3.

#1. Prove the existence of the linear functional $\lambda \colon \mathcal{B}(\mathcal{H}) \to \mathbb{C}$ in the proof of Lemma 13.2.2.

[*Hint:* For $T \in \mathcal{B}(\mathcal{H})$, define $\lambda_n(T) = \langle Tv_n \mid v_n \rangle$, where $\{v_n\}$ is a sequence of unit vectors in $\mathcal{H}$, such that $\|\pi(h)v_n - v_n\| \to 0$ for every $h \in H$. Let λ be an accumulation point of $\{\lambda_n\}$ in an appropriate weak topology.]

#2. Let μ be as defined near the end of the proof of Lemma 13.2.2. Show:
   a) $\mu(\mathbb{R}^2) = 1$.
   b) If A_1 and A_2 are disjoint Borel subsets of $\mathbb{R}^2$, then we have $\mu(A_1 \cup A_2) = \mu(A_1) + \mu(A_2)$.
   c) $\mu(A) \geq 0$ for every Borel set $A \subseteq \mathbb{R}^2$.

#3. Show the finitely additive measure μ in the proof of Lemma 13.2.2 is invariant under the action of H on $\hat{R}$.

[*Hint:* Since

$$\int_{\hat{R}} \tau(r) \, dE(\tau^h) = \int_{\hat{R}} \tau^{h^{-1}}(r) \, dE(\tau) = \int_{\hat{R}} \tau(h^{-1}rh) \, dE(\tau) = \pi(h^{-1}rh)$$

$$= \pi(h^{-1}) \, \pi(r) \, \pi(h) = \int_{\hat{R}} \tau(r) \, (\pi(h^{-1}) \, dE(\tau) \, \pi(h)),$$

we have $E(A^h) = \pi(h^{-1}) \, E(A) \, \pi(h)$ for $A \subseteq \hat{R}$.]

#4. Show that any $\mathrm{SL}(2, \mathbb{R})$-invariant, finitely additive probability measure μ on $\mathbb{R}^2$ is supported on $\{(0,0)\}$.

[*Hint:* Let $V = \{ (x, y) \mid y > |x| \}$ and $h = \begin{bmatrix} 1 & 2 \\ 0 & 1 \end{bmatrix}$. Then $h^i V$ is disjoint from $h^j V$ for $i \neq j \in \mathbb{Z}^+$, so $\mu(V) = 0$. All of $\mathbb{R}^2 \smallsetminus \{(0,0)\}$ is covered by finitely many sets of the form hV with $h \in \mathrm{SL}(2, \mathbb{R})$.]

#5. Show that the natural semidirect product $\mathrm{SL}(3, \mathbb{R}) \ltimes \mathbb{R}^3$ has Kazhdan's property.

> [*Hint:* We know $\mathrm{SL}(3,\mathbb{R})$ has Kazhdan's property, and the proof of Lemma 13.2.2 shows that the pair $(\mathrm{SL}(3,\mathbb{R}) \ltimes \mathbb{R}^3, \mathbb{R}^3)$ has relative property (T).]

#6. Show that the direct product $\mathrm{SL}(3,\mathbb{R}) \times \mathbb{R}^3$ does *not* have Kazhdan's property. (Comparing this with Exercise 5 shows that, for group extensions, Kazhdan's property may depend not only the groups involved, but also on the details of the particular extension.)

§13.4. Lattices in groups with Kazhdan's property

In this section, we will use basic properties of induced representations to prove the following important result:

(13.4.1) **Proposition.** *If G has Kazhdan's property, then Γ has Kazhdan's property.*

Combining this with Theorem 13.2.4, we obtain:

(13.4.2) **Corollary.** *If no simple factor of G is isogenous to $\mathrm{SO}(1,n)$ or $\mathrm{SU}(1,n)$, then Γ has Kazhdan's property.*

By Proposition 13.1.7, this has two important consequences:

(13.4.3) **Corollary.** *If no simple factor of G is isogenous to $\mathrm{SO}(1,n)$ or $\mathrm{SU}(1,n)$, then*

1) *Γ is finitely generated, and*
2) *$\Gamma/[\Gamma,\Gamma]$ is finite.*

(13.4.4) *Remarks.*

1) It was pointed out in Theorem 4.7.10 that (1) remains true without any assumption on the simple factors of G. In fact, Γ is always finitely presented, not merely finitely generated.
2) On the other hand, (2) is not always true, because lattices in $\mathrm{SO}(1,n)$ and $\mathrm{SU}(1,n)$ can have infinite abelian quotients. (In fact, it is conjectured that every lattice in $\mathrm{SO}(1,n)$ has a finite-index subgroup with an infinite abelian quotient, and this is known to be true when $n = 3$.) The good news is that the Margulis Normal Subgroup Theorem implies these are the only examples (modulo multiplying G by a compact factor) if we make the additional assumption that Γ is irreducible (see Exercise 16.1#3 or Exercise 17.1#1).

The proof of Proposition 13.4.1 uses some machinery from the theory of unitary representations.

(13.4.5) **Notation.** Let $(\pi,\mathcal{H})$ and $(\sigma,\mathcal{K})$ be unitary representations of a Lie group H. (In our applications, H will be either G or Γ.)

1) We write $\sigma \le \pi$ if σ is (isomorphic to) a **subrepresentation** of π. This means there exist

- a closed, H-invariant subspace $\mathcal{H}'$ of $\mathcal{H}$, and
- a bijective, linear isometry $T: \mathcal{K} \xrightarrow{\cong} \mathcal{H}'$,

such that $T(\sigma(h)\phi) = \pi(h)\,T(\phi)$, for all $h \in H$ and $\phi \in \mathcal{K}$.

2) We write $\sigma \prec \pi$ if σ is **weakly contained** in π. This means that, for every compact set C in H, every $\epsilon > 0$, and all unit vectors $\phi_1, \ldots, \phi_n \in \mathcal{K}$, there exist unit vectors $\psi_1, \ldots, \psi_n \in \mathcal{H}$, such that, for all $h \in C$ and all $1 \le i, j \le n$, we have

$$\left|\langle \sigma(h)\phi_i \mid \phi_j \rangle - \langle \pi(h)\psi_i \mid \psi_j \rangle\right| < \epsilon.$$

(13.4.6) Remarks.

1) It is obvious that if $\sigma \le \pi$, then $\sigma \prec \pi$.

2) We have:
   - π has invariant vectors if and only if $\mathbb{1} \le \pi$, and
   - π has almost-invariant vectors if and only if $\mathbb{1} \prec \pi$.

   Therefore, Kazhdan's property asserts the converse to (1) in the special case where $\sigma = \mathbb{1}$: for all π, if $\mathbb{1} \prec \pi$, then $\mathbb{1} \le \pi$.

It is not difficult to show that induction preserves weak containment (see Exercise 1):

(13.4.7) Lemma. *If* $\sigma \prec \pi$, *then* $\mathrm{Ind}_\Gamma^G(\sigma) \prec \mathrm{Ind}_\Gamma^G(\pi)$.

This (easily) implies the main result of this section:

Proof of Proposition 13.4.1. Suppose a representation π of Γ has almost-invariant vectors. Then $\pi \succ \mathbb{1}$, so

$$\mathrm{Ind}_\Gamma^G(\pi) \succ \mathrm{Ind}_\Gamma^G(\mathbb{1}) = \mathscr{L}^2(G/\Gamma) \ge \mathbb{1}$$

(see Exercises 2 and 11.3#5). Because G has Kazhdan's property, we conclude that $\mathrm{Ind}_\Gamma^G(\pi) \ge \mathbb{1}$. This implies $\pi \ge \mathbb{1}$ (see Exercise 3), as desired. $\qquad\square$

(13.4.8) Remark. If Γ has Kazhdan's property, and S is any generating set of Γ, then there is some $\epsilon > 0$, such that every unitary representation of Γ with an (ϵ, S)-invariant unit vector must have invariant vectors (see Exercise 13.1#11). Our proof does not provide any estimate on ϵ, but, in many cases, including $\Gamma = \mathrm{SL}(n, \mathbb{Z})$, an explicit value of ϵ can be obtained by working directly with the algebraic structure of Γ (rather than using the fact that Γ is a lattice).

(13.4.9) Remark. For many years, lattices (and some minor modifications of them) were the only discrete groups known to have Kazhdan's property (T), but other constructions are now known. In particular:

1) Groups can be defined by generators and relations. It can be shown that if the relations are selected at random (with respect to a certain

probability distribution), then the resulting group has Kazhdan's property (T) with high probability.

2) An algebraic approach that directly proves Kazhdan's property for $SL(n, \mathbb{Z})$, without using the fact that it is a lattice, has been generalized to allow some other rings, such as polynomial rings, in the place of $\mathbb{Z}$. In particular, $SL(n, \mathbb{Z}[X_1, \ldots, X_k])$ has Kazhdan's property (T) if $n \geq k + 3$.

(13.4.10) *Remark.* We saw in Proposition 12.7.22 that amenability is invariant under quasi-isometry (see Definition 10.1.3). In contrast, this is not true for Kazhdan's property (T). To see this, let

- G be a simple group with Kazhdan's property (T),
- $\widetilde{G}$ be the universal cover of G,
- Γ be a cocompact lattice in G, and
- $\widetilde{\Gamma}$ be the inverse image of Γ in $\widetilde{G}$, so $\widetilde{\Gamma}$ is a lattice in $\widetilde{G}$.

Then $\widetilde{\Gamma}$ has Kazhdan's property (T) (because $\widetilde{G}$ has the property). However, if $G = Sp(4, \mathbb{R})$ (or, more generally, if the fundamental group of G is an infinite cyclic group), then $\widetilde{\Gamma}$ is quasi-isometric to $\Gamma \times \mathbb{Z}$, which obviously does not have Kazhdan's property (because its abelianization is infinite).

Here is a brief explanation of why $\widetilde{\Gamma}$ is quasi-isometric to $\Gamma \times \mathbb{Z}$. Note that $\widetilde{\Gamma}/\mathbb{Z} \cong \Gamma$ yields a 2-cocycle $\alpha \colon \Gamma \times \Gamma \to \mathbb{Z}$ of group cohomology. Since G/Γ is compact, it turns out that α can be chosen to be uniformly bounded, as a function on $\Gamma \times \Gamma$. This implies that the extension $\widetilde{\Gamma}$ is quasi-isometric to the extension corresponding to the trivial cocycle. This extension is $\Gamma \times \mathbb{Z}$.

Exercises for §13.4.

#1. Prove Lemma 13.4.7.

#2. Show $\mathbb{1} \leq \mathscr{L}^2(G/\Gamma)$.

#3. Show that if π is a unitary representation of Γ, and $\mathbb{1} \leq \operatorname{Ind}_\Gamma^G(\pi)$, then $\mathbb{1} \leq \pi$.

#4. Prove the converse of Proposition 13.4.1: Show that if Γ has Kazhdan's property, then G has Kazhdan's property.

[*Hint:* Any Γ-invariant vector v can be averaged over G/Γ to obtain a G-invariant vector. If v is ϵ-invariant for a compact set whose projection to G/Γ has measure $> 1 - \epsilon$, then the average is nonzero.]

§13.5. Fixed points in Hilbert spaces

We now describe an important geometric interpretation of Kazhdan's property.

(13.5.1) **Definition.** Let $\mathcal{H}$ be a Hilbert space. A bijection $T\colon \mathcal{H} \to \mathcal{H}$ is an ***affine isometry*** of $\mathcal{H}$ if there exist a unitary operator U on $\mathcal{H}$, and $b \in \mathcal{H}$, such that

$$T(v) = Uv + b \ \text{ for all } v \in \mathcal{H}.$$

(13.5.2) **Example.** Let w_0 be a nonzero vector in a Hilbert space $\mathcal{H}$. For $t \in \mathbb{R}$, define an affine isometry ϕ^t of $\mathcal{H}$ by $\phi^t(v) = v + tw_0$; this yields an action of $\mathbb{R}$ on $\mathcal{H}$ by affine isometries. Since $\phi^1(v) = v + w_0 \neq v$, we know that the action has no fixed point.

The main theorem of this section shows that the groups that do not have Kazhdan's property are characterized by the existence of a fixed-point-free action as in Example 13.5.2. However, before stating the result, let us introduce some notation, so that we can also state it in cohomological terms.

(13.5.3) **Definition.** Suppose $(\pi, \mathcal{H})$ is a unitary representation of a Lie group H. Define

1) $C(H; \mathcal{H}) = \{\text{ continuous functions } f\colon H \to \mathcal{H} \}$,
2) $\mathscr{L}^1(H; \pi) = \{f \in C(H; \mathcal{H}) \mid \forall g, h \in H,\ f(gh) = f(g) + \pi(g)f(h)\}$,
3) $\mathscr{B}^1(H; \pi) = \{f \in C(H; \mathcal{H}) \mid \exists v \in \mathcal{H},\ \forall h \in H,\ f(h) = v - \pi(h)v\}$,
4) $\mathscr{H}^1(H; \pi) = \mathscr{L}^1(H; \pi)/\mathscr{B}^1(H; \pi)$ (see Exercise 3).

If the representation π on $\mathcal{H}$ is clear from the context, we may write $\mathscr{L}^1(H; \mathcal{H})$, $\mathscr{B}^1(H; \mathcal{H})$, and $\mathscr{H}^1(H; \mathcal{H})$, instead of $\mathscr{L}^1(H; \pi)$, $\mathscr{B}^1(H; \pi)$, and $\mathscr{H}^1(H; \pi)$.

(13.5.4) **Theorem.** *For a Lie group H, the following are equivalent:*

1) *H has Kazhdan's property.*
2) *For every Hilbert space $\mathcal{H}$, every continuous action of H by affine isometries on $\mathcal{H}$ has a fixed point.*
3) *$\mathscr{H}^1(H; \pi) = 0$, for every unitary representation π of H.*

Proof of (2) $\Rightarrow$ (3). Given $f \in \mathscr{L}^1(H; \pi)$, define an action of f on $\mathcal{H}$ via affine isometries by defining

$$hv = \pi(h)v + f(h) \ \text{for } h \in H \text{ and } v \in \mathcal{H}$$

(see Exercise 5). By assumption, this action must have a fixed point v_0. For all $h \in H$, we have $v_0 = hv_0 = \pi(h)v_0 + f(h)$, so $f(h) = v_0 - \pi(h)v_0$. Therefore $f \in \mathscr{B}^1(H; \pi)$. Since f is an arbitrary element of $\mathscr{L}^1(H; \pi)$, this implies $\mathscr{H}^1(H; \pi) = 0$. $\qquad\qquad\square$

Proof of (3) $\Rightarrow$ (1). We prove the contrapositive: assume H does not have Kazhdan's property. This means a unitary representation of H on some Hilbert space $\mathcal{H}$ has almost-invariant vectors, but does not have invariant vectors. We claim $\mathcal{H}^1(H; \pi^\infty) \neq 0$, where π^∞ is the obvious diagonal action of H on the Hilbert space $\mathcal{H}^\infty = \mathcal{H} \oplus \mathcal{H} \oplus \cdots$.

Choose an increasing chain $C_1 \subseteq C_2 \subseteq \cdots$ of compact subsets of G, such that $G = \bigcup_n C_n$. For each n, since $\mathcal{H}$ has almost-invariant vectors, there exists a unit vector $v_n \in \mathcal{H}$, such that

$$\|cv_n - v_n\| < \frac{1}{2^n} \text{ for all } c \in C_n.$$

Now, define $f \colon H \to \mathcal{H}^\infty$ by

$$f(h)_n = n(hv_n - v_n)$$

(see Exercise 9), so $f \in \mathfrak{Z}^1(H; \pi^\infty)$ (see Exercise 10). However, it is easy to see that f is an unbounded function on H (see Exercise 11), so $f \notin \mathfrak{B}^1(H; \mathcal{H}^\infty)$ (see Exercise 4). Therefore f represents a nonzero cohomology class in $\mathcal{H}^1(H; \pi^\infty)$. $\qquad\square$

Alternate proof of (3) $\Rightarrow$ (1). Assume the unitary representation π has no invariant vectors. (We wish to show this implies there are no almost-invariant vectors.) Define a linear map

$$F \colon \mathcal{H} \to \mathfrak{Z}^1(H; \pi) \text{ by } F_v(h) = \pi(h)v - v.$$

Assume, for simplicity, that H is compactly generated (see Exercise 14), so some compact, symmetric set C generates H. By enlarging C, we may assume C has nonempty interior. Then the supremum norm on C turns $\mathfrak{Z}^1(H; \pi)$ into a Banach space (see Exercise 12), and the map F is continuous in this topology (see Exercise 13(a)).

Since there are no invariant vectors in $\mathcal{H}$, we know that F is injective (see Exercise 13(b)). Also, the image of F is obviously $\mathfrak{B}^1(H; \pi)$. Since $\mathcal{H}^1(H; \pi) = 0$, this means that F is surjective. Therefore, F is a bijection. So the Open Mapping Theorem (B7.6(2)) provides a constant $\epsilon > 0$, such that $\|F_v\| > \epsilon$ for every unit vector v. This means there is some $h \in C$, such that $\|\pi(h)v - v\| > \epsilon$, so v is not (C, ϵ)-invariant. Therefore, there are no almost-invariant vectors. $\qquad\square$

Sketch of proof of (1) $\Rightarrow$ (2). We postpone this proof to Section 13.6, where functions of positive type are introduced. They yield an embedding of $\mathcal{H}$ in the unit sphere of a (larger) Hilbert space $\widehat{\mathcal{H}}$. This embedding is nonlinear and non-isometric, but there is a unitary representation $\widehat{\pi}$ on $\widehat{\mathcal{H}}$ for which the embedding is equivariant. Kazhdan's property provides an invariant vector in $\widehat{\mathcal{H}}$, and this pulls back to a fixed point in $\mathcal{H}$. See Section 13.6 for more details. $\qquad\square$

(13.5.5) *Remark.* If H satisfies (2) of Theorem 13.5.4, it is said to have "property (FH)" (because it has **F**ixed points on **H**ilbert spaces).

In Definition 13.5.3, the subspace $\mathcal{B}^1(H; \pi)$ may fail to be closed (see Exercise 16). In this case, the quotient space $\mathcal{H}^1(H; \pi)$ does not have a good topology. Fortunately, it can be shown that Theorem 13.5.4 remains valid even if we replace $\mathcal{B}^1(H; \pi)$ with its closure:

(13.5.6) **Definition.** In the notation of Definition 13.5.3, let:

1) $\overline{\mathcal{B}^1(H; \pi)}$ be the closure of $\mathcal{B}^1(H; \pi)$ in $\mathcal{L}^1(H; \pi)$, and

2) $\overline{\mathcal{H}}^1(H; \pi) = \mathcal{L}^1(H; \pi) / \overline{\mathcal{B}^1(H; \pi)}$. This is called the *reduced* 1st cohomology.

The following result requires the technical condition that H is compactly generated (see Definition 13.1.10 and Exercise 17).

(13.5.7) **Theorem.** *A compactly generated Lie group H has Kazhdan's property if and only if $\overline{\mathcal{H}}^1(H; \pi) = 0$, for every unitary representation π of H.*

Because reduced cohomology behaves well with respect to the direct integral decomposition of a unitary representation (although the unreduced cohomology does not), this theorem implies that it suffices to consider only the irreducible representations of H:

(13.5.8) **Corollary.** *A compactly generated Lie group H has Kazhdan's property if and only if $\overline{\mathcal{H}}^1(H; \pi) = 0$, for every **irreducible** unitary representation π of H.*

(13.5.9) *Remark.* We have seen that a group with Kazhdan's property has bounded orbits whenever it acts isometrically on a Hilbert space. The same conclusion has been proved for isometric actions on some other spaces, including real hyperbolic n-space $\mathfrak{H}^n$, complex hyperbolic n-space $\mathfrak{H}_{\mathbb{C}}^n$, and all "median spaces" (including all $\mathbb{R}$-trees). (In many cases, the existence of a bounded orbit implies the existence of a fixed point.) See Exercise 19 for an example.

Exercises for §13.5.

#1. Let $T: \mathcal{H} \to \mathcal{H}$. Show that if T is an affine isometry, then
   a) $T(v - w) = T(v) - T(w) + T(0)$, and
   b) $\|T(v) - T(w)\| = \|v - w\|$,
   for all $v, w \in \mathcal{H}$.

#2. Prove the converse of Exercise 1.

#3. In the notation of Definition 13.5.3, show that $\mathcal{B}^1(H; \pi) \subseteq \mathcal{L}^1(H; \pi)$ (so the quotient $\mathcal{L}^1(H; \pi) / \mathcal{B}^1(H; \pi)$ is defined).

#4. Suppose $f \in \mathcal{B}^1(H; \pi)$ so $f : H \to \mathcal{H}$. Show f is bounded.

#5. Suppose
   - $(\pi, \mathcal{H})$ is a unitary representation of H, and
   - $\tau : H \to \mathcal{H}$.

   For $h \in H$ and $v \in \mathcal{H}$, let $\alpha(h)v = \pi(h)v + \tau(h)$, so $\alpha(h)$ is an affine isometry of $\mathcal{H}$. Show that α defines a continuous action of H on $\mathcal{H}$ if and only if $\tau \in \mathcal{Z}^1(H; \pi)$ and τ is continuous.

#6. Suppose H acts continuously by affine isometries on the Hilbert space $\mathcal{H}$. Show there is a unitary representation π of H on $\mathcal{H}$, and some $\tau \in \mathcal{Z}^1(H; \pi)$, such that $hv = \pi(h)v + \tau(h)$ for every $h \in H$ and $v \in \mathcal{H}$.

#7. Suppose H acts continuously by affine isometries on the Hilbert space $\mathcal{H}$. Show the following are equivalent:
   a) H has a fixed point in $\mathcal{H}$.
   b) The orbit Hv of each vector v in $\mathcal{H}$ is a bounded subset of $\mathcal{H}$.
   c) The orbit Hv of some vector v in $\mathcal{H}$ is a bounded subset of $\mathcal{H}$.
   [Hint: You may use (without proof) the fact that every nonempty, bounded subset X of a Hilbert space has a unique circumcenter. By definition, the **circumcenter** is a point c, such that, for some $r > 0$, the set X is contained in the closed ball of radius r centered at c, but X is not contained in any ball of radius $< r$ (centered at any point).]

#8. Prove directly that 13.5.4(3) $\Rightarrow$ 13.5.4(2), without using Kazhdan's property.
   [Hint: For each $h \in H$, there is a unique unitary operator $\pi(h)$, such that we have $hv = \pi(h)v + h(0)$ for all $v \in \mathcal{H}$. Fix $v \in \mathcal{H}$ and define $f \in \mathcal{Z}^1(H; \pi)$ by $f(h) = hv - v$. If $f \in \mathcal{B}^1(H; \pi)$, then H has a fixed point.]

#9. In the notation of the proof of 13.5.4(3 $\Rightarrow$ 1), show $f : H \to \mathcal{H}^\infty$.
   [Hint: For each $h \in H$, show the sequence $\{\|f(h)_n\|\}$ is square-summable.]

#10. In the notation of the proof of 13.5.4(3 $\Rightarrow$ 1), show $f \in \mathcal{H}^1(H; \pi^\infty)$.

#11. In the notation of the proof of 13.5.4(3 $\Rightarrow$ 1), show f is unbounded.
   [Hint: You may use (without proof) the fact that every nonempty, bounded subset of a Hilbert space has a unique circumcenter, as in Exercise 7.]

#12. Assume C is a compact, symmetric set that generates H, and has nonempty interior. For each $f \in \mathcal{Z}^1(H; \mathcal{H})$, let $\xi(f)$ be the restriction of f to C. Show that ξ is a bijection from $\mathcal{Z}^1(H; \mathcal{H})$ onto a closed subspace of the Banach space of continuous functions from C to $\mathcal{H}$.

#13. In the notation of the alternate proof of 13.5.4(3 $\Rightarrow$ 1), show:
   a) F is continuous.
   b) F is injective.
   [Hint: (b) If $F_v = F_w$, then what is $\pi(h)(v - w)$?]

#14. Remove the assumption that H is compactly generated from the alternate proof of (3) $\Longrightarrow$ (1).

[*Hint:* The topology of uniform convergence on compact sets makes $\mathscr{Z}^1(H;\pi)$ into a Fréchet space.]

#15. Assume
  - Γ has Kazhdan's property T,
  - V is a vector space,
  - $\mathcal{H}$ is a Hilbert space that is contained in V,
  - $v \in V$, and
  - $\sigma : \Gamma \to GL(V)$ is any homomorphism, such that
    - the restriction $\sigma(\gamma)|_{\mathcal{H}}$ is unitary, for every $\gamma \in \Gamma$, and
    - $\mathcal{H} + v$ is $\sigma(\Gamma)$-invariant.

Show $\sigma(\Gamma)$ has a fixed point in $\mathcal{H} + v$.

[*Hint:* Theorem 13.5.4(2).]

#16. Show that if π has almost-invariant vectors, then $\mathscr{B}^1(H;\pi)$ is not closed in $\mathscr{Z}^1(H;\pi)$.

[*Hint:* See the alternate proof of Theorem 13.5.4(3 $\Longrightarrow$ 1).]

#17. Show the assumption that H is compactly generated cannot be removed from the statement of Theorem 13.5.7.

[*Hint:* Let H be an infinite, discrete group, such that every finitely generated subgroup of H is finite.]

#18. Show H has Kazhdan's property if and only if $\mathcal{H}^1(H;\pi) = 0$, for every **irreducible** unitary representation π of H.

[*Hint:* You may assume Theorem 13.5.7.]

Definition. A *tree* is a contractible, 1-dimensional simplicial complex.

#19. (Watatani) Suppose
  - Λ is a discrete group that has Kazhdan's property, and
  - acts by isometries on a tree T.

Show Λ has a fixed point in T (without assuming Remark 13.5.9).

[*Hint:* Fix an orientation of T, and fix a vertex v in T. For each $\lambda \in \Lambda$, the geodesic path in T from v to $\lambda(v)$ can be represented by a $\{0, \pm 1\}$-valued function P_λ on the set E of edges of T. Verify that $\lambda \mapsto P_\lambda$ is in $\mathscr{Z}^1(\Lambda; \mathscr{L}^2(E))$, and conclude that the orbit of v is bounded.]

#20. Show $SO(1,n)$ and $SU(1,n)$ do not have Kazhdan's property.

[*Hint:* You may assume the facts stated in Remark 13.5.9.]

#21. It is straightforward to verify that all of the results in this chapter remain valid if we require $\mathcal{H}$ to be a **real** Hilbert space (instead of a Hilbert space over $\mathbb{C}$), as in Assumption 13.6.1 below. In this setting, there is no need to restrict attention to *affine* isometries in the statement of Theorem 13.5.4(2), because *all* isometries are affine:

Let $\mathcal{H}$ be a real Hilbert space, and let $\varphi \colon \mathcal{H} \to \mathcal{H}$ be any distance-preserving bijection (so $\|\varphi(v) - \varphi(w)\| = \|v - w\|$ for all $v, w \in \mathcal{H}$). Show that φ is an affine isometry.

[*Hint:* The main problem is to show that if $\varphi(0) = 0$, then φ is $\mathbb{R}$-linear. This is well known (and easy to prove) when $\mathcal{H} = \mathbb{R}^2$. The general case follows from this.]

§13.6. Functions on H that are of positive type

This section completes the proof of Theorem 13.5.4, by showing that affine isometric actions of Kazhdan groups on Hilbert spaces always have fixed points. For this purpose, we develop some of the basic theory of functions of positive type.

(13.6.1) **Assumption.** To simplify some details, Hilbert spaces in this section are assumed to be real, rather than complex. (That is, the field of scalars is $\mathbb{R}$, rather than $\mathbb{C}$.)

(13.6.2) **Definition.**

1) Let A be an $n \times n$ real symmetric matrix.
   (a) A is of *positive type* if $\langle Av \mid v \rangle \geq 0$ for all $v \in \mathbb{R}^n$. Equivalently, this means all of the eigenvalues of A are ≥ 0 (see Exercise 1).
   (b) A is *conditionally of positive type* if
      (i) $\langle Av \mid v \rangle \geq 0$ for all $v = (v_1, \dots, v_n) \in \mathbb{R}^n$, such that we have $v_1 + \cdots + v_n = 0$, and
      (ii) all the diagonal entries of A are 0.
      (The word "conditionally" refers to the fact that the inequality on $\langle Av \mid v \rangle$ is only required to be satisfied when a particular condition is satisfied, namely, when the sum of the coordinates of v is 0.)

2) A continuous, real-valued function φ on a topological group H is said to be of *positive type* (or *conditionally of positive type*, respectively) if, for all n and all $h_1, \dots, h_n \in H$, the matrix $(\varphi(h_i^{-1} h_j))$ is a symmetric matrix of the said type.

(13.6.3) **Warning.** A function that is of positive type is almost never conditionally of positive type. This is because a matrix satisfying (1a) of Definition 13.6.2 will almost never satisfy (1(b)ii) (see Exercise 2).

(13.6.4) **Other terminology.** Functions of positive type are often called *positive definite* or *positive semi-definite*.

Such functions arise naturally from actions of H on Hilbert spaces:

(13.6.5) **Lemma.** *Suppose*
- *H is a topological group,*

- H acts continuously by affine isometries on a Hilbert space $\mathcal{H}$,
- $v \in \mathcal{H}$,
- $\varphi\colon H \to \mathbb{R}$ is defined by $\varphi(h) = -\|hv - v\|^2$ for $h \in H$, and
- $\psi\colon H \to \mathbb{R}$ is defined by $\psi(h) = \langle hv \mid v \rangle$ for $h \in H$.

Then:

1) φ is conditionally of positive type, and

2) ψ is of positive type if $h(0) = 0$ for all $h \in H$.

Proof. Exercises 6 and 7. $\qquad\qquad\qquad\qquad\qquad\qquad\qquad\square$

Conversely, the following result shows that all functions of positive type arise from this construction. (The "GNS" in its name stands for Gelfand, Naimark, and Segal.)

(13.6.6) **Proposition** ("GNS construction"). *If $f\colon H \to \mathbb{R}$ is of positive type, then there exist*

- *a continuous action of H by linear isometries on a Hilbert space $\mathcal{H}$ (so $h(0) = 0$ for all h), and*
- *$v \in \mathcal{H}$,*

such that $f(h) = \langle hv \mid v \rangle$ for all $h \in H$.

Proof. Let $\mathbb{R}[H]$ be the vector space of functions on H with finite support. Since the set of delta functions $\{\, \delta_h \mid h \in H \,\}$ is a basis, there is a unique bilinear form on $\mathbb{R}[H]$, such that

$$\langle \delta_{h_1} \mid \delta_{h_2} \rangle = f(h_1^{-1}h_2) \text{ for all } h_1, h_2.$$

Since f is of positive type, this form is symmetric and satisfies the inequality $\langle w \mid w \rangle \geq 0$ for all w. Let Z be the **radical** of the form, which means

$$Z = \{\, z \in \mathbb{R}[H] \mid \langle z \mid z \rangle = 0 \,\},$$

so $\langle \mid \rangle$ factors through to a well-defined positive-definite, symmetric bilinear form on the quotient $\mathbb{R}[H]/Z$. This makes the quotient into a pre-Hilbert space; let $\mathcal{H}$ be its completion, which is a Hilbert space, and let v be the image of δ_e in $\mathcal{H}$.

The group H acts by translation on $\mathbb{R}[H]$, and it is easy to verify that the action is continuous, and preserves the bilinear form (see Exercise 8). Therefore, the action extends to a unitary representation of H on $\mathcal{H}$. Furthermore, for any $h \in H$, we have

$$f(h) = f(e \cdot h) = \langle \delta_e \mid \delta_h \rangle = \langle \delta_e \mid h\delta_e \rangle = \langle v \mid hv \rangle = \langle hv \mid v \rangle,$$

as desired. $\qquad\qquad\qquad\qquad\qquad\qquad\qquad\qquad\qquad\qquad\square$

We will also use the following important relationship between the two concepts:

(13.6.7) **Lemma** (Schoenberg's Lemma). *If φ is conditionally of positive type, then e^{φ} is of positive type.*

Proof. A function $\kappa\colon H \times H \to \mathbb{R}$ is said to be a ***kernel of positive type*** if the matrix $(\kappa(h_i, h_j))$ is a symmetric matrix of positive type, for all n and all $h_1, \ldots, h_n \in H$.

Define $\kappa\colon H \times H \to \mathbb{R}$ by

$$\kappa(g, h) = \varphi(g^{-1}h) - \varphi(g) - \varphi(h).$$

Then:

- κ is a kernel of positive type (see Exercise 9),
- so e^{κ} is a kernel of positive type (see Exercise 11),
- and $e^{\varphi(g)}e^{\varphi(h)}$ is a kernel of positive type (see Exercise 12),
- so the product $e^{\kappa(g,h)}\left(e^{\varphi(g)}e^{\varphi(h)}\right)$ is a kernel of positive type (see Exercise 10).

This product is $e^{\varphi(g^{-1}h)}$, so e^{φ} is a function of positive type.   □

With these tools, it is not difficult to show that affine isometric actions of Kazhdan groups on Hilbert spaces always have fixed points:

Proof of Theorem 13.5.4 ($1 \Rightarrow 2$). Let α be the given action of H on $\mathcal{H}$ by affine isometries, and let π be the corresponding unitary representation (see Exercise 13.5#6). Therefore, we have

$$\alpha(h)v = \pi(h)v + \tau(h) \text{ for } h \in H \text{ and } v \in \mathcal{H},$$

where $\tau \in \mathscr{L}^1(H; \pi)$.

Let $\widehat{H} = \mathcal{H} \rtimes H$ be the semidirect product of (the additive group of) $\mathcal{H}$ with H, where H acts on $\mathcal{H}$ via π. This means the elements of $\widehat{H}$ are the ordered pairs (v, h), and, for $v_1, v_2 \in \mathcal{H}$ and $h_1, h_2 \in H$, we have

$$(v_1, h_1) \cdot (v_2, h_2) = (v_1 + \pi(h_1)v_2, h_1 h_2).$$

This semidirect product is a topological group, so we can apply the above theory of functions of positive type to it. Define a continuous action $\widehat{\alpha}$ of $\widehat{H}$ on $\mathcal{H}$ by

$$\widehat{\alpha}(v, h)w = \alpha(h)w + v \tag{13.6.8}$$

(see Exercise 3), and define

$$\widehat{\varphi}\colon \widehat{H} \to \mathbb{R} \text{ by } \widehat{\varphi}(v, h) = -\|\widehat{\alpha}(v, h)(0)\|^2.$$

Since $\widehat{\alpha}(v, h)$ is an affine isometry for every v and h, we know $\widehat{\varphi}$ is conditionally of positive type (see Lemma 13.6.5(1)). Therefore $e^{\widehat{\varphi}}$ is of positive type (see Lemma 13.6.7). Hence, the GNS construction (13.6.6) provides a unitary representation $\widehat{\pi}$ of $\widehat{H}$ on a Hilbert space $\widehat{\mathcal{H}}$ and some $\widehat{v} \in \widehat{\mathcal{H}}$,

such that
$$\langle \hat{\pi}(v, h)\hat{v} \mid \hat{v} \rangle = e^{\hat{\varphi}(v,h)} \text{ for all } v \in \mathcal{H} \text{ and } h \in H. \tag{13.6.9}$$
We now define
$$\Phi \colon \mathcal{H} \to \widehat{\mathcal{H}} \text{ by } \Phi(v) = \hat{\pi}(v, e)\hat{v}.$$
We have
$$\Phi(\alpha(h)v) = \hat{\pi}(0, h)\,\Phi(v) \text{ for } h \in H \text{ and } v \in \mathcal{H} \tag{13.6.10}$$
(see Exercise 14), so Φ converts the affine action of H on $\mathcal{H}$ to a linear action on $\widehat{\mathcal{H}}$. Since the linear span of $\Phi(\mathcal{H})$ contains $\hat{v}$ and is invariant under $\hat{\pi}(\hat{H})$ (see Exercise 4), there is no harm in assuming that its closure is all of $\widehat{\mathcal{H}}$.

It is clear from the definition of $\hat{\varphi}$ that $\hat{\varphi}(0, e) = 0$, so we know that $\hat{v}$ is a unit vector. Therefore
$$\begin{aligned} \|\hat{\pi}(v, h)\hat{v} - \hat{v}\|^2 &= \langle \hat{\pi}(v, h)\hat{v} - \hat{v}, \ \hat{\pi}(v, h)\hat{v} - \hat{v} \rangle \\ &= 2(1 - \langle \hat{\pi}(v, h)\hat{v} \mid \hat{v} \rangle) \\ &= 2(1 - e^{\hat{\varphi}(v,h)}). \end{aligned} \tag{13.6.11}$$

Since H has Kazhdan's property, there is a compact subset C of H and some $\epsilon > 0$, such that every unitary representation of H that has a (C, ϵ)-invariant vector must have an invariant vector. There is no harm in multiplying the norm on $\mathcal{H}$ by a small positive scalar, so we may assume $\hat{\varphi}(0, h)$ is as close to 0 as we like, for all $h \in C$. Then (13.6.11) tells us that $\hat{v}$ is (C, ϵ)-invariant, so $\widehat{\mathcal{H}}$ must have a nonzero H-invariant vector $\hat{v}_0$.

Suppose the affine action α does not have any fixed points. (This will lead to a contradiction.) This implies that every H-orbit on $\mathcal{H}$ is unbounded (see Exercise 13.5#7). Hence, for any fixed $v \in \mathcal{H}$, and all $h \in H$, we have
$$\begin{aligned} \langle \Phi(v) \mid \hat{v}_0 \rangle &= \langle \Phi(v) \mid \hat{\pi}(0, h^{-1})\hat{v}_0 \rangle && (\hat{v}_0 \text{ is } H\text{-invariant}) \\ &= \langle \hat{\pi}(0, h)\,\Phi(v) \mid \hat{v}_0 \rangle && (\pi \text{ is unitary}) \\ &= \langle \Phi(\alpha(h)v) \mid \hat{v}_0 \rangle && (13.6.10) \\ &\to 0 \quad \text{as } \|\alpha(h)v\| \to \infty && (\text{Exercise 5}). \end{aligned}$$
So $\hat{v}_0$ is orthogonal to the linear span of $\Phi(\mathcal{H})$, which is dense in $\widehat{\mathcal{H}}$. Therefore $\hat{v}_0 = 0$. This is a contradiction. $\qquad\square$

Exercises for §13.6.

#1. Let A be a real symmetric matrix. Show A is of positive type if and only if all of the eigenvalues of A are ≥ 0.
[*Hint:* A is diagonalizable by an orthogonal matrix.]

#2. Suppose A is an $n \times n$ real symmetric matrix that is of positive type and is also conditionally of positive type. Show $A = 0$.

[*Hint:* What does 13.6.2(1(b)ii) say about the trace of A?]

#3. In the notation of the proof of Theorem 13.5.4, show
$$\hat{\alpha}(v_1, h_1) \cdot \hat{\alpha}(v_2, h_2) = \hat{\alpha}((v_1, h_1) \cdot (v_2, h_2))$$
for all $v_1, v_2 \in \mathcal{H}$ and $h_1, h_2 \in H$.

#4. In the notation of the proof of Theorem 13.5.4, show that the linear span of $\Phi(\mathcal{H})$ is invariant under $\hat{\pi}(\hat{H})$.
[*Hint:* See (13.6.10).]

#5. In the notation of the proof of Theorem 13.5.4, show that if $v \in \mathcal{H}$, and $\{w_n\}$ is a sequence in $\mathcal{H}$, such that $\|w_n\| \to \infty$, then $\Phi(w_n) \to 0$ weakly.
[*Hint:* If $\hat{w} \in \Phi(\mathcal{H})$, then (13.6.9) implies $\langle \Phi(w_n) \mid \hat{w} \rangle \to 0$.]

#6. Prove Lemma 13.6.5(1).
[*Hint:* If $\sum_i t_i = 0$, then $\sum_{i,j} t_i t_j \varphi(h_i^{-1} h_j v) = 2 \|\sum_i t_i h_i v\|^2$.]

#7. Prove Lemma 13.6.5(2).
[*Hint:* $\sum_{i,j} t_i t_j \psi(h_i^{-1} h_j v) = \|\sum_i t_i h_i v\|^2$.]

#8. Prove that the action of H acts on $\mathbb{R}[H]$ by translation is continuous, and preserves the bilinear form defined in the proof of Proposition 13.6.6.

#9. Show that the kernel κ in the proof of Lemma 13.6.7 is of positive type.
[*Hint:* For $v_1, \ldots, v_n \in \mathbb{R}$ and $h_1, \ldots, h_n \in H$, let $v_0 = -\sum v_i$ and $h_0 = e$. Then $\sum_{i,j} v_i v_j \kappa(h_i, h_j) \geq 0$ since φ is conditionally of positive type and $\sum v_i = 0$. However, the terms with either $i = 0$ or $j = 0$ have no net contribution, since $\varphi(e) = 0$.]

#10. Show that if κ and λ are kernels of positive type, then $\kappa\lambda$ is a kernel of positive type.
[*Hint:* Given $h_1, \ldots, h_n \in H$, show there is a matrix L, such that $L^2 = (\lambda(h_i, h_j))$. Note that $\sum_{i,j} v_i v_j \kappa(h_i, h_j) \lambda(h_i, h_j) = \sum_k \sum_{i,j} (L_{i,k} v_i)(L_{k,j} v_j) \kappa(h_i, h_j) \geq 0$.]

#11. If κ is a kernel of positive type, show e^κ is a kernel of positive type.
[*Hint:* Since κ^n is a kernel of positive type for all n, the same is true of $\sum \kappa^n / n!$.]

#12. For every $\varphi \colon H \to \mathbb{R}$, show $\varphi(g) \varphi(h)$ is a kernel of positive type.

#13. Suppose $\varphi \colon H \to \mathbb{R}$. Prove the following converse of Lemma 13.6.7: If $e^{t\varphi}$ is of positive type for all $t > 0$, then φ is conditionally of positive type.
[*Hint:* Verify that $e^{t\varphi} - 1$ is conditionally of positive type. Then $\lim_{t \to 0^+} (e^{t\varphi} - 1)/t$ has the same type.]

#14. Verify (13.6.10).
[*Hint:* Since $\hat{\alpha}(-\tau(h), h)(0) = 0$, we have $\langle \hat{\pi}(e, h)\hat{v} \mid \hat{\pi}(\tau(h), e)\hat{v} \rangle = 1$. Since they are of norm 1, the two vectors must be equal.]

#15. Recall that a Lie group H has the **Haagerup property** if it has a unitary representation, such that there are almost-invariant vectors, and all matrix coefficients decay to 0 at ∞. It is known that H has the Haagerup property if and only if it has a continuous, *proper* action by affine isometries on some Hilbert space. Prove the implication ($\Leftarrow$) of this equivalence.

Notes

The monograph [4] is the standard reference on Kazhdan's property (T). The property was defined by D. Kazhdan in [11], where Propositions 13.1.7 and 13.4.1 and Theorem 13.2.1 were proved.

See [2] for a discussion of the generalization of Kazhdan's property to actions on Banach spaces, including Warning 13.1.3 and Exercise 13.1#1. See [6] for a discussion of the Haagerup property that is mentioned in Exercises 13.1#12 and 13.6#15. See [18] for much more information about expander graphs and their connection with Kazhdan's property, mentioned in Exercise 13.1#16.

Regarding Warning 13.1.11:

- By showing that $SL(3, \mathbf{F}_q[t])$ is not finitely presentable, H. Behr [3] provided the first example of a group with Kazhdan's property that is not finitely presentable.

- The existence of uncountably many Kazhdan groups was proved by M. Gromov [9, Cor. 5.5.E, p. 150]. More precisely, any cocompact lattice in $Sp(1, n)$ has uncountably many different quotients (because it is a "hyperbolic" group), and all of these quotient groups have Kazhdan's property.

- Y. Shalom [20, p. 5] proved that every discrete Kazhdan group is a quotient of a finitely presented Kazhdan group. The proof can also be found in [4, Thm. 3.4.5, p. 187].

Our proof of Theorem 13.2.1 is taken from [4, §1.4].

Theorem 13.2.4 appears in [4, Thm. 3.5.4, p. 177]. It combines work of several people, including Kazhdan [11] and Kostant [14, 15]. See [4, pp. 5–7] for an overview of the various contributions to this theorem.

A detailed solution of Exercise 13.3#1 can be found in [4, Lem. 1.4.1].

See [4, Thm. 1.7.1, p. 60] for a proof of Proposition 13.4.1 and its converse (Exercise 13.4#4).

Regarding Remark 13.4.4(2), see [17] (and its references) for a discussion of W. Thurston's conjecture that lattices in $SO(1, n)$ have finite-index subgroups with infinite abelian quotients. (For $n = 3$, the conjecture was proved by Agol [1].) Lattices in $SU(1, n)$ with an infinite abelian quotient were found by D. Kazhdan [12].

Explicit Kazhdan constants for $SL(n, \mathbb{Z})$ (cf. Remark 13.4.8) were first found by M.Burger [7, Appendix] (or see [4, §4.2]). An approach developed by Y.Shalom (see [21]) applies to more general groups (such as $SL(n, \mathbb{Z}[x])$) that are not assumed to be lattices.

Remark 13.4.9(1) is a theorem of Zuk [23, Thm. 4]. (Or see [16] for a more detailed proof.) Remark 13.4.9(2) is explained in [21].

Remark 13.4.10 is due to S.Gersten (unpublished). A proof (based on the same example, but rather different from our sketch) is in [4, Thm. 3.6.5, p. 182].

Theorem 13.5.4 is due to Delorme [8, Thm. V.1] (for $(1 \Rightarrow 2)$) and Guichardet [10, Thm. 1] (for $(3 \Rightarrow 1)$).

Theorem 13.5.7 was proved for discrete groups by Korevaar and Schoen [13, Cor. 4.1.3]. The general case is due to Shalom [20, Thm. 6.1]. Corollary 13.5.8 and Exercise 13.5#18 are also due to Shalom [20, proof of Thm. 0.2].

The part of Remark 13.5.9 dealing with real or complex hyperbolic spaces is in [4, Cor. 2.7.3]. See [5] for median spaces.

The existence and uniqueness of the circumcenter (mentioned in the hints to Exercises 13.5#7 and 13.5#11) is proved in [4, Lem. 2.2.7].

Exercise 13.5#19 is due to Watatani [22], and can also be found in [4, §2.3]. Serre's book [19] is a very nice exposition of the theory of group actions on trees, but, unfortunately, does not include this theorem.

See [4, §2.10–§2.12 and §C.4] for the material of Section 13.6.

References

[1] I.Agol: The virtual Haken conjecture, *Doc. Math.* 18 (2013), 1045–1087. MR 3104553,
http://www.math.uni-bielefeld.de/documenta/vol-18/33.html

[2] U.Bader, A.Furman, T.Gelander, and N.Monod: Property (T) and rigidity for actions on Banach spaces, *Acta Math.* 198 (2007), no. 1, 57–105. MR 2316269,
http://dx.doi.org/10.1007/s11511-007-0013-0

[3] H.Behr: $SL_3(\mathbf{F}_q[t])$ is not finitely presentable, in: C.T.C.Wall, ed., *Homological Group Theory (Proc. Sympos., Durham, 1977).* Cambridge Univ. Press, Cambridge, 1979, pp. 213–224. ISBN 0-521-22729-1, MR 0564424

[4] B.Bekka, P.de la Harpe, and A.Valette: *Kazhdan's Property (T).* Cambridge U. Press, Cambridge, 2008. ISBN 978-0-521-88720-5, MR 2415834,
http://perso.univ-rennes1.fr/bachir.bekka/KazhdanTotal.pdf

[5] I. Chatterji, C. Druțu, and F. Haglund: Kazhdan and Haagerup properties from the median viewpoint. *Adv. Math.* 225 (2010), no. 2, 882–921. MR 2671183, http://dx.doi.org/10.1016/j.aim.2010.03.012

[6] P.-A. Cherix, M. Cowling, P. Jolissaint, P. Julg, and A. Valette: *Groups with the Haagerup property. Gromov's a-T-menability.* Birkhäuser, Basel, 2001. ISBN 3-7643-6598-6, MR 1852148

[7] P. de la Harpe and A. Valette: *La propriété (T) de Kazhdan pour les groupes localement compacts, Astérisque #175,* Soc. Math. France, 1989. MR 1023471

[8] P. Delorme: 1-cohomologie des représentations unitaires des groupes de Lie semisimples et résolubles. Produits tensoriels continus de représentations, *Bull. Soc. Math. France* 105 (1977), 281–336. MR 0578893, http://www.numdam.org/item?id=BSMF_1977__105__281_0

[9] M. Gromov: Hyperbolic groups, in: S. M. Gersten, ed., *Essays in Group Theory.* Springer, New York, 1987, pp. 75–263. ISBN 0-387-96618-8, MR 0919829

[10] A. Guichardet: Sur la cohomologie des groupes topologiques II, *Bull. Sci. Math.* 96 (1972), 305–332. MR 0340464

[11] D. A. Kazhdan: Connection of the dual space of a group with the structure of its closed subgroups, *Func. Anal. Appl.* 1 (1967) 63–65. MR 0209390, http://dx.doi.org/10.1007/BF01075866

[12] D. A. Kazhdan: Some applications of the Weil representation, *J. Analyse Mat.* 32 (1977), 235–248. MR 0492089, http://dx.doi.org/10.1007/BF02803582

[13] N. J. Korevaar and R. M. Schoen: Global existence theorems for harmonic maps to non-locally compact spaces. *Comm. Anal. Geom.* 5 (1997), no. 2, 333–387. MR 1483983

[14] B. Kostant: On the existence and irreducibility of certain series of representations, *Bull. Amer. Math. Soc.* 75 (1969) 627–642. MR 0245725, http://dx.doi.org/10.1090/S0002-9904-1969-12235-4

[15] B. Kostant: On the existence and irreducibility of certain series of representations, in I. M. Gelfand, ed.: *Lie groups and their representations.* Halsted Press (John Wiley & Sons), New York, 1975, pp. 231–329. MR 0399361

[16] M. Kotowski and M. Kotowski: Random groups and property (T): Zuk's theorem revisited, *J. Lond. Math. Soc.* (2) 88 (2013), no. 2, 396–416. MR 3106728, http://dx.doi.org/10.1112/jlms/jdt024

[17] A. Lubotzky: Free quotients and the first Betti number of some hyperbolic manifolds, *Transform. Groups* 1 (1996), no. 1–2, 71–82. MR 1390750, http://dx.doi.org/10.1007/BF02587736

[18] A. Lubotzky: *Discrete Groups, Expanding Graphs and Invariant Measures.* Birkhäuser, Basel, 2010. ISBN 978-3-0346-0331-7, MR 2569682

[19] J.-P. Serre: *Trees.* Springer, Berlin, 1980 and 2003. ISBN 3-540-44237-5, MR 1954121

[20] Y. Shalom: Rigidity of commensurators and irreducible lattices, *Inventiones Math.* 141 (2000) 1–54. MR 1767270, http://dx.doi.org/10.1007/s002220000064

[21] Y. Shalom: The algebraization of Kazhdan's property (T). *International Congress of Mathematicians, Vol. II,* Eur. Math. Soc., Zürich, 2006, pp. 1283–1310, MR 2275645, http://www.mathunion.org/ICM/ICM2006.2/

[22] Y. Watatani: Property (T) of Kazhdan implies property (FA) of Serre, *Math. Japonica* 27 (1982) 97–103. MR 0649023

[23] A. Zuk: Property (T) and Kazhdan constants for discrete groups, *Geom. Funct. Anal.* 13 (2003), no. 3, 643–670. MR 1995802, http://dx.doi.org/10.1007/s00039-003-0425-8

Ergodic Theory

Ergodic Theory is the study of measure-theoretic aspects of group actions. Topologists and geometers may be more comfortable in the category of continuous functions, but important results in Chapters 16 and 17 will be proved by using measurable properties of actions of Γ, so we will introduce some of the basic ideas.

§14.1. Terminology

The reader is invited to skim through this section, and refer back as necessary.

(14.1.1) **Assumption.**

1) All measures are assumed to be ***σ-finite***. That is, if μ is a measure on a measure space X, then we always assume that X is the union of countably many subsets of finite measure.

2) We have no need for abstract measure spaces, so all measures are assumed to be ***Borel***. That is, when we say μ is a measure on a measure space X, we are assuming that X is a Borel subset of a complete, separable, metrizable space, and the implied σ-algebra

---

Recall: The Standing Assumptions (4.0.0 on page 41) are in effect, so, as always, Γ is a lattice in the semisimple Lie group $G \subseteq \mathrm{SL}(\ell, \mathbb{R})$.

Main prerequisites for this chapter: none.

on X consists of the subsets of X that are equal to a Borel set, modulo a set of measure 0.

(14.1.2) **Definitions.** Let μ be a measure on a measure space X.

1) We say μ is a **finite measure** if $\mu(X) < \infty$.

2) A subset A of X is:
   - **null** if $\mu(A) = 0$,
   - **conull** if the complement of A is null.

3) We often abbreviate "almost everywhere" to "a.e."

4) **Essentially** is a synonym for "almost everywhere." For example, a function f is *essentially constant* iff f is constant (a.e.).

5) Two measures μ and ν on X are in the same **measure class** if they have exactly the same null sets:
$$\mu(A) = 0 \iff \nu(A) = 0.$$
(This defines an equivalence relation.) Note that if $\nu = f\mu$, for some real-valued, measurable function f, such that $f(x) \neq 0$ for a.e. $x \in X$, then μ and ν are in the same measure class (see Exercise 1). The Radon-Nikodym Theorem (B6.13) implies that the converse is true.

(14.1.3) **Definitions.** Suppose H is a Lie group H that acts continuously on a metrizable space X, μ is a measure on X, and A is a subset of X.

1) The set A is **invariant** (or, more precisely, **H-invariant**) if $hA = A$ for all $h \in H$.

2) The measure μ is **invariant** (or, more precisely, **H-invariant**) if $h_*\mu = \mu$ for all $h \in H$. (Recall that the push-forward $h_*\mu$ is defined in (B6.7).)

3) The measure μ is **quasi-invariant** if $h_*\mu$ is in the same measure class as μ, for all $h \in H$.

4) A (measurable) function f on X is **essentially H-invariant** if, for every $h \in H$, we have
$$f(hx) = f(x) \text{ for a.e. } x \in X.$$

The Lebesgue measure on a manifold is not unique, but it determines a well-defined measure class, which is invariant under any smooth action:

(14.1.4) **Lemma** (see Exercise 4). *If X is a manifold, and H acts on X by diffeomorphisms, then Lebesgue measure provides a measure on X that is quasi-invariant for H.*

Exercises for §14.1.

#1. Suppose μ is a measure on a measure space X, and f is a real-valued, measurable function on X, such that $f \geq 0$ for a.e. x. Show that $f\mu$ is in the measure class of μ iff $f(x) \neq 0$ for a.e. $x \in X$.

#2. Suppose a Lie group H acts continuously on a metrizable space X, and μ is a measure on X. Show that μ is quasi-invariant iff the collection of null sets is H-invariant. (This means that if A is a null set, and $h \in H$, then $h(A)$ is a null set.)

#3. Suppose
   - A is a null set in $\mathbb{R}^n$ (with respect to Lebesgue measure), and
   - f is a diffeomorphism of some open subset $\mathcal{O}$ of $\mathbb{R}^n$.
   Show that $f(A \cap \mathcal{O})$ is a null set.
   [*Hint:* Change of variables.]

#4. Suppose X is a (second countable) smooth, n-dimensional manifold. This means that X can be covered by coordinate patches (X_i, φ_i) (where $\varphi_i \colon X_i \to \mathbb{R}^n$, and the overlap maps are smooth).
   a) Show there exists a partition $X = \bigcup_{i=1}^{\infty} \hat{X}_i$ into measurable subsets, such that $\hat{X}_i \subseteq X_i$ for each i.
   b) Define a measure μ on X by $\mu(X) = \lambda(\varphi_i(A \cap \hat{X}_i))$, where λ is the Lebesgue measure on $\mathbb{R}^n$. This measure may depend on the choice of X_i, φ_i, and $\hat{X}_i$, but show that the measure class of μ is independent of these choices.
   [*Hint:* Exercise 3.]

§14.2. Ergodicity

Suppose H acts on a topological space X. If H has a dense orbit on X, then it is easy to see that every continuous, H-invariant function is constant (see Exercise 2). Ergodicity is the much stronger condition that every *measurable* H-invariant function is constant (a.e.):

(14.2.1) **Definition.** Suppose H acts on X with a quasi-invariant measure μ. We say the action of H is **ergodic** (or that μ is an **ergodic** measure for H) if every H-invariant, real valued, measurable function on X is essentially constant.

It is easy to see that transitive actions are ergodic (see Exercise 3). But non-transitive actions can also be ergodic:

(14.2.2) **Example** (Irrational rotation of the circle). For any $\alpha \in \mathbb{R}$, we may define a homeomorphism T_α of the circle $\mathbb{T} = \mathbb{R}/\mathbb{Z}$ by
$$T_\alpha(x) = x + \alpha \pmod{\mathbb{Z}}.$$

By considering Fourier series, it is not difficult to show that if α is irrational, then every T_α-invariant function in $\mathcal{L}^2(\mathbb{T})$ is essentially constant (see Exercise 6). This implies that the $\mathbb{Z}$-action generated by T_α is ergodic (see Exercise 7).

Example 14.2.2 is a special case of the following general result:

(14.2.3) **Proposition.** *If H is any dense subgroup of a Lie group L, then the natural action of H on L by left translation is ergodic (with respect to the Haar measure on L).*

Proof. For any measurable $f: L \to \mathbb{R}$, its *essential stabilizer* in L is defined to be:
$$\mathrm{Stab}_L(f) = \{\, g \in L \mid f(gx) = f(x) \text{ for a.e. } x \in L \,\}.$$
It is not difficult to show that $\mathrm{Stab}_L(f)$ is closed (see Exercise 1). (It is also a subgroup of L, but we do not need this fact.) Hence, if $\mathrm{Stab}_L(f)$ contains a dense subgroup H, then it must be all of L. This implies that f is constant (a.e.) (see Exercise 4).   □

It was mentioned above that transitive actions are ergodic; therefore, G is ergodic on G/Γ. What is not obvious, and leads to important applications for arithmetic groups, is that most subgroups of G are also ergodic on G/Γ:

(14.2.4) **Theorem** (Moore Ergodicity Theorem, see Exercise 11.2#11). *If*
- *H is any noncompact, closed subgroup of G, and*
- *Γ is irreducible,*

then H is ergodic on G/Γ.

If H is ergodic on G/Γ, then Γ is ergodic on G/H (see Exercise 12). Hence:

(14.2.5) **Corollary.** *If H and Γ are as in Theorem 14.2.4, then Γ is ergodic on G/H.*

Exercises for §14.2.

#1. Show $\mathrm{Stab}_L(f)$ is closed, for every Lie group L and measurable $f: L \to \mathbb{R}$.
   [*Hint:* If f is bounded, then, for any $\varphi \in C_c(L)$ and $\{g_n\} \subseteq \mathrm{Stab}_L(f)$, we have $\int_L {}^g f \cdot \varphi \, d\mu = \int_L f \cdot g^{-1}\varphi \, d\mu = \lim \int_L f \cdot g_n^{-1}\varphi \, d\mu = \lim \int_L {}^{g_n}f \cdot \varphi \, d\mu = \int_L f \cdot \varphi \, d\mu.$]

#2. Suppose H acts on a topological space X, and has a dense orbit. Show that every real-valued, continuous, H-invariant function on X is constant.

#3. Show that H is ergodic on H/H_1, for every closed subgroup H_1 of H.
   [*Hint:* Every H-invariant function is constant, not merely essentially constant.]

#4. Suppose H is ergodic on X, and $f: X \to \mathbb{R}$ is measurable and essentially H-invariant. Show that f is essentially constant.

#5. Our definition of ergodicity is not the usual one, but it is equivalent: show that H is ergodic on X iff every H-invariant measurable subset of X is either null or conull.

[*Hint:* The characteristic function of an invariant set is an invariant function. Conversely, the sub-level sets of an invariant function are invariant sets.]

#6. In the notation of Example 14.2.2, show (without using Proposition 14.2.3):
   a) If α is irrational, then every T_α-invariant function in $\mathcal{L}^2(\mathbb{T})$ is essentially constant.
   b) If α is rational, then there exist T_α-invariant functions in $\mathcal{L}^2(\mathbb{T})$ that are not essentially constant.

[*Hint:* Any $f \in \mathcal{L}^2(\mathbb{T})$ can be written as a unique Fourier series: $f = \sum_{n=-\infty}^{\infty} a_n e^{in\theta}$. If f is invariant and α is irrational, then uniqueness implies $a_n = 0$ for $n \neq 0$.]

#7. Suppose μ is an H-invariant, *finite* measure on X. For all $p \in [1, \infty]$, show that H is ergodic iff every H-invariant element of $\mathcal{L}^p(X, \mu)$ is essentially constant.

#8. Let $H = \mathbb{Z}$ act on $X = \mathbb{R}$ by translation, and let μ be Lebesgue measure. Show:
   a) H is not ergodic on X, and
   b) for every $p \in [1, \infty)$, every H-invariant element of $\mathcal{L}^p(X, \mu)$ is essentially constant.
Why is this not a counterexample to Exercise 7?

#9. Let H be a dense subgroup of L. Show that if L is ergodic on X, then H is also ergodic on X.

[*Hint:* Exercise 1.]

#10. Show that if H acts continuously on X, and μ is a quasi-invariant measure on X, then the support of μ is an H-invariant subset of X.

#11. Ergodicity implies that a.e. orbit is dense in the support of μ. More precisely, show that if H is ergodic on X, and the support of μ is all of X (in other words, no open subset of X has measure 0), then a.e. H-orbit in X is dense. (That is, for a.e. $x \in X$, the orbit Hx of x is dense in X.

[*Hint:* The characteristic function of the closure of any orbit is invariant.]

#12. Suppose H is a closed subgroup of G. Show that H is ergodic on G/Γ iff Γ is ergodic on G/H.

[*Hint:* $H \times \Gamma$ acts on G (by letting H act on the left and Γ act on the right). Show H is ergodic on G/Γ iff $H \times \Gamma$ is ergodic on G iff Γ is ergodic on G/H.]

#13. The Moore Ergodicity Theorem has a converse: Assume G is not compact, and show that if H is any compact subgroup of G, then H is not ergodic on G/Γ.

[*Hint:* Γ acts properly discontinuously on G/H, so the orbits are not dense.]

#14. Show that if $n \geq 2$, then
    a) the natural action of $SL(n, \mathbb{Z})$ on $\mathbb{R}^n$ is ergodic, and
    b) the $SL(n, \mathbb{Z})$-orbit of a.e. vector in $\mathbb{R}^n$ is dense in $\mathbb{R}^n$.

[*Hint:* Identify $\mathbb{R}^n$ with a homogeneous space of $G = SL(n, \mathbb{R})$ (a.e.), by noting that G is transitive on the nonzero vectors of $\mathbb{R}^n$.]

#15. Let
- $G = SL(3, \mathbb{R})$,
- Γ be a lattice in G, and
- $P = \begin{bmatrix} * & & \\ * & * & \\ * & * & * \end{bmatrix} \subset G.$

Show:
    a) The natural action of Γ on the homogeneous space G/P is ergodic.
    b) The diagonal action of Γ on $(G/P)^2 = (G/P) \times (G/P)$ is ergodic.
    c) The diagonal action of Γ on $(G/P)^3 = (G/P) \times (G/P) \times (G/P)$ is *not* ergodic.

[*Hint:* G is transitive on a conull subset of $(G/P)^k$, for $k \leq 3$. What is the stabilizer of a generic point in each of these spaces?]

#16. Assume Γ is irreducible, and let H be a closed, noncompact subgroup of G. Show, for a.e. $x \in G/\Gamma$, that Hx is dense in G/Γ.

#17. Suppose H acts ergodically on X, with invariant measure μ. Show that if $\mu(X) < \infty$ and $H \cong \mathbb{R}$, then some cyclic subgroup of H is ergodic on X.

[*Hint:* For each $t \in \mathbb{R}$, choose a nonzero, h^t-invariant function $f_t \in L^2(X)$, such that $f_t \perp 1$. The projection of f_r to the space of h^s-invariant functions is invariant under both h^r and h^s. Therefore, if r and s are linearly independent over $\mathbb{Q}$, then $f_r \perp f_s$. This is impossible, because $\mathcal{L}^2(X, \mu)$ is separable.]

§14.3. Consequences of a finite invariant measure

Measure-theoretic techniques are especially powerful when the action has an invariant measure that is finite. One example of this is the Poincaré Recurrence Theorem (4.6.1). Here is another.

    We know that almost every orbit of an ergodic action is dense (see Exercise 14.2#11). For the case of a $\mathbb{Z}$-action with a finite, invariant measure, the orbits are not only dense, but uniformly distributed:

(14.3.1) **Definition.** Let
- μ be a finite measure on a topological space X, and
- T be a homeomorphism of X.

The $\langle T \rangle$-orbit of a point x in X is **uniformly distributed** with respect to μ if

$$\lim_{n \to \infty} \frac{1}{n} \sum_{k=1}^{n} f(T^k(x)) = \frac{1}{\mu(X)} \int_X f \, d\mu,$$

for every bounded, continuous function f on X.

(14.3.2) **Theorem** (Pointwise Ergodic Theorem). *Suppose*
- *μ is a finite measure on a second countable, metrizable space X, and*
- *T is an ergodic, measure-preserving homeomorphism of X.*

Then a.e. $\langle T \rangle$-orbit in X is uniformly distributed (with respect to μ).

It is tricky to show that $\lim_{n \to \infty} \frac{1}{n} \sum_{k=1}^{n} f(T^k(x))$ converges pointwise (see Exercise 5). Convergence in norm is much easier (see Exercise 3).

(14.3.3) *Remark.* Although the Pointwise Ergodic Theorem was stated only for actions of a cyclic group, it generalizes very nicely to the ergodic actions of any amenable group. (The values of f are averaged over an appropriate Følner set in the amenable group.) See Exercise 8 for actions of $\mathbb{R}$.

Exercises for §14.3.

#1. Suppose the $\langle T \rangle$-orbit of x is uniformly distributed with respect to a finite measure μ on X. Show that if the support of μ is all of X, then the $\langle T \rangle$-orbit of x is dense in X.

#2. Suppose
- U is a unitary operator on a Hilbert space $\mathcal{H}$,
- $v \in \mathcal{H}$, and
- $\langle v \mid w \rangle = 0$, for every vector w that is fixed by U.

Show $\frac{1}{n} \sum_{k=1}^{n} U^k v \to 0$ as $n \to \infty$.

[*Hint:* Apply the Spectral Theorem to diagonalize the unitary operator U.]

#3. (Mean Ergodic Theorem) Assume the setting of the Pointwise Ergodic Theorem (14.3.2). Show that if $f \in \mathscr{L}^2(X, \mu)$, then

$$\frac{1}{n} \sum_{k=1}^{n} f(T^k(x)) \to \frac{1}{\mu(X)} \int_X f \, d\mu \quad \text{in } \mathscr{L}^2.$$

That is, show

$$\lim_{n \to \infty} \left\| \frac{1}{n} \sum_{k=1}^{n} f(T^k(x)) - \frac{1}{\mu(X)} \int_X f \, d\mu \right\|_2 = 0.$$

Do not assume Theorem 14.3.2.

[*Hint:* Exercise 2.]

#4. Assume X, μ, and T are as in Theorem 14.3.2, and that $\mu(X) = 1$. For $f \in L^1(X, \mu)$, define
$$S_n(x) = f(x) + f(T(x)) + \cdots + f(T^{n-1}(x)).$$
Prove the **Maximal Ergodic Theorem**: for every $\alpha \in \mathbb{R}$, if we let
$$E_\alpha = \left\{ x \in X \ \middle| \ \sup_n \frac{S_n(x)}{n} > \alpha \right\},$$
then $\int_E f \, d\mu \geq \alpha \mu(E)$.

[*Hint:* Assume $\alpha = 0$. Let $S_n^+(x) = \max_{0 \leq k \leq n} S_k(x)$, and $E_n = \{x \mid S_n^+ > 0\}$, so $E = \cup_n E_n$. For $x \in E_n$, we have $f(x) \geq S_n^+(x) - S_n^+(T(x))$, so $\int_{E_n} f \, d\mu \geq 0$.]

#5. Prove the Pointwise Ergodic Theorem (14.3.2).

[*Hint:* Either $\{x \mid \lim \sup S_n(x)/n > \alpha\}$ or its complement must have measure 0. If it is the complement, then Exercise 4 implies $\int_X f \, d\mu \geq \alpha$.]

#6. Assume the setting of the Pointwise Ergodic Theorem (14.3.2). For every bounded $\varphi \in \mathcal{L}^1(X, \mu)$, show, for a.e. $x \in X$, that
$$\lim_{n \to \infty} \frac{1}{n} \sum_{k=1}^{n} \varphi(T^k(x)) = \frac{1}{\mu(X)} \int_X \varphi \, d\mu.$$

[*Hint:* You may assume the Pointwise Ergodic Theorem. Use Lusin's Theorem (B6.6) to approximate φ by a continuous function.]

#7. (*harder*) Remove the assumption that φ is bounded in Exercise 6.

#8. Suppose
- $\{a^t\}$ is a (continuous) 1-parameter group of homeomorphisms of a topological space X, and
- μ is an ergodic, a^t-invariant, finite measure on X.

For every bounded, continuous function f on X, show that
$$\lim_{T \to \infty} \frac{1}{T} \int_0^t f(a^t x) \, dt = \frac{1}{\mu(X)} \int_X f \, d\mu \quad \text{for a.e. } x \in X.$$

[*Hint:* Apply Theorem 14.3.2 to $\overline{f}(x) = \int_0^1 f(a^t x) \, dt$ if a^1 is ergodic (cf. Exercise 14.2#17).]

§14.4. Ergodic decomposition

In this section, we briefly explain that every group action (with a quasi-invariant measure) can be decomposed into ergodic actions.

(14.4.1) **Example** (Irrational rotation of the plane). For any irrational $\alpha \in \mathbb{R}$, define a homeomorphism T_α of $\mathbb{C}$ by $T_\alpha(z) = e^{2\pi i \alpha} x$. Then $|T_\alpha(z)| = |z|$, so each circle centered at the origin is invariant under T_α. The restriction of T_α to any such circle is an irrational rotation of the circle, so it is ergodic (see Example 14.2.2). Thus, the entire action can be decomposed as a union of ergodic actions.

A similar decomposition is always possible, as long as we work with nice spaces:

(14.4.2) **Definition.** A topological space X is **Polish** if it is homeomorphic to a complete, separable metric space.

(14.4.3) **Theorem** (Ergodic decomposition). *Suppose a Lie group H acts continuously on a Polish space X, and μ is a quasi-invariant, finite measure on X. Then there exist*

- *a measurable function $\zeta \colon X \to [0,1]$, and*
- *a finite measure μ_z on $\zeta^{-1}(z)$, for each $z \in [0,1]$,*

such that $\mu = \int_{[0,1]} \mu_z \, d\nu(z)$, where $\nu = \zeta_* \mu$. For $f \in C_c(X)$, this means

$$\int_X f \, d\mu = \int_Z \int_{\zeta^{-1}(z)} f \, d\mu_z \, d\nu(z).$$

Furthermore, for a.e. $z \in [0,1]$,

1) *$\zeta^{-1}(z)$ is H-invariant, and*

2) *μ_z is quasi-invariant and ergodic for the action of H.*

(14.4.4) **Remark.** The ergodic decomposition is unique (a.e.). More precisely, if ζ' and μ'_z also satisfy the conclusions, then there is a measurable bijection $\pi \colon [0,1] \to [0,1]$, such that

1) $\zeta' = \pi \circ \zeta$ a.e., and

2) $\mu'_{\pi(z)} = \mu_z$ for a.e. z.

(14.4.5) **Definition.** In the notation of Theorem 14.4.3, each set $\zeta^{-1}(z)$ is called an **ergodic component** of the action.

The remainder of this section sketches two different proofs of Theorem 14.4.3.

§14.4(i). First proof. The main problem is to find the function ζ, because the following general Fubini-like theorem will then provide the required decomposition of μ into an integral of measures μ_z on the fibers of ζ. (In Probability Theory, each μ_z is called a **conditional measure** of μ.)

(14.4.6) **Proposition.** *Suppose*

- *X and Z are Polish spaces,*
- *$\zeta \colon X \to Z$ is a Borel measurable function, and*
- *μ is a probability measure on X.*

Then there is a Borel map $\lambda \colon Z \to \mathrm{Prob}(X)$, such that

1) *$\mu = \int_Z \lambda_z \, d\nu(z)$, where $\nu = \zeta_* \mu$, and*

2) *$\lambda_z(\zeta^{-1}(z)) = 1$, for all $z \in Z$.*

Furthermore, λ is unique (a.e.).

The map ζ is a bit difficult to pin down, since it is not completely well-defined — it can be changed on an arbitrary set of measure zero. We circumvent this difficulty by looking not at the value of ζ on individual points (which is not entirely well-defined), but at its effect on an algebra of functions (which is completely well defined).

(14.4.7) **Definitions.** Assume μ is a finite measure on a Polish space X.

1) Let $\mathcal{B}(X)$ be the collection of all Borel subsets of X, where two sets are identified if they differ by a set of measure 0. This is a σ-algebra.

2) $\mathcal{B}(X)$ is a complete, separable metric space, with respect to the metric $d(A, B) = \mu(A \triangle B)$, where $A \triangle B = (A \smallsetminus B) \cup (B \smallsetminus A)$ is the **symmetric difference** of A and B.

3) If a Lie group H acts continuously on X, we let $\mathcal{B}(X)^H$ be the set of H-invariant elements of $\mathcal{B}(X)$. This is a sub-σ-algebra of $\mathcal{B}(X)$.

The map ζ is constructed by the following result:

(14.4.8) **Lemma.** *Suppose*

- *μ is a finite measure on a Polish space X, and*
- *$\mathcal{B}$ is a sub-σ-algebra of $\mathcal{B}(X)$.*

Then there is a Borel map $\zeta\colon X \to [0, 1]$, such that
$$\mathcal{B} = \{\,\zeta^{-1}(E) \mid E \text{ is a Borel subset of } [0, 1]\,\}.$$

Idea of proof. Let

- $\{E_n\}$ be a countable, dense subset of $\mathcal{B}$,
- χ_n be the characteristic function of E_n, for each n, and
- $\zeta(x) = \sum_{n=1}^{\infty} \dfrac{\chi_n(x)}{3^n}$.

It is clear from the definition of ζ that if I is any open interval in $[0, 1]$, then $\zeta^{-1}(I)$ is a Boolean combination of elements of $\{E_n\}$; therefore, it belongs to $\mathcal{B}$. Since $\mathcal{B}$ is a σ-algebra, this implies that $\zeta^{-1}(E) \in \mathcal{B}$ for every Borel subset E of $[0, 1]$.

Conversely, it is clear from the definition of ζ that each E_n is the inverse image of a (one-point) Borel subset of $[0, 1]$. Since $\{E_n\}$ generates $\mathcal{B}$ as a σ-algebra (see Exercise 2), this implies that every element of $\mathcal{B}$ is the inverse image of a Borel subset of $[0, 1]$.   □

We will use the following very useful fact:

(14.4.9) **Theorem** (von Neumann Selection Theorem). *Let*

- *X and Y be Polish spaces,*
- *μ be a finite measure on X,*
- *$\mathcal{F}$ be a Borel subset of $X \times Y$, and*

- $X_{\mathcal{F}}$ be the projection of $\mathcal{F}$ to X.

Then there is a Borel function $\Phi\colon X \to Y$, such that $(x, \Phi(x)) \in \mathcal{F}$, for a.e. $x \in X_{\mathcal{F}}$.

The first proof of Theorem 14.4.3. We wish to show that μ_z is ergodic (a.e.). If not, then there is a set E of positive measure in $[0,1]$, such that, for each $z \in E$, the action of H on $(\zeta^{-1}(z), \mu_z)$ is not ergodic. This means there exists an H-invariant, measurable, $\{0,1\}$-valued function $\varphi_z \in \mathscr{L}^\infty(\zeta^{-1}(z), \mu_z)$ that is **not** constant (a.e.). There are technical problems that we will ignore, but, roughly speaking, the von Neumann Selection Theorem (14.4.9) implies that the selection of φ_z can be done measurably, so we have a Borel subset A of X, defined by

$$A = \{\, x \in X \mid \zeta(x) \in E \text{ and } \varphi_z(x) = 1 \,\}.$$

Since φ_z is not constant on the fiber $\zeta^{-1}(z)$, we know that A is not of the form $\zeta^{-1}(E)$, for any Borel subset E of $[0,1]$. On the other hand, we have $A \in \mathcal{B}(X)^H$ (since each φ_z is H-invariant). This contradicts the choice of the function ζ. $\square$

§14.4(ii). Second proof. We now describe a different approach. Instead of obtaining the decomposition of μ from the map ζ, we reverse the argument, and obtain the map ζ from a direct-integral decomposition of μ. For simplicity, however, we assume that the space we are acting on is compact. We consider only invariant measures, instead of quasi-invariant measures, so we do not have to keep track of Radon-Nikodym derivatives.

(14.4.10) **Definitions.** Suppose C is a convex subset of a vector space V.

1) A point $c \in C$ is an **extreme point** of C if there do not exist $c_0, c_1 \in C \smallsetminus \{c\}$ and $t \in (0,1)$, such that $c = tc_0 + (1-t)c_1$.

2) Let $\operatorname{ext} C$ be the set of extreme points of C.

(14.4.11) **Example.** Suppose H acts continuously on a compact, separable metric space X, and let

$$\operatorname{Prob}(X)^H = \{\, \mu \in \operatorname{Prob}(X) \mid \mu \text{ is } H\text{-invariant} \,\}.$$

This is a closed, convex subset of $\operatorname{Prob}(X)$, so $\operatorname{Prob}(X)^H$ is a compact, convex subset of $C(X)^*$ (with the weak* topology). The extreme points of this set are precisely the H-invariant probability measures that are ergodic (see Exercise 3).

The well-known Krein-Milman Theorem states that every compact, convex set C is the closure of the convex hull of the set of extreme points of C. (So, in particular, if C is nonempty, then there exists an extreme point.) We will use the following strengthening of this fact:

(14.4.12) **Theorem** (Choquet's Theorem). *Suppose*
- $\mathcal{V}$ *is a locally convex topological vector space over* $\mathbb{R}$,
- C *is a metrizable, compact, convex subset of* $\mathcal{V}$, *and*
- $c_0 \in C$.

Then there is a probability measure v *on* $\mathrm{ext}\, C$, *such that*
$$c_0 = \int_{\mathrm{ext}\, C} x\, dv(x).$$

We will also need a corresponding uniqueness result.

(14.4.13) **Definitions** (Choquet). Suppose $\mathcal{V}$ and C are as in the statement of Theorem 14.4.12.
1) Let $\Sigma C = \{\, tc \mid t \in [0, \infty),\ c \in C \,\} \subseteq \mathcal{V}$.
2) Define a partial order $\le$ on ΣC by $a \le b$ iff $b - a \in \Sigma C$.
3) Two elements $a_1, a_2 \in \Sigma C$ have a **least upper bound** if there exists $b \in \Sigma C$, such that
   - $a_i \le b$ for $i = 1, 2$, and
   - for all $c \in \Sigma C$, such that $a_i \le c$ for $i = 1, 2$, we have $b \le c$.

(14.4.14) **Example.** Any two elements of $\Sigma \mathrm{Prob}(X)$ have a least upper bound (see Exercise 4), so the same is true for $\mathrm{Prob}(X)^H$.

(14.4.15) **Theorem** (Choquet). *Suppose* $\mathcal{V}$, C, *and* c_0 *are as in the statement of Theorem 14.4.12. If every two elements of* ΣC *have a least upper bound, then the measure* v *provided by Theorem 14.4.12 is unique.*

The second proof of Theorem 14.4.3. Assume, for simplicity, that μ is H-invariant, and that X is compact. By normalizing, we may assume $\mu(X) = 1$, so $\mu \in \mathrm{Prob}(X)$. Then Choquet's Theorem (14.4.12) provides a probability measure v, such that
$$\mu = \int_{\mathrm{ext}\, \mathrm{Prob}(X)^H} \sigma\, dv(\sigma).$$
By identifying $\mathrm{ext}\, \mathrm{Prob}(X)$ with a Borel subset of $[0, 1]$, we may rewrite this as:
$$\mu = \int_{[0,1]} \mu_z\, dv(z),$$
where v is a probability measure on $[0, 1]$. Furthermore, Exercise 3 tells us that each $\sigma \in \mathrm{ext}\, \mathrm{Prob}(X)^H$ is ergodic, so μ_z is an ergodic H-invariant measure for a.e. z.

All that remains is to define a map $\zeta \colon X \to [0, 1]$, such that μ_z is concentrated on $\zeta^{-1}(z)$. For each Borel subset E of $\mathrm{ext}\, \mathrm{Prob}(X)^H$, let $\mu_E = \int_E \sigma\, dv(\sigma)$. Then μ_E is absolutely continuous with respect to μ, so we may write $\mu_E = f_E \mu$, for some measurable $f_E \colon X \to [0, \infty)$. Then
$$\psi(E) = \{\, x \in X \mid f_E(x) \ne 0 \,\}$$

is a well-defined element of $\mathcal{B}(X)$. Therefore, we have defined a map $\psi\colon \mathcal{B}([0,1]) \to \mathcal{B}(X)$, and it can be verified that this is a homomorphism of σ-algebras. Hence, there is a measurable function $\zeta\colon X \to [0,1]$, such that $\psi(E) = \zeta^{-1}(E)$, for all E (see Exercise 5). By using the uniqueness of ν, it can be shown that $\mu_z(\zeta^{-1}(z)) = 1$ for a.e. z. $\square$

Exercises for §14.4.

#1. Let $\mathcal{B}$ be a (nonempty) subset of $\mathcal{B}(X)$ that is closed under complements and finite unions. Show that $\mathcal{B}$ is closed under countable unions (so $\mathcal{B}$ is a sub-σ-algebra of $\mathcal{B}(X)$) if and only if $\mathcal{B}$ is a closed set with respect to the topology determined by the metric on $\mathcal{B}(X)$.

[*Hint:* ($\Leftarrow$) If $E = \bigcup_{i=1}^{\infty} E_i$, then $\bigcup_{i=1}^{n} E_i \to E$ in the topology on $\mathcal{B}(X)$.
($\Rightarrow$) If $d(E_i, E) < 2^{-i}$, then $E = \bigcap_{n=1}^{\infty} \bigcup_{i=n}^{\infty} E_i$ (up to a set of measure 0).]

#2. Show that if $\mathcal{E}$ is dense in a sub-σ-algebra $\mathcal{B}$ of $\mathcal{B}(X)$, then $\mathcal{E}$ is not contained in any proper sub-σ-algebra of $\mathcal{B}$.

[*Hint:* If $E_n \to E$, then $\bigcup_{n=1}^{\infty}(E_n \cap E) = E$ (up to a set of measure 0).]

#3. Prove that a measure $\mu \in \mathrm{Prob}(X)^H$ is ergodic iff it is an extreme point.

[*Hint:* If E is an H-invariant set, then μ is a convex combination of the restrictions to E and its complement. Conversely, if $\mu = t\mu_1 + (1-t)\mu_2$, then the Radon-Nikodym Theorem implies $\mu_1 = f\mu$ for some (H-invariant) function f.]

#4. Show that any two elements of $\Sigma\,\mathrm{Prob}(X)$ or $\Sigma\,\mathrm{Prob}(X)^*$ have a least upper bound.

[*Hint:* Write $\nu = f\mu + \nu_s$ (uniquely), where ν_s is concentrated on a set of measure 0.]

#5. Suppose $\psi\colon \mathcal{B}(Z) \to \mathcal{B}(X)$ is a function that respects complements and countable unions (and $\psi(\varnothing) = \varnothing$). Show there is a Borel function $\zeta\colon X \to Z$, such that $\psi(E) = \zeta^{-1}(E)$, for every Borel subset E of Z.

[*Hint:* Assume, for simplicity, that $Z = \{\sum a_k 3^{-k} \mid a_k \in \{0,1\}\} \subset [0,1]$. Then $\zeta = \sum \chi_{E_k} 3^{-k}$ for an appropriate collection $\{E_k\}$ of Borel subsets of X.]

§14.5. Mixing

It is sometimes important to know that a product of group actions is ergodic. To discuss this issue (and related matters), let us fix some notation.

(14.5.1) **Notation.** Throughout this section, we assume:

1) X_i is a Polish space, for every i,

2) H is a Lie group that acts continuously on X_i, for each i, and

3) μ_i is an H-invariant probability measure on X_i, for each i.

Furthermore, we use X and μ as abbreviations for X_0 and μ_0, respectively.

(14.5.2) **Definitions.**

1) The ***product action*** of H on $X_1 \times X_2$ is the H-action defined by $h(x_1, x_2) = (hx_1, hx_2)$. The product measure $\mu_1 \times \mu_2$ is an invariant measure for this action.

2) The action on X is said to be ***weak mixing*** (or ***weakly mixing***) if the product action on $X \times X$ is ergodic.

We have the following very useful characterizations of weakly mixing actions:

(14.5.3) **Theorem.** *The action of H on X is weak mixing if and only if the (one-dimensional) space of constant functions is the only nontrivial, finite-dimensional, H-invariant subspace of $\mathcal{L}^2(X, \mu)$.*

Proof. We prove the contrapositive of each direction.

($\Rightarrow$) Suppose V is a nontrivial, finite-dimensional, H-invariant subspace of $\mathcal{L}^2(X)$. If the functions in V are not all constant (a.e.), then we may assume (by passing to a subspace) that $V \perp 1$. Choose an orthonormal basis $\{\varphi_1, \ldots, \varphi_r\}$ of V, and define $\varphi \colon X \times X \to \mathbb{C}$ by

$$\varphi(x, y) = \sum_{i=1}^{r} \varphi_i(x) \, \overline{\varphi_i(y)}. \tag{14.5.4}$$

Then φ is an H-invariant function that is not constant (see Exercise 2), so H is not ergodic on $X \times X$.

($\Leftarrow$) Suppose φ is a nonconstant, H-invariant, bounded function on $X \times X$. We may assume $\varphi(x, y) = \overline{\varphi(y, x)}$ by replacing φ with either $\varphi(x, y) + \overline{\varphi(y, x)}$ or $\sqrt{-1}\,(\varphi(x, y) - \overline{\varphi(y, x)})$. Therefore, we have a compact, self-adjoint operator on $\mathcal{L}^2(X, \mu)$, defined by

$$(T\psi)(x) = \int_X \varphi(x, y)\, \psi(y)\, d\mu(y).$$

The Spectral Theorem (B7.14) implies T has an eigenspace V that is finite-dimensional (and contains a nonconstant function). This eigenspace is H-invariant, since T commutes with H (because φ is H-invariant). $\square$

We often have the following stronger condition:

(14.5.5) **Definition.** The action of H on X is said to be ***mixing*** (or, alternatively, ***strongly mixing***) if H is noncompact, and, for all $\varphi, \psi \in \mathcal{L}^2(X, \mu)$, such that $\varphi \perp 1$, we have

$$\lim_{h \to \infty} \langle h\varphi \mid \psi \rangle = 0.$$

(We are using 1 to denote the constant function of value 1, and, as usual, $(h\varphi)(x) = \varphi(h^{-1}x)$ (see Example 11.1.3).)

(14.5.6) **Proposition** (see Exercise 3). *If H is not compact, then the following are equivalent:*

1) *The action of H on X is mixing.*

2) *For all $\varphi, \psi \in \mathscr{L}^2(X, \mu)$, we have*
$$\lim_{h \to \infty} \langle h\varphi \mid \psi \rangle = \langle \varphi \mid 1 \rangle \langle 1 \mid \psi \rangle.$$

3) *For all Borel subsets A and B of X, we have*
$$\lim_{h \to \infty} \mu(hA \cap B) = \mu(A)\,\mu(B).$$

(14.5.7) *Remark.* Condition (3) is the motivation for the choice of the term "mixing:" as $h \to \infty$, the space X is getting so stirred up (or well-mixed) that hA is becoming uniformly distributed throughout the entire space.

When G is simple, decay of matrix coefficients (11.2.2) implies that every action of G (with finite invariant measure) is mixing. In fact, we can say more.

(14.5.8) **Definition.** Generalizing Definition 14.5.5, we say that the action of H on X is **mixing of order r** if H is not compact, and, for all Borel subsets $A_1, \ldots, A_r$ of X, we have
$$\lim_{h_i^{-1} h_j \to \infty} \mu\left(\bigcap_{i=1}^{r} h_i A_i \right) = \mu(A_1)\,\mu(A_2) \cdots \mu(A_r).$$
In particular,

- every action of H is mixing of order 1 (if H is not compact), and
- "mixing" is the same as "mixing of order 2."

(14.5.9) **Warning.** Some authors use a different numbering, for which this is "mixing of order $r - 1$," instead of "mixing of order r."

Ledrappier constructed an action of $\mathbb{Z}^2$ that is mixing of order 2, but not of order 3. However, there are no such examples for semisimple groups:

(14.5.10) **Theorem.** *Every mixing action of G (with finite invariant measure) is mixing of all orders.*

In the special case where $H = \mathbb{Z}$, we mention the following additional characterizations, some of which are weaker versions of Proposition 14.5.6(3):

(14.5.11) **Theorem.** *If $H = \langle T \rangle$ is an infinite cyclic group, then the following are equivalent:*

1) *H is weak mixing on X.*

2) *Every eigenfunction of T in $\mathscr{L}^2(X, \mu)$ is constant (a.e.). That is, if $f \in \mathscr{L}^2(X, \mu)$, and there is some $\lambda \in \mathbb{C}$, such that $f(Tx) = \lambda f(x)$ a.e., then f is constant (a.e.).*

3) *The spectral measure of T has no atoms other than 1, and the eigenvalue 1 is simple (that is, the corresponding eigenspace is 1-dimensional).*

4) *The action of H on $X \times X_1$ is ergodic, whenever the action of H on X_1 is ergodic.*

5) *For all Borel subsets A and B of X,*

$$\lim_{n \to \infty} \frac{1}{n} \sum_{k=1}^{n} |\mu(T^k A \cap B) - \mu(A)\mu(B)| = 0.$$

6) *There is a subset $\mathcal{N}$ of full density in $\mathbb{Z}^+$, such that, for all Borel subsets A and B of X, we have*

$$\mu(T^k A \cap B) \longrightarrow \mu(A)\mu(B) \qquad \text{as } k \to \infty \text{ with } k \in \mathcal{N}.$$

*To say $\mathcal{N}$ has **full density** means*

$$\lim_{n \to \infty} \frac{\#(\mathcal{N} \cap \{1, 2, 3, \ldots, n\})}{n} = 1.$$

7) *If $A_0, A_1, \ldots, A_r$ are any Borel subsets of X, then there is a subset $\mathcal{N}$ of full density in $\mathbb{Z}^+$, such that*

$$\lim_{\substack{k \to \infty \\ k \in \mathcal{N}}} \mu(A_0 \cap T^k A_1 \cap T^{2k} A_2 \cap \cdots \cap T^{rk} A_r) = \mu(A_0)\mu(A_1) \cdots \mu(A_r).$$

Sketch of proof. (1 ⇔ 2) By Theorem 14.5.3, it suffices to observe that every finite-dimensional representation contains an irreducible sub-representation, and that the irreducible representations of $\mathbb{Z}$ (or, more generally, of any abelian group) are one-dimensional.

(2 ⇔ 3) These are two different ways of saying the same thing.

(3 ⇒ 4) Since $\mathcal{L}^2(X \times X_1) \cong \mathcal{L}^2(X) \otimes \mathcal{L}^2(X_1)$, the spectral measure v of $\mathcal{L}^2(X \times X_1)$ is the product $v_1 \times v_2$ of the spectral measures of $\mathcal{L}^2(X)$ and $\mathcal{L}^2(X_1)$. Therefore, any point mass in v is obtained by pairing a point mass in v_1 with a point mass in v_2.

(4 ⇒ 1) Take $X_1 = X$.

(1 ⇒ 5) For simplicity, let $a = \mu(A)$ and $b = \mu(B)$. By Exercise 14.3#3 (the Mean Ergodic Theorem), ergodicity on X implies

$$\frac{1}{n} \sum_{k=1}^{n} \mu(T^k A \cap B) \overset{n \to \infty}{\longrightarrow} ab.$$

For the same reason, ergodicity on $X \times X$ implies

$$\sum_{k=1}^{n} (\mu \times \mu)(T^k A \times T^k A) \cap (B \times B)) \overset{n \to \infty}{\longrightarrow} (\mu \times \mu)(A \times A) \cdot (\mu \times \mu)(B \times B).$$

By simplifying both sides, we see that

$$\frac{1}{n} \sum_{k=1}^{n} \mu(T^k A \cap B)^2 \overset{n \to \infty}{\longrightarrow} a^2 b^2.$$

Therefore, simple algebra yields

$$\frac{1}{n} \sum_{k=1}^{n} |\mu(T^k A \cap B) - \mu(A)\,\mu(B)|^2 \overset{n \to \infty}{\longrightarrow} a^2 b^2 - 2(ab)(ab) + (ab)^2 = 0.$$

Exercise 4 implies that we have the same limit without squaring the absolute value.

(5 ⇒ 2) Approximating by linear combinations of characteristic functions implies

$$\lim_{n \to \infty} \frac{1}{n} \sum_{k=1}^{n} |\langle T^k \varphi \mid \varphi \rangle| = 0 \qquad \text{for all } \varphi \perp 1.$$

However, if φ is an eigenfunction for an eigenvalue $\neq 1$, then it is easy to see that the limit is nonzero (either directly, or by applying Exercise 4).

(5 ⇔ 6) Exercise 4 implies that the two assertions are equivalent, up to reversing the order of the quantifiers in (6). To reverse the quantifiers, note that a variant of Cantor diagonalization provides a set $\mathcal{N}$ of full density that works for all A and B in a countable dense subset of $\mathcal{B}(X)$.

(7 ⇒ 6) Take $r = 1$.

(6 ⇒ 7) The proof proceeds by induction on r (with (6) as the starting point), and is nontrivial. We have no need for this result, so we omit the proof.   □

Exercises for §14.5.

#1. Show (directly from the definitions) that if the action of H on X is weak mixing, then it is ergodic.

#2. Let $\varphi \colon X \times X \to \mathbb{C}$ be as in (14.5.4) of the proof of Theorem 14.5.3.
   a) Show φ is H-invariant (a.e.).
   b) Show φ is not constant (a.e.).
   [*Hint:* (a) Write $h\varphi_i = \sum_{i,j} h_{i,j}\varphi_j$, and observe that $[h_{i,j}]$ is a unitary matrix.
   (b) $\varphi(x, x) > 0$, but $\int \varphi(x, y)\, d\mu(y) = 0$.]

#3. Prove Proposition 14.5.6.
   [*Hint:* (1 ⇒ 2) For $c = \langle \varphi \mid 1 \rangle$, we have $\langle (\varphi - c) \mid 1 \rangle = 0$. Now calculate $\lim_{h \to \infty} \langle h(\varphi - c) \mid \psi \rangle$ in two ways.
   (2 ⇒ 3) Let φ and ψ be the characteristic functions of A and B.
   (3 ⇒ 2) Approximate φ and ψ by linear combinations of characteristic functions.]

#4. For every bounded sequence $\{a_k\} \subset [0, \infty)$, show

$$\lim_{n \to \infty} \frac{1}{n} \sum_{k=1}^{n} a_k = 0 \iff a_k \to 0 \text{ as } k \to \infty \text{ in some set of full density.}$$

   [*Hint:* (⇒) For each $m > 0$, the set $A_m = \{k \mid a_k > 1/m\}$ has density 0, so there exists $N_m > N_{m-1}$, such that, for all $n \geq N_m$, we have
   $$N_{m-1} + \#(A_m \cap \{1, 2, \ldots, n\}) < n/m.$$
   Let $\mathcal{N}$ be the complement of $\bigcup_m (A_m \cap [N_m, N_{m+1}))$.]

Notes

The focus of classical Ergodic Theory is on actions of $\mathbb{Z}$ and $\mathbb{R}$ (or other abelian groups). A few of the many introductory books on this subject are [1, 4, 5, 12]. They include proofs of the Poincaré Recurrence Theorem (4.6.1) and the Pointwise Ergodic Theorem (14.3.2).

Some basic results on the Ergodic Theory of noncommutative groups can be found in [13, §2.1].

The Moore Ergodicity Theorem (14.2.4) is due to C.C. Moore [8].

See [7] for a very nice version of the Pointwise Ergodic Theorem that applies to all amenable groups (Remark 14.3.3).

See [3, Thm. 1.1 (and Thm. 5.2)] for a proof of the ergodic decomposition (14.4.3), using Choquet's Theorem as in Subsection 14.4(ii). Proposition 14.4.6 can be found in [10, §3]. See [11, §5.5] for a proof of Theorem 14.4.9.

See [8, Prop. 1, pp. 157–158] for a proof of Proposition 14.5.6.

The standard texts on ergodic theory only prove Proposition 14.5.6 for the special case $H = \mathbb{Z}$, but the same arguments apply in general.

Theorem 14.5.10 is due to S. Mozes [9]. Ledrappier's counterexample for $\mathbb{Z}^2$ is in [6].

Theorem 14.5.11 is in the standard texts on ergodic theory, except for Part (7), which is a "multiple recurrence theorem" that plays a key role in Furstenberg's proof of Szemeredi's theorem that there are arbitrarily long arithmetic progressions in every set of positive density in $\mathbb{Z}^+$. For a proof of $(6 \Rightarrow 7)$, see [1, Prop. 7.13, p. 191] or [2, Thm. 4.10].

A proof of Exercise 14.5#4 is in [1, Lem. 2.41, p. 54].

References

[1] M. Einsiedler and T. Ward: *Ergodic Theory With a View Towards Number Theory.* Springer, London, 2011. ISBN 978-0-85729-020-5, MR 2723325

[2] H. Furstenberg: *Recurrence in Ergodic Theory and Combinatorial Number Theory.* Princeton University Press, Princeton, N.J., 1981. ISBN 0-691-08269-3, MR 0603625

[3] G. Greschonig and K. Schmidt, Ergodic decomposition of quasi-invariant probability measures, *Colloq. Math.* 84/85 (2000), part 2, 495–514. MR 1784210, http://eudml.org/doc/210829

[4] P. R. Halmos: *Lectures on Ergodic Theory.* Chelsea, New York, 1960. MR 0111817

[5] A. Katok and B. Hasselblatt: *Introduction to the Modern Theory of Dynamical Systems.* Cambridge University Press, Cambridge, 1995. ISBN 0-521-34187-6, MR 1326374

[6] F. Ledrappier: Un champ markovien peut être d'entropie nulle et mélangeant, *C. R. Acad. Sci. Paris Sér. A-B* 287 (1978), no. 7, A561–A563. MR 0512106

[7] E. Lindenstrauss: Pointwise theorems for amenable groups, *Invent. Math.* 146 (2001), no. 2, 259–295. MR 1865397, http://dx.doi.org/10.1007/s002220100162

[8] C. C. Moore: Ergodicity of flows on homogeneous spaces, *Amer. J. Math.* 88 1966 154–178. MR 0193188, http://www.jstor.org/stable/2373052

[9] S. Mozes: Mixing of all orders of Lie groups actions, *Invent. Math.* 107 (1992), no. 2, 235–241. (Erratum 119 (1995), no 2, 399.) MR 1144423, MR 1312506, http://eudml.org/doc/143965, http://eudml.org/doc/251560

[10] V. A. Rohlin: On the fundamental ideas of measure theory, *Amer. Math. Soc. Transl.* (1) 10 (1962), 1–54. (Translated from *Mat. Sbornik N.S.* 25(67), (1949), 107–150.) MR 0047744

[11] S. M. Srivastava: *A Course on Borel Sets.* Springer, New York, 1998. ISBN 0-387-98412-7, MR 1619545

[12] P. Walters: *An Introduction to Ergodic Theory.* Springer, New York, 1982 ISBN 0-387-90599-5, MR 0648108

[13] R. J. Zimmer: *Ergodic Theory and Semisimple Groups.* Birkhäuser, Basel, 1984. ISBN 3-7643-3184-4, MR 0776417

Part IV

Major Results

Chapter 15

Mostow Rigidity Theorem

§15.1. Statement of the theorem

In its simplest form, the Mostow Rigidity Theorem says that a single group Γ cannot be a lattice in two different semisimple groups G_1 and G_2 (except for minor modifications involving compact factors, the center, and passing to a finite-index subgroup):

(15.1.1) **Theorem** (Weak version of the Mostow Rigidity Theorem). *Assume*

- *G_1 and G_2 are connected, with trivial center and no compact factors, and*
- *Γ_i is a lattice in G_i, for $i = 1, 2$.*

If $\Gamma_1 \cong \Gamma_2$, then $G_1 \cong G_2$.

In other words, if there is an isomorphism from Γ_1 to Γ_2, then there is also an isomorphism from G_1 to G_2. In fact, it is usually the case that something much stronger is true:

(15.1.2) **Mostow Rigidity Theorem.** *Assume*

- *G_1 and G_2 are connected, with trivial center and no compact factors,*

Recall: The Standing Assumptions (4.0.0 on page 41) are in effect, so, as always, Γ is a lattice in the semisimple Lie group $G \subseteq \mathrm{SL}(\ell, \mathbb{R})$.

Main prerequisites for this chapter: Quasi-isometries (Chapter 10).

- Γ_i *is a lattice in* G_i, *for* $i = 1, 2$, *and*
- *there does not exist a simple factor* N *of* G_1, *such that* $N \cong \mathrm{PSL}(2, \mathbb{R})$
  and $N \cap \Gamma_1$ *is a lattice in* N.

Then any isomorphism from Γ_1 *to* Γ_2 *extends to a continuous isomorphism from* G_1 *to* G_2.

(15.1.3) *Remarks.*

1) Assume G is connected, and has no simple factors that are either compact or isogenous to $\mathrm{SL}(2, \mathbb{R})$. Then the Mostow Rigidity Theorem implies that lattices in G have no nontrivial deformations. More precisely, if Γ_t is a continuous family of lattices in G, then Γ_t is conjugate to Γ_0, for every t (see Exercise 7).

   This is not always true when G is isogenous to $\mathrm{SL}(2, \mathbb{R})$ (see Section 15.3), which explains why the statement of the Mostow Rigidity Theorem must forbid factors that are isogenous to $\mathrm{SL}(2, \mathbb{R})$.

2) In geometric terms, the Mostow Rigidity Theorem tells us that the topological structure of any irreducible finite-volume locally symmetric space of noncompact type completely determines its geometric structure as a Riemannian manifold (up to multiplying the metric by a scalar on each irreducible factor of the universal cover), if the manifold is not 2-dimensional (cf. Exercise 8).

3) Assume Γ is cocompact in G. Then, as a strengthening of Theorem 15.1.1, it can be shown that if Γ is a cocompact lattice in some Lie group H (not assumed to be semisimple), then H must be either G or Γ (modulo the usual minor modifications involving compact groups). In fact, this remains true even if we allow H to be any locally compact group, not necessarily a Lie group.

Exercises for §15.1.

#1. Suppose
- G has trivial center and no compact factors,
- $G \not\cong \mathrm{PSL}(2, \mathbb{R})$, and
- Γ is irreducible.

Show that every automorphism of Γ extends to a continuous automorphism of G.

#2. Show that Theorem 15.1.1 is a corollary of Theorem 15.1.2.

[*Hint:* This is obvious when no simple factor of G_1 is isomorphic to $\mathrm{PSL}(2, \mathbb{R})$. The problem can be reduced to the case where G_1 and G_2 are irreducible.]

#3. Assume G_1 and G_2 are isogenous. Show that if K is any compact group, then some lattice in G_1 is isomorphic to a lattice in $G_2 \times K$

(even though G_1 may not be isomorphic to $G_2 \times K$). This is why Theorem 15.1.1 assumes G_1 and G_2 are connected, with trivial center and no compact factors

[*Hint:* Any torsion-free lattice in G_1° is isomorphic to a lattice in $G_2 \times K$.]

#4. For $i = 1, 2$, suppose
  - Γ_i is a lattice in G_i, and
  - G_i is connected and has no compact factors.

Show that if Γ_1 is isomorphic to Γ_2, then G_1 is isogenous to G_2.

#5. Let Γ_i be a lattice in G_i for $i = 1, 2$. Show that if $\Gamma_1 \cong \Gamma_2$, then there is a compact, normal subgroup K_i of G_i°, for $i = 1, 2$, such that $G_1^\circ / K_1 \cong G_2^\circ / K_2$.

#6. Let $G = \mathrm{PSL}(2, \mathbb{R})$. Find an automorphism φ of some lattice Γ in G, such that φ does not extend to an automorphism of G.

  Why is this not a counterexample to Theorem 15.1.2?

#7. Assume there is a continuous function $\rho \colon \Gamma \times [0, 1] \to G$, such that, if we let $\rho_t(\gamma) = \rho(\gamma, t)$, then:
  - ρ_t is a homomorphism, for all t,
  - $\rho_t(\Gamma)$ is a lattice in G, for all t, and
  - $\rho_0(\gamma) = \gamma$, for all $\gamma \in \Gamma$.

Show that if G is as in the first sentence of Remark 15.1.3(1), then Γ_t is conjugate to Γ, for every t.

[*Hint:* Reduce to the case where G has trivial center. You may use, without proof, the fact that the identity component of the automorphism group of G consists of inner automorphisms (see Remark A6.4).]

#8. (*Requires some familiarity with locally symmetric spaces*) Assume, for $i = 1, 2$:
  - G_i is connected and simple, with trivial center,
  - K_i is a maximal compact subgroup of G_i,
  - Γ_i is a torsion-free, irreducible lattice in G_i,
  - $X_i = K_i \backslash G_i / \Gamma_i$ is the corresponding locally symmetric space of finite volume,
  - the metric on X_i is normalized so that $\mathrm{vol}(X_1) = 1$, and
  - $\dim X_1 \geq 3$.

Show that any homotopy equivalence from X_1 to X_2 is homotopic to an isometry.

[*Hint:* Since the universal cover $K_i \backslash G_i$ is contractible, a homotopy equivalence is determined, up to homotopy, by its effect on the fundamental group.]

§15.2. Sketch of the proof for SO(1, n) (optional)

In most cases, the conclusion of the Mostow Rigidity Theorem (15.1.2) is an easy consequence of the Margulis Superrigidity Theorem, which will be discussed in Chapter 16. More precisely, if we assume, for simplicity, that the lattices Γ_1 and Γ_2 are irreducible (see Exercise 1), then the Margulis Superrigidity Theorem applies unless G_1 and G_2 are isogenous to either $SO(1, n)$ or $SU(1, n)$ (see Exercise 16.2#1). To illustrate the main ideas involved in completing the proof, we discuss a special case:

Proof of Mostow Rigidity Theorem for cocompact lattices in $SO(1, n)$.
Assume, for $i = 1, 2$:

- $G_i \cong PSO(1, n_i)$, for some $n_i \geq 3$,
- Γ_i is cocompact in G_i, and
- $\rho: \Gamma_1 \to \Gamma_2$ is an isomorphism.

In order to show that ρ extends to a continuous homomorphism from G_1 to G_2, we take a geometric approach that uses the action of G_i on its associated symmetric space, which is the hyperbolic space $\mathfrak{H}^{n_i}$. This assumes some understanding of hyperbolic space (and other matters) that is not required elsewhere in this book.

Claim. We have $n_1 = n_2$. By passing to subgroups of finite index, we may assume Γ_1 and Γ_2 are torsion free, so Γ_i acts freely on $\mathfrak{H}^{n_i}$. The action is also properly discontinuous, so, since hyperbolic space is contractible, this implies that $X_i = \Gamma_i \backslash \mathfrak{H}^{n_i}$ is a $K(\Gamma_i, 1)$-space. Since X_i is a compact manifold of dimension n_i, we conclude that the cohomological dimension of Γ_i is n_i. However, the groups Γ_1 and Γ_2 are isomorphic, so they must have the same cohomological dimension. This completes the proof of the claim. $\square$

Therefore, Γ_1 and Γ_2 are two lattices in the same group $G = PSO(1, n)$. To simplify matters, let us assume $n = 3$.

Since Γ_i is cocompact in G_i (so Γ_i is quasi-isometric to $\mathfrak{H}^3$), it is not difficult to construct a quasi-isometry $\varphi: \mathfrak{H}^3 \to \mathfrak{H}^3$, such that

$$\varphi(\gamma x) = \rho(\gamma) \cdot \varphi(x) \quad \text{for all } \gamma \in \Gamma_1 \text{ and } x \in \mathfrak{H}^3 \qquad (15.2.1)$$

(see Exercise 2).

Consider the ball model of $\mathfrak{H}^3$, whose boundary $\partial \mathfrak{H}^3$ is the round 2-sphere S^2. It is easy to see that any isometry ϕ of $\mathfrak{H}^3$ induces a well-defined homeomorphism $\overline{\phi}$ of $\partial \mathfrak{H}^3$. Furthermore, it is well known that this boundary map is **conformal** (i.e., it is angle-preserving). This implies that if C is a very small circle in $\partial \mathfrak{H}^3$, then $\overline{\phi}(C)$ is very close to being a circle; more precisely, if we let $S_r(p)$ be the sphere of radius r around the point p, then

$$\limsup_{r \to 0^+} \frac{\sup_{x \in S_r(p)} d(\overline{\phi}(x), \overline{\phi}(p))}{\inf_{y \in S_r(p)} d(\overline{\phi}(y), \overline{\phi}(p))} = 1.$$

With this background in mind, it should not be difficult to believe (and it is not terribly difficult to prove) that the quasi-isometry φ induces a well-defined homeomorphism $\overline{\varphi}$ of $\partial\mathfrak{H}^3$, such that

$$\overline{\varphi}(\gamma p) = \overline{\rho(\gamma)} \cdot \overline{\varphi}(p) \quad \text{for all } \gamma \in \Gamma_1 \text{ and } p \in \partial\mathfrak{H}^3. \tag{15.2.2}$$

Furthermore, this boundary map is *quasi-conformal*, which means that if C is a very small circle in $\partial\mathfrak{H}^3$, then $\overline{\varphi}(C)$ is approximated by an ellipse of bounded eccentricity; more precisely, there is some constant $\kappa > 0$, such that, for all $p \in \partial\mathfrak{H}^3$, we have

$$\limsup_{r \to 0^+} \frac{\sup_{x \in S_r(p)} d\big(\overline{\varphi}(x), \overline{\varphi}(p)\big)}{\inf_{y \in S_r(p)} d\big(\overline{\varphi}(y), \overline{\varphi}(p)\big)} < \kappa.$$

It is a fundamental, but highly nontrivial, fact that quasi-conformal maps are differentiable almost everywhere. Therefore, if C_p is a circle (centered at the origin) in the tangent plane at almost any point $p \in \partial\mathfrak{H}^3$, then $\overline{\varphi}(C_p)$ is an ellipse in the tangent plane at $\overline{\varphi}(p)$. Furthermore, since multiplying all the vectors in a tangent plane by a scalar does not change the eccentricity of any ellipse in the plane, we see that the eccentricity e_p of this ellipse is independent of the choice of the circle C_p. So e_p is a well-defined, measurable function on (almost all of) $\partial\mathfrak{H}^3$.

Case 1. Assume $e_p = 1$ for almost all $p \in \partial\mathfrak{H}^3$. This implies that the quasi-conformal map $\overline{\varphi}$ is actually conformal. So there is an isometry α of $\mathfrak{H}^3$, such that $\overline{\alpha} = \overline{\phi}$. Then, for $\gamma \in \Gamma_1$ and $p \in \partial\mathfrak{H}^3$, we have

$$\overline{\alpha}(\gamma p) = \overline{\varphi}(\gamma p) = \overline{\rho(\gamma)} \cdot \overline{\varphi}(p) = \overline{\rho(\gamma)} \cdot \overline{\alpha}(p).$$

On the other hand, since G is the identity component of $\mathrm{Isom}(\mathfrak{H}^3)$, it is normalized by α, so there is an automorphism $\hat{\alpha}$ of G, such that, for all $g \in G$ and $p \in \partial\mathfrak{H}^3$, we have

$$\overline{\alpha}(gp) = \overline{\hat{\alpha}(g)} \cdot \overline{\alpha}(p).$$

By comparing the displayed equations (and letting $g = \gamma$), we see that $\hat{\alpha}(\gamma) = \rho(\gamma)$ for all $\gamma \in \Gamma_1$. Therefore, $\hat{\alpha}$ is the desired extension of ρ to an isomorphism defined on all of G.

Case 2. Assume e_p is not almost always equal to 1. Since $\overline{\rho(\gamma)}$ is conformal (because $\rho(\gamma)$ is an isometry), we see from (15.2.2) that $e_{\gamma p} = e_p$ for all $\gamma \in \Gamma_1$ and (almost) all $p \in \partial\mathfrak{H}^3$. However, since $\overline{G}$ is transitive on $\partial\mathfrak{H}^3$ with noncompact point-stabilizers, the Moore Ergodicity Theorem (14.2.5) implies that $\overline{\Gamma_1}$ is ergodic on $\partial\mathfrak{H}^3$, so we conclude that the function e_p is constant (a.e.).

Thus, the assumption of this case implies that, for almost every p, the ellipse $\overline{\varphi}(C_p)$ is not a circle, and therefore has a well-defined major axis ℓ_p, which is a line through the origin in the tangent plane at $\overline{\varphi}(p)$. Hence, $\{\ell_p\}$ is a (measurable) section of a certain bundle $\mathbf{P}\mathfrak{H}^3$, namely, the $\mathbb{R}\mathbf{P}^1$-bundle over $\mathfrak{H}^3$ whose fiber at each point is the projectivization of the tangent space. Furthermore, since $\overline{\Gamma_2} = \overline{\rho(\Gamma_1)}$ is conformal, we see from (15.2.2) that this section is $\overline{\Gamma_2}$-invariant. In fact, if we rotate $\{\ell_p\}$

by any angle θ, then the resulting section $\{\ell_p^\theta\}$ is also invariant (since $\overline{\Gamma_2}$ is conformal). Then $\bigcup_{0 \leq \theta \leq \pi/2}\{\ell_p^\theta\}$ is a measurable, $\overline{\Gamma_2}$-invariant subset of $\mathbf{P}\mathfrak{H}^3$. However, the Moore Ergodicity Theorem (14.2.5) implies there are no such (nontrivial) subsets, since $\overline{G}$ is transitive on $\mathbf{P}\mathfrak{H}^3$ with noncompact point-stabilizers. This contradiction completes the proof, by showing that this case does not occur.   □

(15.2.3) *Remarks.*

1) The above proof makes the simplifying assumption that $n = 3$. Essentially the same proof works for larger values of n, but $\overline{\varphi}(C_p)$ will be an ellipsoid, rather than an ellipse, so the space ℓ_p of major directions may be a higher-dimensional subspace of the tangent space, instead of just being a line.

2) Mostow was able to modify the proof to deal with $SU(1, n)$, instead of $SO(1, n)$, by developing a theory of maps that are quasiconformal over $\mathbb{C}$. (In fact, replacing $\mathbb{C}$ with the quaternions and octonions yields proofs for lattices in the other simple groups of real rank one, namely, $Sp(1, n)$ and $F_{4,1}$. However, this is not necessary, because the Margulis Superrigidity Theorem applies to these groups.)

3) For lattices in $SO(1, n)$ that are **not** cocompact, it is not at all obvious that an isomorphism $\Gamma_1 \cong \Gamma_2$ should yield a quasi-isometry $\mathfrak{H}^n \to \mathfrak{H}^n$. This was proved by G. Prasad, by using the "Siegel set" description of a coarse fundamental domain for the action of Γ_i on $\mathfrak{H}^n$ (cf. Chapter 19). The same method also works for noncocompact lattices in $SU(1, n)$, or, more generally, whenever $\mathrm{rank}_\mathbb{Q}\,\Gamma_1 = 1$.

Exercises for §15.2.

#1. Show that the proof of Theorem 15.1.2 can be reduced to the special case where the lattices Γ_1 and Γ_2 are irreducible in G_1 and G_2, respectively. In other words, assume the conclusion of Theorem 15.1.2 holds whenever Γ_i is irreducible in G_i, for $i = 1, 2$, and show that this additional hypothesis can be eliminated.

#2. Construct a quasi-isometry $\varphi\colon \mathfrak{H}^3 \to \mathfrak{H}^3$ that satisfies (15.2.1).
   [*Hint:* Let $\mathcal{F}$ be a precompact strict fundamental domain for the action of Γ_1 on $\mathfrak{H}^3$, and choose some $x_0 \in \mathfrak{H}^3$. For $x \in \mathfrak{H}^3$, let $\varphi(x) = \rho(\gamma)x_0$, where $x \in \gamma \cdot \mathcal{F}$.]

§15.3. Moduli space of lattices in SL(2, ℝ)

Suppose Γ is a torsion-free, cocompact lattice in $SL(2, \mathbb{R})$. In contrast to the Mostow Rigidity Theorem (15.1.2), we will see that an isomorphism $\Gamma \cong \Gamma'$ need not extend to an automorphism of $SL(2, \mathbb{R})$. In fact, there are uncountably many different embeddings of Γ in $SL(2, \mathbb{R})$ that are not conjugate to each other.

To see this, we take a geometric approach. Since $SL(2, \mathbb{R})$ acts transitively on the hyperbolic plane $\mathfrak{H}^2$ (by isometries), the quotient $\Gamma \backslash \mathfrak{H}^2$ is a compact surface M. We will show there are uncountably many different possibilities for M (up to isometry). The proof is based on the fact that the hyperbolic plane has uncountably many different right-angled hexagons.

(15.3.1) **Lemma.** *The hyperbolic plane $\mathfrak{H}^2$ has uncountably many different right-angled hexagons (no two congruent to each other).*

Proof. Fix a basepoint $p \in \mathfrak{H}^2$, and a starting direction $\vec{v}$. For $\vec{s} \in (\mathbb{R}^+)^6$, construct the piecewise-linear path $L = L(\vec{s})$ in $\mathfrak{H}^2$ determined by:

- The path starts at the point p.
- The path consists of 6 geodesic segments (or "edges") $e_1, \ldots, e_6$, of lengths $s_1, \ldots, s_6$, respectively.
- The first edge e_1 starts at p and heads in the direction $\vec{v}$.
- For $i \geq 2$, edge e_i makes a (clockwise) right angle with e_{i-1}.

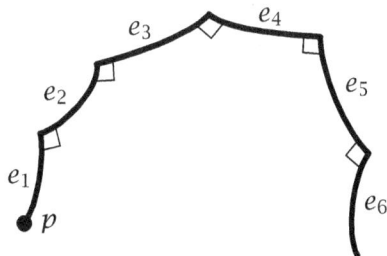

The variables $s_1, \ldots, s_6$ provide 6 degrees of freedom in the construction of L. Requiring that L be a closed path (i.e., that the terminal endpoint of L is equal to p) takes away two degrees of freedom (because $\mathfrak{H}^2$ is 2-dimensional). Then, requiring the angle between e_6 and e_1 to be right angle takes away one more degree of freedom. Hence (from the Implicit Function Theorem), we see that there are 3 degrees of freedom in the construction of a right-angled hexagon in $\mathfrak{H}^2$. □

(15.3.2) *Remark.* In fact, calculations using the trigonometry of triangles in $\mathfrak{H}^2$ yields the much more precise fact that, for any $s_2, s_4, s_6 \in \mathbb{R}^+$, there exists a unique right-angled hexagon, with edges $e_1, \ldots, e_6$, such that the length of edge e_{2i} is exactly s_{2i}, for $i = 1, 2, 3$. (That is, the lengths of the three edges e_2, e_4, and e_6 can be chosen completely arbitrarily, and they uniquely determine the lengths of the other three edges in the right-angled hexagon.)

(15.3.3) **Definition.** A ***hyperbolic surface*** is a compact Riemannian manifold (without boundary) whose universal cover is the hyperbolic plane $\mathfrak{H}^2$.

(15.3.4) **Corollary.** *There are uncountably many non-isometric hyperbolic surfaces of any given genus $g \geq 2$.*

Proof. Choose a right-angled hexagon P in $\mathfrak{H}^2$. Call its edges $e_1, \ldots, e_6$, and let s_i be the length of e_i. Make a copy P' of P, and form a surface $\mathcal{P}$ by gluing e_{2i} to the corresponding edge e'_{2i} of P', for $i = 1, 2, 3$. (Topologically, this surface $\mathcal{P}$ is a disk with two holes, and is usually called a "pair of pants.")

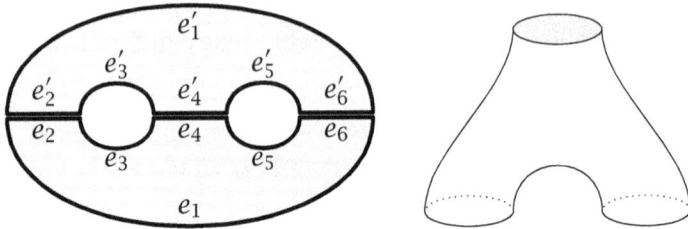

Since P is right-angled, the three boundary curves of $\mathcal{P}$ are geodesics. Their lengths are $2s_1$, $2s_3$, and $2s_5$.

Construct a closed surface M of genus 2 from two copies of $\mathcal{P}$, by gluing corresponding boundary components to each other. (For a discussion of higher genus, see Remark 15.3.5.) Since the only curves that have been glued together are geodesics, it is easy to see that each point in M has a neighborhood that is isometric to an open subset of $\mathfrak{H}^2$. Therefore, the universal cover of M is $\mathfrak{H}^2$ (since M is complete). So M is a hyperbolic surface.

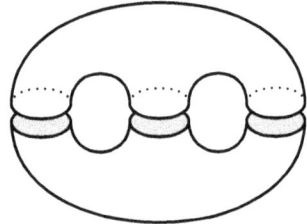

A surface of genus 2 can be made from two pairs of pants.

Furthermore, from the construction, we see that M has a closed geodesic of length $2s_1$. (In fact, there is a geodesic of length $2s_i$, for $1 \leq i \leq 6$.) Since a single closed surface has closed geodesics of only countably many different lengths, but Lemma 15.3.1 implies that there are uncountably many possible values of s_1, this implies there must be uncountably many different isometry classes of surfaces. □

(15.3.5) *Remark.* A hyperbolic surface M of any genus $g \geq 2$ can be constructed by gluing together $2g - 2$ pairs of pants:

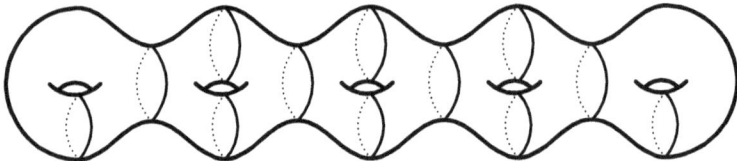

The lengths of the three boundary curves of each pair of pants can be varied independently (cf. Remark 15.3.2), except that curves that will be glued together need to have the same length. The surface can also be modified by rotating any boundary curve through an arbitrary angle θ before it is glued to its mate. This yields $6g - 6$ degrees of freedom in the construction of M. It can be shown that this is precisely the dimension of the space of hyperbolic surfaces of genus g. In other words, $6g - 6$ is the dimension of the **moduli space** of hyperbolic surfaces of genus g.

(15.3.6) **Corollary.** *If Γ is any lattice in* $\mathrm{SL}(2, \mathbb{R})$, *then there are uncountably many nonconjugate embeddings of Γ as a lattice in* $\mathrm{SL}(2, \mathbb{R})$.

Proof. Let us assume $\mathrm{SL}(2, \mathbb{R})/\Gamma$ is compact. (Otherwise, Γ is a free group, so it is easy to find embeddings.) Let us also assume, for simplicity, that Γ is torsion free, so it is the fundamental group of the hyperbolic surface $\Gamma\backslash\mathfrak{H}^2$, which has some genus g. Then Γ is isomorphic to the fundamental group Γ' of any hyperbolic surface $\Gamma'\backslash\mathfrak{H}^2$ of genus g. However, if $\Gamma\backslash\mathfrak{H}^2$ is not isometric to $\Gamma'\backslash\mathfrak{H}^2$, then Γ cannot be conjugate to Γ' in the isometry group of $\mathfrak{H}^2$. Therefore, Corollary 15.3.4 implies that there must be uncountably many different conjugacy classes of subgroups Γ' that are isomorphic to Γ. □

§15.4. Quasi-isometric rigidity

The Mostow Rigidity Theorem's weak form (15.1.1) tells us (under mild hypotheses) that lattices in two different semisimple Lie groups cannot be isomorphic. In fact, they cannot even be quasi-isometric (see Definition 10.1.3):

(15.4.1) **Theorem.** *Assume*

- *G_1 and G_2 are connected, with trivial center and no compact factors, and*
- *Γ_i is an irreducible lattice in G_i, for $i = 1, 2$.*

If $\Gamma_1 \overset{\mathrm{QI}}{\sim} \Gamma_2$, then $G_1 \cong G_2$.

Although nothing more than Theorem 15.4.1 can be said about quasi-isometric lattices that are cocompact (see Corollary 10.1.9), there is a much stronger conclusion for noncocompact lattices. Namely, not only are the Lie groups G_1 and G_2 isomorphic, but the isomorphism can be chosen to make the lattices commensurable (unless $G_1 = G_2 = \mathrm{PSL}(2, \mathbb{R})$):

(15.4.2) **Theorem.** *Assume*

- *G_1 and G_2 are connected, with trivial center and no compact factors, and*
- *Γ_i is an irreducible lattice in G_i, for $i = 1, 2$.*

Then $\Gamma_1 \overset{\mathrm{QI}}{\sim} \Gamma_2$ if and only if $G_1 \cong G_2$ and either

1) *both G_1/Γ_1 and G_2/Γ_2 are compact, or*

2) *there is an isomorphism $\phi\colon G_1 \to G_2$, such that $\phi(\Gamma_1)$ is commensurable to Γ_2, or*

3) G_1 *and* G_2 *are isomorphic to* $\mathrm{PSL}(2, \mathbb{R})$, *and neither* G_1/Γ_1 *nor* G_2/Γ_2 *is compact.*

One of the key ingredients in the proof of this theorem is the fact that any group quasi-isometric to a lattice is isomorphic to a lattice, modulo finite groups:

(15.4.3) **Theorem.** *If Λ is a finitely generated group that is quasi-isometric to an irreducible lattice Γ in G, then there are*

- *a finite-index subgroup Λ' of Λ, and*
- *a finite, normal subgroup N of Λ',*

such that Λ'/N is isomorphic to a lattice in G.

Sketch of proof of a special case. Assume $G = \mathrm{SO}(1,3)$ and Γ is cocompact. The group Λ acts on itself by translation. Since $\Lambda \overset{\mathrm{QI}}{\sim} \Gamma \overset{\mathrm{QI}}{\sim} \mathfrak{H}^3$, this provides an action of Λ by quasi-isometries on $\mathfrak{H}^3$. Let $\overline{\Lambda}$ be the corresponding group of quasiconformal maps on $\partial\mathfrak{H}^3$. Note that the quasiconformality constant κ is uniformly bounded on $\overline{\Lambda}$.

Let $(\partial\mathfrak{H}^3)^3_\circ$ be the space of ordered triples of distinct points in $\partial\mathfrak{H}^3$, and define $p\colon (\partial\mathfrak{H}^3)^3_\circ \to \mathfrak{H}^3$ by letting $p(a,b,c)$ be the point on the geodesic $\overline{ab}$ that is closest to c. It is not difficult to see that p is compact-to-one. Since the action of Λ is cocompact on $\mathfrak{H}^3$, this implies that the action of $\overline{\Lambda}$ is cocompact on $(\partial\mathfrak{H}^3)^3_\circ$.

The above information allows us to apply a theorem of Tukia to conclude that $\overline{\Lambda}$ is quasiconformally conjugate to a subgroup of $\overline{\mathrm{SO}(1,n)}$. Hence, after conjugating the action of Λ by a quasi-isometry, we may assume $\Lambda \subseteq \mathrm{SO}(1,n)$. Furthermore, if we fix a basepoint x_0, then the map $\lambda \mapsto \lambda x_0$ is a quasi-isometry from Λ to $\mathfrak{H}^3$. This implies that Λ is a cocompact lattice in $\mathrm{SO}(1,n)$. $\square$

Many additional ideas are needed to prove Theorem 15.4.2, but this suffices for the weaker version:

Proof of Theorem 15.4.1. Theorem 15.4.3 tells us that Γ_2 is isomorphic to a lattice in G_1 (if we ignore some finite groups). So Γ_2 is a lattice in both G_1 and G_2. Therefore, the Mostow Rigidity Theorem (15.1.1) implies $G_1 \cong G_2$, as desired. $\square$

From the Mostow Rigidity Theorem, we know that every automorphism of Γ extends to an automorphism of G (if G is not isogenous to $\mathrm{SL}(2, \mathbb{R})$ and we ignore compact factors and the center). The following analogue of this result for quasi-isometries is another key ingredient in the proof of Theorem 15.4.2.

(15.4.4) **Theorem.** *Assume*

- *Γ is irreducible, and **not** cocompact,*
- *G has trivial center and no compact factors,*
- *G is not isogenous to* SL(2, ℝ), *and*
- $f: \Gamma \to \Gamma$.

Then f is a quasi-isometry if and only if it is at bounded distance from some automorphism of G.

Exercises for §15.4.

#1. Assume the hypotheses of Theorem 15.1.2, and also assume G_1/Γ_1 is not compact. Show that the conclusion of Theorem 15.1.2 can be obtained by combining Theorem 15.4.2 with Theorem 15.4.4.

Notes

The Mostow Rigidity Theorem (15.1.2) is a combination of (overlapping) special cases proved by three authors:

- Mostow: Γ is cocompact (case where $G = $ SO(1, n) [4], general case [5]),
- Prasad: $\text{rank}_{\mathbb{Q}} \Gamma = 1$ (which includes the case where Γ is *not* cocompact and $\text{rank}_{\mathbb{R}} G = 1$) [6],
- Margulis: $\text{rank}_{\mathbb{R}} G \geq 2$ (and Γ is irreducible) (see Subsection 16.2(i)).

In recognition of this, many authors call Theorem 15.1.2 the Mostow-Prasad Rigidity Theorem, or the Mostow-Prasad-Margulis Rigidity Theorem.

The results described in Remark 15.1.3(3) are due to A. Furman [3].

See the exposition in [7, §5.9, pp. 106–112] for more details of the proof in Section 15.2. A different proof of this special case was found by Gromov, and is described in [7, §6.3, pp. 129–130]. Yet another nice proof (which applies to cocompact lattices in any groups of real rank one) appears in [1, §5.2].

Regarding Remark 15.2.3(2), see [5, §21, esp. (21,18)] for a discussion of the notion of maps that are quasiconformal over ℂ (or ℍ, or 𝕆).

Regarding Remark 15.2.3(3), see [6] for Prasad's proof of Mostow rigidity for lattices of ℚ-rank one.

Regarding Section 15.3, see [7, Thm. 5.3.5] for a more complete discussion of "pairs of pants" and the dimension of the space of hyperbolic metrics on a surface of genus g. (The calculations to justify Remark 15.3.2 can be found in [7, §2.6].)

The results on quasi-isometric rigidity in Section 15.4 include work of A. Eskin, B. Farb, B. Kleiner, B. Leeb, P. Pansu, R. Schwarz, and others.

See the survey [2] for references, discussion of the proofs, and other information.

References

[1] G. Besson, G. Courtois, and S. Gallot: Minimal entropy and Mostow's rigidity theorems, *Ergodic Theory Dynam. Systems* 16 (1996), no. 4, 623–649. MR 1406425, http://dx.doi.org/10.1017/S0143385700009019

[2] B. Farb: The quasi-isometry classification of lattices in semisimple Lie groups, *Math. Res. Letters* 4 (1997), no. 5, 705–717. MR 1484701, http://dx.doi.org/10.4310/MRL.1997.v4.n5.a8

[3] A. Furman: Mostow-Margulis rigidity with locally compact targets, *Geom. Funct. Anal.* 11 (2001), no. 1, 30–59. MR 1829641, http://dx.doi.org/10.1007/PL00001671

[4] G. D. Mostow: Quasi-conformal mappings in n-space and the rigidity of hyperbolic space forms, *Inst. Hautes Études Sci. Publ. Math.* 34 (1968) 53–104. MR 0236383, http://eudml.org/doc/103882

[5] G. D. Mostow: *Strong Rigidity of Locally Symmetric Spaces.* Princeton University Press, Princeton, 1973. ISBN 0-691-08136-0, MR 0385004

[6] G. Prasad: Strong rigidity of $\mathbb{Q}$-rank 1 lattices, *Invent. Math.* 21 (1973) 255–286. MR 0385005, http://eudml.org/doc/142232

[7] W. P. Thurston: *The Geometry and Topology of Three-Manifolds* (unpublished lecture notes). http://library.msri.org/books/gt3m/

Chapter 16

Margulis Superrigidity Theorem

Roughly speaking, the Margulis Superrigidity Theorem tells us that homomorphisms defined on Γ can be extended to be defined on all of G
(unless G is either $SO(1, n)$ or $SU(1, n)$). In cases where it applies, this
fundamental theorem is much stronger than the Mostow Rigidity Theorem (15.1.2). It also implies the Margulis Arithmeticity Theorem (5.2.1
or 16.3.1).

§16.1. Statement of the theorem

It is not difficult to see that every group homomorphism from $\mathbb{Z}^k$ to $\mathbb{R}^n$
can be extended to a continuous homomorphism from $\mathbb{R}^k$ to $\mathbb{R}^n$ (see Exercise 1). Noting that $\mathbb{Z}^k$ is a lattice in $\mathbb{R}^k$, it is natural to hope that,
analogously, homomorphisms defined on Γ can be extended to be de-
fined on all of G. The Margulis Superrigidity Theorem shows this is true
if G has no simple factors isomorphic to $SO(1, m)$ or $SU(1, m)$, except
that the conclusion may only be true modulo finite groups and up to a

Recall: The Standing Assumptions (4.0.0 on page 41) are in effect, so, as
always, Γ is a lattice in the semisimple Lie group $G \subseteq SL(\ell, \mathbb{R})$.

Main prerequisites for this chapter: none, except that the proof of the
main theorem requires real rank (Chapter 8), amenability (Furstenberg's
Lemma (12.6.1)), and the Moore Ergodicity Theorem (Section 11.2).

bounded error. Here is an illustrative special case that is easy to state, because the bounded error does not arise.

(16.1.1) **Theorem** (Margulis). *Assume*
- $G = \mathrm{SL}(k, \mathbb{R})$, *with* $k \geq 3$,
- G/Γ *is **not** compact, and*
- $\varphi \colon \Gamma \to \mathrm{GL}(n, \mathbb{R})$ *is any homomorphism.*

Then there exist:
- *a continuous homomorphism* $\hat{\varphi} \colon G \to \mathrm{GL}(n, \mathbb{R})$, *and*
- *a finite-index subgroup* Γ' *of* Γ,

such that $\hat{\varphi}(\gamma) = \varphi(\gamma)$ *for all* $\gamma \in \Gamma'$.

Proof. See Exercise 16.4#2.   □

Here is a much more general version of the theorem that has a slightly weaker conclusion. To simplify the statement, we preface it with a definition.

(16.1.2) **Definition.** G is *algebraically simply connected* if (for every ℓ) every Lie algebra homomorphism $\mathfrak{g} \to \mathfrak{sl}(\ell, \mathbb{R})$ is the derivative of a well-defined Lie group homomorphism $G \to \mathrm{SL}(\ell, \mathbb{R})$.

(16.1.3) *Remark.* Every simply connected Lie group is algebraically simply connected, but the converse is not true (see Exercise 2). In general, if G is connected, then some finite cover of G is algebraically simply connected. Therefore, assuming that G is algebraically simply connected is just a minor technical assumption that avoids the need to pass to a finite cover.

(16.1.4) **Theorem** (Margulis Superrigidity Theorem). *Assume*
  i) *G is connected, and algebraically simply connected,*
 ii) *G is not isogenous to any group that is of the form* $\mathrm{SO}(1, m) \times K$ *or* $\mathrm{SU}(1, m) \times K$, *where K is compact,*
iii) *Γ is irreducible, and*
 iv) *$\varphi \colon \Gamma \to \mathrm{GL}(n, \mathbb{R})$ is a homomorphism.*

Then there exist:
  1) *a continuous homomorphism* $\hat{\varphi} \colon G \to \mathrm{GL}(n, \mathbb{R})$,
  2) *a compact subgroup C of* $\mathrm{GL}(n, \mathbb{R})$ *that centralizes* $\hat{\varphi}(G)$, *and*
  3) *a finite-index subgroup* Γ' *of* Γ,

such that $\varphi(\gamma) \in \hat{\varphi}(\gamma)\, C$, *for all* $\gamma \in \Gamma'$.

Proof. See Section 16.5.   □

(16.1.5) *Remarks.*

1) Since $\varphi(\gamma) \in \hat{\varphi}(\gamma) C$, we have $\hat{\varphi}(\gamma)^{-1} \varphi(\gamma) \in C$ for all γ. Therefore, although $\hat{\varphi}(\gamma)$ might not be exactly equal to $\varphi(\gamma)$, the error is an element of C, which is a bounded set (because C is compact). Hence, the size of the error is uniformly bounded on all of Γ'.

2) Assumption (ii) cannot be removed. For example, if $G = \mathrm{PSL}(2, \mathbb{R})$, then the lattice Γ can be a free group (see Remark 6.1.6). In this case, there exist many, many homomorphisms from Γ into any group G', and many of them will not extend to G (see Exercise 5).

If we make an appropriate assumption on the range of φ, then there is no need for the compact error term C or the finite-index subgroup Γ':

(16.1.6) **Corollary.** *Assume*

   i) *G is not isogenous to any group that is of the form $\mathrm{SO}(1, m) \times K$ or $\mathrm{SU}(1, m) \times K$, where K is compact,*

  ii) *Γ is irreducible, and*

 iii) *G and G' are connected, with trivial center, and no compact factors.*

If $\varphi \colon \Gamma \to G'$ is any homomorphism, such that $\varphi(\Gamma)$ is Zariski dense in G', then φ extends to a continuous homomorphism $\hat{\varphi} \colon G \to G'$.

Proof. See Exercise 6. □

Because of our standing assumption (4.0.0) that G' is semisimple, Corollary 16.1.6 implicitly assumes that the Zariski closure $\overline{\varphi(\Gamma)} = G'$ is semisimple. In fact, that is automatically the case:

(16.1.7) **Corollary.** *Assume*

   i) *G is not isogenous to any group that is of the form $\mathrm{SO}(1, m) \times K$ or $\mathrm{SU}(1, m) \times K$, where K is compact,*

  ii) *Γ is irreducible, and*

 iii) *$\varphi \colon \Gamma \to \mathrm{GL}(n, \mathbb{R})$ is a homomorphism.*

Then $\overline{\varphi(\Gamma)}$ is semisimple.

Proof. See Exercise 9. □

Exercises for §16.1.

#1. Suppose φ is a homomorphism from $\mathbb{Z}^k$ to $\mathbb{R}^n$. Show that φ extends to a continuous homomorphism from $\mathbb{R}^k$ to $\mathbb{R}^n$.

[*Hint:* Let $\hat{\varphi} \colon \mathbb{R}^k \to \mathbb{R}^n$ be a linear transformation, such that $\hat{\varphi}(\varepsilon_i) = \varphi(\varepsilon_i)$, where $\{\varepsilon_1, \ldots, \varepsilon_k\}$ is the standard basis of $\mathbb{R}^k$.]

#2. Show that $SL(n, \mathbb{R})$ is algebraically simply connected. (On the other hand, $SL(n, \mathbb{R})$ is not simply connected, because its fundamental group is nontrivial.)

[*Hint:* By tensoring with $\mathbb{C}$, any homomorphism $\mathfrak{sl}(n, \mathbb{R}) \to \mathfrak{sl}(\ell, \mathbb{R})$ extends to a homomorphism $\mathfrak{sl}(n, \mathbb{C}) \to \mathfrak{sl}(\ell, \mathbb{C})$, and $SL(n, \mathbb{C})$ is simply connected.]

#3. Assume
  • G is not isogenous to $SO(1, n)$ or $SU(1, n)$, for any n,
  • Γ is irreducible, and
  • G has no compact factors.
Use the Margulis Superrigidity Theorem to show that the abelianization $\Gamma/[\Gamma, \Gamma]$ of Γ is finite. (When G is simple, this was already proved from Kazhdan's property (T) in Corollary 13.4.3(2). We will see yet another proof in Exercise 17.1#1.)

#4. Assume G, Γ, φ, $\hat{\varphi}$, C, and Γ' are as in Theorem 16.1.4. Show there is a homomorphism $\epsilon\colon \Gamma' \to C$, such that $\varphi(\gamma) = \hat{\varphi}(\gamma) \cdot \epsilon(\gamma)$, for all $\gamma \in \Gamma'$.

#5. Suppose $G = PSL(2, \mathbb{R})$ and Γ is a free group. Construct a homomorphism $\varphi\colon \Gamma \to GL(n, \mathbb{R})$ (for some n), such that, for every continuous homomorphism $\hat{\varphi}\colon G \to GL(n, \mathbb{R})$, and every finite-index subgroup Γ' of Γ, the set $\{\hat{\varphi}(\gamma)^{-1} \varphi(\gamma) \mid \gamma \in \Gamma'\}$ is not precompact.

[*Hint:* φ may have an infinite kernel.]

#6. Prove Corollary 16.1.6 from Theorem 16.1.4.

#7. Show that the extension $\hat{\varphi}$ in Corollary 16.1.6 is unique.

[*Hint:* Borel Density Theorem.]

#8. In each case, find
  • a lattice Γ in G and
  • a homomorphism $\varphi\colon \Gamma \to G'$,
such that
  • $\varphi(\Gamma)$ is Zariski dense in G', and
  • φ does not extend to a continuous homomorphism $\hat{\varphi}\colon G \to G'$.
Also explain why they are not counterexamples to Corollary 16.1.6.
  a) $G = G' = PSL(2, \mathbb{R}) \times PSL(2, \mathbb{R})$.
  b) $G = PSL(4, \mathbb{R})$ and $G' = SL(4, \mathbb{R})$.
  c) $G = SO(2, 3)$ and $G' = SO(2, 3) \times SO(5)$.

#9. Prove Corollary 16.1.7 from Theorem 16.1.4.

#10. Derive Theorem 16.1.4 from the combination of Corollary 16.1.6 and Corollary 16.1.7. (This is a converse to Exercises 6 and 9.)

§16.2. Applications

We briefly describe a few important consequences of the Margulis Super-rigidity Theorem.

§16.2(i). Mostow Rigidity Theorem. The special case of the Margulis Superrigidity Theorem (16.1.6) in which the homomorphism φ is assumed to be an isomorphism onto a lattice Γ' in G' is very important:

(16.2.1) Theorem (Mostow Rigidity Theorem, cf. (15.1.2)). *Assume*

- G_1 *and* G_2 *are connected, with trivial center and no compact factors,*
- $G_1 \not\approx \mathrm{PSL}(2, \mathbb{R})$,
- Γ_i *is an irreducible lattice in* G_i, *for* $i = 1, 2$, *and*
- $\varphi \colon \Gamma_1 \to \Gamma_2$ *is a group isomorphism.*

Then φ *extends to a continuous isomorphism from* G_1 *to* G_2.

This theorem has already been discussed in Chapter 15. In most cases, it follows easily from the Margulis Superrigidity Theorem (see Exercise 1). However, since the superrigidity theorem does not apply when G_1 is either $\mathrm{SO}(1, m)$ or $\mathrm{SU}(1, m)$, a different argument is needed for those cases; see Section 15.2 for a sketch of the proof.

§16.2(ii). Triviality of flat vector bundles over G/Γ.

(16.2.2) Definition. For any homomorphism $\varphi \colon \Gamma \to \mathrm{GL}(n, \mathbb{R})$, there is a diagonal action of Γ on $G \times \mathbb{R}^n$, defined by
$$(x, v) \cdot y = (xy, \varphi(y^{-1})v).$$
Let $\mathcal{E}_\varphi = (G \times \mathbb{R}^n)/\Gamma$ be the space of orbits of this action. Then there is a well-defined map
$$\pi \colon \mathcal{E}_\varphi \to G/\Gamma, \text{ defined by } \pi([x, v]) = x\Gamma,$$
and this makes $\mathcal{E}_\varphi$ into a vector bundle over G/Γ (with fiber $\mathbb{R}^n$) (see Exercise 3). A vector bundle defined from a homomorphism in this way is said to be a ***flat vector bundle***.

The Margulis Superrigidity Theorem implies (in some cases) that every flat vector bundle over G/Γ is nearly trivial. Here is an example:

(16.2.3) Proposition. *Let* $G = \mathrm{SL}(n, \mathbb{R})$ *and* $\Gamma = \mathrm{SL}(n, \mathbb{Z})$. *If* $\mathcal{E}_\varphi$ *is any flat vector bundle over* G/Γ, *then there is a finite-index subgroup* Γ' *of* Γ, *such that the lift of* $\mathcal{E}_\varphi$ *to the finite cover* G/Γ' *is trivial.*

In other words, if we let φ' *be the restriction of* φ *to* Γ', *then the vector bundle* $\mathcal{E}_{\varphi'}$ *is isomorphic to the trivial vector bundle* $(G/\Gamma') \times \mathbb{R}^n$.

Proof. From Theorem 16.1.1, we may choose Γ' so that the restriction φ' extends to a homomorphism $\hat{\varphi} \colon G \to \mathrm{GL}(n, \mathbb{R})$. Define a continuous function $T \colon G \times \mathbb{R}^n \to G \times \mathbb{R}^n$ by

$$T(g, v) = (g, \hat{\varphi}(g)v).$$

Then, for any $\gamma \in \Gamma'$, a straightforward calculation shows

$$T((g, v) \cdot \gamma) = T(g, v) * \gamma, \text{ where } (g, v) * \gamma = (g\gamma, v) \qquad (16.2.4)$$

(see Exercise 4). Therefore T factors through to a well-defined bundle isomorphism $\mathcal{E}_{\varphi'} \xrightarrow{\cong} (G/\Gamma') \times \mathbb{R}^n$. $\qquad\square$

§16.2(iii). Embeddings of locally symmetric spaces from embeddings of lattices. Let $M = \Gamma\backslash G/K$ and $M' = \Gamma'\backslash G'/K'$. Roughly speaking, the Mostow Rigidity Theorem (15.1.2) tells us that if Γ is isomorphic to Γ', then M is isometric to M'. More generally, superrigidity implies that if Γ is isomorphic to a subgroup of Γ', then M is isometric to a submanifold of M' (modulo finite covers).

(16.2.5) **Proposition.** *Suppose*

- *$M = \Gamma\backslash G/K$ and $M' = \Gamma'\backslash G'/K'$ are irreducible locally symmetric spaces with no compact factors,*
- *Γ is isomorphic to a subgroup of Γ', and*
- *the universal cover of M is neither the real hyperbolic space $\mathfrak{H}^n$ nor the complex hyperbolic space $\mathbb{C}\mathfrak{H}^n$.*

Then some finite cover of $\Gamma\backslash G/K$ embeds as a totally geodesic submanifold of a finite cover of $\Gamma'\backslash G'/K'$.

Idea of proof. There is no harm in assuming that G and G' have trivial center and no compact factors. After passing to a finite-index subgroup of Γ (and ignoring a compact group C), the Margulis Superrigidity Theorem tells us that the embedding $\Gamma \hookrightarrow \Gamma'$ extends to a continuous embedding $\varphi \colon G \hookrightarrow G'$. Conjugate φ by an element of G', so that $\varphi(K) \subseteq K'$, and $\varphi(G)$ is invariant under the Cartan involution of G' corresponding to the maximal compact subgroup K'. Then φ induces an embedding $\Gamma\backslash G/K \to \Gamma'\backslash G'/K'$ whose image is a totally geodesic submanifold. $\qquad\square$

Exercises for §16.2.

#1. Prove the Mostow Rigidity Theorem (16.2.1) under the additional assumption that G_1 is neither $\mathrm{PSO}(1, n)$ nor $\mathrm{PSU}(1, n)$.

#2. The statement of the Mostow Rigidity Theorem in Theorem 15.1.2 is slightly different from Theorem 16.2.1. Show that these two theorems are corollaries of each other.
[*Hint:* Exercise 15.2#1.]

#3. In the notation of Definition 16.2.2:
   a) Show the map π is well defined.
   b) Show $\mathcal{E}_\varphi$ is a vector bundle over G/Γ with fiber $\mathbb{R}^n$.

#4. Verify (16.2.4) for all $\gamma \in \Gamma'$.

§16.3. Why superrigidity implies arithmeticity

Recall the following major theorem that was stated without proof in Theorem 5.2.1:

(16.3.1) Margulis Arithmeticity Theorem. *Every irreducible lattice in G is arithmetic, except, perhaps, when G is isogenous to* $\mathrm{SO}(1, m) \times K$ *or* $\mathrm{SU}(1, m) \times K$, *for some compact group K.*

This important fact is a consequence of the Margulis Superrigidity Theorem, but the implication is not at all obvious. In this section, we will explain the main ideas that are involved.

In addition to our usual assumption that $G \subseteq \mathrm{SL}(\ell, \mathbb{R})$, let us also assume, for simplicity:

• $G \cong \mathrm{SL}(3, \mathbb{R})$ (or, more generally, G is algebraically simply connected; see Definition 16.1.2), and

• G/Γ is not compact.

We wish to show that Γ is arithmetic. It suffices to show $\Gamma \subseteq G_\mathbb{Z}$, that is, that every matrix entry of every element of Γ is an integer, for then Γ is commensurable to $G_\mathbb{Z}$ (see Exercise 4.1#10).

Here is a loose description of the 4 steps of the proof:

1) The Margulis Superrigidity Theorem (16.1.1) implies that every matrix entry of every element of Γ is an algebraic number.

2) By restriction of scalars, we may assume that these algebraic numbers are rational; that is, $\Gamma \subseteq G_\mathbb{Q}$.

3) For every prime p, a "p-adic" version of the Margulis Superrigidity Theorem provides a natural number N_p, such that no element of Γ has a matrix entry whose denominator is divisible by p^{N_p}.

4) This implies that some finite-index subgroup Γ' of Γ is contained in $G_\mathbb{Z}$.

Step 1. Every matrix entry of every element of Γ is an algebraic number.
Suppose some $\gamma_{i,j}$ is transcendental. Then, for any transcendental number α, there is a field automorphism ϕ of $\mathbb{C}$ with $\phi(\gamma_{i,j}) = \alpha$. Applying ϕ to all the entries of a matrix induces an automorphism $\tilde{\phi}$ of $\mathrm{SL}(\ell, \mathbb{C})$. Let

$$\varphi \text{ be the restriction of } \tilde{\phi} \text{ to } \Gamma,$$

so φ is a homomorphism from Γ to $\mathrm{SL}(\ell, \mathbb{C})$. The Margulis Superrigidity Theorem implies there is a continuous homomorphism $\hat{\varphi} \colon G \to \mathrm{SL}(\ell, \mathbb{C})$, such that $\hat{\varphi} = \varphi$ on a finite-index subgroup of Γ (see Exercise 16.4#2). By passing to this finite-index subgroup, we may assume $\hat{\varphi} = \varphi$ on all of Γ.

Since there are uncountably many transcendental numbers α, there are uncountably many different choices of ϕ, so there must be uncountably many different n-dimensional representations $\hat{\varphi}$ of G. However, it is well known from the the theory of "roots and weights" that G (or, more generally, any connected, simple Lie group) has only finitely many non-isomorphic representations of any given dimension, so this is a contradiction.[3]

Step 2. We have $\Gamma \subseteq \mathrm{SL}(\ell, \mathbb{Q})$. Let F be the subfield of $\mathbb{C}$ generated by the matrix entries of the elements of Γ, so $\Gamma \subseteq \mathrm{SL}(\ell, F)$. From Step 1, we know that this is an algebraic extension of $\mathbb{Q}$. Furthermore, because Γ is finitely generated (see Theorem 4.7.10), we see that this field extension is finitely generated. Therefore, F is finite-degree field extension of $\mathbb{Q}$ (in other words, F is an algebraic number field). This means that F is almost the same as $\mathbb{Q}$, so it is only a slight exaggeration to say that we have proved $\Gamma \subseteq \mathrm{SL}(\ell, \mathbb{Q})$.

Indeed, restriction of scalars (5.5.8) provides a way to change F into $\mathbb{Q}$: there is a representation $\rho \colon G \to \mathrm{SL}(r, \mathbb{C})$, for some r, such that $\rho(G \cap \mathrm{SL}(\ell, F)) \subseteq \mathrm{SL}(r, \mathbb{Q})$ (see Exercise 1). Therefore, after replacing G with $\rho(G)$, we have the desired conclusion (without any exaggeration).

Step 3. For every prime p, there is a natural number N_p, such that no element of Γ has a matrix entry whose denominator is divisible by p^{N_p}. The fields $\mathbb{R}$ and $\mathbb{C}$ are complete (that is, every Cauchy sequence converges), and they obviously contain $\mathbb{Q}$. For any prime p, the p-adic numbers $\mathbb{Q}_p$ are another field that has these same properties.

As we have stated it, the Margulis Superrigidity Theorem deals with homomorphisms into $\mathrm{SL}(\ell, \mathbb{F})$, where $\mathbb{F} = \mathbb{R}$, but Margulis also proved a version of the theorem that applies when $\mathbb{F}$ is a p-adic field (see Theorem 16.3.2). Now G is connected, but p-adic fields are totally disconnected, so every continuous homomorphism from G to $\mathrm{SL}(\ell, \mathbb{Q}_p)$ is trivial. Therefore, superrigidity tells us that φ is trivial, after we mod out a compact group (cf. Theorem 16.1.4). In other words, the closure of $\varphi(\Gamma)$ is compact in $\mathrm{SL}(\ell, \mathbb{Q}_p)$.

This conclusion can be rephrased in more elementary terms, without any mention of p-adic numbers. Namely, it says that there is a bound on

---

[3] Actually, this is not quite a contradiction, because it is possible that two different choices of φ yield the same representation of Γ, up to isomorphism; that is, after a change of basis. The trace of a matrix is independent of the basis, so the preceding argument really shows that the trace of $\varphi(y)$ must be algebraic, for every $y \in \Gamma$. Then one can use some algebraic methods to construct some other matrix representation φ' of Γ, such that the matrix entries of $\varphi'(y)$ are algebraic, for every $y \in \Gamma$.

the highest power of p that divides the denominator of any matrix entry of any element of Γ. This is what we wanted.

Step 4. Some finite-index subgroup Γ' of Γ is contained in $\mathrm{SL}(\ell, \mathbb{Z})$. Let $D \subseteq \mathbb{N}$ be the set consisting of the denominators of the matrix entries of the elements of $\varphi(\Gamma)$.

We claim there exists $N \in \mathbb{N}$, such that every element of D is less than N. Since Γ is known to be finitely generated, some finite set of primes $\{p_1, \ldots, p_r\}$ contains all the prime factors of every element of D. (If p is in the denominator of some matrix entry of $y_1 y_2$, then it must appear in a denominator somewhere in either y_1 or y_2.) Therefore, every element of D is of the form $p_1^{m_1} \cdots p_r^{m_r}$, for some $m_1, \ldots, m_r \in \mathbb{N}$. From Step 3, we know $m_i < N_{p_i}$, for every i. Thus, every element of D is less than $p_1^{N_{p_1}} \cdots p_r^{N_{p_r}}$. This establishes the claim.

From the preceding paragraph, we see that $\Gamma \subseteq \frac{1}{N!} \mathrm{Mat}_{\ell \times \ell}(\mathbb{Z})$. Note that if $N = 1$, then $\Gamma \subseteq \mathrm{SL}(\ell, \mathbb{Z})$. In general, N is a finite distance from 1, so it should not be hard to believe (and it can indeed be shown) that some finite-index subgroup of Γ must be contained in $\mathrm{SL}(\ell, \mathbb{Z})$ (see Exercise 2). Therefore, a finite-index subgroup of Γ is contained in $G_{\mathbb{Z}}$, as desired. $\square$

For ease of reference, we officially record the key fact used in Step 3:

(16.3.2) **Theorem** (Margulis superrigidity over p-adic fields). *Assume*

i) *G is not isogenous to any group that is of the form* $\mathrm{SO}(1, m) \times K$ *or* $\mathrm{SU}(1, m) \times K$, *where K is compact,*

ii) *Γ is irreducible,*

iii) *$\mathbb{Q}_p$ is the field of p-adic numbers, for some prime p, and*

iv) *$\varphi \colon \Gamma \to \mathrm{GL}(n, \mathbb{Q}_p)$ is a homomorphism.*

Then $\overline{\varphi(\Gamma)}$ is compact.

In other words, there is some $N \in \mathbb{Z}$, such that every matrix entry of every element of $\varphi(\Gamma)$ is in $p^N \mathbb{Z}_p$, where $\mathbb{Z}_p$ is the ring of p-adic integers.

The Margulis Arithmeticity Theorem (16.3.1) does not apply to lattices in $\mathrm{SO}(1, n)$ or $\mathrm{SU}(1, n)$, but, for those groups, Margulis proved the following characterization of the lattices that are arithmetic:

(16.3.3) **Commensurability Criterion for Arithmeticity** (Margulis). *Assume*

- *G is connected, with no compact factors, and*

- *Γ is irreducible.*

Then Γ is arithmetic if and only if the commensurator $\mathrm{Comm}_G(\Gamma)$ of Γ is dense in G.

As was already mentioned in Remark 5.2.5(1), the direction ($\Rightarrow$) follows from the simple observation that $\mathrm{Comm}_G(G_{\mathbb{Z}})$ contains $G_{\mathbb{Q}}$.

The proof of ($\Leftarrow$) is more difficult. It is the same as the proof of the Margulis Arithmeticity Theorem, but replacing the Margulis Superrigidity Theorem (16.1.6) with the following superrigidity theorem (and also replacing the p-adic superrigidity theorem with a suitable commensurator analogue):

(16.3.4) **Theorem** (Commensurator Superrigidity). *Assume*

  i) *Γ is irreducible,*

 ii) *$\mathrm{Comm}_G(\Gamma)$ is dense in G, and*

 iii) *G and G' are connected, with trivial center, and no compact factors.*

If $\varphi\colon \mathrm{Comm}_G(\Gamma) \to G'$ is any homomorphism whose image is Zariski dense in G', then φ extends to a continuous homomorphism $\hat{\varphi}\colon G \to G'$.

Exercises for §16.3.

#1. Suppose
  - $G \subseteq \mathrm{SL}(\ell, \mathbb{C})$,
  - $\Gamma \subseteq \mathrm{SL}(\ell, F)$, for some algebraic number field F, and
  - G has no compact factors.

Show there is a continuous homomorphism $\rho\colon G \to \mathrm{SL}(r, \mathbb{C})$, for some r, such that $\rho(G \cap \mathrm{SL}(\ell, F)) \subseteq \mathrm{SL}(r, \mathbb{Q})$.
[*Hint:* Apply restriction of scalars (§5.5) after noting that the Borel Density Theorem (§4.5) implies G is defined over F.]

#2. Show that if Λ is a subgroup of $\mathrm{SL}(\ell, \mathbb{Q})$, and $\Lambda \subseteq \frac{1}{N}\mathrm{Mat}_{\ell \times \ell}(\mathbb{Z})$, for some $N \in \mathbb{N}$, then $\mathrm{SL}(\ell, \mathbb{Z})$ contains a finite-index subgroup of Λ.
[*Hint:* The additive group of $\mathbb{Q}^\ell$ contains a Λ-invariant subgroup V, such that we have $\mathbb{Z}^\ell \subseteq V \subseteq \frac{1}{N}\mathbb{Z}^\ell$. Choose $g \in \mathrm{GL}(\ell, \mathbb{Q})$, such that $g(V) = \mathbb{Z}^\ell$. Then g commensurates $\mathrm{SL}(\ell, \mathbb{Z})$ and we have $g\Lambda g^{-1} \subseteq \mathrm{SL}(\ell, \mathbb{Z})$.]

#3. Assume, as usual, that
  - G is not isogenous to any group that is of the form $\mathrm{SO}(1,m) \times K$ or $\mathrm{SU}(1,m) \times K$, where K is compact, and
  - Γ is irreducible.

Use the proof of the Margulis Arithmeticity Theorem to show that if $\varphi\colon \Gamma \to \mathrm{SL}(n, \mathbb{C})$ is any homomorphism, then every eigenvalue of every element of $\varphi(\Gamma)$ is an algebraic integer.

§16.4. Homomorphisms into compact groups

The Margulis Superrigidity Theorem (16.1.4) does not say anything about homomorphisms whose image is contained in a compact subgroup of $\mathrm{GL}(n, \mathbb{R})$. (This is because all of $\varphi(\Gamma)$ can be put into the error term C, so the homomorphism $\hat{\alpha}$ can be taken to be trivial.) Fortunately, there is a different version that completely eliminates the error term (and applies very generally). Namely, from the Margulis Arithmeticity Theorem (5.2.1), we know that the lattice Γ must be arithmetic (if no simple factors of G are $\mathrm{SO}(1, m)$ or $\mathrm{SU}(1, m)$). This means that if we add some compact factors to G, then we can assume that Γ is commensurable to $G_{\mathbb{Z}}$. In this situation, there is no need for the error term C:

(16.4.1) Corollary. *Assume*
- *G is connected, algebraically simply connected, and defined over $\mathbb{Q}$,*
- *G is not isogenous to any group that is of the form $\mathrm{SO}(1, m) \times K$ or $\mathrm{SU}(1, m) \times K$, where K is compact,*
- *Γ is irreducible, and*
- *$\varphi \colon \Gamma \to \mathrm{GL}(n, \mathbb{R})$ is a homomorphism.*

If Γ is commensurable to $G_{\mathbb{Z}}$, then there exist:
1) *a continuous homomorphism $\hat{\varphi} \colon G \to \mathrm{GL}(n, \mathbb{R})$, and*
2) *a finite-index subgroup Γ' of Γ,*

such that $\varphi(\gamma) = \hat{\varphi}(\gamma)$, for all $\gamma \in \Gamma'$.

Here is a less precise version of Corollary 16.4.1 that may be easier to apply in situations where the lattice Γ is not explicitly given as the $\mathbb{Z}$-points of G. However, it only applies to the homomorphism into each simple component of $\overline{\alpha(\Gamma)}$, not to the entire homomorphism all at once.

(16.4.2) Corollary. *Assume*
- *G is connected, and algebraically simply connected,*
- *G is not isogenous to any group that is of the form $\mathrm{SO}(1, m) \times K$ or $\mathrm{SU}(1, m) \times K$, where K is compact,*
- *Γ is irreducible,*
- *$\varphi \colon \Gamma \to \mathrm{GL}(n, \mathbb{C})$ is a homomorphism, and*
- *$\overline{\varphi(\Gamma)}$ is simple.*

Then there exist:
1) *a continuous homomorphism $\hat{\varphi} \colon G \to \mathrm{GL}(n, \mathbb{C})$,*
2) *a finite-index subgroup Γ' of Γ, and*
3) *a Galois automorphism σ of $\mathbb{C}$,*

such that $\varphi(\gamma) = \sigma(\hat{\varphi}(\gamma))$, for all $\gamma \in \Gamma'$.

Proof. We may assume $\overline{\varphi(\Gamma)}$ is compact, for otherwise Corollary 16.1.6 applies (after modding out the centers of G and $\overline{\alpha(\Gamma)}$). Then every element of $\overline{\varphi(\Gamma)}$ is semisimple.

Choose some $h \in \varphi(\Gamma)$, such that h has infinite order (see Exercise 4.8#12). Then the conclusion of the preceding paragraph implies that some eigenvalue λ of h is not a root of unity. On the other hand, if λ is algebraic, then p-adic superrigidity (16.3.2) implies that λ is an algebraic integer (see Exercise 16.3#3). So there is a Galois automorphism σ

of $\mathbb{C}$, such that $|\sigma(\lambda)| \neq 1$ (see Exercise 4). Then $\{ \sigma(\lambda)^k \mid k \in \mathbb{Z} \}$ is an unbounded subset of $\mathbb{C}$, so $\langle \sigma(h) \rangle$ is not contained in any compact subgroup of $\mathrm{GL}(n, \mathbb{C})$.

Now, let

- φ' be the composition $\sigma \circ \varphi$, and
- G' be the Zariski closure of $\varphi'(\Gamma)$.

Then G' is simple, and the conclusion of the preceding paragraph implies that G' is not compact (since $\sigma(h) \in G'$). After passing to a finite-index subgroup (so G' is connected), Corollary 16.1.6 provides a continuous homomorphism $\tilde{\varphi}\colon G \to G'$, such that $\varphi'(\gamma) = \hat{\varphi}(\gamma)$, for all γ in some finite-index subgroup of Γ.　　　　□

(16.4.3) **Warning.** Assume Γ is irreducible, and G is not isogenous to any group of the form $\mathrm{SO}(1, m) \times K$ or $\mathrm{SU}(1, m) \times K$. Corollary 16.4.2 implies that if there exists a homomorphism φ from Γ to a compact Lie group (and $\varphi(\Gamma)$ is infinite), then G/Γ must be compact (see Exercise 1). However, **the converse is not true.** Namely, Corollary 16.4.1 tells us that if Γ is commensurable to $G_{\mathbb{Z}}$, where G is defined over $\mathbb{Q}$, and $G_{\mathbb{R}}$ has no compact factors, then Γ does not have any homomorphisms to compact groups (with infinite image). It does not matter whether G/Γ is compact or not.

Exercises for §16.4.

#1. Assume, as usual, that the lattice Γ is irreducible, that G is not isogenous to any group of the form $\mathrm{SO}(1, m) \times K$ or $\mathrm{SU}(1, m) \times K$, and that $\varphi\colon \Gamma \to \mathrm{GL}(n, \mathbb{R})$ is a homomorphism. If G/Γ is **not** compact, show the semisimple group $\overline{\varphi(\Gamma)}$ has no compact factors.
[*Hint:* Godement's Criterion (5.3.1).]

#2. Assume
- G is algebraically simply connected,
- G is not isogenous to any group that is of the form $\mathrm{SO}(1, m) \times K$ or $\mathrm{SU}(1, m) \times K$, where K is compact,
- Γ is irreducible,
- G/Γ is not compact, and
- $\varphi\colon \Gamma \to \mathrm{SL}(n, \mathbb{R})$ is a homomorphism.

Show there is a continuous homomorphism $\hat{\varphi}\colon G \to \mathrm{SL}(n, \mathbb{R})$, such that $\varphi(\gamma) = \hat{\varphi}(\gamma)$ for all γ in some finite-index subgroup of Γ.
[*Hint:* Theorem 16.1.4, Corollary 16.1.7, and Exercise 1.]

#3. Assume Γ is irreducible, and G has no factors isogenous to $\mathrm{SO}(1, m)$ or $\mathrm{SU}(1, m)$. Show that if N is an infinite normal subgroup of Γ, such that Γ/N is linear (i.e., isomorphic to a subgroup of $\mathrm{GL}(\ell, \mathbb{C})$, for some ℓ), then Γ/N is finite.

#4. (*Kronecker's Theorem*) Assume λ is an algebraic integer. Show that if $|\sigma(\lambda)| = 1$ for every Galois automorphism σ of $\mathbb{C}$, then λ is a root of unity.

[*Hint:* The powers of λ form a set that (by restriction of scalars) is discrete in $\times_{\sigma \in S^\infty} F_\sigma^\times$. Alternate proof: there are only finitely many polynomials of degree n with integer coefficients that are all $\leq C$ in absolute value.]

§16.5. Proof of the Margulis Superrigidity Theorem

In order to establish Corollary 16.1.6, it suffices to prove the following special case (see Exercise 1):

(16.5.1) Theorem. *Suppose*

- *G is connected, and it is not isogenous to any group that is of the form $\mathrm{SO}(1, m) \times K$ or $\mathrm{SU}(1, m) \times K$, where K is compact,*
- *the lattice Γ is irreducible in G,*
- *H is a connected, noncompact, simple subgroup of $\mathrm{SL}(n, \mathbb{R})$, for some n (and H has trivial center),*
- *$\varphi : \Gamma \to H$ is a homomorphism, and*
- *$\varphi(\Gamma)$ is Zariski dense in H.*

Then φ extends to a continuous homomorphism $\hat{\varphi} : G \to H$.

Although it does result in some loss of generality, we assume:

(16.5.2) Assumption. $\mathrm{rank}_\mathbb{R} G \geq 2$.

The case where $\mathrm{rank}_\mathbb{R} G = 1$ requires quite different methods — see Section 16.8 for a very brief discussion.

§16.5(i). Geometric reformulation.

To set up the proof of Theorem 16.5.1, let us translate the problem into a geometric setting, by replacing the homomorphism φ with the corresponding flat vector bundle $\mathcal{E}_\varphi$ over G/Γ (see Definition 16.2.2).

(16.5.3) Remark. The sections of the vector bundle $\mathcal{E}_\varphi$ are in natural one-to-one correspondence with the right Γ-equivariant maps from G to $\mathbb{R}^n$ (see Exercise 2).

(16.5.4) Lemma. *φ extends to a homomorphism $\tilde{\varphi} : G \to \mathrm{GL}(n, \mathbb{R})$ if and only if there exists a G-invariant subspace $V \subseteq \mathrm{Sect}(\mathcal{E}_\varphi)$, such that the evaluation map $V \to V_{[e]}$ is bijective.*

Proof. ($\Leftarrow$) Since V is G-invariant, we have a representation of G on V; let us say $\pi : G \to \mathrm{GL}(V)$. Therefore, the isomorphism $V \to V_{[e]} = \mathbb{R}^n$ yields a representation $\hat{\pi}$ of G on $\mathbb{R}^n$. It is not difficult to verify that $\hat{\pi}$ extends φ (see Exercise 3).

($\Rightarrow$) For $v \in \mathbb{R}^n$ and $g \in G$, let

$$\xi_v(g) = \tilde{\varphi}(g^{-1})v.$$

It is easy to verify that $\xi_v \colon G \to \mathbb{R}^n$ is right Γ-equivariant (see Exercise 4), so we may think of ξ_v as a section of $\mathcal{E}_\varphi$ (see Remark 16.5.3). Let

$$V = \{\, \xi_v \mid v \in \mathbb{R}^n \,\} \subseteq \mathrm{Sect}(\mathcal{E}_\varphi).$$

Now the map $v \mapsto \xi_v$ is linear and G-equivariant (see Exercise 5), so V is a G-invariant subspace of $\mathrm{Sect}(\mathcal{E}_\varphi)$. Since

$$\xi_v([e]) = \tilde{\varphi}(e)v = v,$$

it is obvious that the evaluation map is bijective. $\square$

In fact, if we assume the representation φ is irreducible, then it is not necessary to have the evaluation map $V \to V_{[e]}$ be bijective. Namely, in order to show that φ extends, it suffices to have V be finite dimensional (and nonzero):

(16.5.5) **Lemma** (see Exercise 6). *Assume that the representation φ is irreducible. If there exists a (nontrivial) G-invariant subspace V of $\mathrm{Sect}(\mathcal{E}_\varphi)$ that is finite dimensional, then φ extends to a continuous homomorphism $\tilde{\varphi} \colon G \to \mathrm{GL}(n, \mathbb{R})$.*

§16.5(ii). The need for higher real rank. We now explain how Assumption 16.5.2 comes into play.

(16.5.6) **Notation.** Let A be a maximal $\mathbb{R}$-split torus of G. For example, if $G = \mathrm{SL}(3, \mathbb{R})$, we let

$$A = \begin{bmatrix} * & 0 & 0 \\ 0 & * & 0 \\ 0 & 0 & * \end{bmatrix}.$$

By definition, the assumption that $\mathrm{rank}_\mathbb{R} G \geq 2$ means $\dim A \geq 2$.

It is the following result that relies on our assumption $\mathrm{rank}_\mathbb{R} G \geq 2$. It is easy to prove if G has more than one noncompact simple factor (see Exercise 7), and is not difficult to verify for the case $G = \mathrm{SL}(\ell, \mathbb{R})$ (cf. Exercise 8). Readers familiar with the structure of semisimple groups (including the theory of real roots) should have little difficulty in generalizing to any semisimple group of real rank ≥ 2 (see Exercise 9).

(16.5.7) **Lemma.** *If $\mathrm{rank}_\mathbb{R} G \geq 2$, then, for some $r \in \mathbb{N}$, there exist closed subgroups $L_1, L_2, \ldots, L_r$ of G, such that*

1) $G = L_r L_{r-1} \cdots L_1$, *and*
2) *both H_i and $H_i^\perp$ are noncompact, where*
   - $H_i = L_i \cap A$, *and*
   - $H_i^\perp = C_A(L_i)$ *(so L_i centralizes $H_i^\perp$).*

§16.5(iii). **Outline of the proof.** The idea for proving Theorem 16.5.1 is quite simple. We begin by finding a (nonzero) A-invariant section of $\mathcal{E}_\varphi$; this section spans a (1-dimensional) subspace V_0 of $\mathcal{E}_\varphi$ that is invariant under A. Since (by definition) the subgroup H_1 of Lemma 16.5.7 is contained in A, we know that V_0 is invariant under H_1, so Lemma 16.5.8 below provides a subspace of $\mathrm{Sect}(\mathcal{E}_\varphi)$ that is invariant under a larger subgroup of G, but is still finite dimensional. Applying the lemma repeatedly yields finite-dimensional subspaces that are invariant under more and more of G. Eventually, the lemma yields a finite-dimensional subspace that is invariant under all of G. Then Lemma 16.5.5 implies that φ extends to a homomorphism that is defined on G, as desired.

(16.5.8) **Lemma.** *If*

- *H is a closed, noncompact subgroup of A, and*

- *V is an H-invariant subspace of $\mathrm{Sect}(\mathcal{E}_\varphi)$ that is finite dimensional,*

then $\langle C_G(H) \cdot V \rangle$ is finite dimensional.

Idea of proof. To illustrate the idea of the proof, let us assume that $V = \mathbb{R}\sigma$ is the span of an H-invariant section (see Exercise 13). Since H is noncompact, the Moore Ergodicity Theorem (14.2.4) tells us that H has a dense orbit on G/Γ (see Exercise 14.2#16). (In fact, almost every orbit is dense.) This implies that any continuous H-invariant section of $\mathcal{E}_\varphi$ is determined by its value at a single point (see Exercise 10), so the space of H-invariant sections is finite-dimensional (see Exercise 11). Since this space contains $\langle C_G(H) \cdot V \rangle$ (see Exercise 12), the desired conclusion is immediate.   □

Here is a more detailed outline:

Idea of the proof of Theorem 16.5.1. Assume there exists a nonzero A-invariant section σ of $\mathcal{E}_\varphi$. Let

$$H_0 = A \text{ and } V_0 = \langle \sigma \rangle.$$

Thus, V_0 is a 1-dimensional subspace of $\mathrm{Sect}(\mathcal{E}_\varphi)$ that is H_0-invariant.
Now, for $i = 1, \ldots, r$, let

$$V_i = \langle L_i \cdot A \cdot L_{i-1} \cdot A \cdots L_1 \cdot A \cdot V_0 \rangle.$$

Since $L_r L_{r-1} \cdots L_1 = G$, it is clear that V_r is G-invariant. Therefore, it will suffice to show (by induction on i) that each V_i is finite dimensional.
Since $H_{i-1} \subseteq L_{i-1}$, it is clear that V_{i-1} is H_{i-1}-invariant. Therefore, since A centralizes H_{i-1}, Lemma 16.5.8 implies that $\langle A \cdot V_{i-1} \rangle$ is finite dimensional. Now, since $H_i^\perp \subseteq A$, we know that $\langle A \cdot V_{i-1} \rangle$ is $H_i^\perp$-invariant. Then, since L_i centralizes $H_i^\perp$, Lemma 16.5.8 implies that the subspace $V_i = \langle L_i \cdot A \cdot V_{i-1} \rangle$ is finite dimensional.   □

Therefore, the key to proving Theorem 16.5.1 is finding a nonzero A-invariant section σ of $\mathcal{E}_\varphi$. Unfortunately, the situation is a bit more complicated than the above would indicate, because we will not find a *continuous* A-invariant section, but only a *measurable* one (see Key Fact 16.6.1). Then the proof appeals to Lemma 16.5.10 below, instead of Lemma 16.5.8. We leave the details to the reader (see Exercise 16).

(16.5.9) **Definition.** Let $\mathrm{Sect}_{\mathrm{meas}}(\mathcal{E}_\varphi)$ be the vector space of measurable sections of $\mathcal{E}_\varphi$, where two sections are identified if they agree almost everywhere.

(16.5.10) **Lemma** (see Exercises 14 and 15). *If*
- *H is a closed, noncompact subgroup of A, and*
- *V is a finite-dimensional, H-invariant subspace of $\mathrm{Sect}_{\mathrm{meas}}(\mathcal{E}_\varphi)$,*

then $\langle C_G(H) \cdot V \rangle$ is finite dimensional.

Exercises for §16.5.

#1. Derive Corollary 16.1.6 as a corollary of Theorem 16.5.1.

#2. Suppose $\xi \colon G \to \mathbb{R}^n$. Show that $\overline{\xi} \colon G/\Gamma \to \mathcal{E}_\varphi$, defined by
$$\overline{\xi}(g\Gamma) = [(g, \xi(g))],$$
is a well-defined section of $\mathcal{E}_\varphi$ if and only if ξ is right Γ-equivariant; i.e., $\xi(g\gamma) = \varphi(\gamma^{-1})\,\xi(g)$.

#3. In the notation of the proof of Lemma 16.5.4($\Leftarrow$), show $\widehat{\pi}(\gamma) = \varphi(\gamma)$ for every $\gamma \in \Gamma$.

#4. In the notation of the proof of Lemma 16.5.4($\Rightarrow$), show that we have $\xi_v(gh) = \widetilde{\varphi}(h^{-1})\,\xi_v(g)$. Since $\widetilde{\varphi}(\gamma^{-1}) = \varphi(\gamma^{-1})$ for all $\gamma \in \Gamma$, this implies that ξ_v is right Γ-equivariant.

#5. In the notation of the proof of Lemma 16.5.4($\Rightarrow$), show that we have $\xi_{\widetilde{\varphi}(g)v} = g \cdot \xi_v$, where the action of G on $\mathrm{Sect}(\mathcal{E}_\varphi)$ is defined by $(g \cdot \xi_v)(x) = \xi_v(g^{-1}x)$, as usual.

#6. Prove Lemma 16.5.5.
   [*Hint:* By choosing V of minimal dimension, we may assume it is an irreducible G-module, so the evaluation map is either 0 or injective. It cannot be 0, and then it must also be surjective, since φ is irreducible.]

#7. Prove Lemma 16.5.7 under that additional assumption that we have $G = G_1 \times G_2$, where G_1 and G_2 are noncompact (and semisimple).
   [*Hint:* Let $L_i = G_i$ for $i = 1, 2$.]

#8. Prove the conclusion of Lemma 16.5.7 for $G = \mathrm{SL}(3, \mathbb{R})$.
   [*Hint:* A **unipotent elementary matrix** is a matrix with 1's on the diagonal and only one nonzero off-diagonal entry. Every element of $\mathrm{SL}(3, \mathbb{R})$ is a product of ≤ 10 unipotent elementary matrices, and any such matrix is contained in a subgroup isogenous to $\mathrm{SL}(2, \mathbb{R})$ that has a 1-dimensional intersection with A.]

#9. Prove Lemma 16.5.7.

#10. Let H be a subgroup of G. Show that if σ_1 and σ_2 are H-invariant, continuous sections of $\mathcal{E}_\varphi$, and there is some $x \in G/\Gamma$, such that
   - Hx is dense in G/Γ and
   - $\sigma_1(x) = \sigma_2(x)$,

   then $\sigma_1 = \sigma_2$.

#11. Let H be a subgroup of G, and assume H has a dense orbit in G/Γ. Show the space of H-invariant, continuous sections of $\mathcal{E}_\varphi$ has finite dimension.

#12. Let H be a subgroup of G. Show that if σ is an H-invariant section of $\mathcal{E}_\varphi$, and c is an element of G that centralizes H, then $\sigma \cdot c$ is also H-invariant.

#13. Prove Lemma 16.5.8 without assuming that V is 1-dimensional.
   [*Hint:* Fix $x \in G$. For $c \in C_G(H)$ and $\sigma \in V$, define $T: V \to \mathbb{R}^n$ by $T(\xi) = \xi(x)$, and note that $(c\sigma)(hx) = T(h^{-1} \cdot \sigma)$ for all $h \in H$. If $Hx\Gamma$ is dense, this implies that $c\sigma$ is determined by σ and T. So $\dim(C_G(H) \cdot V) \leq (\dim V) \cdot (\dim \operatorname{Hom}(V, \mathbb{R}^n))$.]

#14. Prove Lemma 16.5.10 in the special case where $V = \mathbb{R}\sigma$ is the span of an H-invariant measurable section.
   [*Hint:* This is similar to Lemma 16.5.8, but use the fact that H is ergodic on G/Γ.]

#15. Prove Lemma 16.5.10 (without assuming $\dim V = 1$).
   [*Hint:* This is similar to Exercise 13.]

#16. Prove Theorem 16.5.1.

§16.6. An *A*-invariant section

This section sketches the proof of the following result, which completes the proof of Theorem 16.5.1 (under the assumption that $\operatorname{rank}_\mathbb{R} G \geq 2$).

(16.6.1) **Key Fact.** *For some n, there is an embedding of H in $\mathrm{SL}(n, \mathbb{R})$, such that*

1) *the associated representation $\varphi: \Gamma \to H \subseteq \mathrm{SL}(n, \mathbb{R})$ is irreducible, and*

2) *there exists a nonzero A-invariant $\sigma \in \mathrm{Sect}_{\mathrm{meas}}(\mathcal{E}_\varphi)$.*

Remark 16.5.3 allows us to restate this as follows:

(16.6.1′) **Key Fact.** *For some embedding of H in $\mathrm{SL}(n, \mathbb{R})$,*

1) *H acts irreducibly on $\mathbb{R}^n$, and*

2) *there exists a Γ-equivariant, measurable function $\xi: G/A \to \mathbb{R}^n$ (and ξ is nonzero).*

In this form, the result is closely related to the following consequence of amenability (from Chapter 12). For simplicity, it is stated only for the case $G = \mathrm{SL}(3, \mathbb{R})$.

(12.6.2′) **Proposition** (Furstenberg). *If*

- $G = \mathrm{SL}(3, \mathbb{R})$,

- $P = \begin{bmatrix} * & * & * \\ & * & * \\ & & * \end{bmatrix} \subset G$, *and*

- Γ *acts continuously on a compact metric space X,*

then there is a Borel measurable map $\psi \colon G/P \to \mathrm{Prob}(X)$, such that ψ is essentially Γ-equivariant.

For convenience, let $W = \mathbb{R}^n$. There are 3 steps in the proof of Key Fact 16.6.1′:

1) (amenability) Letting X be the projective space $\mathbb{P}(W)$, which is compact, Proposition 12.6.2′ provides a Γ-equivariant, measurable map $\hat{\xi} \colon G/P \to \mathrm{Prob}(\mathbb{P}(W))$.

2) (proximality) The representation of Γ on W induces a representation of Γ on any exterior power $\bigwedge^k W$. By replacing W with an appropriate subspace of such an exterior power, we may assume there is some $\gamma \in \Gamma$, such that γ has a unique eigenvalue of maximal absolute value (see Exercise 1). Therefore, the action of γ on $\mathbb{P}(W)$ is "proximal" (see Lemma 16.7.3). The theory of proximality (discussed in Section 16.7) now tells us that the Γ-equivariant random map $\hat{\xi}$ must actually be a well-defined map into $\mathbb{P}(W)$ (see Corollary 16.7.10).

3) (algebra trick) We have a Γ-equivariant map $\hat{\xi} \colon G/P \to \mathbb{P}(W)$. By the same argument, there is a Γ-equivariant map $\hat{\xi}^* \colon G/P \to \mathbb{P}(W^*)$, where W^* is the dual of W. Combining these yields a Γ-equivariant map
$$\overline{\xi} \colon G/P \times G/P \to \mathbb{P}(W \otimes W^*) \cong \mathbb{P}(\mathrm{End}(W)).$$
We can lift $\overline{\xi}$ to a well-defined map
$$\xi \colon G/P \times G/P \to \mathrm{End}(W),$$
by specifying that $\mathrm{trace}(\xi(x)) = 1$ (see Exercise 2). Since the action of Γ on $\mathrm{End}(W)$ is by conjugation (see Exercise 3) and the trace of conjugate matrices are equal, we see that ξ is Γ-equivariant (see Exercise 4).

Finally, note that there is a G-orbit in $G/P \times G/P$ whose complement is a set of measure 0, and the stabilizer of a point is (conjugate to) the group A of diagonal matrices (see Exercises 5 and 6). Therefore, after discarding a set of measure 0, we may identify $G/P \times G/P$ with G/A, so $\xi \colon G/A \to \mathrm{End}(W)$.

Exercises for §16.6.

#1. Let
- y be a semisimple element of Γ, such that some eigenvalue of y is not of absolute value 1.
- $\lambda_1, \ldots, \lambda_k$ be the eigenvalues of y (with multiplicity) that have maximal absolute value.
- $W' = \bigwedge^k W$.

Show that, in the representation of Γ on W', the element y has a unique eigenvalue of maximal absolute value.

#2. Let $\tilde{\xi}\colon G/P \to W$ and $\tilde{\xi}^*\colon G/P \to W^*$ be well-defined, measurable lifts of $\hat{\xi}$ and $\hat{\xi}^*$.

a) Show, for a.e. $x, y \in G/P$, that $\tilde{\xi}(x)$ is not in the kernel of the linear functional $\tilde{\xi}^*(y)$.

b) Show, for a.e. $x, y \in G/P$, that, under the natural identification of $W \otimes W^*$ with $\text{End}(W)$, we have

$$\text{trace}\big(\tilde{\xi}(x) \otimes \tilde{\xi}^*(y)\big) \neq 0.$$

c) Show $\overline{\xi}$ can be lifted to a well-defined measurable function $\xi\colon G/P \times G/P \to \text{End}(W)$, such that $\text{trace}(\xi(x, y)) = 1$, for a.e. $x, y \in G/P$.

[*Hint:* Γ acts irreducibly on W, and ergodically on $G/P \times G/P$.]

#3. Show that the action of Γ on $\text{End}(W) \cong W \otimes W^*$ is given by conjugation: $\overline{\varphi}(y)T = \varphi(y)\, T\, \varphi(y)^{-1}$.

#4. Show that ξ is Γ-equivariant.

#5. Recall that a *flag* in $\mathbb{R}^3$ is a pair (ℓ, Π), where
- ℓ is a line through the origin (in other words, a 1-dimensional linear subspace), and
- Π is a plane through the origin (in other words, a 2-dimensional linear subspace), such that
- $\ell \subset \Pi$.

Show:

a) $\text{SL}(3, \mathbb{R})$ acts transitively on the set of all flags in $\mathbb{R}^3$, and

b) the stabilizer of any flag is conjugate to the subgroup P of Proposition 12.6.2′.

Therefore, the set of flags can be identified with G/P.

#6. Two flags (ℓ_1, Π_1) and (ℓ_2, Π_2) are in *general position* if

$$\ell_1 \notin \Pi_2, \quad \text{and} \quad \ell_2 \notin \Pi_1 .$$

Letting G be the subset of $G/P \times G/P$ corresponding to the pairs of flags that are in general position, show:

a) $\text{SL}(3, \mathbb{R})$ is transitive on G,

  b) the stabilizer of any point in G is conjugate to the group of diagonal matrices, and
  c) the complement of G has measure zero in $G/P \times G/P$.

  [*Hint:* For (a) and (b), identify G with the set of triples (ℓ_1, ℓ_2, ℓ_3) of lines that are in general position, by letting $\ell_3 = \Pi_1 \cap \Pi_2$.]

§16.7. A quick look at proximality

(16.7.1) **Assumption.** Assume
  1) $\Gamma \subset \mathrm{SL}(\ell, \mathbb{R})$,
  2) every finite-index subgroup of Γ is irreducible on $\mathbb{R}^\ell$, and
  3) there exists a semisimple element $\overline{y} \in \Gamma$, such that $\overline{y}$ has a unique eigenvalue $\overline{\lambda}$ of maximal absolute value (and the eigenvalue is simple, which means the corresponding eigenspace is 1-dimensional).

(16.7.2) **Notation.**
  1) Let $\overline{v}$ be an eigenvector associated to the eigenvalue $\overline{\lambda}$.
  2) For convenience, let $W = \mathbb{R}^\ell$.

(16.7.3) **Lemma** (Proximality). *The action of Γ on $\mathbb{P}(W)$ is **proximal**. This means that, for every $[w_1], [w_2] \in \mathbb{P}(W)$, there exists a sequence $\{y_n\}$ in Γ, such that $d([y_n(w_1)], [y_n(w_2)]) \to 0$ as $n \to \infty$.*

Proof. Assume, to simplify the notation, that all of the eigenspaces of $\overline{y}$ are orthogonal to each other. Then, for any $w \in W \smallsetminus \overline{v}^\perp$, we have $\overline{y}^n[w] \to [\overline{v}]$, as $n \to \infty$ (see Exercise 1). Since the finite-index subgroups of Γ act irreducibly, there is some $y \in \Gamma$, such that $y(w_1), y(w_2) \notin \overline{v}^\perp$ (see Exercise 2). Therefore,

$$d(\overline{y}^n y([w_1]), \overline{y}^n y([w_2])) \to d([\overline{v}], [\overline{v}]) = 0,$$

as desired.   □

  In the above proof, it is easy to see that the convergence $\overline{y}^n[w] \to [\overline{v}]$ is uniform on compact subsets of $W \smallsetminus \overline{v}^\perp$ (see Exercise 3). This leads to the following stronger assertion (see Exercise 4):

(16.7.4) **Proposition** (Measure proximality). *Let μ be any probability measure on $\mathbb{P}(W)$. Then there is a sequence $\{y_n\}$ in Γ, such that $(y_n)_*\mu$ converges to a delta-mass supported at a single point of $\mathbb{P}(W)$.*

  It is obvious from Proposition 16.7.4 that there is no Γ-invariant probability measure on $\mathbb{P}(W)$. However, it is easy to see that there does exist a probability measure that is invariant "on average," in the following sense (see Exercise 5):

(16.7.5) **Definition.**

1) Fix a finite generating set S of Γ, such that $S^{-1} = S$. A probability measure μ on $\mathbb{P}(W)$ is **stationary** for S if

$$\frac{1}{\#S} \sum_{\gamma \in S} \gamma_* \mu = \mu.$$

2) More generally, let ν be a probability measure on Γ. A probability measure μ on $\mathbb{P}(W)$ is **ν-stationary** if $\nu * \mu = \mu$. More concretely, this means

$$\sum_{\gamma \in \Gamma} \nu(\gamma) \, \gamma_* \mu = \mu.$$

(Some authors call μ "harmonic," rather than "stationary.")

(16.7.6) *Remark.* A random walk on $\mathbb{P}(W)$ can be defined as follows: Choose a sequence $\gamma_1, \gamma_2, \ldots$ of elements of Γ, independently and with distribution ν. Also choose a random $x_0 \in \mathbb{P}(W)$, with respect to some probability distribution μ on $\mathbb{P}(W)$. Then $x_n \in \mathbb{P}(W)$ is defined by

$$x_n = \gamma_1 \gamma_2 \cdots \gamma_n(x_0),$$

so $\{x_n\}$ is a random walk on $\mathbb{P}(W)$. A stationary measure represents a "stationary state" (or equilibrium distribution) for this random walk. Hence the terminology.

If the initial distribution μ is stationary, then a basic result of probability (the "Martingale Convergence Theorem") implies, for almost every sequence $\{\gamma_n\}$, that the resulting random walk $\{x_n\}$ has a limiting distribution; that is,

for a.e. $\{\gamma_n\}$, $(\gamma_1 \gamma_2 \cdots \gamma_n)_* \mu$ converges in $\mathrm{Prob}(\mathbb{P}(W))$.

This theorem applies to stationary measures on any space, with no need for Assumption 16.7.1. By using measure proximality, we will now show that the limiting distribution is almost always a point mass.

(16.7.7) **Definition.** A closed, nonempty, Γ-invariant subset of $\mathbb{P}(W)$ is **minimal** if it does not have any nonempty, proper, closed, Γ-invariant subsets. (Since $\mathbb{P}(W)$ is compact, the finite-intersection property implies that every nonempty, closed, Γ-invariant subset of $\mathbb{P}(W)$ contains a minimal set.)

(16.7.8) **Theorem** (Mean proximality). *Assume*

- *ν is a probability measure on Γ, such that $\nu(\gamma) > 0$ for all $\gamma \in \Gamma$,*
- *C is a minimal closed, Γ-invariant subset of $\mathbb{P}(W)$, and*
- *μ is a ν-stationary probability measure on C.*

Then, for a.e. $\{\gamma_n\} \in \Gamma^\infty$, there exists $c \in \mathbb{P}(W)$, such that

$$(\gamma_1 \gamma_2 \cdots \gamma_n)_* (\mu) \to \delta_c \text{ as } n \to \infty.$$

Proof. It was mentioned above that the Martingale Convergence Theorem implies $(\gamma_1 \gamma_2 \cdots \gamma_n)_*(\mu)$ has a limit (almost surely), so it suffices to show there is (almost surely) a subsequence that converges to a measure of the form δ_c.

Proposition 16.7.4 provides a sequence $\{g_k\}$ of elements of Γ, such that $(g_k)_*\mu \to \delta_{c_0}$, for some $c_0 \in C$. To extend this conclusion to a.e. sequence $\{\gamma_n\}$, we use equicontinuity: we may write Γ is the union of finitely many sets $E_1, \ldots, E_r$, such that each E_i is equicontinuous on some nonempty open subset U_i of C (see Exercise 6).

The minimality of C implies $\Gamma U_i = C$ for every i. Then, by compactness, there is a finite subset $F = \{f_1, \ldots, f_s\}$ of Γ, such that $FU_i = C$ for each i. Since $\nu(\gamma) > 0$ for every $\gamma \in \Gamma$, there is (almost surely) a subsequence $\{\gamma_{n_k}\}$ of $\{\gamma_n\}$, such that, for every k, we have

$$\gamma_{n_k+1} \gamma_{n_k+2} \cdots \gamma_{n_k+j} = f_j^{-1} g_k \text{ for } 1 \leq j \leq s.$$

By passing to a subsequence, we may assume there is some i, such that

$$\gamma_1 \gamma_2 \cdots \gamma_{n_k} \in E_i, \text{ for all } k.$$

To simplify the notation, let us assume $i = 1$.

Since $FU_1 = C$, we may write $c_0 = f_j u$, for some $f_j \in F$ and $u \in U_1$. Then

$$(\gamma_{n_k+1} \gamma_{n_k+2} \cdots \gamma_{n_k+j})_*\nu = (f_j^{-1} g_k)_*\nu \to (f_j^{-1})_*\delta_{c_0} = \delta_{f_j^{-1} c_0} = \delta_u.$$

By passing to a subsequence, we may assume $(\gamma_1 \gamma_2 \cdots \gamma_{n_k})u$ converges to some $c \in C$. Then, since $\gamma_1 \gamma_2 \cdots \gamma_{n_k} \in E_1$, and E_1 is equicontinuous on U_1, this implies

$$(\gamma_1 \gamma_2 \cdots \gamma_{n_k+j})_*\nu = (\gamma_1 \gamma_2 \cdots \gamma_{n_k})_*((\gamma_{n_k+1} \cdots \gamma_{n_k+j})_*\nu) \to \delta_c. \quad \square$$

In order to apply this theorem, we need a technical result, whose proof we omit:

(16.7.9) Lemma. *There exist:*

- *a probability measure ν on Γ, and*
- *a ν-stationary probability measure μ on G/P,*

such that

1) *the support of ν generates Γ, and*
2) *μ is in the class of Lebesgue measure. (That is, μ has exactly the same sets of measure 0 as Lebesgue measure does.)*

Also note that if C is any nonempty, closed, Γ-invariant subset of $\mathbb{P}(W)$, then $\text{Prob}(C)$ is a nonempty, compact, convex Γ-space, so Furstenberg's Lemma (12.6.1) provides a Γ-equivariant map $\overline{\xi}: G/P \to \text{Prob}(C)$. This observation allows us to replace $\mathbb{P}(W)$ with a minimal subset.

We can now fill in the missing part of the proof of Key Fact 16.6.1′:

(16.7.10) **Corollary.** *Suppose*

- *C is a minimal closed, Γ-invariant subset of $\mathbb{P}(W)$, and*
- *$\overline{\xi}\colon G/P \to \mathrm{Prob}(C)$ is Γ-equivariant.*

Then $\overline{\xi}(x)$ is a point mass, for a.e. $x \in G/P$.

Hence, there exists $\hat{\xi}\colon G/P \to \mathbb{P}(W)$, such that $\overline{\xi}(x) = \delta_{\hat{\xi}(x)}$, for a.e. $x \in G/P$.

Proof. Let

- $\delta_{\mathbb{P}(W)} = \{\, \delta_x \mid x \in \mathbb{P}(W) \,\}$ be the set of all point masses in the space $\mathrm{Prob}(\mathbb{P}(W))$, and
- μ be a ν-stationary probability measure on G/P that is in the class of Lebesgue measure (see Lemma 16.7.9).

We wish to show $\overline{\xi}(x) \in \delta_{\mathbb{P}(W)}$, for a.e. $x \in G/P$. In other words, we wish to show that $\overline{\xi}_*(\mu)$ is supported on $\delta_{\mathbb{P}(W)}$.

Note that:

- $\delta_{\mathbb{P}(W)}$ is a closed, Γ-invariant subset of $\mathrm{Prob}(\mathbb{P}(W))$, and
- because $\overline{\xi}$ is Γ-equivariant, we know that $\overline{\xi}_*(\mu)$ is a ν-stationary probability measure on $\mathrm{Prob}(\mathbb{P}(W))$.

Roughly speaking, the idea of the proof is that almost every trajectory of the random walk on $\mathrm{Prob}(\mathbb{P}(W))$ converges to a point in $\delta_{\mathbb{P}(W)}$ (see 16.7.8). On the other hand, being stationary, $\overline{\xi}_*(\mu)$ is invariant under the random walk. Therefore, we conclude that $\overline{\xi}_*(\mu)$ is supported on $\delta_{\mathbb{P}(W)}$, as desired.

We now make this rigorous. Let

$$\mu_{\mathbb{P}(W)} = \int_{G/P} \overline{\xi}(x)\, d\mu(x),$$

so $\mu_{\mathbb{P}(W)}$ is a stationary probability measure on $\mathbb{P}(W)$. By mean proximality (16.7.8), we know, for a.e. $(\gamma_1, \gamma_2, \ldots) \in \Gamma^\infty$, that

$$d\big((\gamma_1 \gamma_2 \cdots \gamma_n)_*(\mu_{\mathbb{P}(W)}),\, \delta_{\mathbb{P}(W)}\big) \overset{n \to \infty}{\longrightarrow} 0.$$

For any $\epsilon > 0$, this implies, by using the definition of $\mu_{\mathbb{P}(W)}$, that

$$\mu\Big(\big\{\, x \in G/P \mid d(\gamma_1 \gamma_2 \cdots \gamma_n(\overline{\xi}(x)), \delta_{\mathbb{P}(W)}) > \epsilon \,\big\}\Big) \overset{n \to \infty}{\longrightarrow} 0.$$

Since $\overline{\xi}$ is Γ-equivariant, we may

replace $\gamma_1 \gamma_2 \cdots \gamma_n(\overline{\xi}(x))$ with $\overline{\xi}(\gamma_1 \gamma_2 \cdots \gamma_n x)$.

Then, since the measure μ on G/P is stationary, we can delete $\gamma_1 \gamma_2 \cdots \gamma_n$, and conclude that

$$\mu\big\{\, x \in G/P \mid d(\overline{\xi}(x), \delta_{\mathbb{P}(W)}) > \epsilon \,\big\} \overset{n \to \infty}{\longrightarrow} 0 \qquad (16.7.11)$$

(see Exercise 7). Since the left-hand side does not depend on n, but tends to 0 as $n \to \infty$, it must be 0. Since $\epsilon > 0$ is arbitrary, we conclude that $\overline{\xi}(x) \in \delta_{\mathbb{P}(W)}$ for a.e. x, as desired. $\qquad\square$

Exercises for §16.7.

#1. In the notation of Lemma 16.7.3, show, for every $w \in W \smallsetminus \overline{v}^{\perp}$, that $\overline{y}^n[w] \to [\overline{v}]$, as $n \to \infty$.

#2. Show, for any nonzero $w_1, w_2 \in W$, that there exists $y \in \Gamma$, such that neither yw_1 nor yw_2 is orthogonal to $\overline{v}$.

[*Hint:* Let H be the Zariski closure of Γ in $\mathrm{SL}(\ell, \mathbb{R})$, and assume, by passing to a finite-index subgroup, that H is connected. Then $W_i = \{h \in H \mid hw_i \in \overline{v}^{\perp}\}$ is a proper, Zariski-closed subset. Since Γ is Zariski dense in H, it must intersection the complement of $W_1 \cup W_2$.]

#3. Show that the convergence in Exercise 1 is uniform on compact subsets of $W \smallsetminus \overline{v}^{\perp}$.

#4. Prove Proposition 16.7.4.

[*Hint:* Show $\max_{w \in \mathbb{P}(W) \, v \in \overline{\Gamma\mu}} v(w) = 1$.]

#5. Show there exists a stationary probability measure on $\mathbb{P}(W)$.

[*Hint:* Kakutani-Markov Fixed-Point Theorem (cf. 12.2.1).]

#6. Let C be a subset of $\mathbb{P}(\mathbb{R}^n)$, and assume that C is not contained in any $(n - 1)$-dimensional hyperplane. Prove that $\mathrm{GL}(n, \mathbb{R})$ is the union of finitely many sets $E_1, \ldots, E_r$, such that each E_i is equicontinuous on some nonempty open subset U_i of C.

[*Hint:* Each matrix $T \in \mathrm{Mat}_{n \times n}(\mathbb{R})$ induces a well-defined, continuous function $\overline{T} \colon (\mathbb{P}(\mathbb{R}^n) \smallsetminus \mathbb{P}(\ker T)) \to \mathbb{P}(\mathbb{R}^n)$. If B_T is a small ball around $\overline{T}$ in $\mathbb{P}(\mathrm{Mat}_{n \times n}(\mathbb{R}))$, then B_T is equicontinuous on an open set. A compact set can be covered by finitely many balls.]

#7. Establish (16.7.11).

[*Hint:* Since μ is stationary, the map
$$\Gamma^{\infty} \times G/P \to G/P \colon ((y_1, y_2, \ldots), x) \mapsto y_1 y_2 \cdots y_n x$$
is measure preserving.]

#8. Show that if $\mathbb{P}(W)$ is minimal, then the Γ-equivariant measurable map $\xi \colon G/P \to \mathbb{P}(W)$ is unique (a.e.).

[*Hint:* If ψ is another Γ-equivariant map, then define $\overline{\xi} \colon G/P \to \mathrm{Prob}(\mathbb{P}(W))$ by $\overline{\xi}(x) = \frac{1}{2}(\delta_{\xi(x)} + \delta_{\psi(x)})$.]

§16.8. Groups of real rank one

The Margulis Superrigidity Theorem (16.1.4) was proved for groups of real rank at least two in Section 16.5. Suppose, now, that $\mathrm{rank}_{\mathbb{R}} G = 1$ (and G has no compact factors). The classification of simple Lie groups tells us that G is isogenous to the isometry group of either:

- real hyperbolic space $\mathfrak{H}^n$,
- complex hyperbolic space $\mathbb{C}\mathfrak{H}^n$,
- quaternionic hyperbolic space $\mathbb{H}\mathfrak{H}^n$, or

- the Cayley hyperbolic plane $\mathbb{O}\mathfrak{H}^2$ (where $\mathbb{O}$ is the ring of "Cayley numbers" or "octonions")

(cf. Theorem 8.3.1). Assumption 16.1.4(ii) rules out $\mathfrak{H}^n$ and $\mathbb{C}\mathfrak{H}^n$, so, from the connection of superrigidity with totally geodesic embeddings (cf. Subsection 16.2(iii)), the following result completes the proof:

(16.8.1) **Theorem.** *Assume*

- $X = \mathbb{H}\mathfrak{H}^n$ *or* $\mathbb{O}\mathfrak{H}^2$,
- Γ *is a torsion-free, discrete group of isometries of X, such that $\Gamma \backslash X$ has finite volume,*
- X' *is an irreducible symmetric space of noncompact type, and*
- $\varphi: \Gamma \to \mathrm{Isom}(X')^{\circ}$ *is a homomorphism whose image is Zariski dense.*

Then there is a map $f: X \to X'$, such that

1) *$f(X)$ is totally geodesic, and*
2) *f is φ-equivariant, which means $f(\gamma x) = \varphi(\gamma) \cdot f(x)$.*

Brief outline of proof. Choose a (nice) fundamental domain $\mathcal{F}$ for the action of Γ on X. For any φ-equivariant map $f: X \to X'$, define the **energy** of f to be the L^2-norm of the derivative of f over $\mathcal{F}$. Since f is φ-equivariant, and the groups Γ and $\varphi(\Gamma)$ act by isometries, this is independent of the choice of the fundamental domain $\mathcal{F}$.

It can be shown that this energy functional attains its minimum at some function f. The minimality implies that f is harmonic. Then, by using the geometry of X and the negative curvature of X', it can be shown that f must be totally geodesic. $\qquad\square$

Notes

This chapter is largely based on [6, Chaps. 6 and 7]. (However, we usually replace the assumption that $\mathrm{rank}_{\mathbb{R}} G \geq 2$ with the weaker assumption that G is not $\mathrm{SO}(1, m) \times K$ or $\mathrm{SU}(1, m) \times K$. (See [10, Thm. 5.1.2, p. 86] for a different exposition that proves version (16.1.6) for $\mathrm{rank}_{\mathbb{R}} G \geq 2$.) In particular:

- For $\mathrm{rank}_{\mathbb{R}} G \geq 2$, our statement of the Margulis Superrigidity Theorem (16.1.6) is a special case of [6, Thm. 7.5.6, p. 228].
- For $\mathrm{rank}_{\mathbb{R}} G \geq 2$, Corollary 16.1.7 is stated in [6, Thm. 9.6.15(i)(a), p. 332].
- For $\mathrm{rank}_{\mathbb{R}} G \geq 2$, Theorem 15.1.2 is stated in [6, Thm. 7.7.5, p. 254]. (See [8, Thm. B] for the general case, which does not follow from superrigidity.)
- Lemma 16.5.5 is a version of [6, Prop. 4.6, p. 222]
- Lemma 16.5.7 is a version of [6, Lem. 7.5.5, p. 227].

- Lemma 16.5.10 is [6, Prop. 7.3.6, p. 219].
- Key Fact 16.6.1′ is adapted from [6, Thm. 6.4.3(b)2, p. 209].
- Theorem 16.7.8 is based on [6, Prop. 6.2.13, pp. 202-203].
- Lemma 16.7.9 is taken from [6, Prop. 6.4.2, p. 209].
- Corollary 16.7.10 is based on [6, Prop. 6.2.9, p. 200].
- Exercise 16.7#6 is [6, Lem. 6.3.2, p. 203].

Long before the general theorem of Margulis for groups of real rank ≥ 2, it was proved by Bass, Milnor, and Serre [2, Thm. 61.2] that the Congruence Subgroup Property implies $SL(n, \mathbb{Z})$ is superrigid in $SL(n, \mathbb{R})$.

"Geometric superrigidity" is the study of differential geometric versions of the Margulis Superrigidity Theorem, such as Proposition 16.2.5. (See, for example, [7].)

Details of the derivation of arithmeticity from superrigidity (Section 16.3) appear in [6, Chap. 9] and [10, §6.1].

Proofs of the Commensurability Criterion (16.3.3) and Commensurator Superrigidity (16.3.4) can be found in [1], [6, §9.2.11, pp. 305ff, and Thm. 7.5.4, pp. 226-227], and [10, §6.2].

Much of the material in Section 16.7 is due to Furstenberg [4].

The superrigidity of lattices in the isometry groups of $\mathbb{H}\mathfrak{H}^n$ and $\mathbb{O}\mathfrak{H}^2$ (see Section 16.8) was proved by Corlette [3]. The p-adic version (16.3.2) for these groups was proved by Gromov and Schoen [5].

References

[1] N. A'Campo and M. Burger: Réseaux arithmétiques et commensurateur d'après G. A. Margulis, *Invent. Math.* 116 (1994) 1-25. MR 1253187, http://eudml.org/doc/144182

[2] H. Bass, J. Milnor, and J.-P. Serre: Solution of the congruence subgroup problem for SL_n ($n \geq 3$) and Sp_{2n} ($n \geq 2$), *Inst. Hautes Études Sci. Publ. Math.* 33 (1967) 59-137. MR 0244257, http://www.numdam.org/item?id=PMIHES_1967__33__59_0

[3] K. Corlette: Archimedean superrigidity and hyperbolic geometry, *Ann. Math.* (2) 135 (1992), no. 1, 165-182. MR 1147961, http://www.jstor.org/stable/2946567

[4] H. Furstenberg: Boundary theory and stochastic processes on homogeneous spaces, in C. C. Moore, ed.: *Harmonic Analysis on Homogeneous Spaces (Williamstown, Mass., 1972)*. Amer. Math. Soc., Providence, R.I., 1973, pp. 193-229. MR 0352328

[5] M. Gromov and R. Schoen: Harmonic maps into singular spaces and p-adic superrigidity for lattices in groups of rank one, *Inst. Hautes*

Études Sci. Publ. Math. 76 (1992) 165–246. MR 1215595,
http://www.numdam.org/item?id=PMIHES_1992__76__165_0

[6] G. A. Margulis: *Discrete Subgroups of Semisimple Lie Groups.*
Springer, New York, 1991. ISBN 3-540-12179-X, MR 1090825

[7] N. Mok, Y. T. Siu, and S.-K. Yeung: Geometric superrigidity, *Invent.
Math.* 113 (1993) 57–83. MR 1223224, http://eudml.org/doc/144122

[8] G. Prasad: Strong rigidity of $\mathbb{Q}$-rank 1 lattices, *Invent. Math.* 21 (1973)
255–286. MR 0385005, http://eudml.org/doc/142232

[9] Y. T. Siu: Geometric super-rigidity, in *Geometry and Analysis
(Bombay, 1992)*, Oxford U. Press, Oxford, 1996, pp. 299–312. ISBN
0-19-563740-2, MR 1351514

[10] R. J. Zimmer: *Ergodic Theory and Semisimple Groups.* Birkhäuser,
Basel, 1984. ISBN 3-7643-3184-4, MR 0776417

Chapter 17

Normal Subgroups of Γ

This chapter presents a contrast between the lattices in groups of real rank 1 and those of higher real rank:

- If $\operatorname{rank}_{\mathbb{R}} G = 1$, then Γ has many, many normal subgroups, so Γ is very far from being simple.

- If $\operatorname{rank}_{\mathbb{R}} G > 1$ (and Γ is irreducible), then Γ is simple modulo finite groups. More precisely, if N is any normal subgroup of Γ, then either N is finite, or Γ/N is finite.

§17.1. Normal subgroups in lattices of real rank ≥ 2

(17.1.1) **Theorem** (Margulis Normal Subgroups Theorem). *Assume*

- $\operatorname{rank}_{\mathbb{R}} G \geq 2$,
- Γ *is an irreducible lattice in G, and*
- N *is a normal subgroup of* Γ.

Then either N is finite, or Γ/N *is finite.*

Recall: The Standing Assumptions (4.0.0 on page 41) are in effect, so, as always, Γ is a lattice in the semisimple Lie group $G \subseteq \mathrm{SL}(\ell, \mathbb{R})$.

Main prerequisites for this chapter: amenability (Furstenberg's Lemma (12.6.1)) and Kazhdan's Property (T) (Chapter 13). *Also used:* the σ-algebra of Borel sets modulo sets of measure 0 (Section 14.4) and manifolds of negative curvature.

(17.1.2) **Example.** Every lattice in $SL(3, \mathbb{R})$ is simple, modulo finite groups. In particular, this is true of $SL(3, \mathbb{Z})$.

(17.1.3) *Remarks.*

  1) The hypotheses on G and Γ are essential:
       (a) If $rank_{\mathbb{R}} G = 1$, then every lattice in G has an infinite normal subgroup of infinite index (see Theorem 17.2.1).
       (b) If Γ is reducible (and G has no compact factors), then Γ has an infinite normal subgroup of infinite index (see Exercise 2).
  2) The finite normal subgroups of Γ are easy to understand (if Γ is irreducible): the Borel Density Theorem implies that they are the subgroups of the finite abelian group $\Gamma \cap Z(G)$ (see Corollary 4.5.4).
  3) If Γ is infinite, then Γ has infinitely many normal subgroups of finite index (see Exercise 5), so Γ is *not* simple.
  4) In most cases, the subgroups of finite index are described by the "Congruence Subgroup Property." For example, if $\Gamma = SL(3, \mathbb{Z})$, then the principal congruence subgroups are obvious subgroups of finite index (see Exercise 4.8#3). More generally, any subgroup of Γ that contains a principal congruence subgroup obviously has finite index. The Congruence Subgroup Property is the assertion that every finite-index subgroup is one of these obvious ones. It is true for $SL(n, \mathbb{Z})$, whenever $n \geq 3$, and a similar (but slightly weaker) statement is conjectured to be true whenever $rank_{\mathbb{R}} G \geq 2$ and Γ is irreducible.

The remainder of this section presents the main ideas in the proof of Theorem 17.1.1. In a nutshell, we will show that if N is an infinite, normal subgroup of Γ, then

  1) Γ/N has Kazhdan's property (T), and
  2) Γ/N is amenable.

This implies that Γ/N is finite (see Corollary 13.1.5).

In most cases, it is easy to see that Γ/N has Kazhdan's property (because Γ has the property), so the main problem is to show that Γ/N is amenable. This amenability follows easily from an ergodic-theoretic result that we will now describe.

(17.1.4) **Assumption.** To minimize the amount of Lie theory needed, let us assume
$$G = SL(3, \mathbb{R}).$$

(17.1.5) **Notation.** Let
$$P = \begin{bmatrix} * & & \\ * & * & \\ * & * & * \end{bmatrix} \subset SL(3, \mathbb{R}) = G.$$

Hence, P is a (minimal) parabolic subgroup of G.

Note that if Q is any closed subgroup of G that contains P, then the natural map $G/P \to G/Q$ is G-equivariant, so we may say that G/Q is a G-equivariant quotient of G/P. Conversely, it is easy to see that spaces of the form G/Q are the only G-equivariant quotients of G/P. In fact, these are the only quotients even if we only assume that quotient map is equivariant *almost* everywhere (see Exercise 6).

Furthermore, since Γ is a subgroup of G, it is obvious that every G-equivariant map is Γ-equivariant. Conversely, the following surprising result shows that every Γ-equivariant quotient of G/P is G-equivariant (up to a set of measure 0):

(17.1.6) Theorem (Margulis). *Suppose*

- $\mathrm{rank}_{\mathbb{R}}\, G \geq 2$,
- *P is a minimal parabolic subgroup of G,*
- *Γ is irreducible,*
- *Γ acts by homeomorphisms on a compact, metrizable space Z, and*
- *$\psi\colon G/P \to Z$ is essentially Γ-equivariant (and measurable).*

Then the action of Γ on Z is measurably isomorphic to the natural action of Γ on G/Q (a.e.), for some closed subgroup Q of G that contains P.

(17.1.7) Remark.

1) Perhaps we should clarify the choice of measures in the statement of Theorem 17.1.6. (A measure class on G/P is implicit in the assumption that ψ is *essentially* Γ-equivariant. Measure classes on Z and G/Q are implicit in the "(a.e.)" in the conclusion of the theorem.)

   (a) Because G/P and G/Q are C^{∞} manifolds, Lebesgue measure supplies a measure class on each of these spaces. The Lebesgue class is invariant under all diffeomorphisms, so, in particular, it is G-invariant.

   (b) There is a unique measure class on Z for which ψ is measure-class preserving (see Exercise 7).

2) The proof of Theorem 17.1.6 will be presented in Section 17.3. It may be skipped on a first reading.

Proof of Theorem 17.1.1. Let N be a normal subgroup of Γ, and assume N is infinite. We wish to show Γ/N is finite. Let us assume, for simplicity, that Γ has Kazhdan's Property (T). (For example, this is true if $G = \mathrm{SL}(3, \mathbb{R})$, or, more generally, if G is simple (see Corollary 13.4.2).) Then Γ/N also has Kazhdan's Property (T) (see Proposition 13.1.7), so it suffices to show that Γ/N is amenable (see Corollary 13.1.5).

Suppose Γ/N acts by homeomorphisms on a compact, metrizable space X. In order to show that Γ/N is amenable, it suffices to find an invariant probability measure on X (see Theorem 12.3.1(3)). In other words, we wish to show that Γ has a fixed point in $\mathrm{Prob}(X)$.

- Because P is amenable, there is an (essentially) Γ-equivariant measurable map $\psi \colon G/P \to \mathrm{Prob}(X)$ (see Corollary 12.6.2).
- From Theorem 17.1.6, we know there is a closed subgroup Q of G, such that the action of Γ on $\mathrm{Prob}(X)$ is measurably isomorphic (a.e.) to the natural action of Γ on G/Q.

Since N acts trivially on X, we know it acts trivially on $\mathrm{Prob}(X) \cong G/Q$. Hence, the kernel of the G-action on G/Q is infinite (see Exercise 10). However, G is simple (modulo its finite center), so this implies that the action of G on G/Q is trivial (see Exercise 11). (It follows that G/Q is a single point, so $Q = G$, but we do not need quite such a strong conclusion.) Since $\Gamma \subseteq G$, then the action of Γ on G/Q is trivial. In other words, every point in G/Q is fixed by Γ. Since $G/Q \cong \mathrm{Prob}(X)$ (a.e.), we conclude that almost every point in $\mathrm{Prob}(X)$ is fixed by Γ; therefore, Γ has a fixed point in $\mathrm{Prob}(X)$, as desired.   □

(17.1.8) *Remark.* The proof of Theorem 17.1.1 concludes that "almost every point in $\mathrm{Prob}(X)$ is fixed by Γ," so it may seem that the proof provides not just a single Γ-invariant measure, but many of them. This is not the case: The proof implies that ψ is essentially constant (see Exercise 12). This means that the Γ-invariant measure class $[\psi_*\mu]$ is supported on a single point of $\mathrm{Prob}(X)$, so "a.e." means only one point.

Exercises for §17.1.

#1. Assume
- G is not isogenous to $\mathrm{SO}(1,n)$ or $\mathrm{SU}(1,n)$, for any n,
- Γ is irreducible, and
- G has no compact factors.

In many cases, Kazhdan's property (T) implies that the abelianization $\Gamma/[\Gamma,\Gamma]$ of Γ is finite (see Corollary 13.4.3(2)). Use Theorem 17.1.1 to prove this in the remaining cases. (We saw a different proof of this in Exercise 16.1#3.)

#2. Verify Remark 17.1.3(1b).
[*Hint:* Proposition 4.3.3.]

#3. Suppose Γ is a lattice in $\mathrm{SL}(3,\mathbb{R})$. Show that Γ has no nontrivial, finite, normal subgroups.

#4. Suppose Γ is an irreducible lattice in G. Show that Γ has only finitely many finite, normal subgroups.

#5. Show that if Γ is infinite, then it has infinitely many normal subgroups of finite index.

[*Hint:* Exercise 4.8#9.]

#6. Suppose
  - H is a closed subgroup of G,
  - G acts continuously on a metrizable space Z, and
  - $\psi: G/H \to Z$ is essentially G-equivariant (and measurable).

Show the action of G on Z is measurably isomorphic to the action of G on G/Q (a.e.), for some closed subgroup Q of G that contains H. More precisely, show there is a measurable $\phi: Z \to G/Q$, such that:
  a) ϕ is measure-class preserving (i.e., a subset A of G/Q has measure 0 if and only if its inverse image $\phi^{-1}(A)$ has measure 0),
  b) ϕ is one-to-one (a.e.) (i.e., ϕ is one-to-one on a conull subset of Z), and
  c) ϕ is essentially G-equivariant.

[*Hint:* See Remark 17.1.7(1) for an explanation of the measure classes to be used on G/H, G/Q, and Z. For each $g \in G$, the set $\{x \in G/H \mid \psi(gx) = g \cdot \psi(x)\}$ is conull. By Fubini's Theorem, there is some $x_0 \in G/H$, such that $\psi(gx_0) = g \cdot \psi(x_0)$ for a.e. g. Show the G-orbit of $\psi(x_0)$ is conull in Z, and let $Q = \text{Stab}_G(\psi(x_0))$.]

#7. Suppose
  - $\psi: Y \to Z$ is measurable, and
  - μ_1 and μ_2 are measures on Y that are in the same measure class.

Show:
  a) The measures $\psi_*(\mu_1)$ and $\psi_*(\mu_2)$ on Z are in the same measure class.
  b) For any measure class on Y, there is a unique measure class on Z for which ψ is measure-class preserving.

#8. In the setting of Theorem 17.1.6, show that ψ is essentially onto. That is, the image $\psi(G/P)$ is a conull subset of Z.

[*Hint:* By choice of the measure class on Z, we know that ψ is measure-class preserving.]

#9. Let $G = \text{SL}(3, \mathbb{R})$ and $\Gamma = \text{SL}(3, \mathbb{Z})$. Show that the natural action of Γ on $\mathbb{R}^3/\mathbb{Z}^3 = \mathbb{T}^3$ is a Γ-equivariant quotient of the action on $\mathbb{R}^3$, but is not a G-equivariant quotient.

#10. In the proof of Theorem 17.1.1, we know that $\text{Prob}(X) \cong G/Q$ (a.e.), so each element of N fixes a.e. point in G/Q. Show that N acts trivially on G/Q (everywhere, not only a.e.).

[*Hint:* The action of N is continuous.]

#11. In the notation of the proof of Theorem 17.1.1, show that the action of G on G/Q is trivial.

[*Hint:* Show that the kernel of the action of G on G/Q is closed. You may assume, without proof, that G is an almost simple Lie group. This means that every proper, closed, normal subgroup of G is finite.]

#12. In the setting of the proof of Theorem 17.1.1, show that ψ is constant (a.e.).

[*Hint:* The proof shows that a.e. point in the image of ψ is fixed by G. Because ψ is G-equivariant, and G is transitive on G/P, this implies that ψ is constant (a.e.).]

§17.2. Normal subgroups in lattices of rank one

Theorem 17.1.1 assumes $\mathrm{rank}_{\mathbb{R}} G \geq 2$. The following result shows that this condition is necessary:

(17.2.1) **Theorem.** *If* $\mathrm{rank}_{\mathbb{R}} G = 1$, *then* Γ *has a normal subgroup* N, *such that neither* N *nor* Γ/N *is finite.*

Proof (assumes familiarity with manifolds of negative curvature). For simplicity, assume:

- Γ is torsion free, so it is the fundamental group of the locally symmetric space $M = \Gamma \backslash G/K$ (where K is a maximal compact subgroup of G).
- M is compact.
- The locally symmetric metric on M has been normalized to have sectional curvature ≤ -1.
- The injectivity radius of M is ≥ 2.
- There are closed geodesics y and λ in M, such that $\mathrm{length}(\lambda) > 2\pi$ and $\mathrm{dist}(y, \lambda) > 2$.

The geodesics y and λ represent (conjugacy classes of) nontrivial elements $\hat{y}$ and $\hat{\lambda}$ of the fundamental group Γ of M. Let N be the smallest normal subgroup of Γ that contains $\hat{\lambda}$.

It suffices to show that $\hat{y}^n$ is nontrivial in Γ/N, for every $n \in \mathbb{Z}^+$ (see Exercise 1). Construct a CW complex $\overline{M}$ by gluing the boundary of a 2-disk D_λ to M along the curve λ, so the fundamental group of $\overline{M}$ is Γ/N.

We wish to show that y^n is not null-homotopic in $\overline{M}$. Suppose there is a continuous map $f: D^2 \to \overline{M}$, such that the restriction of f to the boundary of D^2 is y^n. Let

$$D_0^2 = f^{-1}(M),$$

so D_0^2 is a surface of genus 0 with some number k of boundary curves. We may assume f is minimal (i.e., the area of D^2 under the pull-back metric is minimal). Then D_0^2 is a surface of curvature $\kappa(x) \leq -1$ whose boundary curves are geodesics. Note that f maps

- one boundary geodesic onto y^n, and
- the other $k - 1$ boundary geodesics onto multiples of λ.

This yields a contradiction:

$$2\pi(k-2) = -2\pi\,\chi(D_0^2) \qquad \text{(see Exercise 2)}$$

$$= -\int_{D_0^2} \kappa(x)\,dx \qquad \text{(Gauss-Bonnet Theorem)}$$

$$\geq \int_{D_0^2} 1\,dx$$

$$\geq (k-1)\,\text{length}(\lambda) \qquad \text{(see Exercise 3)}$$

$$> 2\pi(k-1). \qquad\qquad\qquad \square$$

(17.2.2) *Remark.* Perhaps the simplest example of Theorem 17.2.1 is when $G = \mathrm{SL}(2,\mathbb{R})$ and Γ is a free group (see Remark 6.1.6). In this case, it is easy to find a normal subgroup N, such that N and Γ/N are both infinite. (For example, we could take $N = [\Gamma,\Gamma]$.)

There are numerous strengthenings of Theorem 17.2.1 that provide infinite quotients of Γ with various interesting properties (if $\mathrm{rank}_\mathbb{R} G = 1$). We will conclude this section by briefly describing just one such example.

A classical theorem of Higman, Neumann, and Neumann states that every countable group can be embedded in a 2-generated group. Since 2-generated groups are precisely the quotients of the free group F_2 on 2 generators, this means that F_2 is "SQ-universal" in the following sense:

(17.2.3) **Definition.** Γ is *SQ-universal* if every countable group is iso-morphic to a subgroup of a quotient of Γ. (The letters "SQ" stand for "subgroup-quotient.")

More precisely, the SQ-universality of Γ means that if Λ is any count-able group, then there exists a normal subgroup N of Γ, such that Λ is isomorphic to a subgroup of Γ/N.

(17.2.4) **Example.** F_n is SQ-universal, for any $n \geq 2$ (see Exercise 4).

SQ-universality holds not only for free groups, which are lattices in $\mathrm{SL}(2,\mathbb{R})$ (see Remark 6.1.6), but for any other lattice of real rank one:

(17.2.5) **Theorem.** *If $\mathrm{rank}_\mathbb{R} G = 1$, then Γ is SQ-universal.*

(17.2.6) *Remark.* Although the results in this section have been stated only for Γ, which is a lattice, the theorems are valid for a much more gen-eral class of groups. This is because normal subgroups can be obtained from an assumption of negative curvature (as is illustrated by the proof of Theorem 17.2.1). Indeed, Theorems 17.2.1 and 17.2.5 remain valid when Γ is replaced with any group that is Gromov hyperbolic (see Defini-tion 10.2.1), or even "relatively" hyperbolic (and not commensurable to a cyclic group).

Exercises for §17.2.

#1. Suppose
   - y and λ are nontrivial elements of Γ,
   - Γ is torsion free,
   - N is a normal subgroup of Γ,
   - $\lambda \in N$, and
   - $y^n \notin N$, for every positive integer n.

   Show that neither N nor Γ/N is infinite.

#2. Show that the Euler characteristic of a 2-disk with $k - 1$ punctures is $2 - k$.

#3. In the notation of the proof of Theorem 17.2.1, show
   $$\int_{D_0^2} 1 \, dx \geq (k - 1) \, \text{length}(\lambda).$$

   [*Hint:* All but one of the boundary components are at least as long as λ, and a boundary collar of width 1 is disjoint from the collar around any other boundary component.]

#4. Justify Example 17.2.4.

   [*Hint:* You may assume the theorem of Higman, Neumann, and Neumann on embedding countable groups in 2-generated groups.]

§17.3. Γ-equivariant quotients of G/P (optional)

In this section, we explain how to prove Theorem 17.1.6. However, we will assume $G = \text{SL}(2, \mathbb{R}) \times \text{SL}(2, \mathbb{R})$, for simplicity.

The space Z is not known explicitly, so it is difficult to study directly. Instead, as in the proof of the ergodic decomposition in Section 14.4, we will look at the σ-algebra $\mathcal{B}(Z)$ of Borel sets, modulo the sets of measure 0. (We will think of this as the set of $\{0, 1\}$-valued functions in $\mathcal{L}^\infty(Z)$, by identifying each set with its characteristic function.) Note that ψ induces a Γ-equivariant inclusion

$$\psi^* \colon \mathcal{B}(Z) \hookrightarrow \mathcal{B}(G/P)$$

(see Exercise 1). Via the inclusion ψ^*, we can identify $\mathcal{B}(Z)$ with a sub-σ-algebra of $\mathcal{B}(G/P)$:

$$\mathcal{B}(Z) \subseteq \mathcal{B}(G/P).$$

In order to establish that Z is a G-equivariant quotient of G/P, we wish to show that $\mathcal{B}(Z)$ is G-invariant (see Exercise 2). Therefore, Theorem 17.1.6 can be reformulated as follows:

(17.1.6′) **Theorem.** *If $\mathcal{B}$ is any Γ-invariant sub-σ-algebra of $\mathcal{B}(G/P)$, then $\mathcal{B}$ is G-invariant.*

To make things easier, let us settle for a lesser goal temporarily:

(17.3.1) Definition. The *trivial* Boolean sub-σ-algebra of $\mathcal{B}(G/P)$ is $\{0,1\}$ (the set of constant functions).

(17.3.2) Proposition. *If $\mathcal{B}$ is any nontrivial, Γ-invariant sub-σ-algebra of $\mathcal{B}(G/P)$, then $\mathcal{B}$ contains a nontrivial G-invariant Boolean algebra.*

(17.3.3) Remark.

  1) To establish Proposition 17.3.2, we will find a characteristic function $\overline{f} \in \mathcal{B}(G/P) \setminus \{0,1\}$, such that $G\,\overline{f} \subseteq \mathcal{B}$.

  2) The proof of Theorem 17.1.6′ is similar: let $\mathcal{B}_G$ be the (unique) maximal G-invariant Boolean subalgebra of $\mathcal{B}$. If $\mathcal{B}_G \neq \mathcal{B}$, we will find some $\overline{f} \in \mathcal{B}(G/P) \setminus \mathcal{B}_G$, such that $G\,\overline{f} \subseteq \mathcal{B}$. (This is a contradiction.)

(17.3.4) Assumption. To simplify the algebra in the proof of Proposition 17.3.2, let us assume $G = \mathrm{SL}(2, \mathbb{R}) \times \mathrm{SL}(2, \mathbb{R})$.

(17.3.5) Notation.

  • $G = G_1 \times G_2$, where $G_1 = G_2 = \mathrm{SL}(2, \mathbb{R})$,

  • $P = P_1 \times P_2$, where $P_i = \begin{bmatrix} * & \\ * & * \end{bmatrix} \subset G_i$,

  • $U = U_1 \times U_2$, where $U_i = \begin{bmatrix} 1 & \\ * & 1 \end{bmatrix} \subset P_i$,

  • $V = V_1 \times V_2$, where $V_i = \begin{bmatrix} 1 & * \\ & 1 \end{bmatrix} \subset G_i$,

  • $\Gamma =$ some irreducible lattice in G, and

  • $\mathcal{B} =$ some Γ-invariant sub-σ-algebra of $\mathcal{B}(G/P)$.

(17.3.6) Remark. We have $G/P = (G_1/P_1) \times (G_2/P_2)$. Here are two useful, concrete descriptions of this space:

  • $G/P = \mathbb{R}P^1 \times \mathbb{R}P^1 \cong \mathbb{R}^2$ (a.e.), and

  • $G/P \cong V_1 \times V_2$ (a.e.) (see Exercise 4).

Note that, if we identify G/P with $\mathbb{R}^2$ (a.e.), then, for the action of G_1 on G/P, we have

  • $\begin{bmatrix} k & \\ & k^{-1} \end{bmatrix} (x,y) = (k^2 x, y)$, and

  • $\begin{bmatrix} 1 & t \\ & 1 \end{bmatrix} (x,y) = (x+t, y)$

(see Exercise 3).

The proof of Proposition 17.3.2 employs two preliminary results. The first is based on a standard fact from first-year analysis:

(17.3.7) Lemma (Lebesgue Differentiation Theorem). *Let*

  • $f \in \mathscr{L}^1(\mathbb{R}^n)$,

  • λ *be the Lebesgue measure on $\mathbb{R}^n$, and*

  • $B_r(p)$ *be the ball of radius r centered at p.*

For a.e. $p \in \mathbb{R}^n$, we have

$$\lim_{r \to 0} \frac{1}{\lambda(B_r(p))} \int_{B_r(p)} f \, d\lambda = f(p). \qquad (17.3.8)$$

Letting $n = 1$ and applying Fubini's Theorem yields:

(17.3.9) Corollary. *Let*

- $f \in \mathcal{L}^\infty(\mathbb{R}^2)$,

- $a = \begin{bmatrix} k & \\ & k^{-1} \end{bmatrix} \in G_1$, *for some $k > 1$, and*

- $\pi_2 \colon \mathbb{R}^2 \to \{0\} \times \mathbb{R}$ *be the projection onto the y-axis.*

Then, for a.e. $v \in V_1$,

$$a^n v f \text{ converges in measure to } (v f) \circ \pi_2 \text{ as } n \to \infty.$$

Proof. Exercise 6. $\qquad\qquad\qquad\qquad\qquad\qquad\qquad\qquad \Box$

The other result to be used in the proof of Proposition 17.3.2 is a consequence of the Moore Ergodicity Theorem:

(17.3.10) Proposition. *For a.e. $v \in V_1$, $\Gamma v^{-1} a^{-\mathbb{N}}$ is dense in G.*

Proof. Taking inverses, we wish to show $\overline{a^{\mathbb{N}} v \Gamma} = G$; i.e., the (forward) a-orbit of $v\Gamma$ is dense in G/Γ, for a.e. $v \in V_1$. We will show that

$$\overline{a^{\mathbb{N}} g \Gamma} = G, \text{ for a.e. } g \in G,$$

and leave the remainder of the proof to the reader (see Exercise 7).

Given a nonempty open subset $\mathcal{O}$ of G/Γ, let

$$E = \bigcup_{n > 0} a^{-n} \mathcal{O}.$$

Clearly, $a^{-1} E \subseteq E$. Since $\mu(a^{-1} E) = \mu(E)$ (because the measure on G/Γ is G-invariant), this implies E is a-invariant (a.e.). Since the Moore Ergodicity Theorem (14.2.4) tells us that a is ergodic on G/Γ, we conclude that $E = G/\Gamma$ (a.e.). This means that, for a.e. $g \in G$, the forward a-orbit of g intersects $\mathcal{O}$.

Since $\mathcal{O}$ is an arbitrary open subset, and G/Γ is second countable, we conclude that the forward a-orbit of a.e. g is dense. $\qquad\qquad\qquad \Box$

Proof of Proposition 17.3.2 for $G = \mathrm{SL}(2, \mathbb{R}) \times \mathrm{SL}(2, \mathbb{R})$. Identify G/P with $\mathbb{R}^2$, as in Remark 17.3.6. Since $\mathcal{B}$ is nontrivial, it contains some nonconstant f. Now f cannot be essentially constant both on almost every vertical line and on almost every horizontal line (see Exercise 8), so we may assume there is a non-null set of vertical lines on which it is not constant. This means that

$$\left\{ v \in V_1 \;\middle|\; \begin{array}{c} (v f) \circ \pi_2 \text{ is not} \\ \text{essentially constant} \end{array} \right\} \text{ has positive measure.}$$

Corollary 17.3.9 and Proposition 17.3.10 tell us we may choose v in this set, with the additional properties that

- $a^n v f \to (vf) \circ \pi_2$, and
- $\Gamma v^{-1} a^{-\mathbb{N}}$ is dense in G.

Let $\overline{f} = (vf) \circ \pi_2$, so
$$a^n v f \to \overline{f}.$$
Now, for any $g \in G$, there exist $\gamma_i \in \Gamma$ and $n_i \to \infty$, such that
$$g_i := \gamma_i v^{-1} a^{-n_i} \to g.$$
Then we have
$$g_i a^{n_i} v = \gamma_i \in \Gamma,$$
so the Γ-invariance of $\mathcal{B}$ implies
$$\mathcal{B} \ni \gamma_i f = g_i a^{n_i} v f \to g \overline{f}$$
(see Exercise 12). Since $\mathcal{B}$ is closed (see Exercise 11), we conclude that $g\overline{f} \in \mathcal{B}$. Since g is an arbitrary element of G, this means $G\overline{f} \subseteq \mathcal{B}$. Also, from the choice of v, we know that $\overline{f} = (vf) \circ \pi_2$ is not essentially constant. □

Combining the above argument with a list of the G-invariant Boolean subalgebras of $\mathcal{B}(G/P)$ yields Theorem 17.1.6′:

Proof of Theorem 17.1.6′ for $G = \mathrm{SL}(2, \mathbb{R}) \times \mathrm{SL}(2, \mathbb{R})$. Let $\mathcal{B}_G$ be the largest G-invariant subalgebra of $\mathcal{B}$, and suppose $\mathcal{B} \neq \mathcal{B}_G$. (This will lead to a contradiction.)

It is shown in Exercise 10 that the only G-invariant subalgebras of $\mathcal{B}(G/P) = \mathcal{B}(\mathbb{R}^2)$ are

- $\mathcal{B}(\mathbb{R}^2)$,
- { functions constant on horizontal lines (a.e.) },
- { functions constant on vertical lines (a.e.) }, and
- $\{0, 1\}$.

So $\mathcal{B}_G$ must be one of these 4 subalgebras.

We know $\mathcal{B}_G \neq \mathcal{B}(\mathbb{R}^2)$ (otherwise $\mathcal{B} = \mathcal{B}_G$). Also, we know $\mathcal{B}$ is nontrivial (otherwise $\mathcal{B} = \{0, 1\} = \mathcal{B}_G$), so Proposition 17.3.2 tells us that $\mathcal{B}_G \neq \{0, 1\}$. Hence, we may assume, by symmetry, that
$$\mathcal{B}_G = \{ \text{functions constant on vertical lines (a.e.)} \}. \tag{17.3.11}$$
Since $\mathcal{B} \neq \mathcal{B}_G$, there is some $f \in \mathcal{B}$, such that f is *not* essentially constant on vertical lines. Applying the proof of Proposition 17.3.2 yields $\overline{f}$, such that

- $G\overline{f} \subseteq \mathcal{B}$, so $\overline{f} \in \mathcal{B}_G$, and
- $\overline{f}$ is *not* essentially constant on vertical lines.

This contradicts (17.3.11). □

Very similar ideas yield the general case of Theorem 17.1.6, if one is familiar with real roots and parabolic subgroups. To illustrate this, without using extensive Lie-theoretic language, let us explicitly describe the setup for $G = \mathrm{SL}(3, \mathbb{R})$.

Modifications for SL(3, ℝ).

- $P = \begin{bmatrix} * \\ * & * \\ * & * & * \end{bmatrix}$, $V = \begin{bmatrix} 1 & * & * \\ & 1 & * \\ & & 1 \end{bmatrix}$, $V_1 = \begin{bmatrix} 1 & * \\ & 1 \\ & & 1 \end{bmatrix}$, $V_2 = \begin{bmatrix} 1 \\ & 1 & * \\ & & 1 \end{bmatrix}$.

  Note that $V = \langle V_1, V_2 \rangle$.

- There are exactly four subgroups containing P, namely,

  $$P, \quad G, \quad P_1 = \begin{bmatrix} * & * \\ * & * \\ * & * & * \end{bmatrix} = \langle V_1, P \rangle, \quad P_2 = \begin{bmatrix} * \\ * & * & * \\ * & * & * \end{bmatrix} = \langle V_2, P \rangle.$$

  Hence, there are precisely four G-invariant subalgebras of $\mathcal{B}(G/P)$. Namely, if we identify $\mathcal{B}(G/P)$ with $\mathcal{B}(V)$, then the G-invariant subalgebras of $\mathcal{B}(V)$ are
  - $\mathcal{B}(V)$,
  - $\{0, 1\}$,
  - right V_1-invariant functions,
  - right V_2-invariant functions.

  (17.3.12) *Remark.* The homogeneous spaces G/P_1 and G/P_2 are $\mathbb{R}P^2$ and the Grassmannian $G_{2,3}$ of 2-planes in $\mathbb{R}^3$ (see Exercise 13). Hence, in geometric terms, the G-invariant Boolean subalgebras of $\mathcal{B}(G/P)$ are $\mathcal{B}(G/P)$, $\{0, 1\}$, $\mathcal{B}(\mathbb{R}P^2)$, and $\mathcal{B}(G_{2,3})$.

- Let π_2 be the projection onto V_2 in the natural semidirect product $V = V_2 \ltimes V_2^\perp$, where $V_2^\perp = \begin{bmatrix} 1 & * \\ & 1 & * \\ & & 1 \end{bmatrix}$.

- For $a = \begin{bmatrix} k \\ & k \\ & & 1/k^2 \end{bmatrix} \in G$, Exercise 14 tells us

  $$a \begin{bmatrix} 1 & x & z \\ & 1 & y \\ & & 1 \end{bmatrix} P = \begin{bmatrix} 1 & x & k^3 z \\ & 1 & k^3 y \\ & & 1 \end{bmatrix} P. \qquad (17.3.13)$$

- A generalization of the Lebesgue Differentiation Theorem tells us, for $f \in \mathcal{B}(G/P) = \mathcal{B}(V)$ and a.e. $v \in V_2^\perp$, that

  $$a^n v f \text{ converges in measure to } (vf) \circ \pi_2.$$

With these facts in hand, it is not difficult to prove Theorem 17.1.6′ under the assumption that $G = \mathrm{SL}(3, \mathbb{R})$ (see Exercise 15).

Exercises for §17.3.

#1. In the setting of Theorem 17.1.6, define $\psi^*\colon \mathcal{B}(Z) \to \mathcal{B}(G/P)$ by $\psi^*(f) = f \circ \psi$. Show that ψ^* is injective and Γ-equivariant.

[*Hint:* Injectivity relies on the fact that ψ is measure-class preserving.]

#2. In the setting of Theorem 17.1.6, show that if the sub-σ-algebra $\psi^*(\mathcal{B}(Z))$ of $\mathcal{B}(G/P)$ is G-invariant, then Z is a G-equivariant quotient of G/P (a.e.).

[*Hint:* To reduce problems of measurability, you may pretend that G is countable. More precisely, use Exercise 14.4#5 to show that if H is any countable subgroup of G that contains Γ, then the Γ-action can be extended to an action of H on Z by Borel maps, such that, for each $h \in H$, we have $\psi(hx) = h\psi(x)$ for a.e. $x \in G/P$.]

#3. Let G_i and P_i be as in Notation 17.3.5. Show that choosing appropriate coordinates on $\mathbb{R}P^1 = \mathbb{R} \cup \{\infty\}$ identifies the action of G_i on G_i/P_i with the action of $G_i = \mathrm{SL}(2, \mathbb{R})$ on $\mathbb{R} \cup \{\infty\}$ by linear-fractional transformations:

$$\begin{bmatrix} a & b \\ c & d \end{bmatrix}(x) = \frac{ax + b}{cx + d}.$$

In particular,

$$\begin{bmatrix} k & \\ & k^{-1} \end{bmatrix}(x) = k^2 x \qquad \text{and} \qquad \begin{bmatrix} 1 & t \\ & 1 \end{bmatrix}(x) = x + t.$$

[*Hint:* Map a nonzero vector $(x_1, x_2) \in \mathbb{R}^2$ to its reciprocal slope $x_1/x_2 \in \mathbb{R} \cup \{\infty\}$.]

#4. Let G_i, P_i, and V_i be as in Notation 17.3.5. Show that the map $V_i \to G_i/P_i\colon v \mapsto vP_i$ injective and measure-class preserving.

[*Hint:* Exercise 3.]

#5. Show that Equation (17.3.8) is equivalent to

$$\lim_{k \to \infty} \frac{1}{\lambda(B_1(0))} \int_{B_1(0)} f\left(p + \frac{x}{k}\right) d\lambda(x) = f(p).$$

[*Hint:* A change of variables maps $B_1(0)$ onto $B_r(p)$ with $r = 1/k$.]

#6. Prove Corollary 17.3.9.

[*Hint:* Exercise 5.]

#7. Complete the proof of Proposition 17.3.10: assume, for a.e. $g \in G$, that $a^{\mathbb{N}} g\Gamma$ is dense in G, and show, for a.e. $v \in V_1$, that $a^{\mathbb{N}} v\Gamma$ is dense in G.

[*Hint:* If $a^{\mathbb{N}} g\Gamma$ is dense, then the same is true when g is replaced by any element of $C_G(a) U_1 g$.]

#8. Let $f \in \mathcal{B}(\mathbb{R}^2)$. Show that if f is essentially constant on a.e. vertical line and on a.e. horizontal line, then f is constant (a.e.).

#9. Assume Notation 17.3.5. Show that the only subgroups of G containing P are P, $G_1 \times P_2$, $P_1 \times G_2$, and G.

[*Hint:* P is the stabilizer of a point in $\mathbb{R}P^1 \times \mathbb{R}P^1$, and has only 4 orbits.]

#10. Assume Notation 17.3.5. Show that the only G-equivariant quotients of G/P are G/P, G_2/P_2, G_1/P_1, and G/G.
[*Hint:* Exercise 9.]

#11. Suppose $\mathcal{B}$ is a sub-σ-algebra of $\mathcal{B}(G/P)$. Show that $\mathcal{B}$ is closed under convergence in measure.

More precisely, fix a probability measure μ in the Lebesgue measure class on G/P, and show that $\mathcal{B}$ is a closed in the topology corresponding to the metric on $\mathcal{B}(G/P)$ that is defined by $d(A_1, A_2) = \mu(A_1 \bigtriangleup A_2)$.

#12. Show that the action of G on $\mathcal{B}(G/P)$ is continuous.
[*Hint:* Suppose $g_n \to e$ and $\mu(A_n \bigtriangleup A) \to 0$. The Radon-Nikodym derivative $d(g_n)_* \mu/d\mu$ tends uniformly to 1, so $\mu(g_n A_n \bigtriangleup g_n A) \to 0$. To bound $\mu(g_n A \bigtriangleup A)$, note that $\int_{g_n A} \varphi \, d\mu \to \int_A \varphi d\mu$, for every $\varphi \in C_c(G/P)$.]

#13. In the notation of Remark 17.3.12, show that G/P_1 and G/P_2 are G-equivariantly diffeomorphic to $\mathbb{R}P^2$ and $G_{2,3}$, respectively.
[*Hint:* Verify that the stabilizer of a point in $\mathbb{R}P^2$ is P_1, and the stabilizer of a point in $G_{2,3}$ is P_2.]

#14. Verify Equation (17.3.13).
[*Hint:* Since $a \in P$, we have $agP = (aga^{-1})P$, for any $g \in G$.]

#15. Prove Theorem 17.1.6' under the assumption that $G = \mathrm{SL}(3, \mathbb{R})$.
[*Hint:* You may assume (without proof) the facts stated in the "Modifications for $\mathrm{SL}(3, \mathbb{R})$."]

Notes

The Normal Subgroups Theorem (17.1.1) is due to G. A. Margulis [5, 6, 7]. Expositions of the proof appear in [8, Chap. 4] and [12, Chap. 8]. (However, the proof in [12] assumes that G has Kazhdan's property (T).)

When Γ is not cocompact, the Normal Subgroups Theorem can be proved by algebraic methods derived from the proof of the Congruence Subgroup Problem (see [9, Thms. A and B, p. 109] and [10, Cor. 1, p. 75]). On the other hand, it seems that the ergodic-theoretic approach of Margulis provides the only known proof in the cocompact case.

Regarding Remark 17.1.3(4), see [11] for an introduction to the Congruence Subgroup Property.

Theorem 17.1.6 is stated for general G of real rank ≥ 2 in [8, Cor. 2.13] and [12, Thm. 8.1.4]. Theorem 17.1.6' is in [8, Thm. 4.2.11] and [12, Thm. 8.1.3]. See [12, §8.2 and §8.3] and [8, §4.2] for expositions of the proof.

The proof of Theorem 17.2.1 is adapted from [3, 5.5.F, pp. 150–152].

The Higman-Neumann-Neumann Theorem on SQ-universality of F_2 (see p. 353) was proved in [4]. A very general version of Theorem 17.2.5 that applies to all relatively hyperbolic groups was proved in [1]. (The notion of a relatively hyperbolic group was introduced in [3], and generalized in [2].)

References

[1] G. Arzhantseva, A. Minasyan, and D. Osin: The SQ-universality and residual properties of relatively hyperbolic groups, *J. Algebra* 315 (2007), no. 1, pp. 165–177. MR 2344339, http://dx.doi.org/10.1016/j.jalgebra.2007.04.029

[2] B. Farb: Relatively hyperbolic groups, *Geom. Funct. Anal.* 8 (1998), no. 5, 810–840. MR 1650094, http://dx.doi.org/10.1007/s000390050075

[3] M. Gromov: Hyperbolic groups, in: S. M. Gersten, ed., *Essays in Group Theory*. Springer, New York, 1987, pp. 75–263. ISBN 0-387-96618-8, MR 0919829

[4] G. Higman, B. H. Neumann and H. Neumann: Embedding theorems for groups, *J. London Math. Soc.* 24 (1949), 247–254. MR 0032641, http://dx.doi.org/10.1112/jlms/s1-24.4.247

[5] G. A. Margulis: Factor groups of discrete subgroups, *Soviet Math. Doklady* 19 (1978), no. 5, 1145–1149 (1979). MR 0507138

[6] G. A. Margulis: Quotient groups of discrete subgroups and measure theory, *Func. Anal. Appl.* 12 (1978), no. 4, 295–305 (1979). MR 0515630, http://dx.doi.org/10.1007/BF01076383

[7] G. A. Margulis: Finiteness of quotient groups of discrete subgroups, *Func. Anal. Appl.* 13 (1979), no. 3, 178–187 (1979). MR 0545365, http://dx.doi.org/10.1007/BF01077485

[8] G. A. Margulis: *Discrete Subgroups of Semisimple Lie Groups*. Springer, Berlin Heidelberg New York, 1991. ISBN 3-540-12179-X, MR 1090825

[9] M. S. Raghunathan: On the congruence subgroup problem, *Publ. Math. IHES* 46 (1976) 107–161. MR 0507030 http://www.numdam.org/item?id=PMIHES_1976__46__107_0

[10] M. S. Raghunathan: On the congruence subgroup problem, II, *Invent. Math.* 85 (1986), no. 1, 73–117. MR 0842049, http://eudml.org/doc/143360

[11] B. Sury: *The Congruence Subgroup Problem*. Hindustan Book Agency, New Delhi, 2003. ISBN 81-85931-38-0, MR 1978430

[12] Robert J. Zimmer: *Ergodic Theory and Semisimple Groups*. Birkhäuser, Basel, 1984. ISBN 3-7643-3184-4, MR 0776417

Chapter 18

Arithmetic Subgroups of Classical Groups

This chapter will give a quite explicit description (up to commensurability) of all the arithmetic subgroups of almost every classical Lie group G (see Theorem 18.5.3). (Recall that a simple Lie group G is "classical" if it is either a special linear group, an orthogonal group, a unitary group, or a symplectic group (see Definition A2.1).) The key point is that all the $\mathbb{Q}$-forms of G are also classical, not exceptional, so they are fairly easy to understand. However, there is an exception to this rule: some 8-dimensional orthogonal groups have $\mathbb{Q}$-forms of so-called "triality type" that are not classical and will not be discussed in any detail here (see Remark 18.5.10).

Given G, which is a Lie group over $\mathbb{R}$, we would like to know all of its $\mathbb{Q}$-forms (because, by definition, arithmetic groups are made from $\mathbb{Q}$-forms). However, we will start with the somewhat simpler problem that replaces the fields $\mathbb{Q}$ and $\mathbb{R}$ with the fields $\mathbb{R}$ and $\mathbb{C}$: finding the $\mathbb{R}$-forms of the classical Lie groups over $\mathbb{C}$.

Recall: The Standing Assumptions (4.0.0 on page 41) are in effect, so, as always, Γ is a lattice in the semisimple Lie group $G \subseteq \mathrm{SL}(\ell, \mathbb{R})$.

Main prerequisites for this chapter: Restriction of Scalars (Section 5.5) and examples of arithmetic subgroups (Chapter 6).

§18.1. ℝ-forms of classical simple groups over ℂ

To set the stage, let us recall the classical result that almost all complex simple groups are classical:

(18.1.1) **Theorem** (Cartan, Killing). *All but finitely many of the simple Lie groups over ℂ are isogenous to either* $\mathrm{SL}(n,\mathbb{C})$, $\mathrm{SO}(n,\mathbb{C})$, *or* $\mathrm{Sp}(2n,\mathbb{C})$, *for some n.*

(18.1.2) *Remark.* Up to isogeny, there are exactly five simple Lie groups over ℂ that are not classical. They are the "exceptional" simple groups, and are called E_6, E_7, E_8, F_4, and G_2.

Now, we would like to describe the ℝ-forms of each of the classical groups. For example, finding all the ℝ-forms of $\mathrm{SL}(n,\mathbb{C})$ would mean making a list of the (simple) Lie groups G, such that the "complexification" of G is $\mathrm{SL}(n,\mathbb{C})$. This is not difficult, but we should perhaps begin by explaining more clearly what it means.

It has already been mentioned that, intuitively, the complexification of G is the complex Lie group that is obtained from G by replacing real numbers with complex numbers. For example, the complexification of $\mathrm{SL}(n,\mathbb{R})$ is $\mathrm{SL}(n,\mathbb{C})$. In general, G is (isogenous to) the set of real solutions of a certain set of equations, and we let $G_{\mathbb{C}}$ be the set of complex solutions of the same set of equations:

(18.1.3) **Notation.** Assume $G \subseteq \mathrm{SL}(\ell,\mathbb{R})$, for some ℓ. Since G is almost Zariski closed (see Theorem A4.9), there is a certain subset $\mathcal{Q}$ of $\mathbb{R}[x_{1,1},\ldots,x_{\ell,\ell}]$, such that $G^{\circ} = \mathrm{Var}(\mathcal{Q})^{\circ}$. Let

$$G_{\mathbb{C}} = \mathrm{Var}_{\mathbb{C}}(\mathcal{Q}) = \{\, g \in \mathrm{SL}(\ell,\mathbb{C}) \mid Q(g) = 0, \text{ for all } Q \in \mathcal{Q} \,\}.$$

Then $G_{\mathbb{C}}$ is a (complex, semisimple) Lie group.

(18.1.4) **Example.**

1) $\mathrm{SL}(n,\mathbb{R})_{\mathbb{C}} = \mathrm{SL}(n,\mathbb{C})$.
2) $\mathrm{SO}(n)_{\mathbb{C}} = \mathrm{SO}(n,\mathbb{C})$.
3) $\mathrm{SO}(m,n)_{\mathbb{C}} \cong \mathrm{SO}(m+n,\mathbb{C})$ (see Exercise 1).

(18.1.5) **Definition.** If $G_{\mathbb{C}}$ is isomorphic to H, then we say that

- H is the *complexification* of G, and that
- G is an *ℝ-form* of H.

The following result lists the complexification of each classical group. It is not difficult to memorize the correspondence. For example, it is obvious from the notation that the complexification of $\mathrm{Sp}(m,n)$ should be symplectic. Indeed, the only case that really requires memorization is the complexification of $\mathrm{SU}(m,n)$ (see Proposition 18.1.6(Aiv)).

(18.1.6) **Proposition.** *Here is the complexification of each classical Lie group.*

   A) *Real forms of special linear groups:*
      (i) $\mathrm{SL}(n, \mathbb{R})_{\mathbb{C}} = \mathrm{SL}(n, \mathbb{C})$,
      (ii) $\mathrm{SL}(n, \mathbb{C})_{\mathbb{C}} \cong \mathrm{SL}(n, \mathbb{C}) \times \mathrm{SL}(n, \mathbb{C})$,
      (iii) $\mathrm{SL}(n, \mathbb{H})_{\mathbb{C}} \cong \mathrm{SL}(2n, \mathbb{C})$,
      (iv) $\mathrm{SU}(m, n)_{\mathbb{C}} \cong \mathrm{SL}(m + n, \mathbb{C})$.

   B) *Real forms of orthogonal groups:*
      (i) $\mathrm{SO}(m, n)_{\mathbb{C}} \cong \mathrm{SO}(m + n, \mathbb{C})$,
      (ii) $\mathrm{SO}(n, \mathbb{C})_{\mathbb{C}} \cong \mathrm{SO}(n, \mathbb{C}) \times \mathrm{SO}(n, \mathbb{C})$,
      (iii) $\mathrm{SO}(n, \mathbb{H})_{\mathbb{C}} \cong \mathrm{SO}(2n, \mathbb{C})$.

   C) *Real forms of symplectic groups:*
      (i) $\mathrm{Sp}(n, \mathbb{R})_{\mathbb{C}} = \mathrm{Sp}(n, \mathbb{C})$,
      (ii) $\mathrm{Sp}(n, \mathbb{C})_{\mathbb{C}} \cong \mathrm{Sp}(n, \mathbb{C}) \times \mathrm{Sp}(n, \mathbb{C})$,
      (iii) $\mathrm{Sp}(m, n)_{\mathbb{C}} \cong \mathrm{Sp}(2(m + n), \mathbb{C})$.

Some parts of this proposition are more-or-less obvious (such as $\mathrm{SL}(n, \mathbb{R})_{\mathbb{C}} = \mathrm{SL}(n, \mathbb{C})$). A few other examples appear in Section 18.2 below, and the methods used there can be applied to all of the cases. In fact, all of the calculations are straightforward adaptations of the examples, except perhaps the determination of $\mathrm{SO}(n, \mathbb{H})_{\mathbb{C}}$ (see Exercise 18.2#4).

Nothing in Proposition 18.1.6 is very surprising. What is not at all obvious is that this list of real forms is complete:

(18.1.7) **Theorem** (É. Cartan). *Every real form of* $\mathrm{SL}(n, \mathbb{C})$, $\mathrm{SO}(n, \mathbb{C})$, *or* $\mathrm{Sp}(n, \mathbb{C})$ *appears in Proposition 18.1.6 (up to isogeny).*

We will discuss a proof of this theorem in Section 18.3.

(18.1.8) *Remarks.*

   1) From Proposition 18.1.6, we see that a single complex group may have several different real forms. However, there are always only finitely many (even for exceptional groups).

   2) The Lie algebra of $G_{\mathbb{C}}$ is the tensor product $\mathfrak{g} \otimes \mathbb{C}$ (see Exercise 2). This is independent of the embedding of G in $\mathrm{SL}(\ell, \mathbb{C})$, so, up to isogeny, $G_{\mathbb{C}}$ is independent of the embedding of G in $\mathrm{SL}(\ell, \mathbb{C})$.

   3) We ignored a technical issue in Notation 18.1.3: there may be many different choices of Q (having the same set of real solutions), and it may be the case that different choices yield different sets of complex solutions. (In fact, a bad choice of Q can yield a set of complex solutions that is not a group.) To eliminate this problem, we should insist that Q be maximal; that is,

$$Q = \{ Q \in \mathbb{R}[x_{1,1}, \ldots, x_{\ell,\ell}] \mid Q(g) = 0, \text{ for all } g \in G \}.$$

Then $G_{\mathbb{C}}$ is the Zariski closure of G (over the field $\mathbb{C}$), from which it follows that $G_{\mathbb{C}}$, like G, is a semisimple Lie group.

(18.1.9) **Example.** Because the center of $\mathrm{SL}(3, \mathbb{R})$ is trivial, we see that $\mathrm{SL}(3, \mathbb{R})$ is the same Lie group as $\mathrm{PSL}(3, \mathbb{R})$. On the other hand, we have

$$\mathrm{SL}(3, \mathbb{R})_{\mathbb{C}} = \mathrm{SL}(3, \mathbb{C}) \not\cong \mathrm{PSL}(3, \mathbb{C}) = \mathrm{PSL}(3, \mathbb{R})_{\mathbb{C}}.$$

This is a concrete illustration of the fact that different embeddings of G can yield different complexifications. Note, however, that $\mathrm{SL}(3, \mathbb{C})$ is isogenous to $\mathrm{PSL}(3, \mathbb{C})$, so the difference between the complexifications is negligible (cf. Remark 18.1.8(2)).

Exercises for §18.1.

#1. Show that $\mathrm{SO}(m, n)_{\mathbb{C}} \cong \mathrm{SO}(m + n, \mathbb{C})$.
   [*Hint:* $\mathrm{SO}(m, n)_{\mathbb{C}}$ is conjugate to $\mathrm{SO}(m + n, \mathbb{C})$ in $\mathrm{SL}(m + n, \mathbb{C})$, because -1 is a square in $\mathbb{C}$.]

#2. Show that the Lie algebra of $G_{\mathbb{C}}$ is $\mathfrak{g} \otimes \mathbb{C}$.

§18.2. Calculating the complexification of G

This section justifies Proposition 18.1.6, by calculating the complexification of each classical group.

Let us start with $\mathrm{SL}(n, \mathbb{C})$. This is already a complex Lie group, but we can think of it as a real Lie group of twice the dimension. As such, it has a complexification:

(18.2.1) **Proposition.** $\mathrm{SL}(n, \mathbb{C})_{\mathbb{C}} \cong \mathrm{SL}(n, \mathbb{C}) \times \mathrm{SL}(n, \mathbb{C})$.

Proof. We should embed $\mathrm{SL}(n, \mathbb{C})$ as a subgroup of $\mathrm{SL}(2n, \mathbb{R})$, find the corresponding set $\mathcal{Q}$ of defining polynomials, and determine the complex solutions. However, it is more convenient to sidestep some of these calculations by using restriction of scalars, the method described in §5.5.

Define $\Delta: \mathbb{C} \to \mathbb{C} \oplus \mathbb{C}$ by $\Delta(z) = (z, \overline{z})$. Then the vectors $\Delta(1) = (1, 1)$ and $\Delta(i) = (i, -i)$ are linearly independent (over $\mathbb{C}$), so they form a basis of $\mathbb{C} \oplus \mathbb{C}$. Thus, $\Delta(\mathbb{C})$ is the $\mathbb{R}$-span of a basis, so it is a $\mathbb{R}$-form of $\mathbb{C} \oplus \mathbb{C}$. Therefore, letting $V = \mathbb{C}^{2n}$, we see that

$$V_{\mathbb{R}} = \Delta(\mathbb{C}^n) = \{ (v, \overline{v}) \mid v \in \mathbb{C}^n \}$$

is a real form of V. Let

$$(\mathrm{SL}(n, \mathbb{C}) \times \mathrm{SL}(n, \mathbb{C}))_{\mathbb{R}} = \{ g \in \mathrm{SL}(n, \mathbb{C}) \times \mathrm{SL}(n, \mathbb{C}) \mid g(V_{\mathbb{R}}) = V_{\mathbb{R}} \}.$$

Then we have an isomorphism

$$\tilde{\Delta}: \mathrm{SL}(n, \mathbb{C}) \xrightarrow{\cong} (\mathrm{SL}(n, \mathbb{C}) \times \mathrm{SL}(n, \mathbb{C}))_{\mathbb{R}},$$

defined by $\tilde{\Delta}(g) = (g, \overline{g})$, so

$$\mathrm{SL}(n, \mathbb{C})_{\mathbb{C}} \cong ([\mathrm{SL}(n, \mathbb{C}) \times \mathrm{SL}(n, \mathbb{C})]_{\mathbb{R}})_{\mathbb{C}} = \mathrm{SL}(n, \mathbb{C}) \times \mathrm{SL}(n, \mathbb{C}). \qquad \square$$

(18.2.2) *Remarks.*

1) Generalizing Proposition 18.2.1, one can show that if G is isogenous to a complex Lie group, then $G_{\mathbb{C}}$ is isogenous to $G \times G$.

2) From Proposition 18.2.1, we see that $G_{\mathbb{C}}$ need not be simple, even if G is simple. However, this only happens when G is complex: if G is simple, and G is not isogenous to a complex Lie group, then $G_{\mathbb{C}}$ is simple.

Although not stated explicitly there, the proof of Proposition 18.2.1 is based on the fact that $\mathbb{C} \otimes_{\mathbb{R}} \mathbb{C} \cong \mathbb{C} \oplus \mathbb{C}$. Namely, the map

$$\mathbb{C} \otimes_{\mathbb{R}} \mathbb{C} \to \mathbb{C} \oplus \mathbb{C} \text{ defined by } v \otimes \lambda \mapsto \Delta(v)\lambda$$

is an isomorphism of $\mathbb{C}$-algebras. Analogously, understanding the complexification of a group defined from the algebra $\mathbb{H}$ of quaternions will be based on a calculation of $\mathbb{H} \otimes_{\mathbb{R}} \mathbb{C}$.

(18.2.3) **Lemma.** *The tensor product $\mathbb{H} \otimes_{\mathbb{R}} \mathbb{C}$ is isomorphic to $\mathrm{Mat}_{2 \times 2}(\mathbb{C})$.*

Proof. Define an $\mathbb{R}$-linear map $\phi \colon \mathbb{H} \to \mathrm{Mat}_{2 \times 2}(\mathbb{C})$ by

$$\phi(1) = \mathrm{Id}, \quad \phi(i) = \begin{bmatrix} i & 0 \\ 0 & -i \end{bmatrix}, \quad \phi(j) = \begin{bmatrix} 0 & 1 \\ -1 & 0 \end{bmatrix}, \quad \phi(k) = \begin{bmatrix} 0 & i \\ i & 0 \end{bmatrix}.$$

It is straightforward to verify that ϕ is an injective ring homomorphism. Furthermore, $\phi(\{1, i, j, k\})$ is a $\mathbb{C}$-basis of $\mathrm{Mat}_{2 \times 2}(\mathbb{C})$. Therefore, the map $\hat{\phi} \colon \mathbb{H} \otimes \mathbb{C} \to \mathrm{Mat}_{2 \times 2}(\mathbb{C})$ defined by $\hat{\phi}(v \otimes \lambda) = \phi(v)\lambda$ is a ring isomorphism (see Exercise 1). $\qquad\square$

(18.2.4) **Proposition.** $\mathrm{SL}(n, \mathbb{H})_{\mathbb{C}} \cong \mathrm{SL}(2n, \mathbb{C})$.

Proof. From Lemma 18.2.3, we have

$$\mathrm{SL}(n, \mathbb{H})_{\mathbb{C}} \cong \mathrm{SL}(n, \mathrm{Mat}_{2 \times 2}(\mathbb{C})) \cong \mathrm{SL}(2n, \mathbb{C})$$

(see Exercises 2 and 3). $\qquad\square$

As additional examples, let us look at the complexifications of the classical simple Lie groups that are compact, namely, $\mathrm{SO}(n)$, $\mathrm{SU}(n)$, and $\mathrm{Sp}(n)$. As observed in Example 18.1.4(2), we have $\mathrm{SO}(n)_{\mathbb{C}} = \mathrm{SO}(n, \mathbb{C})$. The other cases are not as obvious.

(18.2.5) **Proposition.** $\mathrm{SU}(n)_{\mathbb{C}} = \mathrm{SL}(n, \mathbb{C})$.

Proof. Let

- $\sigma \colon \mathbb{C} \to \mathbb{C}$,
- $\vec{\sigma} \colon \mathbb{C}^n \to \mathbb{C}^n$, and
- $\tilde{\sigma} \colon \mathrm{SL}(n, \mathbb{C}) \to \mathrm{SL}(n, \mathbb{C})$

be the usual complex conjugations $\sigma(z) = \bar{z}$, $\vec{\sigma}(v) = \bar{v}$, and $\tilde{\sigma}(g) = \bar{g}$. We have

$$SU(n) = \{\, g \in SL(n, \mathbb{C}) \mid g^* g = \mathrm{Id} \,\}$$
$$= \{\, g \in SL(n, \mathbb{C}) \mid \tilde{\sigma}(g^T) g = \mathrm{Id} \,\},$$

so, in order to calculate $SU(n)_{\mathbb{C}}$, we should determine the map $\tilde{\eta}$ on $SL(n, \mathbb{C}) \times SL(n, \mathbb{C})$ that corresponds to $\tilde{\sigma}$ when we identify $\mathbb{C}^n$ with $(\mathbb{C}^n \oplus \mathbb{C}^n)_{\mathbb{R}}$ under the map $\vec{\Delta}$.

First, let us determine $\vec{\eta}$. That is, we wish to identify $\mathbb{C}^n$ with $\mathbb{R}^{2n}$, and extend $\vec{\sigma}$ to a $\mathbb{C}$-linear map on $\mathbb{C}^{2n}$. However, as usual, we use the $\mathbb{R}$-form $\vec{\Delta}(\mathbb{C}^n)$, in place of $\mathbb{R}^{2n}$. It is obvious that if we

define $\vec{\eta}: \mathbb{C}^n \oplus \mathbb{C}^n \to \mathbb{C}^n \oplus \mathbb{C}^n$ by $\vec{\eta}(x, y) = (y, x)$,

then $\vec{\eta}$ is $\mathbb{C}$-linear, and the following diagram commutes:

$$
\begin{array}{ccc}
\mathbb{C}^n & \overset{\vec{\Delta}}{\to} & \mathbb{C}^n \oplus \mathbb{C}^n \\
\downarrow{\vec{\sigma}} & & \downarrow{\vec{\eta}} \\
\mathbb{C}^n & \overset{\vec{\Delta}}{\to} & \mathbb{C}^n \oplus \mathbb{C}^n.
\end{array}
$$

Thus, it is fairly clear that $\tilde{\eta}(g, h) = (h, g)$. Hence

$$SU(n)_{\mathbb{C}} = \{\, (g, h) \in SL(n, \mathbb{C}) \times SL(n, \mathbb{C}) \mid \tilde{\eta}(g^T, h^T)(g, h) = (\mathrm{Id}, \mathrm{Id}) \,\}$$
$$= \{\, (g, h) \in SL(n, \mathbb{C}) \times SL(n, \mathbb{C}) \mid (h^T, g^T)(g, h) = (\mathrm{Id}, \mathrm{Id}) \,\}$$
$$= \{\, (g, (g^T)^{-1}) \mid g \in SL(n, \mathbb{C}) \,\}$$
$$\cong SL(n, \mathbb{C}). \qquad \square$$

(18.2.6) **Proposition.** $Sp(n)_{\mathbb{C}} = Sp(2n, \mathbb{C})$.

Proof. Let

- $\phi: \mathbb{H} \to \mathrm{Mat}_{2 \times 2}(\mathbb{C})$ be the embedding that is described in the proof of Lemma 18.2.3,

- τ be the usual conjugation on $\mathbb{H}$,

- $J = \begin{bmatrix} 0 & 1 \\ -1 & 0 \end{bmatrix}$, and

- $\eta: \mathrm{Mat}_{2 \times 2}(\mathbb{C}) \to \mathrm{Mat}_{2 \times 2}(\mathbb{C})$ be defined by $\eta(x) = J^{-1} x^T J$.

Then η is $\mathbb{C}$-linear, and the following diagram commutes:

$$
\begin{array}{ccc}
\mathbb{H} & \overset{\phi}{\to} & \mathrm{Mat}_{2 \times 2}(\mathbb{C}) \\
\downarrow{\tau} & & \downarrow{\eta} \\
\mathbb{H} & \overset{\phi}{\to} & \mathrm{Mat}_{2 \times 2}(\mathbb{C}).
\end{array}
$$

Hence, because
$$\mathrm{Sp}(2) = \{\, g \in \mathrm{SL}(2, \mathbb{H}) \mid g^* g = \mathrm{Id} \,\}$$
$$= \left\{\, \begin{bmatrix} a & b \\ c & d \end{bmatrix} \in \mathrm{SL}(2, \mathbb{H}) \; \middle| \; \begin{bmatrix} \tau(a) & \tau(c) \\ \tau(b) & \tau(d) \end{bmatrix} \begin{bmatrix} a & b \\ c & d \end{bmatrix} = \mathrm{Id} \,\right\},$$

we see that
$$\mathrm{Sp}(2)_{\mathbb{C}} = \left\{\, \begin{bmatrix} a & b \\ c & d \end{bmatrix} \in \mathrm{SL}(2, \mathrm{Mat}_{2\times2}(\mathbb{C})) \; \middle| \; \begin{bmatrix} \eta(a) & \eta(c) \\ \eta(b) & \eta(d) \end{bmatrix} \begin{bmatrix} a & b \\ c & d \end{bmatrix} = \mathrm{Id} \,\right\}$$
$$= \left\{\, \begin{bmatrix} a & b \\ c & d \end{bmatrix} \in \mathrm{SL}(2, \mathrm{Mat}_{2\times2}(\mathbb{C})) \; \middle| \; J^{-1} \begin{bmatrix} a^T & c^T \\ b^T & d^T \end{bmatrix} J \begin{bmatrix} a & b \\ c & d \end{bmatrix} = \mathrm{Id} \,\right\}$$
$$= \left\{\, g \in \mathrm{SL}(4, \mathbb{C}) \; \middle| \; J^{-1} g^T J g = \mathrm{Id} \,\right\}$$
$$= \left\{\, g \in \mathrm{SL}(4, \mathbb{C}) \; \middle| \; g^T J g = J \,\right\}$$
$$= \mathrm{Sp}(4, \mathbb{C}).$$

Similarly, letting
$$\hat{J}_n = \begin{bmatrix} J & & & \\ & J & & \\ & & \ddots & \\ & & & J \end{bmatrix} \in \mathrm{SL}(2n, \mathbb{C}),$$

the same calculations show that
$$\mathrm{Sp}(n)_{\mathbb{C}} = \{\, g \in \mathrm{SL}(2n, \mathbb{C}) \mid g^T \hat{J}_n g = \hat{J}_n \,\} \cong \mathrm{Sp}(2n, \mathbb{C}). \qquad \square$$

Exercises for §18.2.

#1. In the proof of Lemma 18.2.3, verify:
   a) ϕ is an injective ring homomorphism,
   b) $\phi(\{1, i, j, k\})$ is a $\mathbb{C}$-basis of $\mathrm{Mat}_{2\times2}(\mathbb{C})$, and
   c) $\hat{\phi}$ is an isomorphism of $\mathbb{C}$-algebras.

#2. Show $\mathrm{SL}(n, \mathbb{H})_{\mathbb{C}} \cong \mathrm{SL}(n, \mathrm{Mat}_{2\times2}(\mathbb{C}))$.
   [*Hint:* Define ϕ as in the proof of Lemma 18.2.3. Use the proof of Proposition 18.2.1, with ϕ in the place of Δ.]

#3. Show $\mathrm{SL}(n, \mathrm{Mat}_{d\times d}(\mathbb{C})) \cong \mathrm{SL}(dn, \mathbb{C})$.

#4. Show that $\mathrm{SO}(n, \mathbb{H})_{\mathbb{C}} \cong \mathrm{SO}(2n, \mathbb{C})$.
   [*Hint:* Similar to (18.2.6). To calculate $\tau_r \otimes \mathbb{C}$, note that $\tau_r(x) = j^{-1} \tau(x) j$, for $x \in \mathbb{H}$.]

§18.3. How to find the real forms of complex groups

In this section, we will explain how to find all of the possible $\mathbb{R}$-forms of $\mathrm{SL}(n, \mathbb{C})$. (Similar techniques can be used to justify the other cases of Theorem 18.1.7, but additional calculations are needed, and we omit the details.) We take an algebraic approach, based on Galois theory, and we first review the most basic terminology from the theory of (nonabelian) group cohomology.

§18.3(i). Definition of the first cohomology of a group.

(18.3.1) **Definitions.** Suppose a group X acts (on the left) by automorphisms on a group M. (For $x \in X$ and $m \in M$, we write $^x m$ for the image of m under x.)

1) A function $\alpha\colon X \to M$ is a **1-cocycle** (or "crossed homomorphism") if
$$\alpha(xy) = \alpha(x) \cdot {}^x\alpha(y) \text{ for all } x, y \in X.$$

2) Two 1-cocycles α and β are equivalent (or "cohomologous") if there is some $m \in M$, such that
$$\alpha(x) = m^{-1} \cdot \beta(x) \cdot {}^x m \text{ for all } x \in X.$$

3) $\mathcal{H}^1(X; M)$ is the set of equivalence classes of all 1-cocycles. It is called the **first cohomology** of X with coefficients in M.

4) A 1-cocycle is a **coboundary** if it is cohomologous to the trivial 1-cocycle defined by $\tau(x) = e$ for all $x \in X$.

(18.3.2) **Warning.** In our applications, the coefficient group M is sometimes **non**abelian. In this case, $\mathcal{H}^1(X; M)$ is a set with no obvious algebraic structure. However, if M is an abelian group (as is often assumed in textbooks on group cohomology), then $\mathcal{H}^1(X; M)$ is an abelian group.

§18.3(ii). How Galois cohomology comes into the picture.

For convenience, let $G_{\mathbb{C}} = \mathrm{SL}(n, \mathbb{C})$. Suppose $\rho\colon G_{\mathbb{C}} \to \mathrm{SL}(N, \mathbb{C})$ is an embedding, such that $\rho(G_{\mathbb{C}})$ is defined over $\mathbb{R}$. We wish to find all the possibilities for the group $\rho(G_{\mathbb{C}})_{\mathbb{R}} = \rho(G_{\mathbb{C}}) \cap \mathrm{SL}(N, \mathbb{R})$ that can be obtained by considering all the possible choices of ρ.

Let σ denote complex conjugation, the nontrivial Galois automorphism of $\mathbb{C}$ over $\mathbb{R}$. Since $\mathbb{R} = \{ z \in \mathbb{C} \mid \sigma(z) = z \}$, we have
$$\mathrm{SL}(N, \mathbb{R}) = \{ g \in \mathrm{SL}(N, \mathbb{C}) \mid \sigma(g) = g \},$$
where we apply σ to a matrix by applying it to each of the matrix entries. Therefore
$$\rho(G_{\mathbb{C}})_{\mathbb{R}} = \rho(G_{\mathbb{C}}) \cap \mathrm{SL}(N, \mathbb{R}) = \{ g \in \rho(G_{\mathbb{C}}) \mid \sigma(g) = g \}.$$
Since $\rho(G_{\mathbb{C}})$ is defined over $\mathbb{R}$, we know that it is invariant under σ, so we have
$$G_{\mathbb{C}} \xrightarrow{\rho} \rho(G_{\mathbb{C}}) \xrightarrow{\sigma} \rho(G_{\mathbb{C}}) \xrightarrow{\rho^{-1}} G_{\mathbb{C}}.$$

Let $\tilde{\sigma} = \rho^{-1}\sigma\rho\colon G_{\mathbb{C}} \to G_{\mathbb{C}}$ be the composition. Then the real form corresponding to ρ is

$$G_{\mathbb{R}} = \rho^{-1}\left(\rho(G_{\mathbb{C}}) \cap \mathrm{SL}(N,\mathbb{R})\right) = \{\, g \in G_{\mathbb{C}} \mid \tilde{\sigma}(g) = g \,\}.$$

To summarize, the obvious $\mathbb{R}$-form of $G_{\mathbb{C}}$ is the set of fixed points of the usual complex conjugation, and any other $\mathbb{R}$-form is the set of fixed points of some other automorphism of $G_{\mathbb{C}}$.

Now let

$$\alpha(\sigma) = \tilde{\sigma}\,\sigma^{-1}\colon G_{\mathbb{C}} \to G_{\mathbb{C}}. \tag{18.3.3}$$

It is not difficult to see that

- $\alpha(\sigma)$ is an automorphism of $G_{\mathbb{C}}$ (as an abstract group), and
- $\alpha(\sigma)$ is holomorphic (since ρ^{-1} and $\sigma\rho\sigma^{-1}$ are holomorphic — in fact, they can be represented by polynomials in local coordinates).

So $\alpha(\sigma) \in \mathrm{Aut}(G_{\mathbb{C}})$. Thus, by defining $\alpha(1)$ to be the trivial automorphism, we obtain a function $\alpha\colon \mathrm{Gal}(\mathbb{C}/\mathbb{R}) \to \mathrm{Aut}(G_{\mathbb{C}})$.

Let $\mathrm{Gal}(\mathbb{C}/\mathbb{R})$ act on $\mathrm{Aut}(G_{\mathbb{C}})$, by defining

$${}^{\sigma}\varphi = \sigma\varphi\sigma^{-1} \quad \text{for } \varphi \in \mathrm{Aut}(G_{\mathbb{C}}).$$

Then $\alpha(\sigma) = \varphi^{-1}\,{}^{\sigma}\varphi$, so $\alpha(\sigma) \cdot {}^{\sigma}\alpha(\sigma) = \alpha(1)$ (since $\sigma^2 = 1$). This means that α is 1-cocycle of group cohomology, and therefore defines an element of the cohomology set $\mathcal{H}^1(\mathrm{Gal}(\mathbb{C}/\mathbb{R}), \mathrm{Aut}(G_{\mathbb{C}}))$. In fact:

This construction provides a one-to-one correspondence between $\mathcal{H}^1(\mathrm{Gal}(\mathbb{C}/\mathbb{R}), \mathrm{Aut}(G_{\mathbb{C}}))$ and the set of $\mathbb{R}$-forms of $G_{\mathbb{C}}$ (18.3.4)

(see Exercise 1). Thus, finding all of the $\mathbb{R}$-forms of $G_{\mathbb{C}}$ amounts to calculating the cohomology of a Galois group, or, in other words, "Galois cohomology."

(18.3.5) **Observation.** The above discussion is an example of a fairly general principle: if X is an algebraic object that is defined over $\mathbb{R}$, then $\mathcal{H}^1(\mathrm{Gal}(\mathbb{C}/\mathbb{R}), \mathrm{Aut}(X_{\mathbb{C}}))$ is in one-to-one correspondence with the set of $\mathbb{R}$-isomorphism classes of $\mathbb{R}$-defined objects whose $\mathbb{C}$-points are isomorphic to $X_{\mathbb{C}}$.

(18.3.6) **Example.** Suppose V_1 and V_2 are two vector spaces over $\mathbb{R}$, and they are isomorphic over $\mathbb{C}$. (I.e., $V_1 \otimes \mathbb{C} \cong V_2 \otimes \mathbb{C}$.) Then the two vector spaces have the same dimension, so elementary linear algebra tells us that they are isomorphic over $\mathbb{R}$. This means that the $\mathbb{R}$-form of any complex vector space $V_{\mathbb{C}}$ is unique (up to isomorphism), so the general principle (18.3.5) tells us

$$\mathcal{H}^1(\mathrm{Gal}(\mathbb{C}/\mathbb{R}), \mathrm{Aut}(V_{\mathbb{C}})) = 0.$$

In other words, we have

$$\mathcal{H}^1(\mathrm{Gal}(\mathbb{C}/\mathbb{R}), \mathrm{GL}(n,\mathbb{C})) = 0.$$

A similar argument shows $\mathcal{H}^1(\mathrm{Gal}(\mathbb{C}/\mathbb{R}), \mathrm{SL}(n,\mathbb{C})) = 0$ (see Exercise 2).

(18.3.7) **Warning.** The "fairly general principle" (18.3.5) is not completely general. Although almost nothing needs to be assumed in order to construct a well-defined, injective map from the set of $\mathbb{Q}$-forms to the cohomology set (cf. Exercise 1), this map might not be surjective. That is, there might be cohomology classes that do not come from $\mathbb{Q}$-forms, unless some (fairly mild) hypotheses are imposed on the class of algebraic objects.

(18.3.8) **Warning.** Up to now, we have usually ignored finite groups in this book: an answer up to isogeny or commensurability was good enough. However, such sloppiness is unacceptable when calculating Galois cohomology groups. For example, even though $SL(n, \mathbb{C})$ is isogenous to $PSL(n, \mathbb{C})$, the two groups have completely different cohomology. Namely:

- we saw in Example 18.3.6 that $\mathcal{H}^1(Gal(\mathbb{C}/\mathbb{R}), SL(n, \mathbb{C}))$ is trivial, but
- Subsection 18.3(iii) will show $\mathcal{H}^1(Gal(\mathbb{C}/\mathbb{R}), PSL(n, \mathbb{C}))$ is infinite.

§18.3(iii). Constructing explicit $\mathbb{R}$-forms from cohomology classes.
Given $\alpha \in \mathcal{H}^1(Gal(\mathbb{C}/\mathbb{R}), Aut(G_\mathbb{C}))$, we will now see how to find the corresponding $\mathbb{R}$-form $G_\mathbb{R}$.

It is known that the **outer** automorphism group of $G_\mathbb{C} = SL(n, \mathbb{C})$ has only one nontrivial element, namely, the "transpose-inverse" automorphism, defined by $w(g) = (g^T)^{-1}$. So

$$Aut(G_\mathbb{C}) = PSL(n, \mathbb{C}) \rtimes \langle w \rangle.$$

We consider two cases.

Case 1. Assume $\alpha \in \mathcal{H}^1(Gal(\mathbb{C}/\mathbb{R}), PSL(n, \mathbb{C}))$. It is a fundamental fact in the theory of finite-dimensional algebras that every $\mathbb{C}$-linear automorphism of the matrix algebra $Mat_{n \times n}(\mathbb{C})$ is inner (see Exercise 5). Since the center acts trivially, this means $Aut(Mat_{n \times n}(\mathbb{C})) = PSL(n, \mathbb{C})$. Therefore,

$$\mathcal{H}^1(Gal(\mathbb{C}/\mathbb{R}), PSL(n, \mathbb{C})) = \mathcal{H}^1(Gal(\mathbb{C}/\mathbb{R}), Aut(Mat_{n \times n}(\mathbb{C}))),$$

so, by the general principle (18.3.5), we can identify this cohomology set with the set of $\mathbb{R}$-forms of $Mat_{n \times n}(\mathbb{C})$. More precisely, it is the set of algebras A over $\mathbb{R}$, such that $A \otimes \mathbb{C} \cong Mat_{n \times n}(\mathbb{C})$. Such an algebra must be simple (since $Mat_{n \times n}(\mathbb{C})$ is simple), so, by Wedderburn's Theorem (6.8.5), it is a matrix algebra over a division algebra: $A \cong Mat_k(D)$, where D is a division algebra over $\mathbb{R}$. The corresponding $\mathbb{R}$-form $G_\mathbb{R}$ is $SL(k, D)$. It is well known that the only division algebras over $\mathbb{R}$ are $\mathbb{R}$, $\mathbb{C}$, and $\mathbb{H}$ (see Exercise 6), so the real form must be either $SL(k, \mathbb{R})$, $SL(k, \mathbb{C})$, or $SL(k, \mathbb{H})$, all of which are on the list in Proposition 18.1.6(A).

Case 2. Assume the image of α is **not** contained in $PSL(n, \mathbb{C})$. In this case, we have $\alpha(\sigma) = $ (conjugation by A) w for some $A \in GL(n, \mathbb{R})$. Hence, for every $g \in G_{\mathbb{R}}$,

$$g = \tilde{\sigma}(g) = (\alpha(\sigma)\,\sigma)(g) = A\,\omega\,(\sigma(g))\,A^{-1} = A\left(({}^{\sigma}g)^T\right)^{-1} A^{-1},$$

which means $g\,A\,({}^{\sigma}g)^T = A$. In other words, g is in the unitary group $\mathrm{SU}(A,\sigma)$ corresponding to the Hermitian form on $\mathbb{C}^n$ that is defined by the matrix A. Since every Hermitian form on $\mathbb{C}^n$ is determined (up to isometry) by the number of positive and negative eigenvalues of A, we conclude that $G_{\mathbb{R}} \cong \mathrm{SU}(m,n)$ for some m and n. So $G_{\mathbb{R}}$ is listed in Proposition 18.1.6(A).

Exercises for §18.3.

#1. Suppose $\rho_1(G_{\mathbb{C}})_{\mathbb{R}}$ and $\rho_2(G_{\mathbb{C}})_{\mathbb{R}}$ are two $\mathbb{R}$-forms of $\mathrm{SL}(n,\mathbb{C})$, with corresponding 1-cocycles α_1 and α_2.
   a) Show that if $\rho_1(G_{\mathbb{C}})_{\mathbb{R}} \cong \rho_2(G_{\mathbb{C}})_{\mathbb{R}}$, then α_1 and α_2 are cohomologous. (So the correspondence in (18.3.4) is well-defined.)
   b) Conversely, show that if α_1 is cohomologous to α_2, then we have $\rho_1(G_{\mathbb{C}})_{\mathbb{R}} \cong \rho_2(G_{\mathbb{C}})_{\mathbb{R}}$. (So the correspondence in (18.3.4) is one-to-one.)
   In Subsection 18.3(iii), a real form of $\mathrm{SL}(n,\mathbb{C})$ is constructed for each cohomology class α. This shows that the correspondence is onto, and therefore completes the proof of (18.3.4).
   [Hint: In (a), you may assume, without proof, that every isomorphism from $\rho_1(G_{\mathbb{C}})_{\mathbb{R}}$ to $\rho_2(G_{\mathbb{C}})_{\mathbb{R}}$ extends to an isomorphism from $\rho_1(G_{\mathbb{C}})$ to $\rho_2(G_{\mathbb{C}})$.]

#2. Show $\mathcal{H}^1(\mathrm{Gal}(\mathbb{C}/\mathbb{R}), \mathrm{SL}(n,\mathbb{C})) = 0$, by identifying $\mathrm{SL}(n,\mathbb{C})$ with the automorphism group of a pair (V, ξ), where V is an n-dimensional vector space and ξ is a nonzero element of the exterior power $\bigwedge^n V$.

#3. The short exact sequence

$$1 \to \mathrm{SL}(n,\mathbb{C}) \hookrightarrow \mathrm{GL}(n,\mathbb{C}) \xrightarrow{\det} \mathbb{C}^{\times} \to 1$$

gives rise to the following long exact sequence of cohomology:

$$\mathcal{H}^0(\mathrm{Gal}(\mathbb{C}/\mathbb{R}), \mathrm{GL}(n,\mathbb{C})) \to \mathcal{H}^0(\mathrm{Gal}(\mathbb{C}/\mathbb{R}), \mathbb{C}^{\times})$$
$$\to \mathcal{H}^1(\mathrm{Gal}(\mathbb{C}/\mathbb{R}), \mathrm{SL}(n,\mathbb{C})) \to \mathcal{H}^1(\mathrm{Gal}(\mathbb{C}/\mathbb{R}), \mathrm{GL}(n,\mathbb{C})).$$

Show that the first map in this sequence is surjective, and combine this with the vanishing of the last term to provide another proof that $\mathcal{H}^1(\mathrm{Gal}(\mathbb{C}/\mathbb{R}), \mathrm{SL}(n,\mathbb{C})) = 0$.
   [Hint: The 0th cohomology group is the set of fixed points of the action.]

#4. Show that if n is odd, then every $\mathbb{R}$-form of $\mathrm{SO}(n,\mathbb{C})$ is isogenous to $\mathrm{SO}(p,q)$, for some p and q.
   [Hint: You may assume, without proof, that every automorphism of $\mathrm{SO}(n,\mathbb{C})$ is inner. Also note that $\mathrm{SO}(n,\mathbb{C}) = \mathrm{PSO}(n,\mathbb{C})$ (why?). Both of these observations require the assumption that n is odd.]

#5. Show that if α is any $\mathbb{C}$-linear automorphism of the ring $\mathrm{Mat}_{n \times n}(\mathbb{C})$, then there exists $T \in \mathrm{GL}(n, \mathbb{C})$, such that $\varphi(X) = TXT^{-1}$ for all $X \in \mathrm{Mat}_{n \times n}(\mathbb{C})$.

[*Hint:* For $A = \mathrm{Mat}_{n \times n}(\mathbb{C})$, make $\mathbb{C}^n$ into a simple A-module via $a * v = \alpha(a)v$. However, the usual action on $\mathbb{C}^n$ is the unique simple A-module (up to isomorphism), because A is a direct sum of submodules that are isomorphic to $\mathbb{C}^n$.]

#6. Show:

a) $\mathbb{C}$ is the only finite field extension of $\mathbb{R}$ (other than $\mathbb{R}$ itself).

b) $\mathbb{H}$ is the only division algebra over $\mathbb{R}$ that is not commutative.

[*Hint:* (a) You may assume, without proof, that $\mathbb{C}$ is algebraically closed. This implies that every irreducible real polynomial is either linear or quadratic. (b) If $x \in D \smallsetminus \mathbb{R}$, then $\mathbb{R}[x]$ is a field extension of $\mathbb{R}$; identify it with $\mathbb{C}$. Then conjugation by i is a $\mathbb{C}$-linear map on D. Choose j to be in the -1-eigenspace, and let $b = j^2$. Show $b \in \mathbb{R}$ and $D \cong \mathbb{H}_{\mathbb{R}}^{-1,b}$.]

§18.4. The $\mathbb{Q}$-forms of SL$(n, \mathbb{R})$

To illustrate how the method of the preceding section is used to find $\mathbb{Q}$-forms, instead of $\mathbb{R}$-forms, we prove the following result that justifies the claims made in Chapter 6 about arithmetic subgroups of SL$(n, \mathbb{R})$:

(18.4.1) **Theorem** (cf. Section 6.8). *Every $\mathbb{Q}$-form $G_{\mathbb{Q}}$ of SL$(n, \mathbb{R})$ is either a special linear group or a unitary group (perhaps over a division algebra).*

(18.4.2) *Remark.* More precisely, $G_{\mathbb{Q}}$ is isomorphic to either:

1) SL(m, D), for some m and some division algebra D over $\mathbb{Q}$, or

2) SU$(A, \tau; D) = \{ g \in \mathrm{SL}(k, D) \mid gA(^{\tau}g)^T = A \}$, where
   - D is a division algebra over $\mathbb{Q}$,
   - τ is an anti-involution of D that acts nontrivially on the center of D, and
   - A is a matrix in $\mathrm{Mat}_{k \times k}(D)$ that is Hermitian (i.e., $(^{\tau}A)^T = A$).

The proof is based on the following connection with Galois cohomology. We will work with $G_{\mathbb{C}}$, instead of G, because algebraically closed fields are much more amenable to Galois Theory. (That is, we are replacing SL$(n, \mathbb{R})$ with SL$(n, \mathbb{C})$ to avoid technical issues.)

(18.4.3) **Proposition.** *There is a one-to-one correspondence between the $\mathbb{Q}$-forms of $G_{\mathbb{C}}$ and the Galois cohomology set $\mathcal{H}^1(\mathrm{Gal}(\mathbb{C}/\mathbb{Q}), \mathrm{Aut}(G_{\mathbb{C}}))$.*

Proof. We assume familiarity with the proof in Section 18.3, and highlight the changes that need to be made.

Suppose we have an embedding $\rho \colon G_{\mathbb{C}} \to \mathrm{SL}(N, \mathbb{C})$, such that $\rho(G_{\mathbb{C}})$ is defined over $\mathbb{Q}$. The main difference from Section 18.3 is that, unlike $\mathrm{Gal}(\mathbb{C}/\mathbb{R})$, the Galois group $\mathrm{Gal}(\mathbb{C}/\mathbb{Q})$ has infinitely many nontrivial elements, and we need to consider all of them: since

$$\mathbb{Q} = \{ z \in \mathbb{C} \mid \sigma(z) = z, \ \forall \sigma \in \mathrm{Gal}(\mathbb{C}/\mathbb{Q}) \},$$

we have
$$\rho(G_\mathbb{C})_\mathbb{Q} = \{\, g \in \rho_\mathbb{C}(G_\mathbb{C}) \mid \sigma(g) = g, \ \forall \sigma \in \mathrm{Gal}(\mathbb{C}/\mathbb{Q}) \,\}.$$
For each $\sigma \in \mathrm{Gal}(\mathbb{C}/\mathbb{Q})$, let
$$\tilde{\sigma} = \rho^{-1}\sigma\rho\colon G_\mathbb{C} \to G_\mathbb{C} \quad \text{and} \quad \alpha(\sigma) = \tilde{\sigma}\,\sigma^{-1}\colon G_\mathbb{C} \to G_\mathbb{C}.$$
Then
$$G_\mathbb{Q} = \{\, g \in G_\mathbb{C} \mid \tilde{\sigma}(g) = g, \ \forall \sigma \in \mathrm{Gal}(\mathbb{C}/\mathbb{Q}) \,\},$$
and $\alpha(\sigma) \in \mathrm{Aut}(G_\mathbb{C})$. Furthermore, since $\alpha(\sigma) = \rho^{-1}\sigma\rho\sigma^{-1} = \rho^{-1}\,{}^\sigma\rho$ is formally a 1-coboundary, it is easily seen to be a 1-cocycle, and therefore represents a cohomology class in $\mathcal{H}^1(\mathrm{Gal}(\mathbb{C}/\mathbb{Q}), \mathrm{Aut}(G_\mathbb{C}))$.

This defines the desired map from the set of $\mathbb{Q}$-forms to the Galois cohomology set. It can be proved to be well-defined and injective by replacing $\mathbb{R}$ with $\mathbb{Q}$ in Exercise 18.3#1. That the map is surjective will be established in the proof of Theorem 18.4.1 below, where we explicitly describe the $\mathbb{Q}$-form corresponding to each cohomology class. $\qquad\square$

More generally, we have the following natural analogue of Observation 18.3.5:

(18.4.4) **Observation.** If X is an algebraic object that is defined over $\mathbb{Q}$ (and satisfies mild hypotheses; cf. Warning 18.3.7), then the Galois cohomology set $\mathcal{H}^1(\mathrm{Gal}(\mathbb{C}/\mathbb{Q}), \mathrm{Aut}(X_\mathbb{C}))$ is in one-to-one correspondence with the set of $\mathbb{Q}$-isomorphism classes of $\mathbb{Q}$-defined objects whose $\mathbb{C}$-points are isomorphic to $X_\mathbb{C}$.

(18.4.5) **Corollary** (cf. Example 18.3.6).
$$\mathcal{H}^1(\mathrm{Gal}(\mathbb{C}/\mathbb{Q}), \mathrm{GL}(n, \mathbb{C})) = 0 \ \textit{and} \ \mathcal{H}^1(\mathrm{Gal}(\mathbb{C}/\mathbb{Q}), \mathrm{SL}(n, \mathbb{C})) = 0.$$

Proof of Theorem 18.4.1. Let $G_\mathbb{C} = \mathrm{SL}(n, \mathbb{C})$. As in Subsection 18.3(iii), we have
$$\mathrm{Aut}(G_\mathbb{C}) = \mathrm{PSL}(n, \mathbb{C}) \rtimes \langle \omega \rangle,$$
where $\omega(g) = (g^T)^{-1}$. Given $\alpha \in \mathcal{H}^1(\mathrm{Gal}(\mathbb{C}/\mathbb{Q}), \mathrm{Aut}(G_\mathbb{C}))$, corresponding to a $\mathbb{Q}$-form $G_\mathbb{Q}$, we consider two cases.

Case 1. Assume $\alpha \in \mathcal{H}^1(\mathrm{Gal}(\mathbb{C}/\mathbb{Q}), \mathrm{PSL}(n, \mathbb{C}))$. By arguing exactly as in Case 1 of Subsection 18.3(iii) (but with $\mathbb{Q}$ in the place of $\mathbb{R}$), we see that $G_\mathbb{Q} \cong \mathrm{SL}(k, D)$, for some k and some division algebra D over $\mathbb{Q}$.

Case 2. Assume the image of α is **not** contained in $\mathrm{PSL}(n, \mathbb{C})$. Since the outer automorphism group $\mathrm{Out}(G_\mathbb{C})$ is of order 2, it has no nontrivial automorphisms. Therefore, the action of the Galois group $\mathrm{Gal}(\mathbb{C}/\mathbb{Q})$ on $\mathrm{Out}(G_\mathbb{C})$ must be trivial. Hence, if we let $\overline{\alpha}\colon \mathrm{Gal}(\mathbb{C}/\mathbb{Q}) \to \mathrm{Out}(G_\mathbb{C})$ be the 1-cocycle obtained from α by modding out $\mathrm{PSL}(n, \mathbb{C})$, then $\overline{\alpha}$ is an actual homomorphism (not merely a "crossed homomorphism").

By the assumption of this case (and the fact that $|\mathrm{Out}(G_\mathbb{C})| = 2$), the kernel of $\overline{\alpha}$ is a subgroup of index 2 in $\mathrm{Gal}(\mathbb{C}/\mathbb{Q})$. This means that the fixed field of $\ker \overline{\alpha}$ is a quadratic extension $L = \mathbb{Q}[\sqrt{r}]$ of $\mathbb{Q}$. Then, by construction, we have $\mathrm{Gal}(\mathbb{C}/L) = \ker \overline{\alpha}$.

For any $\sigma \in \mathrm{Gal}(\mathbb{C}/L)$, the conclusion of the preceding paragraph tells us that $\overline{\alpha}(\sigma)$ is trivial. For simplicity, let us assume that the bar can be removed, so $\alpha(\sigma)$ is trivial (see Correction 18.4.6(1)). Since, by definition, we have $\alpha(\sigma) = \tilde{\sigma}\,\sigma^{-1}$ (see (18.3.3)), this implies $\sigma = \tilde{\sigma}$. Therefore, for any $g \in G_{\mathbb{Q}}$, we have $g^{\sigma} = g^{\tilde{\sigma}} = g$. Since this holds for all $\sigma \in \mathrm{Gal}(\mathbb{C}/L)$, we conclude that $g \in \mathrm{SL}(n, L)$.

Now, for the unique nontrivial $\tau \in \mathrm{Gal}(L/\mathbb{Q})$, we have $\tau \notin \ker \overline{\alpha}$, so $\alpha(\tau) = (\text{conj by } A)\, \omega$ for some $A \in \mathrm{GL}(n, \mathbb{R})$. Hence, for any $g \in G_{\mathbb{Q}}$, we have

$$g = \tilde{\tau}(g) = (\alpha(\tau)\,\tau)(g) = A\,\omega(\tau(g))\,A^{-1} = A\,\big(({}^{\tau}g)^{T}\big)^{-1} A^{-1},$$

so $g\, A\,({}^{\tau}g)^{T} = A$, which means $g \in \mathrm{SU}(A, \tau; L)$. Furthermore, the equation $\tilde{\tau}^2 = 1$ provides an equation that can be used to show A is Hermitian (or, more precisely, can be chosen to be Hermitian) (see Exercise 1). $\qquad\square$

(18.4.6) Corrections.

1) *Mixed case.* We seem to have shown that all $\mathbb{Q}$-forms of $\mathrm{SL}(n, \mathbb{R})$ can be constructed from either division algebras (Case 1) or unitary groups (Case 2). However, the discussion in Case 2 assumes that $\alpha(\sigma)$ is trivial, when all we actually know is that $\overline{\alpha}(\sigma)$ is trivial. Removing this assumption means that α can map a part of the Galois group into $\mathrm{PSL}(n, \mathbb{C})$. In other words, in addition to the homomorphism $\overline{\alpha}$, there is a nontrivial cocycle from $\mathrm{Gal}(\mathbb{C}/L)$ to $\mathrm{PSL}(n, \mathbb{C})$. By the argument of Case 1, this cocycle yields a division algebra D over L. The resulting $\mathbb{Q}$-form $G_{\mathbb{Q}} = \mathrm{SU}(A, \tau; D)$ is obtained by combining division algebras with unitary groups.

2) $\mathbb{C}$ *vs.* $\overline{\mathbb{Q}}$. We should really be using the algebraic closure $\overline{\mathbb{Q}}$ of $\mathbb{Q}$, instead of $\mathbb{C}$. The Galois cohomology set $\mathcal{H}^1(\mathrm{Gal}(\overline{\mathbb{Q}}/\mathbb{Q}), \mathrm{Aut}(G_{\overline{\mathbb{Q}}}))$ is defined to be the natural limit of the sets $\mathcal{H}^1(\mathrm{Gal}(F/\mathbb{Q}), \mathrm{Aut}(G_{\overline{\mathbb{Q}}}))$, where F ranges over all finite Galois extensions of $\mathbb{Q}$.

Exercises for §18.4.

#1. In Case 2 of the proof of Theorem 18.4.1, show that the matrix A can be chosen to be Hermitian.

[*Hint:* A must be a scalar multiple λ of a Hermitian matrix (since $\tilde{\tau}^2 = 1$). Use the fact that $\mathcal{H}^1(\mathrm{Gal}(\mathbb{C}/L); \mathbb{C}^{\times})$ is trivial (why?) to replace A with a scalar multiple of itself that makes $\lambda = 1$.]

§18.5. $\mathbb{Q}$-forms of classical groups

By arguments similar to the ones applied to $\mathrm{SL}(n, \mathbb{R})$ in Section 18.4, it can be shown that the $\mathbb{Q}$-forms of almost any classical group come from special linear groups, unitary groups, orthogonal groups, or symplectic groups. However, the special linear groups and unitary groups may involve division algebras, and restriction of scalars (5.5.8) implies that the groups may be over an extension F of $\mathbb{Q}$. (Recall that unitary groups over division algebras were defined in Definition 6.8.13, and the involutions τ_c and τ_r on the quaternion algebra $\mathbb{H}_F^{a,b}$ were defined in Example 6.8.12.) Here is a list of the groups that arise:

(18.5.1) **Definition.** For any algebraic number field F, and any n, the following groups are said to be of *classical type*:

1) $SL(n, D)$, where D is a division algebra whose center is F.

2) $Sp(2n, F)$.

3) $SO(A; F)$, where A is an invertible, symmetric matrix in $\mathrm{Mat}_{n \times n}(F)$.

4) $SU(A, \tau_c; \mathbb{H}_F^{a,b})$, where $\mathbb{H}_F^{a,b}$ is a quaternion division algebra over F, and A is an invertible, τ_c-Hermitian matrix in $\mathrm{Mat}_{n \times n}(\mathbb{H}_F^{a,b})$.

5) $SU(A, \tau_r; \mathbb{H}_F^{a,b})$, where $\mathbb{H}_F^{a,b}$ is a quaternion division algebra, and A is an invertible, τ_r-Hermitian matrix in $\mathrm{Mat}_{n \times n}(\mathbb{H}_F^{a,b})$.

6) $SU(A, \tau; D)$, where
   - D is a division algebra whose center is a quadratic extension L of F,
   - τ is an anti-involution whose restriction to L is the Galois automorphism of L over F, and
   - A is an invertible, τ-Hermitian matrix in $\mathrm{Mat}_{n \times n}(D)$.

(18.5.2) *Remark.* Definition 18.5.1 is directly analogous to the list of classical simple Lie groups (see Examples A2.3 and A2.4). Specifically:

1) $SL(n, D)$ is the analogue of $SL(n, \mathbb{R})$, $SL(n, \mathbb{C})$, and $SL(n, \mathbb{H})$.

2) $Sp(2n, F)$ is the analogue of $Sp(2n, \mathbb{R})$ and $Sp(2n, \mathbb{C})$.

3) $SO(A; F)$ is the analogue of $SO(m, n)$ and $SO(n, \mathbb{C})$.

4) $SU(A, \tau_c; \mathbb{H}_F^{a,b})$ is the analogue of $Sp(m, n)$.

5) $SU(A, \tau_r; \mathbb{H}_F^{a,b})$ is the analogue of $SO(n, \mathbb{H})$.

6) $SU(A, \tau; D)$ (with τ nontrivial on the center) is the analogue of $SU(m, n)$.

(18.5.3) **Theorem.** *Suppose*
   - *G is classical, and*
   - *no simple factor of $G_{\mathbb{C}}$ is isogenous to $SO(8, \mathbb{C})$.*

Then every irreducible, arithmetic lattice in G is commensurable to the integer points of some group (of classical type) listed in Definition 18.5.1.

(18.5.4) *Remark.* To state the conclusion of Theorem 18.5.3 more explicitly, let us assume, for simplicity, that the center of G is trivial. Then Theorem 18.5.3 states that there exist:

- algebraic number field F, with places S^∞ and ring of integers $\mathcal{O}$,
- a group $\hat{G}_F$ listed in Definition 18.5.1, with corresponding semisimple Lie group $\hat{G}$ that is defined over F, and
- a homomorphism $\varphi\colon \prod_{\sigma\in S^\infty} \hat{G}^\sigma \to G$, with compact kernel,

such that $\varphi(\Delta(G_\mathcal{O}))$ is commensurable to Γ (cf. Proposition 5.5.8).

(18.5.5) **Warning.** Although $\varphi(\Delta(G_\mathcal{O}))$ is commensurable to Γ, this does **not** imply that $\varphi(\Delta(G_F))$ is commensurable to $G_\mathbb{Q}$. For example, the image of $\mathrm{SL}(2,\mathbb{Q})$ in $\mathrm{PSL}(2,\mathbb{Q})$ has infinite index (cf. Exercise 5.2#1).

Each of the groups in Definition 18.5.1 has a corresponding semisimple Lie group G that is defined over F. Before determining which Lie group corresponds to each F-group, we first find the complexification of G. This is similar to calculations that we have already seen, so we omit the details.

(18.5.6) **Proposition** (cf. Section 18.2). *The notation of each part of this proposition is taken from the corresponding part of Definition 18.5.1. We use d to denote the degree of the central division algebra D, and the matrix A is assumed to be $n \times n$.*

1) $\mathrm{SL}(n, D \otimes_F \mathbb{C}) \cong \mathrm{SL}(dn, \mathbb{C})$.
2) $\mathrm{Sp}(2n, \mathbb{C}) = \mathrm{Sp}(2n, \mathbb{C})$ *(obviously!)*.
3) $\mathrm{SO}(A;\mathbb{C}) \cong \mathrm{SO}(n, \mathbb{C})$.
4) $\mathrm{SU}(A, \tau_c; \mathbb{H}_F^{a,b} \otimes_F \mathbb{C}) \cong \mathrm{Sp}(2n, \mathbb{C})$.
5) $\mathrm{SU}(A, \tau_r; \mathbb{H}_F^{a,b} \otimes_F \mathbb{C}) \cong \mathrm{SO}(2n, \mathbb{C})$.
6) $\mathrm{SU}(A, \tau; D \otimes_F \mathbb{C}) \cong \mathrm{SL}(dn, \mathbb{C})$.

If $F \not\subset \mathbb{R}$, then the semisimple Lie group G corresponding to G_F is the complex Lie group in the corresponding line of the above proposition. However, if $F \subset \mathbb{R}$, then G is some $\mathbb{R}$-form of that complex group. The following result lists the correct $\mathbb{R}$-form for each of the groups of classical type.

(18.5.7) **Proposition.** *The notation of each part of this proposition is taken from the corresponding part of Definition 18.5.1. We use d to denote the degree of the central division algebra D, and the matrix A is assumed to be $n \times n$. Assume F is an algebraic number field, and that $F \subset \mathbb{R}$. Then:*

1) $\mathrm{SL}(n, D \otimes_F \mathbb{R}) \cong \begin{cases} \mathrm{SL}(dn, \mathbb{R}) & \text{if } D \otimes_F \mathbb{R} \cong \mathrm{Mat}_{d\times d}(\mathbb{R}), \\ \mathrm{SL}(dn/2, \mathbb{H}) & \text{otherwise.} \end{cases}$
2) $\mathrm{Sp}(2n, \mathbb{R}) = \mathrm{Sp}(2n, \mathbb{R})$ *(obviously!)*.

3) $SO(A, \mathbb{R}) \cong SO(p, n - p)$.

4) $SU(A, \tau_c; \mathbb{H}_F^{a,b} \otimes_F \mathbb{R}) \cong \begin{cases} Sp(2n, \mathbb{R}) & \text{if } \mathbb{H}_\mathbb{R}^{a,b} \cong Mat_{2\times2}(\mathbb{R}), \\ Sp(p, n - p) & \text{if } \mathbb{H}_\mathbb{R}^{a,b} \cong \mathbb{H}. \end{cases}$

5) $SU(A, \tau_r; \mathbb{H}_F^{a,b} \otimes_F \mathbb{R}) \cong \begin{cases} SO(p, 2n - p) & \text{if } \mathbb{H}_\mathbb{R}^{a,b} \cong Mat_{2\times2}(\mathbb{R}), \\ SO(n, \mathbb{H}) & \text{if } \mathbb{H}_\mathbb{R}^{a,b} \cong \mathbb{H}. \end{cases}$

6) $SU(A, \tau; D \otimes_F \mathbb{R}) \cong \begin{cases} SU(p, dn - p) & \text{if } L \not\subset \mathbb{R}, \\ SL(dn, \mathbb{R}) & \begin{array}{l} \text{if } L \subset \mathbb{R} \text{ and} \\ \quad D \otimes_F \mathbb{R} \cong Mat_{d\times d}(\mathbb{R}), \end{array} \\ SL(dn/2, \mathbb{H}) & \text{otherwise.} \end{cases}$

(18.5.8) *Remark.* Proposition 18.5.7 does not specify the value of p, where it appears. However, it can be calculated for any particular matrix A. For example, to calculate p in (6), note that, because $L \not\subset \mathbb{R}$, we have

$$D \otimes_F \mathbb{R} \cong D \otimes_L \mathbb{C} \cong Mat_{d\times d}(\mathbb{C}),$$

so we may think of $A \in Mat_{n\times n}(D)$ as a $(dn) \times (dn)$ Hermitian matrix. Then p is the number of positive eigenvalues of this Hermitian matrix (and $dn - p$ is the number of negative eigenvalues). We have already seen this type of consideration in Notation 6.4.10 and Proposition 6.4.11.

(18.5.9) *Remark.* The table on page 380 summarizes the above results in a format that makes it easy to find the arithmetic subgroups of any given simple Lie group G (or, by restriction of scalars, to find the irreducible arithmetic subgroups of any semisimple Lie group that has G as a simple factor), except that (as indicated by "?") the list is not complete for groups whose complexification is isogenous to $SO(8, \mathbb{C})$.

The arithmetic group Γ that corresponds to a given F-form G_F is obtained by:

- replacing F with its ring of integers $\mathcal{O}$, or
- replacing D with an order $\mathcal{O}_D$ (see Lemma 6.8.7).

By restriction of scalars (5.5.8), Γ is an arithmetic subgroup of $\prod_{\sigma \in S^\infty} G^\sigma$.

A parenthetical reference indicates the corresponding part of Definition 18.5.1, and also of Proposition 18.5.6 (for $F_\sigma = \mathbb{C}$) and Proposition 18.5.7 (for $F_\sigma = \mathbb{R}$). The reference column (combined with the "m or $p + q$" column) also lists additional conditions that determine $G_F \otimes_F F_\infty$ is the desired simple Lie group G (except that the parameters p and q will need to be calculated, if they arise).

The $\mathbb{Q}$-rank of the corresponding arithmetic group Γ is either given explicitly (as a function of n), or is the dimension of a maximal isotropic subspace (of the associated vector space over either the field F or the division algebra D, as indicated).

Lie group G	F-form G_F	reference	m or $p+q$	rank$_\mathbb{Q}\,\Gamma$
$SL(m,\mathbb{R})$	$SL(n,D)$	(1), $F\subset\mathbb{R}$, D split/$\mathbb{R}$	$m=dn$	$n-1$
	$SU(B,\tau;D)$	(6), $F\subset L\subset\mathbb{R}$, D split/$\mathbb{R}$	$m=dn$	D-subspace
$SL(m,\mathbb{C})$	$SL(n,D)$	(1), $F\not\subset\mathbb{R}$	$m=dn$	$n-1$
	$SU(B,\tau;D)$	(6), $F\not\subset\mathbb{R}$ (so $L\not\subset\mathbb{R}$)	$m=dn$	D-subspace
$SL(m,\mathbb{H})$	$SL(n,D)$	(1), $F\subset\mathbb{R}$, D not split/$\mathbb{R}$	$m=dn/2$, d even	$n-1$
	$SU(B,\tau;D)$	(6), $F\subset L\subset\mathbb{R}$, D not split/$\mathbb{R}$	$m=dn/2$, d even	D-subspace
$SU(p,q)$	$SU(B,\tau;D)$	(6), $F\subset\mathbb{R}$, $L\not\subset\mathbb{R}$	$p+q=dn$	D-subspace
$SO(p,q)$	$SO(B;F)$	(3), $F\subset\mathbb{R}$	$p+q=n$	F-subspace
	$SU(B,\tau_r;D)$	(5), $F\subset\mathbb{R}$, D split/$\mathbb{R}$	$p+q=2n$, $d=2$	D-subspace
	?	Remark 18.5.10	$p+q=8$	?
$SO(m,\mathbb{C})$	$SO(B;F)$	(3), $F\not\subset\mathbb{R}$	$m=n$	F-subspace
	$SU(B,\tau_r;D)$	(5), $F\not\subset\mathbb{R}$	$m=2n$, $d=2$	D-subspace
	?	Remark 18.5.10	$m=8$	?
$SO(m,\mathbb{H})$	$SU(B,\tau_r;D)$	(5), $F\subset\mathbb{R}$, D not split/$\mathbb{R}$	$m=n$, $d=2$	D-subspace
	?	Remark 18.5.10	$m=4$	?
$Sp(2m,\mathbb{R})$	$Sp(2n,F)$	(2), $F\subset\mathbb{R}$	$m=n$	n
	$SU(B,\tau_c;D)$	(4), $F\subset\mathbb{R}$, D split/$\mathbb{R}$	$m=n$, $d=2$	D-subspace
$Sp(2m,\mathbb{C})$	$Sp(2n,F)$	(2), $F\not\subset\mathbb{R}$	$m=n$	n
	$SU(B,\tau_c;D)$	(4), $F\not\subset\mathbb{R}$	$m=n$, $d=2$	D-subspace
$Sp(p,q)$	$SU(B,\tau_c;D)$	(4), $F\subset\mathbb{R}$, D not split/$\mathbb{R}$	$p+q=n$, $d=2$	D-subspace

See Remark 18.5.9 on page 379 for an explanation of this table.

(18.5.10) *Remark* ("triality"). Perhaps we should explain why the statement of Theorem 18.5.3 assumes no simple factor of $G_{\mathbb{C}}$ is isogenous to $SO(8, \mathbb{C})$. Fundamentally, the reason $PSO(8, \mathbb{C})$ is special is that, unlike all the other simple Lie groups over $\mathbb{C}$, it has an outer automorphism ϕ of order 3, called "triality." For all of the other simple groups, the outer automorphism group is either trivial or has order 2.

Here is how the triality automorphism ϕ can be used to construct $\mathbb{Q}$-forms that are not listed in Theorem 18.5.3. We first choose any homomorphism $\alpha \colon \mathrm{Gal}(\mathbb{C}/\mathbb{Q}) \xrightarrow{\text{onto}} \langle \phi \rangle$ (so the kernel of α is a cubic, Galois extension of $\mathbb{Q}$). The triality automorphism can be chosen so that it commutes with the action of the Galois group (in other words, ϕ is "defined over $\mathbb{Q}$"), so the homomorphism α is a 1-cocycle into $\mathrm{Aut}(PSO(8, \mathbb{C}))$. Therefore, by the correspondence between cohomology and $\mathbb{Q}$-forms (18.4.3), there is a corresponding $\mathbb{Q}$-form $G_{\mathbb{Q}}$. This $\mathbb{Q}$-form is not any of the groups listed in Theorem 18.5.3, because, for all those groups, the image of the induced homomorphism $\overline{\alpha} \colon \mathrm{Gal}(\mathbb{C}/\mathbb{Q}) \to \mathrm{Out}(G_{\mathbb{C}})$ has order 1 or 2, not 3.

Mathematicians who understand the triality automorphism can construct the corresponding $\mathbb{Q}$-form explicitly, by reversing the steps in the proof of Proposition 18.4.3. Namely, for each $\sigma \in \mathrm{Gal}(\mathbb{C}/\mathbb{Q})$, let

$$\tilde{\sigma} = \alpha(\sigma) \cdot \sigma \in \mathrm{Aut}(PSO(8, \mathbb{C})).$$

Then

$$G_{\mathbb{Q}} = \{ g \in PSO(8, \mathbb{C}) \mid \tilde{\sigma}(g) = g, \ \forall \sigma \in \mathrm{Gal}(\mathbb{C}/\mathbb{Q}) \}.$$

§18.6. Applications of the classification of arithmetic groups

Several results that were stated without proof in previous chapters are easy consequences of the above classification of F-forms.

(18.6.1) **Corollary** (cf. Proposition 6.4.5). *Suppose Γ is an arithmetic subgroup of $SO(m, n)$, and $m + n \geq 5$ is odd. Then there is a finite extension F of $\mathbb{Q}$, with ring of integers $\mathcal{O}$, such that Γ is commensurable to $SO(A; \mathcal{O})$, for some invertible, symmetric matrix A in $\mathrm{Mat}_{n \times n}(F)$.*

Proof. Let $G = SO(m, n)$. Restriction of scalars (5.5.16) implies there is a group $\hat{G}$ that is defined over an algebraic number field F and has a simple factor that is isogenous to G, such that Γ is commensurable to $\hat{G}_{\mathcal{O}}$. By inspection, we see that a group of the form $SO(m, n)$ never appears in Proposition 18.5.6, and appears at two places in Proposition 18.5.7. However, in our situation, we know that $m + n$ is odd, so the only possibility for $\hat{G}_F$ is $SO(A; F)$. Therefore, Γ is commensurable to $SO(A; \mathcal{O})$. $\square$

(18.6.2) **Corollary** (cf. Remark 9.1.7(3b)). *If* $G = \mathrm{SO}(2, n)$, *with* $n \geq 5$, *and* n *is odd, then* $\mathrm{rank}_{\mathbb{R}}\, G = 2$, *but there is no lattice* Γ *in* G, *such that* $\mathrm{rank}_{\mathbb{Q}}\, \Gamma = 1$.

Proof. We have $\mathrm{rank}_{\mathbb{R}}\, \mathrm{SO}(2, n) = \min\{2, n\} = 2$ (see Proposition 8.1.8).

From the Margulis Arithmeticity Theorem (5.2.1), we know that Γ is arithmetic, so Corollary 18.6.1 tells us that Γ is of the form $\mathrm{SO}(B; \mathcal{O})$, where

- $\mathcal{O}$ is the ring of integers of some algebraic number field F, and
- B is a symmetric bilinear form on F^{n+2}.

If $\mathrm{rank}_{\mathbb{Q}}\, \Gamma = 1$, then G/Γ is not compact, so Corollary 5.3.2 tells us that we may take $F = \mathbb{Q}$; therefore $\mathcal{O} = \mathbb{Z}$. We see that:

1) B has signature $(2, n)$ on $\mathbb{R}^{n+2}$ (because $G = \mathrm{SO}(2, n)$), and
2) no 2-dimensional $\mathbb{Q}$-subspace of $\mathbb{Q}^{n+2}$ is totally isotropic (because we have $\mathrm{rank}_{\mathbb{Q}}\, \Gamma < 2$).

Recall the following important fact that was used in the proof of Proposition 6.4.1:

> **Meyer's Theorem.** *If* $B_0(x, y)$ *is any nondegenerate, symmetric bilinear form on* $\mathbb{R}^d$, *such that*
> - B *is defined over* $\mathbb{Q}$,
> - $d \geq 5$, *and*
> - B_0 *is isotropic over* $\mathbb{R}$ (*that is,* $B(v, v) = 0$ *for some nonzero* $v \in \mathbb{R}^d$),
> *then* B_0 *is also isotropic over* $\mathbb{Q}$ (*that is,* $B(v, v) = 0$ *for some nonzero* $v \in \mathbb{Q}^d$).

This theorem, tells us there is a nontrivial isotropic vector $v \in \mathbb{Q}^{n+2}$. Then, because B is nondegenerate, there is a vector $w \in \mathbb{Q}^{n+2}$, such that $B(v, w) = 1$ and $B(w, w) = 0$. Let V be the $\mathbb{R}$-span of $\{v, w\}$. Because the restriction of B to V is nondegenerate, we have $\mathbb{R}^{n+2} = V \oplus V^\perp$. This direct sum is obviously orthogonal (with respect to B), and the restriction of B to V has signature $(1, 1)$, so we conclude that the restriction of B to $V^\perp$ has signature $(1, n-1)$. Hence, there is an isotropic vector in $V^\perp$. By applying Meyer's Theorem again, we conclude that there is an isotropic vector z in $(V^\perp)_{\mathbb{Q}}$. Then $\langle v, z \rangle$ is a 2-dimensional totally isotropic subspace of $\mathbb{Q}^{n+2}$. This is a contradiction. $\qquad\square$

(18.6.3) **Corollary** (cf. Proposition 6.4.2). *If* $n \notin \{3, 7\}$, *then every noncompact, arithmetic subgroup of* $\mathrm{SO}(1, n)$ *is commensurable to a conjugate of* $\mathrm{SO}(A; \mathbb{Z})$, *for some invertible, symmetric matrix* $A \in \mathrm{Mat}_{n \times n}(\mathbb{Q})$.

Proof. Assume, for simplicity, that $n = 5$, and let Γ be a noncocompact, arithmetic subgroup of $G = \mathrm{SO}(1, 5)$. Since Γ is not cocompact, there is no need for restriction of scalars: Γ corresponds to a $\mathbb{Q}$-form $G_{\mathbb{Q}}$ on G itself (see Corollary 5.3.2). We may assume the $\mathbb{Q}$-form is not $SO(A; \mathbb{Q})$; otherwise, Γ is as described. Therefore, by inspection of Proposition 18.5.6 and Proposition 18.5.7, we see that $G_{\mathbb{Q}}$ must be of the form $SU(A, \tau_r; \mathbb{H}_{\mathbb{Q}}^{a,b})$, where $A \in \mathrm{Mat}_{3\times 3}(\mathbb{H}_{\mathbb{Q}}^{a,b})$, and $\mathbb{H}_{\mathbb{Q}}^{a,b} \otimes_{\mathbb{Q}} \mathbb{R} \cong \mathrm{Mat}_{2\times 2}(\mathbb{R})$.

Because G/Γ is not compact, there is a vector $v \in (\mathbb{H}_{\mathbb{Q}}^{a,b})^3$, such that $\tau_r(v)^T A v = 0$ (see Exercise 18.7#4). Hence, it is not difficult to see that, by making a change of basis, we may assume

$$A = \begin{bmatrix} 1 & 0 & 0 \\ 0 & -1 & 0 \\ 0 & 0 & p \end{bmatrix}, \quad \text{for some } p \in \mathbb{H}_{\mathbb{Q}}^{a,b}.$$

Since the identity matrix $\mathrm{Id}_{2\times 2}$ is the image of $1 \in \mathbb{H}_{\mathbb{Q}}^{a,b}$ under any isomorphism $\mathbb{H}_{\mathbb{Q}}^{a,b} \otimes_{\mathbb{Q}} \mathbb{R} \to \mathrm{Mat}_{2\times 2}(\mathbb{R})$, this means

$$G = SU(A; \mathbb{H}_{\mathbb{Q}}^{a,b} \otimes_{\mathbb{Q}} \mathbb{R}) \cong SO(A_{\mathbb{R}}; \mathbb{R}), \quad \text{where } A_{\mathbb{R}} = \begin{bmatrix} 1 & & & & \\ & 1 & & & \\ & & -1 & & \\ & & & -1 & \\ & & & & * & * \\ & & & & * & * \end{bmatrix}.$$

Therefore, G is isomorphic to either $SO(2, 4)$ or $SO(3, 3)$; it is not isogenous to $SO(1, 5)$. This is a contradiction. □

(18.6.4) Proposition (see Proposition 6.6.5). *Every noncocompact, arithmetic subgroup of* $SL(3, \mathbb{R})$ *is commensurable to a conjugate of either* $SL(3, \mathbb{Z})$ *or a subgroup of the form* $SU(J_3, \sigma; \mathcal{O})$, *where*

- $J_3 = \begin{bmatrix} 0 & 0 & 1 \\ 0 & 1 & 0 \\ 1 & 0 & 0 \end{bmatrix}$,
- *L is a real quadratic extension of $\mathbb{Q}$, with Galois automorphism σ, and*
- *$\mathcal{O}$ is the ring of integers of L.*

Proof. Let Γ be an arithmetic subgroup of $G = SL(3, \mathbb{R})$, such that G/Γ is not compact. We know, from the Margulis Arithmeticity Theorem (5.2.1), that Γ is arithmetic. Since G/Γ is not compact, there is no need for restriction of scalars (see Corollary 5.3.2), so there is a $\mathbb{Q}$-form $G_{\mathbb{Q}}$ of G, such that Γ is commensurable to $G_{\mathbb{Z}}$. By inspection of Propositions 18.5.6 and 18.5.7, we see that there are only two possibilities for $G_{\mathbb{Q}}$. We consider them individually, as separate cases.

Case 1. Assume $G_{\mathbb{Q}} = SL(n, D)$, *for some central division algebra D of degree d over $\mathbb{Q}$, with $dn = 3$.* Because 3 is prime, there are only two possibilities for n and d.

Subcase 1.1. Assume $n = 3$ and $d = 1$. Since $d = 1$, we have $\dim_{\mathbb{Q}} D = 1$, so $D = \mathbb{Q}$. Therefore, $G_{\mathbb{Q}} = SL(3, \mathbb{Q})$. So $\Gamma \approx SL(3, \mathbb{Z})$.

Subcase 1.2. Assume $n = 1$ and $d = 3$. We have $G_{\mathbb{Q}} = SL(1, D)$. Therefore $\Gamma \approx SL(1, \mathcal{O}_D)$ is cocompact (see Proposition 6.8.8(2)). This is a contradiction.

Case 2. Assume $G_{\mathbb{Q}} = SU(A, \tau; D)$, for A, D, σ as in 18.5.1(6), with $F = \mathbb{Q}$, $L \subset \mathbb{R}$, and $dn = 3$. If $n = 1$, then $G_{\mathbb{Q}} \subseteq SL(1, D)$, so it has no unipotent elements, which contradicts the fact that G/Γ is not compact. Thus, we may assume that $n = 3$ and $d = 1$.

Since $d = 1$, we have $D = L$, so $G_{\mathbb{Q}} = SU(A, \sigma; L)$, where σ is the (unique) Galois automorphism of L over $\mathbb{Q}$, and B is a σ-Hermitian form on L^3.

Since Γ is not cocompact, we know $\mathrm{rank}_{\mathbb{Q}} \Gamma \geq 1$, so there is some nonzero $v \in L^3$ with $v^T A v = 0$ (cf. Example 9.1.5(4)). From this, it is not difficult to construct a basis of L^3 in which A is a scalar multiple of J_3 (see Exercise 3). □

Exercises for §18.6.

#1. Assume that $F \subset \mathbb{R}$, and that A is an invertible, symmetric matrix in $\mathrm{Mat}_{n \times n}(F)$. Show that if exactly p of the eigenvalues of A are positive, then $SU(A; \mathbb{R}) \cong SO(p, n - p)$.

#2. Suppose
   - $F \subset \mathbb{R}$, $\mathbb{H}_F^{a,b}$ is a quaternion division algebra over F, and
   - A is an invertible τ_c-Hermitian matrix in $\mathrm{Mat}_{n \times n}(D)$ that is diagonal.

   Show:
   a) every entry of the matrix A belongs to F (and, hence, to $\mathbb{R}$), and
   b) if exactly p of the diagonal entries of A are positive, then we have $SU(A, \tau_c; \mathbb{H}_F^{a,b} \otimes_F \mathbb{R}) \cong Sp(p, n - p)$.

#3. Complete the proof of Proposition 18.6.4, by showing that we may assume $A = J$.

   [*Hint:* Assume v_1 and v_3 are isotropic, and v_2 is orthogonal to both v_1 and v_3. Multiply A by a scalar in $\mathbb{Q}$, so $v_2^* A v_2 = 1$. Then normalize v_3, so $v_1^* A v_3 = 1$.]

#4. (B. Farb) For each $n \geq 2$, find a *cocompact* lattice Γ_n in $SL(n, \mathbb{R})$, such that $\Gamma_2 \subseteq \Gamma_3 \subseteq \Gamma_4 \subseteq \cdots$. (If we did not require Γ_n to be cocompact, we could let $\Gamma_n = SL(n, \mathbb{Z})$.)

#5. Show that if $G = SU(A, \tau_r; \mathbb{H}_F^{a,b})$, as in Definition 18.5.1(5), then there exists $A' \in \mathrm{Mat}_{n \times n}(\mathbb{H}_F^{a,b})$, such that A' is skew-Hermitian with respect to the standard anti-involution τ_c, and $G = SU(A', \tau_c; \mathbb{H}_F^{a,b})$.

   [*Hint:* Use the fact that $\tau_r(x) = j^{-1} \tau_c(x) j$.]

§18.7. G has a cocompact lattice

We have already seen that if G is not compact, then it has a noncocompact lattice (see Corollary 5.1.17). In this section, we will show there is also a lattice that is cocompact:

(18.7.1) **Theorem.** *G has a cocompact, arithmetic lattice.*

To illustrate the main idea, we briefly recall the prototypical case, which is a generalization of Example 5.5.4.

(18.7.2) **Proposition.** $SO(m,n)$ *has a cocompact, arithmetic lattice.*

Proof. Let

- $F = \mathbb{Q}(\sqrt{2})$,
- σ be the Galois automorphism of F over $\mathbb{Q}$,
- $\mathcal{O} = \mathbb{Z}[\sqrt{2}]$,
- $B(x,y) = \sum_{j=1}^{p} x_j y_j - \sqrt{2} \sum_{j=1}^{q} x_{p+j} y_{p+j}$, for $x, y \in F^{p+q}$,
- $G = SO(B)^\circ$,
- $\Gamma = G_\mathcal{O}$, and
- $\Delta \colon G_F \to G \times G^\sigma$ defined by $\Delta(g) = (g, \sigma(g))$.

We know (from restriction of scalars) that $\Delta(\Gamma)$ is an irreducible, arithmetic lattice in $G \times G^\sigma$ (see Proposition 5.5.8). Since $G^\sigma \cong SO(p+q)$ is compact, we may mod it out, to conclude that Γ is an arithmetic lattice in $G \cong SO(m,n)^\circ$. Also, since G^σ is compact, we know that Γ is cocompact (see Corollary 5.5.10). □

More generally, the same idea can be used to prove that any simple group G has a cocompact, arithmetic lattice. Namely, start by letting K be a compact group that has the same complexification as G. (For classical groups, the correct choice of K can be found by looking at Proposition 18.1.6.) Then show that $G \times K$ has an irreducible, arithmetic lattice. More precisely, construct

- an extension F of $\mathbb{Q}$ that has exactly two places σ and τ, and
- a group $\hat{G}$ that is defined over F,

such that $\hat{G}^\sigma$ and $\hat{G}^\tau$ are isogenous to G and K, respectively.

Although we could do this explicitly for the classical groups, we will save ourselves a lot of work (and also be able to handle the exceptional groups at the same time) by quoting the following powerful theorem.

(18.7.3) **Theorem** (Borel-Harder). *Suppose*

- *F is an algebraic number field,*
- S^∞ *is the set of places of F,*
- *G is defined over F,*

- $G_{\mathbb{C}}$ is connected and simple, and has trivial center, and
- for each $\sigma \in S^\infty$, we are given some F_σ-form G_σ of $G_{\mathbb{C}}$.

Then there exists a group $\hat{G}$ that is defined over F, such that $\hat{G}^\sigma \cong G_\sigma$, for each $\sigma \in S^\infty$.

Remark on the proof. For any place σ of F, there is a natural map

$$\sigma^*: \mathcal{H}^1(\mathrm{Gal}(\mathbb{C}/F), \mathrm{Aut}(G_{\mathbb{C}})) \to \mathcal{H}^1(\mathrm{Gal}(\mathbb{C}/F_\sigma), \mathrm{Aut}(G_{\mathbb{C}})).$$

Namely, any element of the domain corresponds to an F-form $\hat{G}$ of $G_{\mathbb{C}}$. The twisted group $\hat{G}^\sigma$ is defined over $\sigma(F)$, and is therefore also defined over F_σ. Hence, it determines an element of the range. (The map can also be defined directly, in terms of 1-cocycles, by restricting a cocycle $\alpha: \mathrm{Gal}(\mathbb{C}/F) \to \mathrm{Aut}(G_{\mathbb{C}})$ to the subgroup $\mathrm{Gal}(\mathbb{C}/F_\sigma)$.)

However, we should replace $\mathbb{C}$ with $\overline{\mathbb{Q}}$ in the domain (see Correction 18.4.6(2)). Making this correction, and putting together the maps for the various choices of σ, we obtain a map

$$\mathcal{H}^1(\mathrm{Gal}(\overline{\mathbb{Q}}/F), \mathrm{Aut}(G_{\overline{\mathbb{Q}}})) \to \underset{\sigma \in S^\infty}{\times} \mathcal{H}^1(\mathrm{Gal}(\mathbb{C}/F_\sigma), \mathrm{Aut}(G_{\mathbb{C}})).$$

The theorem is proved by showing that this map is surjective. $\square$

(18.7.4) **Corollary.** *If G is isotypic (see Definition 5.6.1), then G has a cocompact, irreducible, arithmetic lattice.*

Proof. Assume G has trivial center (by passing to an isogenous group), and write $G = G^1 \times \cdots \times G^n$, where each G^i is simple. Let $\hat{G} = G^1$, and assume, for simplicity, that $G^i_{\mathbb{C}}$ is simple, for every i (see Exercise 2). Then, by assumption, $G^i_{\mathbb{C}} \cong \hat{G}_{\mathbb{C}}$ for every i, which means that G^i is an $\mathbb{R}$-form of $\hat{G}_{\mathbb{C}}$.

Let F be an extension of $\mathbb{Q}$, such that F has exactly n places $v_1, \ldots, v_n$, and all of them are real (see Lemma 18.7.8). The Borel-Harder Theorem (18.7.3) provides some group $\hat{G}$ that is defined over F, such that $\hat{G}_{F_{v_i}} \cong G_i$, for each i. Then restriction of scalars (5.5.8) tells us that $\hat{G}_\mathcal{O}$ is (isomorphic to) an irreducible, arithmetic lattice in $\prod_{i=1}^n G_{F_{v_i}} \cong G$.

We may assume that G^n is compact (by replacing G with $G \times K$ for a compact group K, such that $G \times K$ is isotypic). Then Corollary 5.5.10 tells us that every irreducible, arithmetic lattice in G is cocompact. $\square$

Proof of Theorem 18.7.1. We may assume G is simple. (If Γ_1 and Γ_2 are cocompact, arithmetic lattices in G_1 and G_2, then $\Gamma_1 \times \Gamma_2$ is a cocompact, arithmetic lattice in $G_1 \times G_2$.) Then G is isotypic, so Corollary 18.7.4 applies. $\square$

The converse of Corollary 18.7.4 is true (even without assuming cocompactness):

(18.7.5) **Proposition** (cf. Proposition 5.5.12). *If G has an irreducible, arithmetic lattice, then G is isotypic.*

By the Margulis Arithmeticity Theorem (5.2.1), there is usually no need to assume that the lattice is arithmetic:

(18.7.6) **Corollary.** *G is isotypic if*
- *it has an irreducible lattice, and*
- *it is not isogenous to a group of the form* $SO(1, n) \times K$ *or* $SU(1, n) \times K$, *where K is compact.*

We know that if *G* has an irreducible, arithmetic lattice that is not cocompact, then *G* is isotypic (see Proposition 18.7.5) and has no compact factors (see Corollary 5.5.10). However, the converse is not true:

(18.7.7) **Example.** Every irreducible lattice in $SO(3, \mathbb{H}) \times SO(1, 5)$ is cocompact.

Proof. Suppose Γ is an irreducible lattice in $SO(3, \mathbb{H}) \times SO(1, 5)$, such that G/Γ is **not** compact. This will lead to a contradiction.

The Margulis Arithmeticity Theorem (5.2.1) implies that Γ is arithmetic, so Corollary 5.5.15 implies that Γ can be obtained by restriction of scalars. Hence, there exist:
- an algebraic number field *F* with exactly two places 1 and σ, and
- a (connected) group $\hat{G}$ that is defined over *F*,

such that
- $\hat{G}$ is isogenous to $SO(3, \mathbb{H})$,
- $\hat{G}^\sigma$ is isogenous to $SO(1, 5)$, and
- Γ is commensurable to $\Delta(G_{\mathcal{O}})$ in $\hat{G} \times \hat{G}^\sigma$.

Since $SO(n, \mathbb{H})$ occurs only once in Propositions 18.5.6 and 18.5.7 combined, $\hat{G}_F$ must be of the form $SU(A, \tau_r; \mathbb{H}_F^{a,b})$, for some $A \in \text{Mat}_{3 \times 3}(\mathbb{H}_F^{a,b})$.

However, since G/Γ is not compact, the proof of Corollary 18.6.3 implies that $\hat{G}^\sigma$ is isomorphic to either $SO(2, 4)$ or $SO(3, 3)$; it is not isogenous to $SO(1, 5)$. This is a contradiction. □

We close with a stronger version of a fact that was used in the proof of Corollary 18.7.4:

(18.7.8) **Lemma.** *For any natural numbers r and s, not both 0, there is an algebraic number field F with exactly r real places and s complex places.*

Proof. Let $n = r + 2s$. It suffices to find an irreducible polynomial $f(x) \in \mathbb{Q}[x]$ of degree *n*, such that $f(x)$ has exactly *r* real roots. (Then we may let $F = \mathbb{Q}(\alpha)$, where α is any root of $f(x)$.)

Choose a monic polynomial $g(x) \in \mathbb{Z}[x]$, such that

- $g(x)$ has degree n,
- $g(x)$ has exactly r real roots, and
- all of the real roots of $g(x)$ are simple.

For example, choose distinct integers $a_1, \ldots, a_r$, and let
$$g(x) = (x - a_1) \cdots (x - a_r)(x^{2s} + 1).$$

Fix a prime p. Exercise 5 allows us to assume

1) $g(x) \equiv x^n \pmod{p^2}$, and
2) $\min\{ g(t) \mid g'(t) = 0 \} > p$,

by replacing $g(x)$ with $k^n g(x/k)$, for an appropriate integer k.

Let $f(x) = g(x) - p$. From (1), we know that $f(x) \equiv x^n - p \pmod{p^2}$, so the Eisenstein Criterion (B4.6) implies that f is irreducible. From (2), we know that $f(x)$ has the same number of real roots as $g(x)$ (see the figure below). Therefore $f(x)$ has exactly r real roots.   □

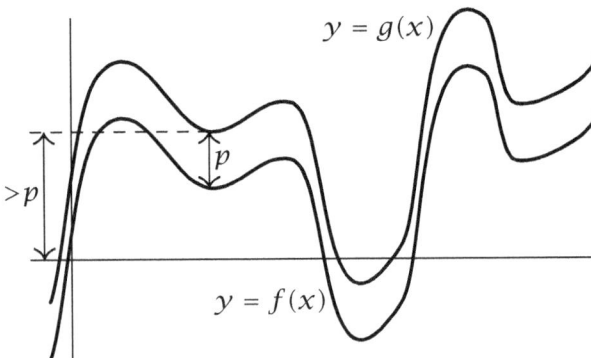

Exercises for §18.7.

#1. Use restriction of scalars to construct cocompact arithmetic subgroups of $\mathrm{SU}(m, n)$ and $\mathrm{Sp}(m, n)$ for all m and n.
[*Hint:* See the proof of Proposition 18.7.2.]

#2. The proof of Corollary 18.7.4 assumes that $G_{\mathbb{C}}^i$ is simple for every i. Remove this assumption.
[*Hint:* You may use Remark 18.2.2 (without proof), and you will need the full strength of Lemma 18.7.8.]

#3. Construct a noncocompact, irreducible lattice in $\mathrm{SL}(2, \mathbb{R}) \times \mathrm{SO}(3)$.
[*Hint:* The free group F_2 is a noncocompact lattice in $\mathrm{SL}(2, \mathbb{R})$. Let Γ be the graph of a homomorphism $F_2 \to \mathrm{SO}(3)$ that has dense image.]

#4. In the proof of Example 18.7.7, show there exists $v \in (\mathbb{H}_F^{a,b})^3$, such that $\tau_r(v)^T A v = 0$.

[*Hint:* If g is a nontrivial unipotent element of G_F, then there is some w, such that g fixes the nonzero vector $v = gw - w$.]

#5. a) Suppose $g(x)$ is a monic polynomial of degree n, and assume $k \in \mathbb{Z} \smallsetminus \{0\}$, such that $k\,g(x) \in \mathbb{Z}[x]$. Show $k^n g(x/k) \in \mathbb{Z}[x]$.

   b) Suppose $g(x)$ is a monic, integral polynomial of degree n, and p is a prime. Show that $p^{2n} g(x/p^2) \equiv x^n \pmod{p^2}$.

   c) Suppose $g(x)$ and $h(x)$ are monic polynomials, and k and n are nonzero integers, such that $h(x) = k^n g(x/k)$. Show that

   $$\min\{\,|h(t)| \mid h'(t) = 0\,\} = k^n \min\{\,|g(t)| \mid g'(t) = 0\,\}.$$

Notes

A proof of the classification of complex semisimple Lie algebras (18.1.1, 18.1.2) can be found in standard texts, such as [6, Thm. 11.4, pp. 57–58, and Thm. 18.4, p. 101].

The classification of real simple Lie algebras (Theorem 18.1.7) was obtained by É. Cartan [4]. (The intervening decades have led to enormous simplifications in the proof.)

In Sections 18.3 and 18.4, our cohomological approach to the classification of F-forms of the classical groups is based on [10, §2.3], where full details can be found. See [12] for a list of all F-forms (mostly without proof), including exceptional groups (intended for readers familiar with root systems). The special case of real forms (including exceptional groups) is proved, by a somewhat different approach, in [5, Chap. 10].

Theorem 18.5.3 is due to Weil [13]. A proof (together with Propositions 18.1.6 and 18.5.6) is in [10, §2.3, pp. 78–92]. We copied (18.1.6), (18.5.1), and (18.5.6) from [10, p. 92], except that [10] uses a different description of the groups in 18.5.1(5) (see Exercise 18.6#5).

Exercise 18.3#5 is an easy special case of the Skolem-Noether Theorem, which can be found in texts on ring theory, such as [9, §12.6, p. 230].

Exercise 18.3#6, the classification of division algebras over $\mathbb{R}$, is due to Frobenius (1878), and a proof can be found in [7, pp. 452–453].

Theorem 18.7.3 is due to A. Borel and G. Harder [1]. See [8] for an explicit construction of $\widehat{G}$ in the special case where the simple factors of G are classical.

G. Prasad (personal communication) supplied Example 18.7.7. It is a counterexample to the noncocompact part of [8, Thm. C], which erroneously states that isotypic groups with no compact factors have both cocompact and noncocompact irreducible lattices.

Remark 18.2.2 is a consequence of [3, Prop. 1 of App. 2, p. 385], since a connected Lie group is simple if and only if its adjoint representation has no nonzero, proper, invariant subspaces.

Meyer's Theorem (used in the proof of Corollary 18.6.2) can be found in [2, Thm. 1 of §1.7 and Thm. 5 of §1.6, pp. 61 and 51] or [11, Cor. 2 of §4.3.2, p. 43].

References

[1] A. Borel and G. Harder: Existence of discrete cocompact subgroups of reductive groups over local fields, *J. Reine Angew. Math.* 298 (1978) 53–64. MR 0483367, http://eudml.org/doc/151965

[2] Z. I. Borevich and I. R. Shafarevich: *Number Theory,* Academic Press, New York, 1966. MR 0195803

[3] N. Bourbaki: *Lie Groups and Lie Algebras, Chapters 7–9.* Springer, Berlin, 2005. ISBN 3-540-43405-4, MR 2109105

[4] É. Cartan: Les groupes réels simples finis et continus, *Ann. Sci. École Norm. Sup.* 31 (1914) 263–355. Zbl 45.1408.03, http://www.numdam.org/item?id=ASENS_1914_3_31__263_0

[5] S. Helgason: *Differential Geometry, Lie Groups, and Symmetric Spaces.* Academic Press, New York, 1978. ISBN 0-12-338460-5, MR 0514561

[6] J. E. Humphreys: *Introduction to Lie Algebras and Representation Theory.* Springer, Berlin Heidelberg New York, 1972. MR 0323842

[7] N. Jacobson: *Basic Algebra I, 2nd ed..* Freeman, New York, 1985. ISBN 0-7167-1480-9, MR 0780184

[8] F. E. A. Johnson: On the existence of irreducible discrete subgroups in isotypic Lie groups of classical type, *Proc. London Math. Soc.* (3) 56 (1988) 51–77. MR 0915530, http://dx.doi.org/10.1112/plms/s3-56.1.51

[9] R. S. Pierce: *Associative Algebras.* Springer, New York, 1982. ISBN 0-387-90693-2, MR 0674652

[10] V. Platonov and A. Rapinchuk: *Algebraic Groups and Number Theory.* Academic Press, Boston, 1994. ISBN 0-12-558180-7, MR 1278263

[11] J.-P. Serre: *A Course in Arithmetic,* Springer, New York 1973. MR 0344216

[12] J. Tits: Classification of algebraic semisimple groups, in A. Borel and G. D. Mostow, eds.: *Algebraic Groups and Discontinuous Subgroups (Boulder, Colo., 1965),* Amer. Math. Soc., Providence, R.I., 1966, pp. 33–62. MR 0224710

[13] A. Weil: Algebras with involution and the classical groups, *J. Indian Math. Soc.* 24 (1960) 589–623. MR 0136682

Chapter 19

Construction of a Coarse Fundamental Domain

The ordinary 2-torus is often depicted as a square with opposite sides identified, and it would be useful to have a similar representation of $\Gamma \backslash G$, so we would like to construct a fundamental domain for Γ in G. Unfortunately, it is usually not feasible to do this explicitly, so, as in Chapter 7, where we showed that $\mathrm{SL}(n, \mathbb{Z})$ is a lattice in $\mathrm{SL}(n, \mathbb{R})$, we will make do with a nice set that is close to being a fundamental domain:

(19.0.1) **Definition** (cf. Definition 4.7.2). A subset $\mathcal{F}$ of G is called a *coarse fundamental domain* for Γ in G if

1) $\Gamma \mathcal{F} = G$, and
2) $\{\gamma \in \Gamma \mid \mathcal{F} \cap \gamma \mathcal{F} \neq \varnothing\}$ is finite.

The main result is Theorem 19.2.2, which states that the desired set $\mathcal{F}$ can be constructed as a finite union of (translates of) "Siegel sets" in G. Applications of the construction are described in Section 19.3.

Recall: The Standing Assumptions (4.0.0 on page 41) are in effect, so, as always, Γ is a lattice in the semisimple Lie group $G \subseteq \mathrm{SL}(\ell, \mathbb{R})$.

Main prerequisites for this chapter: $\mathbb{Q}$-rank (Chapter 9). *Recommended:* Siegel sets for $\mathrm{SL}(n, \mathbb{Z})$ (sections 7.1 to 7.3).

§19.1. What is a Siegel set?

Before defining Siegel sets in every semisimple group, we recall the following special case:

(19.1.1) **Definition** (cf. Definition 7.2.4). A *Siegel set* for $SL(n, \mathbb{Z})$ is a set of the form $\mathfrak{S}_{\overline{N},c} = \overline{N} A_c K \subseteq SL(n, \mathbb{R})$, where

- $\overline{N}$ is a compact subset of the group N of upper-triangular unipotent matrices,
- $A_c = \{ a \in A \mid a_{i-1,i-1} \geq c\, a_{i,i}$ for $i = 1, \ldots, n-1 \}$, where A is the group of positive-definite diagonal matrices (and $c > 0$), and
- $K = SO(n)$.

In this section, we generalize this notion by replacing $SL(n, \mathbb{Z})$ with any arithmetic subgroup (or, more generally, any lattice) in any semisimple Lie group G. To this end, note that the subgroups N, A, and K above are the components of the Iwasawa decomposition $G = KAN$ (or $G = NAK$), which can be defined for any semisimple group (see Theorem 8.4.9):

- N is a maximal unipotent subgroup of G.
- A is a maximal $\mathbb{R}$-split torus of G that normalizes N, and
- K is a maximal compact subgroup of G.

Now, to construct Siegel sets in the general case, we will do two things. First, we rephrase Definition 19.1.1 in a way that does not refer to any specific realization of G as a matrix group. To this end, recall that, for $G = SL(n, \mathbb{R})$, the *positive Weyl chamber* is

$$A^+ = \{ a \in A \mid a_{i,i} > a_{i+1,i+1} \text{ for } i = 1, \ldots, n-1 \}.$$

Therefore, in the notation of Definition 19.1.1, we have $A^+ = A_1$, and, for any $c > 0$, it is not difficult to see that there exists some $a \in A$, such that $A_c = aA^+$ (see Exercise 7.2#11). Therefore, letting $C = \overline{N}a$, we see that

$$\mathfrak{S}_{\overline{N},c} = CA^+K, \text{ and } C \text{ is a compact subset of } NA.$$

This description of Siegel sets can be generalized in a natural way to any semisimple group G.

However, all of the above is based entirely on the structure of G, with no mention of Γ, but a coarse fundamental domain $\mathcal{F}$ needs to be constructed with a particular arithmetic subgroup Γ in mind. For example, if $\Gamma \backslash G$ is compact (or, in other words, if $\text{rank}_{\mathbb{Q}} \Gamma = 0$), then our coarse fundamental domain needs to be compact, so none of the factors in the definition of a Siegel set can be unbounded. Therefore, we need to replace the maximal $\mathbb{R}$-split torus A with a smaller torus S that reflects the choice of a particular subgroup Γ. In fact, S will be the trivial torus when G/Γ is compact. In general, S is a maximal $\mathbb{Q}$-split torus of G (hence, S is compact if and only if $\Gamma \backslash G$ is compact).

Now, if S is properly contained in A, then NSK is not all of G. Hence, NS will usually not be the appropriate replacement for the subgroup NA. Instead, if we note that NA is the identity component of a minimal parabolic subgroup of $\mathrm{SL}(n, \mathbb{R})$ (see Example 8.4.4(1)), and that NA is obviously defined over $\mathbb{Q}$, then it is natural to replace NA with a minimal parabolic $\mathbb{Q}$-subgroup P of G.

The following definition implements these considerations.

(19.1.2) Definition. Assume

- G is defined over $\mathbb{Q}$,
- Γ is commensurable to $G_{\mathbb{Z}}$,
- P is a minimal parabolic $\mathbb{Q}$-subgroup of G,
- S is a maximal $\mathbb{Q}$-split torus that is contained in P,
- S^+ is the positive Weyl chamber in S (with respect to P),
- K is a maximal compact subgroup of G, and
- C is any nonempty, compact subset of P.

Then $\mathfrak{S} = \mathfrak{S}_C = C\,S^+K$ is a **Siegel set** for Γ in G.

(19.1.3) Warning. Our definition of a Siegel set is slightly more general than what is usually found in the literature, because other authors place some restrictions on the compact set C. For example, it is often assumed that C has nonempty interior.

Exercises for §19.1.

#1. Show that if $\mathfrak{S} = C\,S^+K$ is a Siegel set, then there is a compact subset C' of G, such that $\mathfrak{S} \subseteq S^+C'$.

[*Hint:* Conjugation by any element of S^+ centralizes MS and contracts N (where $P = MSN$ is the Langlands decomposition).]

#2. For every compact subset C of G, show there is a Siegel set that contains C.

[*Hint:* Exercise 9.3#4.]

§19.2. Coarse fundamental domains made from Siegel sets

(19.2.1) Example. Let

- $G = \mathrm{SL}(2, \mathbb{R})$,
- $\mathfrak{S}$ be a Siegel set that is a coarse fundamental domain for $\mathrm{SL}(2, \mathbb{Z})$ in G (see Figure 7.2A(b)), and
- Γ be a subgroup of finite index in $\mathrm{SL}(2, \mathbb{Z})$.

Then $\mathfrak{S}$ may not be a coarse fundamental domain for Γ, because $\Gamma\mathfrak{S}$ may not be all of G. In fact, if the hyperbolic surface $\Gamma\backslash\mathfrak{H}$ has more than one cusp, then no Siegel set is a coarse fundamental domain for Γ.

However, if we let F be a set of coset representatives for Γ in $\mathrm{SL}(2,\mathbb{Z})$, then $F\mathfrak{S}$ is a coarse fundamental domain for Γ (see Exercise 7.2#7(b)).

From the above example, we see that a coarse fundamental domain can sometimes be the union of several translates of a Siegel set, even in cases where it cannot be a single Siegel set. In fact, this construction always works (if Γ is arithmetic):

(19.2.2) Reduction Theory for Arithmetic Groups. *If Γ is commensurable to $G_\mathbb{Z}$, then there exist a Siegel set $\mathfrak{S}$ and a finite subset F of $G_\mathbb{Q}$, such that $\mathcal{F} = F\mathfrak{S}$ is a coarse fundamental domain for Γ in G.*

The proof will be given in Section 19.4.

Although the statement of this result only applies to arithmetic lattices, it can be generalized to the non-arithmetic case. However, this extension requires a notion of Siegel sets in groups that are not defined over $\mathbb{Q}$. The following definition reduces this problem to the case where Γ is irreducible.

(19.2.3) Definition. If $\mathfrak{S}_i$ is a Siegel set for Γ_i in G_i, for $i = 1, 2, \ldots, n$, then

$$\mathfrak{S}_1 \times \mathfrak{S}_2 \times \cdots \times \mathfrak{S}_n$$

is a **Siegel set** for the lattice $\Gamma_1 \times \cdots \times \Gamma_n$ in $G_1 \times \cdots \times G_n$.

Then, by the Margulis Arithmeticity Theorem (5.2.1), all that remains is to define Siegel sets for lattices in $\mathrm{SO}(1,n)$ and $\mathrm{SU}(1,n)$, but we can use the same definition for all simple groups of real rank one:

(19.2.4) Definition. Assume G is simple, $\mathrm{rank}_\mathbb{R} G = 1$, and K is a maximal compact subgroup of G.

0) If $\mathrm{rank}_\mathbb{Q} \Gamma = 0$, and C is any compact subset of G, then $\mathfrak{S} = CK$ is a **Siegel set** in G.

1) Assume now that $\mathrm{rank}_\mathbb{Q} \Gamma = 1$. Let P be a minimal parabolic subgroup of G, with Langlands decomposition $P = MAN$, such that

$$\Gamma \cap N \text{ is a maximal unipotent subgroup of } \Gamma. \qquad (19.2.4N)$$

If
  - C is any compact subset of P, and
  - A^+ is the positive Weyl chamber of A (with respect to P),
  then $\mathfrak{S} = CA^+K$ is a **generalized Siegel set** in G.

(19.2.5) Remark. If Γ is commensurable to $G_\mathbb{Z}$ (and G is defined over $\mathbb{Q}$), then (19.2.4N) holds if and only if P is defined over $\mathbb{Q}$ (and is therefore a minimal parabolic $\mathbb{Q}$-subgroup).

We can now state a suitable generalization of Theorem 19.2.2:

(19.2.6) **Theorem.** *If G has no compact factors, then there exist a generalized Siegel set $\mathfrak{S}$ and a finite subset F of G, such that $\mathcal{F} = F\mathfrak{S}$ is a coarse fundamental domain for Γ in G.*

The proof is essentially the same as for Theorem 19.2.2.

Exercises for §19.2.

#1. Without using any of the results in this chapter (other than the definitions of "Siegel set" and "coarse fundamental domain"), show that if $\mathrm{rank}_{\mathbb{Q}}\,\Gamma = 0$, then some Siegel set is a coarse fundamental domain for Γ in G.

#2. Suppose $\mathcal{F}_1$ and $\mathcal{F}_2$ are coarse fundamental domains for Γ_1 and Γ_2 in G_1 and G_2, respectively. Show that $\mathcal{F}_1 \times \mathcal{F}_2$ is a coarse fundamental domain for $\Gamma_1 \times \Gamma_2$ in $G_1 \times G_2$.

#3. Suppose N is a compact, normal subgroup of G, and let $\overline{\Gamma}$ be the image of Γ in $\overline{G} = G/N$. Show that if $\overline{\mathcal{F}}$ is a coarse fundamental domain for $\overline{\Gamma}$ in $\overline{G}$, then
$$\mathcal{F} = \{\, g \in G \mid gN \in \overline{\mathcal{F}} \,\}$$
is a coarse fundamental domain for Γ in G.

#4. If G is simple, $\mathrm{rank}_{\mathbb{R}}\,G = 1$, and G is defined over $\mathbb{Q}$, then Definitions 19.1.2 and 19.2.4 give two different definitions of the Siegel sets for $G_{\mathbb{Z}}$. Show that Definition 19.2.4 is more general: any Siegel set according to Definition 19.1.2 is also a Siegel set by the other definition.
[*Hint:* Remark 19.2.5.]

§19.3. Applications of reduction theory

Having a coarse fundamental domain is very helpful for understanding the geometry and topology of $\Gamma \backslash G$. Here are a few examples of this (with only sketches of the proofs).

§19.3(i). Γ is finitely presented. Proposition 4.7.7 tells us that if Γ has a coarse fundamental domain that is a connected, open subset of G, then Γ is finitely presented. The coarse fundamental domains constructed in Theorems 19.2.2 and 19.2.6 are closed, rather than open, but it is easy to deal with this minor technical issue:

(19.3.1) **Definition.** A subset $\overset{\circ}{\mathfrak{S}}$ of G is an *open Siegel set* if $\overset{\circ}{\mathfrak{S}} = \mathcal{O}S^{+}K$, where $\mathcal{O}$ is a nonempty, precompact, open subset of P.

Choose a maximal compact subgroup K of G that contains a maximal compact subgroup of $C_G(S)$. Then we may let:

- $\mathcal{F} = F\mathfrak{S}$ be a coarse fundamental domain, with $\mathfrak{S} = C\,S^+K$, such that $C \subseteq P^\circ$ and $\mathcal{F}$ is connected (see Exercise 4),
- $\mathcal{O}$ be a connected, open, precompact subset of P° that contains C,
- $\mathring{\mathfrak{S}} = \mathcal{O}S^+K$ be the corresponding open Siegel set, and
- $\mathring{\mathcal{F}} = F\mathring{\mathfrak{S}}$.

Then $\mathring{\mathcal{F}}$ is a coarse fundamental domain for Γ (see Exercise 5), and $\mathring{\mathcal{F}}$ is both connected and open.

This establishes Theorem 4.7.10, which stated (without proof) that Γ is finitely presented.

§19.3(ii). Mostow Rigidity Theorem.

When $\mathrm{rank}_\mathbb{Q}\,\Gamma_1 = 1$, G. Prasad constructed a quasi-isometry $\varphi\colon G_1/K_1 \to G_2/K_2$ from an isomorphism $\rho\colon \Gamma_1 \to \Gamma_2$, by using the Siegel-set description of the coarse fundamental domain for $\Gamma_i \backslash G_i$. This completed the proof of the Mostow Rigidity Theorem (15.1.2).

§19.3(iii). Divergent torus orbits.

(19.3.2) Definition. Let T be an $\mathbb{R}$-split torus in G, and let $x \in G/\Gamma$. We say the T-orbit of x is ***divergent*** if the natural map $T \to Tx$ is proper.

(19.3.3) Theorem. $\mathrm{rank}_\mathbb{Q}\,\Gamma$ *is the maximal dimension of an $\mathbb{R}$-split torus that has a divergent orbit on G/Γ.*

We start with the easy half of the proof:

(19.3.4) Lemma (cf. Exercise 2). *Assume G is defined over $\mathbb{Q}$, and let S be a maximal $\mathbb{Q}$-split torus in G (so $\dim S = \mathrm{rank}_\mathbb{Q}\,G_\mathbb{Z}$). Then the S-orbit of $eG_\mathbb{Z}$ is divergent.*

Now, the other half:

(19.3.5) Theorem. *If T is an $\mathbb{R}$-split torus, and $\dim T > \mathrm{rank}_\mathbb{Q}\,\Gamma$, then no T-orbit in G/Γ is divergent.*

Proof (assuming $\mathrm{rank}_\mathbb{Q}\,\Gamma = 1$). Let T be a 2-dimensional, $\mathbb{R}$-split torus T of G, and define $\pi\colon T \to G/\Gamma$, by $\pi(t) = t\Gamma$. Suppose π is proper. (This will lead to a contradiction.)

Let P be a minimal parabolic $\mathbb{Q}$-subgroup of G, and let S be a maximal $\mathbb{Q}$-split torus in P. For simplicity, let us assume that $\Gamma = G_\mathbb{Z}$, and also that a single open Siegel set $\mathring{\mathfrak{S}} = KS^-\mathcal{O}$ provides a coarse fundamental domain for Γ in G. (Note that, since we are considering G/Γ, instead of $\Gamma\backslash G$, we have reversed the order of the factors in the definition of the Siegel set, and we use the opposite Weyl chamber.)

Choose a large, compact subset C of G/Γ, and let T_R be a large circle in T that is centered at e. Since π is proper, we may assume T_R is so large that $\pi(T_R)$ is disjoint from C. Since T_R is connected, this implies $\pi(T_R)$ is contained in a connected component of the complement of C. So there exists $y \in \Gamma$, such that $T_R \subseteq \mathfrak{S}P_{\mathbb{Z}}y$ (cf. Remark 19.4.10 below).

Let $t \in T_R$, and assume, for simplicity, that $y = e$. Then $t \in \mathfrak{S}P_{\mathbb{Z}}$, and, since T_R is closed under inverses, we see that $\mathfrak{S}P_{\mathbb{Z}}$ also contains t^{-1}. However, it is not difficult to see that conjugation by any large element of $\mathfrak{S}P_{\mathbb{Z}}$ expands the volume form on P (see Exercise 7). Since the inverse of an expanding element is a contracting element, not an expanding element, this is a contradiction. $\square$

Theorem 19.3.3 can be restated in the following geometric terms:

(19.3.6) Theorem (see Theorem 2.2.1). $\mathrm{rank}_{\mathbb{Q}} \Gamma$ *is the largest natural number* r, *such that some finite cover of the locally symmetric space* $\Gamma \backslash G/K$ *contains a closed, simply connected, r-dimensional flat.*

§19.3(iv). The large-scale geometry of locally symmetric spaces. If we let $\pi\colon G \to \Gamma \backslash G/K$ be the natural map, then it is not difficult to see that the restriction of π to any Siegel set is proper (see Exercise 6). In fact, with much more work (which we omit), it can be shown that the restriction of π is very close to being an isometry:

(19.3.7) Theorem. *If* $\mathfrak{S} = C\,S^+K$ *is any Siegel set, and*
$$\pi\colon G \to \Gamma \backslash G/K \text{ is the natural map,}$$
then there exists $c \in \mathbb{R}^+$, *such that, for all* $x, y \in \mathfrak{S}$, *we have*
$$d(\pi(x), \pi(y)) \le d(x, y) \le d(\pi(x), \pi(y)) + c.$$

This allows us to describe the precise shape of the the locally symmetric space associated to Γ, up to quasi-isometry:

(19.3.8) Theorem. *Let*
- $X = \Gamma \backslash G/K$ *be the locally symmetric space associated to* Γ, *and*
- $r = \mathrm{rank}_{\mathbb{Q}} \Gamma$.

Then X is quasi-isometric to the cone on a certain $(r - 1)$-dimensional simplicial complex at ∞.

Idea of proof. Modulo quasi-isometry, any features of bounded size in X can be completely ignored. Note that:
- Theorem 19.3.7 tells us that, up to a bounded error, $\mathfrak{S}$ looks the same as its image in X.
- There is a compact subset C' of G, such that $\mathfrak{S} \subseteq S^+C'$ (see Exercise 19.1#1), so every element of $\mathfrak{S}$ is within a bounded distance of S^+. Therefore, $\mathfrak{S}$ and S^+ are indistinguishable, up to quasi-isometry.

Then, since $\mathcal{F} = F \mathfrak{S}$ covers all of X, we conclude that X is quasi-isometric to $\bigcup_{f \in F} f S^+$.

The Weyl chamber S^+ is a cone; more precisely, it is the cone on an $(r - 1)$-simplex at ∞. Therefore, up to quasi-isometry, X is the union of these finitely many cones, so it is the cone on some $(r - 1)$-dimensional simplicial complex at ∞. $\square$

(19.3.9) *Remarks.*

1) The same argument shows that we get the same picture if, instead of looking at X modulo quasi-isometry, we look at it from farther and farther away, as in the definition of the asymptotic cone of X in Definition 2.2.6. Therefore, the asymptotic cone of X is the cone on a certain $(r - 1)$-dimensional simplicial complex at ∞. This establishes Theorem 2.2.8.

2) For a reader familiar with "Tits buildings," the proof (and the construction of $\mathcal{F}$) shows that this simplicial complex at ∞ can be constructed by taking the Tits building of parabolic $\mathbb{Q}$-subgroups of G, and modding out by the action of Γ.

Exercises for §19.3.

#1. Show that Γ has only finitely many conjugacy classes of finite subgroups.

[*Hint:* If H is a finite subgroup of Γ, then $H^g \subseteq K$, for some $g \in G$. Write $g = \gamma x$, with $\gamma \in \Gamma$ and $x \in \mathcal{F}$. Then $H^\gamma x = x \cdot H^g \subseteq \mathcal{F}$, so H is conjugate to a subset of $\{\gamma \in \Gamma \mid \mathcal{F} \cap \gamma \mathcal{F} \neq \emptyset\}$.]

#2. Let $G = \mathrm{SL}(n, \mathbb{R})$, $\Gamma = \mathrm{SL}(n, \mathbb{Z})$, and S be the group of positive-definite diagonal matrices. Show the S-orbit of Γe is proper.

[*Hint:* If $s_{j,j}/s_{i,i}$ is large, then conjugation by s contracts a unipotent matrix y whose only off-diagonal entry is $y_{i,j}$.]

#3. Show that every open Siegel set is an open subset of G (so the terminology is consistent).

#4. Assume
   - K contains a maximal compact subgroup of $C_G(S)$,
   - C is a compact subset of P, and
   - F is a finite subset of G.

   Show there is a compact subset $C_\circ$ of P°, such that $C S^+ K \subseteq C_\circ S^+ K$ and $F C_\circ S^+ K$ is connected.

[*Hint:* Show $P^\circ (K \cap C_P(S)) = P$.]

#5. Show the set $\overset{\circ}{\mathcal{F}}$ constructed in Subsection 19.3(i) is indeed a coarse fundamental domain for Γ in G.

[*Hint:* Exercise 7.2#2.]

#6. Let $\pi \colon G \to \Gamma \backslash G$ be the natural map. Show that if $\mathfrak{S} = CS^+K$ is a Siegel set for Γ, then the restriction of π to $\mathfrak{S}$ is proper.

[*Hint:* Let v be a nontrivial element of $N \cap \Gamma$. If g is a large element of $\mathfrak{S}$, then $v^g \approx e$.]

#7. Show that if $\mathcal{O}$ is contained in a compact subset of P, then conjugation by any large element of $KS^- \mathcal{O} P_{\mathbb{Z}}$ expands the Haar measure on P.

[*Hint:* Conjugation by any element of $M \cup N$ preserves the measure, conjugation by an element of $\mathcal{O}$ is bounded, and S^- centralizes SM and expands N. Also note that $P_{\mathbb{Z}} \doteq M_{\mathbb{Z}} N_{\mathbb{Z}}$ (see Exercise 19.4#4).]

§19.4. Outline of the proof of reduction theory

(19.4.1) **Notation.** Throughout this section, we assume

- G is defined over $\mathbb{Q}$,
- Γ is commensurable to $G_{\mathbb{Z}}$, and
- P is a minimal parabolic $\mathbb{Q}$-subgroup of G, with Langlands decomposition $P = MSN$.

In order to use Siegel sets to construct a coarse fundamental domain, a bit of care needs to be taken when choosing a maximal compact subgroup K. Before stating the precise condition, we recall that the **Cartan involution** corresponding to K is an automorphism τ of G, such that τ^2 is the identity, and K is the set of fixed points of τ. (For example, if $G = \mathrm{SL}(n, \mathbb{R})$ and $K = \mathrm{SO}(n)$, then $\tau(g) = (g^T)^{-1}$ is the transpose-inverse of g.)

(19.4.2) **Definition.** A Siegel set $\mathfrak{S} = CS^+K$ is **normal** if S is invariant under the Cartan involution corresponding to K.

Fix a normal Siegel set $\mathfrak{S} = CS^+K$, and some finite $F \subseteq G_{\mathbb{Q}}$. Then, letting $\mathcal{F} = F\mathfrak{S}$, the proof of Theorem 19.2.2 has two parts, corresponding to the two conditions in the definition of coarse fundamental domain (19.0.1):

i) $\mathfrak{S}$ and F can be chosen so that $\Gamma \mathcal{F} = G$ (see Theorems 19.4.3 and 19.4.4), and

ii) for all choices of $\mathfrak{S}$ and F, the set $\{\, y \in \Gamma \mid \mathcal{F} \cap y\mathcal{F} \neq \varnothing \,\}$ is finite (see Theorem 19.4.8).

We will sketch proofs of both parts (assuming $\mathrm{rank}_{\mathbb{Q}} \Gamma = 1$). However, as a practical matter, the methods of proof are not as important as understanding the construction of the coarse fundamental domain as a union of Siegel sets (see Section 19.2), and being able to use this in applications (as in Section 19.3).

§19.4(i). Proof that $\Gamma \mathcal{F} = G$. Here is the rough idea: Fix a base point in $\Gamma \backslash G$. A Siegel set can easily cover all of the nearby points (see Exercise 19.1#2), so consider a point Γg that is far away. Godement's Criterion (5.3.1) implies there is some nontrivial unipotent $v \in \Gamma$, such that $v^g \approx e$. Replacing g with a different representative of the coset replaces v with a conjugate element. If we assume all the maximal unipotent $\mathbb{Q}$-subgroups of G are conjugate under Γ, this implies that we may assume $v \in N$. If we furthermore assume, for simplicity, that the maximal $\mathbb{Q}$-split torus S is actually a maximal $\mathbb{R}$-split torus, then the Iwasawa decomposition (8.4.9) tells us $G = NSK$. The compact group K is contained in our Siegel set $\mathfrak{S}$, and the subgroup N is contained in $\Gamma \mathfrak{S}$ if $\mathfrak{S}$ is sufficiently large, so let us assume $g \in S$. Since g contracts the element v of N, and, by definition, S^+ consists of the elements of S that contract N, we conclude that $g \in S^+ \subseteq \mathfrak{S}$.

We now explain how to turn this outline into a proof.

Recall that all minimal parabolic $\mathbb{Q}$-subgroups of G are conjugate under $G_\mathbb{Q}$ (see Proposition 9.3.6(1)). The following technical result from the algebraic theory of arithmetic groups asserts that there are only finitely many conjugacy classes under the much smaller group $G_\mathbb{Z}$. In geometric terms, it is a generalization of the fact that hyperbolic manifolds of finite volume have only finitely many cusps.

(19.4.3) Theorem. *There is a finite subset F of $G_\mathbb{Q}$, such that $\Gamma F P_\mathbb{Q} = G_\mathbb{Q}$.*

The finite subset F provided by the theorem can be used to construct the coarse fundamental domain $\mathcal{F}$:

(19.4.4) Theorem. *If F is a finite subset of $G_\mathbb{Q}$, such that $\Gamma F P_\mathbb{Q} = G_\mathbb{Q}$, then there is a (normal) Siegel set $\mathfrak{S} = C S^+ K$, such that $\Gamma F \mathfrak{S} = G$.*

Idea of proof (assuming $\operatorname{rank}_\mathbb{Q} \Gamma \leq 1$). For simplicity, assume $\Gamma = G_\mathbb{Z}$, and that $F = \{e\}$ has only one element (see Exercise 6), so

$$\Gamma P_\mathbb{Q} = G_\mathbb{Q}. \tag{19.4.5}$$

The theorem is trivial if Γ is cocompact (see Exercise 19.2#1), so let us assume $\operatorname{rank}_\mathbb{Q} \Gamma = 1$.

From the proof of the Godement Compactness Criterion (5.3.1), we have a compact subset C_0 of G, such that, for each $g \in G$, either $g \in \Gamma C_0$, or there is a nontrivial unipotent element v of Γ, such that $v^g \approx e$. By choosing C large enough, we may assume $C_0 \subseteq \mathfrak{S}$ (see Exercise 19.1#2).

Now suppose some element g of G is not in $\Gamma \mathfrak{S}$. Then $g \notin \Gamma C_0$, so there is a nontrivial unipotent element v of Γ, such that

$$v^g \approx e. \tag{19.4.6}$$

From (19.4.5), we see that we may assume $v \in N$, after multiplying g on the left by an element of Γ (see Exercise 2).

We have $G = PK$ (cf. Exercise 9.3#4). Furthermore, $P = MSN$, and Γ intersects both M and N in a cocompact lattice (see Example 9.1.5(3), Theorem 9.3.3(3), and Exercise 9.3#6). Therefore, if we multiply g on the left by an element of $\Gamma \cap P$, and ignore a bounded error, we may assume $g \in S$ (see Exercise 3). Then, since $\mathrm{rank}_{\mathbb{Q}} \Gamma = 1$, we have either $g \in S^+$ or $g^{-1} \in S^+$ (see Exercise 5). From (19.4.6), we conclude it is g that is in S^+. So $g \in \mathfrak{S}$, which contradicts the fact that $g \notin \Gamma\mathfrak{S}$.   □

(19.4.7) *Remark.* The above proof overlooks a technical issue: in the Langlands decomposition $P = MSN$, the subgroup M may be reductive, rather than semisimple. However, the maximality of S implies that the central torus T of M has no $\mathbb{R}$-split subtori (cf. Theorem 9.3.3(3)), so it can be shown that this implies $T/T_{\mathbb{Z}}$ is compact. Therefore $M/M_{\mathbb{Z}}$ is compact, even if M is not semisimple.

§**19.4(ii). Proof that** $\mathfrak{S}$ **intersects only finitely many Γ-translates.** We know that Γ is commensurable to $G_{\mathbb{Z}}$. Therefore, if we make the minor assumption that $G_{\mathbb{C}}$ has trivial center, then $\Gamma \subseteq G_{\mathbb{Q}}$ (see Exercise 5.2#4). Hence, the following result establishes Condition 19.0.1(2) for $\mathcal{F} = F\mathfrak{S}$:

(19.4.8) **Theorem** ("Siegel property"). *If*

- $\mathfrak{S} = CS^+K$ *is a normal Siegel set, and*
- $q \in G_{\mathbb{Q}}$,

then $\{ y \in G_{\mathbb{Z}} \mid q\mathfrak{S} \cap y\mathfrak{S} \neq \varnothing \}$ *is finite.*

Proof (assuming $\mathrm{rank}_{\mathbb{Q}} \Gamma \leq 1$). The desired conclusion is obvious if $\mathfrak{S}$ is compact, so we may assume $\mathrm{rank}_{\mathbb{Q}} \Gamma = 1$. To simplify matters, let $\Gamma = G_{\mathbb{Z}}$, and

$$\text{assume } q = e \text{ is trivial.}$$

The proof is by contradiction: assume

$$\sigma = y\sigma',$$

for some large element y of Γ, and some $\sigma, \sigma' \in \mathfrak{S}$. Since y is large, we may assume σ is large (by interchanging σ with σ' and replacing y with y^{-1}, if necessary). Let

$$u \text{ be an element of } N_{\mathbb{Z}} \text{ of bounded size.}$$

Since $\sigma \in \mathfrak{S} = CS^+K$, we may write

$$\sigma = csk \text{ with } c \in C, s \in S^+, \text{ and } k \in K.$$

Then s must be large (since K and C are compact), so conjugation by s performs a large contraction on N. Since u^c is an element of N of bounded size, and K is compact, this implies that $u^\sigma \approx e$. In other words,

$$u^{y\sigma'} \approx e.$$

In addition, we know that $u^y \in G_{\mathbb{Z}}$. Since $\sigma' \in \mathfrak{S}$, we conclude that $u^y \in N$ (see Exercise 7).

Now we use the assumption that $\operatorname{rank}_{\mathbb{Q}} \Gamma = 1$: since

$$u^y \in N \cap N^y,$$

Lemma 9.2.5 tells us that $N = N^y$, so Proposition 9.3.6(3) implies

$$y \in \mathcal{N}_G(N) = P.$$

Then, since $(MN)_{\mathbb{Z}}$ has finite index in $P_{\mathbb{Z}}$ (see Exercise 4), we may assume $y \in (MN)_{\mathbb{Z}}$.

This implies that we may work inside of MN: if we choose a compact subset $\overline{C} \subseteq MN$, such that $CS^+ \subseteq \overline{C}S$, then we have

$$\overline{C}(K \cap P) \cap y\overline{C}(K \cap P) \neq \varnothing \qquad (19.4.9)$$

(see Exercise 9). Since $\overline{C}(K \cap P)$ is compact (and Γ is discrete), we conclude that there are only finitely many possibilities for y.   □

(19.4.10) *Remark.* When $\operatorname{rank}_{\mathbb{Q}} \Gamma = 1$, the first part of the proof establishes the useful fact that there is a compact subset C_0 of G, such that if $y \in \Gamma$, and $\mathfrak{S} \cap y\mathfrak{S} \not\subseteq C_0$, then $y \in P$.

Exercises for §19.4.

#1. Show that every $\mathbb{Q}$-split torus of G is invariant under some Cartan involution of G. (Therefore, for any maximal $\mathbb{Q}$-split torus S, there exists a maximal compact subgroup K, such that the resulting Siegel sets CS^+K are normal.)

   [*Hint:* If τ is any Cartan involution, then there is a maximal $\mathbb{R}$-split torus A, such that $\tau(a) = a^{-1}$ for all $a \in A$. Any $\mathbb{R}$-split torus is contained in some conjugate of A.]

#2. In the proof of Theorem 19.4.4, explain why it may be assumed that $v \in N$.

   [*Hint:* Being unipotent, v is contained in the unipotent radical of some minimal parabolic $\mathbb{Q}$-subgroup (see Proposition 9.3.6(2)). Since $\Gamma P_{\mathbb{Q}} = G_{\mathbb{Q}}$, we know that all minimal parabolic $\mathbb{Q}$-subgroups are conjugate under Γ.]

#3. In the proof of Theorem 19.4.4, complete the proof without assuming that $g \in S$.

   [*Hint:* Write $g = pk \in PK$. If C is large enough that $MN \subseteq (\Gamma \cap P)C$, then we have $g \in \Gamma csk \subseteq \Gamma CSK$, so $v^s \approx e$.]

#4. Show that if $P = MAN$ is a Langlands decomposition of a parabolic $\mathbb{Q}$-subgroup of G, then $(MN)_{\mathbb{Z}}$ contains a finite-index subgroup of $P_{\mathbb{Z}}$.

   [*Hint:* A $\mathbb{Q}$-split torus can have only finitely many integer points.]

#5. Show that if $\operatorname{rank}_{\mathbb{Q}} \Gamma = 1$, then $S = S^+ \cup (S^+)^{-1}$.

   [*Hint:* If $s \in S^+$, then $s^t \in S^+$ for all $t \in \mathbb{Q}^+$ (and, hence, for all $t \in \mathbb{R}^+$).]

#6. Prove Theorem 19.4.4 in the case where $\mathrm{rank}_{\mathbb{Q}}\,\Gamma = 1$.

[*Hint:* Replace the imprecise arguments of the text with rigorous statements, and do not assume F is a singleton. (In order to assume $v \in N$, multiply g on the left by an element x of $F\Gamma$. Then $x\Gamma x^{-1}$ contains a finite-index subgroup of $\Gamma \cap P$.)]

#7. Let $\mathfrak{S} = CS^+K$ be a Siegel set, and let P be the minimal parabolic $\mathbb{Q}$-subgroup corresponding to S^+. Show there is a neighborhood W of e in G, such that if $y \in G_{\mathbb{Z}}$ and $y^\sigma \in W$, for some $\sigma \in \mathfrak{S}$, then $y \in \mathrm{unip}\,P$.

#8. Suppose
- τ is the Cartan involution corresponding to the maximal compact subgroup K, and
- S is a τ-invariant.

Show $K \cap P \subseteq M \subseteq C_G(S)$.

[*Hint:* Since $C_G(S)$ is τ-invariant, the restriction of τ to the semisimple part of M is a Cartan involution. Therefore $K \cap M$ contains a maximal compact subgroup of M, which is a maximal compact subgroup of P. The second inclusion is immediate from the definition of the Langlands decomposition.]

#9. Establish (19.4.9).

[*Hint:* $\mathfrak{S} \cap P \cap y(\mathfrak{S} \cap P) \ne \varnothing$ (since $y \in P$) and $\mathfrak{S} \cap P = CS^+(K \cap P) = C(K \cap P)S^+$ (see Exercise 8).]

#10. Give a complete proof of Theorem 19.4.8.

#11. Show that Γ has only finitely many conjugacy classes of maximal unipotent subgroups.

[*Hint:* You may assume, for simplicity, that $\Gamma = G_{\mathbb{Z}}$ is arithmetic. Use Proposition 9.3.6(2) and Theorem 19.4.3.]

Notes

The main results of this chapter were obtained for many classical groups by L. Siegel (see, for example, [8]), and the general results are due to A. Borel and Harish-Chandra [2].

The book of A. Borel [1] is the standard reference for this material; see [1, Thm. 13.1, p. 90] for the construction of a fundamental domain for $G_{\mathbb{Z}}$ as a union of Siegel sets. The proof there does not assume $G_{\mathbb{Z}}$ is a lattice, so this provides a proof of the fundamental fact that every arithmetic subgroup of G is a lattice (see Theorem 5.1.11). See [6, §4.6] for an exposition of Borel and Harish-Chandra's original proof of this fact (using the Siegel set $\mathfrak{S}_{c_1,c_2,c_3}$ for $SL(n,\mathbb{Z})$ from Definition 7.2.4).

Theorem 19.3.7 was conjectured by C. L. Siegel in 1959, and was proved by L. Ji [4, Thm. 7.6]. Another proof is in E. Leuzinger [5, Thm. B].

The construction of coarse fundamental domains for non-arithmetic lattices in groups of real rank one is due to Garland and Raghunathan [3]. (An exposition appears in [7, Chap. 13].) Combining this result with Theorem 19.2.2 yields Theorem 19.2.6.

Exercise 19.3#1 can be found in [6, Thm. 4.3, p. 203].

Regarding Remark 19.4.7, see [1, Prop. 8.5, p. 55] or [6, Thm. 4.11, p. 208] for a proof that if T is a $\mathbb{Q}$-torus that has no $\mathbb{Q}$-split subtori, then $T/T_{\mathbb{Z}}$ is compact.

References

[1] A. Borel: *Introduction aux Groupes Arithmétiques*. Hermann, Paris, 1969. MR 0244260

[2] A. Borel and Harish-Chandra: Arithmetic subgroups of algebraic groups, *Ann. Math.* (2) 75 (1962) 485–535. MR 0147566, http://www.jstor.org/stable/1970210

[3] H. Garland and M. S. Raghunathan: Fundamental domains for lattices in (ℝ-)rank 1 semisimple Lie groups *Ann. Math.* (2) 92 (1970) 279–326. MR 0267041, http://www.jstor.org/stable/1970838

[4] L. Ji: Metric compactifications of locally symmetric spaces, *Internat. J. Math.* 9 (1998), no. 4, 465–491. MR 1635185, http://dx.doi.org/10.1142/S0129167X98000208

[5] E. Leuzinger: Tits geometry, arithmetic groups, and the proof of a conjecture of Siegel, *J. Lie Theory* 14 (2004), no. 2, 317–338. MR 2066859, http://www.emis.de/journals/JLT/vol.14_no.2/6.html

[6] V. Platonov and A. Rapinchuk: *Algebraic Groups and Number Theory*. Academic Press, Boston, 1994. ISBN 0-12-558180-7, MR 1278263

[7] M. S. Raghunathan: *Discrete Subgroups of Lie Groups*. Springer, New York, 1972. ISBN 0-387-05749-8, MR 0507234

[8] C. L. Siegel: Discontinuous groups, *Ann. of Math.* (2) 44 (1943) 674–689. MR 0009959, http://www.jstor.org/stable/1969104

Chapter 20

Ratner's Theorems on Unipotent Flows

This chapter presents three theorems that strengthen and vastly generalize the well-known and useful observation that if V is any straight line
in the Euclidean plane $\mathbb{R}^2$, then the closure of the image of L in $\mathbb{T}^2$ is a
very nice submanifold (see Example 20.1.1). The plane can be replaced
with any Lie group H, and V can be any subgroup of H that is generated
by unipotent elements.

§20.1. Statement of Ratner's Orbit-Closure Theorem

(20.1.1) **Example.** Let

- V be any 1-dimensional subspace of the vector space $\mathbb{R}^2$,
- $\mathbb{T}^2 = \mathbb{R}^2/\mathbb{Z}^2$ be the ordinary 2-torus,
- $x \in \mathbb{T}^2$, and
- $\pi: \mathbb{R}^2 \to \mathbb{T}^2$ be the natural covering map.

Geometrically, $V + x$ is a straight line in the plane, and it is classical (and
not difficult to prove) that:

Recall: The Standing Assumptions (4.0.0 on page 41) are in effect, so, as
always, Γ is a lattice in the semisimple Lie group $G \subseteq \mathrm{SL}(\ell, \mathbb{R})$.

Main prerequisites for this chapter: none.

1) If the slope of the line $V + x$ is rational, then $\pi(V + x)$ is closed, and it is homeomorphic to the circle $\mathbb{T}^1$ (see Exercise 1).

2) If the slope of the line $V + x$ is irrational, then the closure of $\pi(V + x)$ is the entire torus $\mathbb{T}^2$ (see Exercise 2).

An analogous result holds in higher dimensions: if we take any vector subspace of $\mathbb{R}^\ell$, the closure of its image in $\mathbb{T}^\ell$ will always be a nice submanifold of $\mathbb{T}^\ell$. Indeed, the closure will be a subtorus of $\mathbb{T}^\ell$:

(20.1.2) Example (see Exercise 3). Let

- V be a vector subspace of $\mathbb{R}^\ell$,
- $\mathbb{T}^\ell = \mathbb{R}^\ell / \mathbb{Z}^\ell$ be the ordinary ℓ-torus,
- $x \in \mathbb{T}^\ell$, and
- $\pi \colon \mathbb{R}^\ell \to \mathbb{T}^\ell$ be the natural covering map.

Then the closure of $\pi(V + x)$ in $\mathbb{T}^\ell$ is homeomorphic to a torus $\mathbb{T}^k$ (with $0 \le k \le \ell$).

More precisely, there is a vector subspace L of $\mathbb{R}^\ell$, such that

- the closure of $\pi(V + x)$ is $\pi(L + x)$, and
- L is defined over $\mathbb{Q}$ (or, in other words, $L \cap \mathbb{Z}^\ell$ is a $\mathbb{Z}$-lattice in L), so $\pi(L + x) \cong L/(L \cap \mathbb{Z}^\ell)$ is a torus.

The above observation about tori generalizes in a natural way to much more general homogeneous spaces, by replacing:

- $\mathbb{R}^3$ with any connected Lie group H,
- $\mathbb{Z}^3$ with a lattice Λ in H,
- the vector subspace V of $\mathbb{R}^3$ with any subgroup of H that is generated by unipotent elements,
- $x + V$ with the coset xV,
- the map $\pi \colon \mathbb{R}^\ell \to \mathbb{T}^\ell$ with the natural covering map $\pi \colon H \to H/\Lambda$, and
- the vector subspace L of $\mathbb{R}^\ell$ with a closed subgroup L of H.

Because it suffices for our purposes, we state only the case where H is semisimple (so we call the group G, instead of H):

(20.1.3) Ratner's Orbit-Closure Theorem. *Suppose*

- *V is a subgroup of G that is generated by unipotent elements, and*
- *$x \in G/\Gamma$.*

Then there is a closed subgroup L of G, such that the closure of Vx is Lx. Furthermore, L can be chosen so that:

1) *L contains the identity component of V,*
2) *L has only finitely many connected components, and*
3) *there is an L-invariant, finite measure on Lx.*

(Also note that Lx is closed in G/Γ, since it is the closure of Vx.)

(20.1.4) *Remark.* Write $x = g\Gamma$, for some $g \in G$, and let $\Lambda = (g\Gamma g^{-1}) \cap L$.

1) The theorem tells us that the closure of Vx is a very nice submanifold of G/Γ. Indeed, the closure is homeomorphic to the homogeneous space L/Λ.

2) Conclusion (3) of the theorem is equivalent to the assertion that Λ is a lattice in L.

(20.1.5) **Warning.** The assumption that V is generated by unipotent elements cannot be eliminated. For example, it is known that if V is the group of diagonal matrices in $G = \mathrm{SL}(2, \mathbb{R})$, then there are points $x \in G/\mathrm{SL}(2, \mathbb{Z})$, such that the closure of Vx is a fractal. This means that the closure of Vx can be a very bad set that is not anywhere close to being a submanifold.

Unfortunately, the known proofs of Ratner's Orbit-Closure Theorem are rather long. One of the paramount ideas in the proof will be described in Section 20.4, but, first, we will present a few of the theorem's applications (in Section 20.2) and state two other variants of the theorem (in Section 20.3).

Exercises for §20.1.

#1. Verify Example 20.1.1(1).

#2. Verify Example 20.1.1(2).

#3. Verify Example 20.1.2.

#4. Show that if V is connected, then the subgroup L in the conclusion of Theorem 20.1.3 can also be taken to be connected.

#5. (Non-divergence of unipotent flows) Suppose
   - $\{u^t\}$ is a unipotent one-parameter subgroup of G, and
   - $x \in G/\Gamma$.

Use Theorem 20.1.3 to show there is a compact subset K of G, such that

$$\{\, t \in \mathbb{R} \mid u^t x \in K \,\} \text{ is unbounded.}$$

(Hence, Theorem 7.4.6 is logically a corollary of Theorem 20.1.3. However, in practice, Theorem 7.4.6 is used in the proof of Theorem 20.1.3.)

[*Hint:* Conclusion (3) of Theorem 20.1.3 is crucial.]

#6. Show, by providing an explicit counterexample, that the assumption that $\{u^t\}$ is unipotent cannot be eliminated in Exercise 5.

[*Hint:* Consider the one-parameter group of diagonal matrices in $\mathrm{SL}(2, \mathbb{R})$, and let $\Gamma = \mathrm{SL}(2, \mathbb{Z})$.]

§20.2. Applications

We will briefly describe just a few of the many diverse applications of Ratner's Orbit-Closure Theorem (20.1.3).

§20.2(i). Closures of totally geodesic subspaces.

(20.2.1) **Example.** Let M be a compact, hyperbolic n-manifold (with $n \geq 2$).

- There is a covering map $\pi : \mathfrak{H}^n \to M$ that is a local isometry.
- There is a natural embedding $\iota : \mathfrak{H}^1 \hookrightarrow \mathfrak{H}^n$.
- Let f_1 be the composition $\pi \circ \iota$, so $f_1 : \mathfrak{H}^1 \to M$.

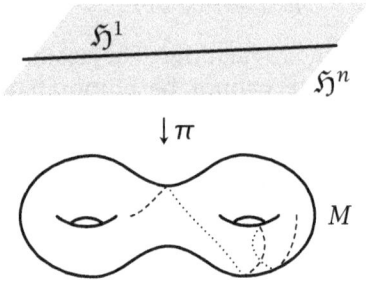

Then the image of f_1 is a curve in M. It is well known (though not at all obvious) that the closure of this curve can be a very bad set; in fact, even though $\mathfrak{H}^1$ and M are nice, smooth manifolds, this closure can be a fractal. (This is a higher-dimensional analogue of the example in Warning 20.1.5. In the literature, it is the fact that the closure of a geodesic in a compact manifold of negative curvature can be a fractal.)

It is a consequence of Ratner's Theorem that this pathology never occurs if we replace $\mathfrak{H}^1$ with a higher-dimensional hyperbolic space:

(20.2.2) **Corollary.** *Let:*

- *$m, n \in \mathbb{N}$, with $m \leq n$,*
- *M be a compact, hyperbolic n-manifold,*
- *$\pi : \mathfrak{H}^n \to M$ be a covering map that is a local isometry,*
- *$\iota : \mathfrak{H}^m \hookrightarrow \mathfrak{H}^n$ be a totally geodesic embedding, and*
- *$f_m = \pi \circ \iota$, so $f_m : \mathfrak{H}^m \to M$.*

If $m \geq 2$, then the closure $\overline{f_m(\mathfrak{H}^m)}$ of the image of f_m is a (totally geodesic) immersed submanifold of M.

Proof. We prove only that the closure is a submanifold, not that it is totally geodesic. Let
$$V = \mathrm{SO}(1, m), \quad G = \mathrm{SO}(1, n), \quad \text{and } x \in \iota(\mathfrak{H}^m),$$

so

- G acts by isometries on $\mathfrak{H}^n$,
- $M = \Gamma \backslash \mathfrak{H}^n$, for some lattice Γ in G, and
- $\iota(\mathfrak{H}^m) = Vx$.

From Ratner's Orbit-Closure Theorem (20.1.3), we know there is a subgroup L of G, such that $\overline{\Gamma V} = \Gamma L$. So

$$\overline{f_m(\mathfrak{H}^m)} = \overline{\pi(\iota(\mathfrak{H}^m))} = \overline{\pi(Vx)} = \pi(Lx)$$

is an immersed submanifold of M.   □

(20.2.3) *Remarks.*

1) The same conclusion holds (with the same proof) when $\mathfrak{H}^m$ and $\mathfrak{H}^n$
   are replaced with much more general symmetric spaces $\tilde{X}$ and $\tilde{Y}$
   that have no compact factors, except that the closure may not be
   totally geodesic if rank $\tilde{X}$ < rank $\tilde{Y}$.

2) When rank $\tilde{X}$ = rank $\tilde{Y}$, one proves that the submanifold is totally
   geodesic by showing that the subgroup L in the above proof is invariant under the appropriate Cartan involution of G.

§20.2(ii). Values of quadratic forms. Many of the most impressive
applications of Ratner's Orbit-Closure Theorem (and the related results
that will be described in Section 20.3) are in Number Theory. As an example, we present a famous result on values of quadratic forms. It is now an
easy corollary of Ratner's Orbit-Closure Theorem, but, historically, it was
proved by Margulis before this major theorem was available (by proving
the relevant special case of the general theorem).
   Let

$$Q(\vec{x}) = Q(x_1, x_2, \ldots, x_n) \text{ be a quadratic form in } n \text{ variables}$$

(in other words, $Q(\vec{x})$ is a homogeneous polynomial of degree 2).
   Classical number theorists were interested in determining the values
of c for which the equation $Q(\vec{x}) = c$ has an integer solution; that is, a
solution with $\vec{x} \in \mathbb{Z}^n$. For example:

1) Lagrange's 4-Squares Theorem tells us that if
   $$Q(x_1, x_2, x_3, x_4) = x_1^2 + x_2^2 + x_3^2 + x_4^2,$$
   then $Q(\vec{x}) = c$ has a solution iff $c \in \mathbb{Z}^{\geq 0}$.

2) Fermat's 2-Squares Theorem tells us that if $Q(x_1, x_2) = x_1^2 + x_2^2$, and
   p is an odd prime, then $Q(\vec{x}) = p$ has a solution iff $p \equiv 1 \pmod 4$.

These very classical results consider only forms whose coefficients are
integers, but we can also look at forms with *irrational* coefficients, such
as

$$Q(x_1, x_2, x_3, x_4) = 3x_1^2 - \sqrt{2}x_2x_3 + \pi x_4^2.$$

For a given quadratic form $Q(\vec{x})$, it is clear that

the equation $Q(\vec{x}) = c$ does not have an integral solution,
for most real values of c,

for the simple reason that there are only countably many possible integer values of the variables $x_1, x_2, \ldots, x_n$, but there are uncountably many possible choices of c. Therefore, instead of trying to solve the equation *exactly*, we must be content with solving the equation *approximately*. That is, we will be satisfied with knowing that we can find a value of $Q(\vec{x})$ that is within ϵ of c, for every ϵ (and every c). In other words, we would like to know that $Q(\mathbb{Z}^n)$ is *dense* in $\mathbb{R}$.

(20.2.4) **Examples.** There are some simple reasons that $Q(\mathbb{Z}^n)$ may fail to be dense in $\mathbb{R}$:

1) Suppose all of the coefficients of $Q(\vec{x})$ are integers. Then we have $Q(\mathbb{Z}^n) \subseteq \mathbb{Z}$, so $Q(\mathbb{Z}^n)$ is obviously not dense in $\mathbb{R}$. More generally, if $Q(\vec{x})$ is a scalar multiple of a form with integer coefficients, then $Q(\mathbb{Z}^n)$ is not dense in $\mathbb{R}$.

2) Suppose all values of $Q(\vec{x})$ are ≥ 0 (or all are ≤ 0). (In this case, we say that $Q(\vec{x})$ is *positive-definite* (or *negative-definite*, respectively). For example, this is the case if
$$Q(\vec{x}) = a_1 x_1^2 + a_2 x_2^2 + a_3 x_3^2 + \cdots + a_n x_n^2,$$
with all coefficients a_i of the same sign. Then it is clear that $Q(\mathbb{Z}^n)$ is not dense in all of $\mathbb{R}$.

3) Let $Q(x_1, x_2) = x_1^2 - \alpha x_2^2$, where $\alpha = 3 + 2\sqrt{2}$. Then, although it is not obvious, one can show that $Q(\mathbb{Z}^2)$ is not dense in $\mathbb{R}$ (see Exercise 3). Certain other choices of α also provide examples where $Q(\mathbb{Z}^2)$ is not dense (see Exercise 2), so having only 2 variables in the quadratic form can cause difficulties.

4) Even if a form has many variables, there may be a linear change of coordinates that turns it into a form with fewer variables. (For example, letting $z = x + \sqrt{2}y$ transforms $x^2 + 2\sqrt{2}xy + 2y^2$ into z^2.) A form that admits such a change of coordinates is said to be *degenerate*. Therefore, a degenerate form with more than 2 variables could merely be a disguised version of a form with 2 variables whose image is not dense in $\mathbb{R}$.

The following result shows that any quadratic form avoiding these simple obstructions does have values that are dense in $\mathbb{R}$. It is often called the "Oppenheim Conjecture," because it was an open problem under that name for more than 50 years, but that terminology is no longer appropriate, since it is now a theorem.

(20.2.5) **Corollary** (Margulis' Theorem on Values of Quadratic Forms). *Let $Q(\vec{x})$ be a quadratic form in $n \geq 3$ variables, and assume $Q(\vec{x})$ is:*

- *not a scalar multiple of a form with integer coefficients,*
- *neither positive-definite nor negative-definite, and*

- *nondegenerate.*

Then $Q(\mathbb{Z}^n)$ *is dense in* $\mathbb{R}$.

Proof. For simplicity, assume $n = 3$. Let

- $G = \mathrm{SL}(3, \mathbb{R})$, and
- $H = \mathrm{SO}(Q)^\circ = \{\, h \in G \mid Q(h\vec{x}) = Q(\vec{x}) \text{ for all } \vec{x} \in \mathbb{R}^3 \,\}^\circ$.

Since $Q(\vec{x})$ is nondegenerate, and neither positive-definite nor negative-definite, we have $H \cong \mathrm{SO}(1,2)^\circ \cong \mathrm{PSL}(2, \mathbb{R})$, so H is generated by unipotent elements. Furthermore, calculations in Lie theory (which we omit) show that the only connected subgroups of G containing H are the obvious ones: H and G. Therefore, Ratner's Orbit-Closure Theorem (20.1.3) tells us that either:

- $HG_{\mathbb{Z}}$ is closed, and $G_{\mathbb{Z}} \cap H$ is a lattice in H, or
- the closure of $HG_{\mathbb{Z}}$ is all of G.

However, if $H_{\mathbb{Z}} = G_{\mathbb{Z}} \cap H$ is a lattice in H, then the Borel Density Theorem (4.5.6) implies that H is defined over $\mathbb{Q}$ (see Exercise 5.1#5). Then, since $H = \mathrm{SO}(Q)^\circ$, a bit of algebra shows that $Q(\vec{x})$ is a scalar multiple of a form with integer coefficients (see Exercise 4). This is a contradiction.

Therefore, we conclude that the closure of $HG_{\mathbb{Z}}$ is all of G. In other words,

$$HG_{\mathbb{Z}} \text{ is dense in } G,$$

so

$$HG_{\mathbb{Z}}(1,0,0) \text{ is dense in } G(1,0,0).$$

Since $G_{\mathbb{Z}}(1,0,0) \subseteq \mathbb{Z}^3$, and $G(1,0,0) = \mathbb{R}^3 \smallsetminus \{0\}$, this tells us that

$$H\mathbb{Z}^3 \text{ is dense in } \mathbb{R}^3.$$

Then, since $Q(\vec{x})$ is continuous, we conclude that

$$Q(H\mathbb{Z}^3) \text{ is dense in } Q(\mathbb{R}^3).$$

We also know:

- $Q(H\mathbb{Z}^3) = Q(\mathbb{Z}^3)$, by the definition of H, and
- $Q(\mathbb{R}^3) = \mathbb{R}$, because $Q(\vec{x})$ is neither positive-definite nor negative-definite (see Exercise 5).

Therefore $Q(\mathbb{Z}^3)$ is dense in $\mathbb{R}$. $\qquad\square$

§20.2(iii). Products of lattices.

(20.2.6) Corollary (see Exercise 6)**.** *If* Γ_1 *and* Γ_2 *are any two lattices in* G, *and* G *is simple, then either*

1) Γ_1 *and* Γ_2 *are commensurable, so the product* $\Gamma_1 \Gamma_2$ *is discrete, or*
2) $\Gamma_1 \Gamma_2$ *is dense in* G.

Exercises for §20.2.

#1. Suppose β is a quadratic irrational. (This means that α is irrational, and that α is a root of a quadratic polynomial with integer coefficients.) Show that β is **badly approximable**: i.e., there exists $\epsilon > 0$, such that if p/q is any rational number, then

$$\left| \frac{p}{q} - \beta \right| > \frac{\epsilon}{q^2}.$$

#2. Let $Q(x_1, x_2) = x_1^2 - \beta^2 x_2^2$, where β is any badly approximable number (cf. Exercise 1). Show that $Q(\mathbb{Z}^2)$ is not dense in $\mathbb{R}$.
[*Hint:* There exists $\delta > 0$, such that $|Q(p, q)| \geq \delta$ for $p, q \in \mathbb{Z} \smallsetminus \{0\}$.]

#3. Let $Q(x_1, x_2) = x_1^2 - \alpha x_2^2$, where $\alpha = 3 + 2\sqrt{2}$. Show $Q(\mathbb{Z}^2)$ is not dense in $\mathbb{R}$.
[*Hint:* Use previous exercises, and note that $3 + 2\sqrt{2} = (1 + \sqrt{2})^2$.]

#4. Suppose $Q(\overline{x})$ is a nondegenerate quadratic form in n variables. Show that if $\mathrm{SO}(Q)^\circ$ is defined over $\mathbb{Q}$, then $Q(\overline{x})$ is a scalar multiple of a form with integer coefficients.
[*Hint:* Up to scalar multiples, there is a unique quadratic form that is invariant under $\mathrm{SO}(Q)^\circ$, and the uniqueness implies that it is invariant under the Galois group $\mathrm{Gal}(\mathbb{C}/\mathbb{Q})$.]

#5. Suppose $Q(\overline{x})$ is a quadratic form in n variables that is neither positive-definite nor negative-definite. Show $Q(\mathbb{R}^n) = \mathbb{R}$.
[*Hint:* $Q(\lambda \overline{x}) = \lambda^2 Q(\overline{x})$.]

#6. Prove Corollary 20.2.6.
[*Hint:* Let $\Gamma = \Gamma_1 \times \Gamma_2 \subset G \times G$, and $H = \{ (g, g) \mid g \in G \}$. Show the only connected subgroups of $G \times G$ that contain H are the two obvious ones: H and $G \times G$. Therefore, Ratner's Theorem implies that either $H \cap \Gamma$ is a lattice in H, or ΓH is dense in $G \times G$.]

§20.3. Two measure-theoretic variants of the theorem

Ratner's Orbit-Closure Theorem (20.1.3) is purely topological, or qualitative. In some situations, it is important to have quantitative information.

(20.3.1) **Example.** We mentioned earlier that if V is a line with irrational slope in $\mathbb{R}^2$, then the image $\pi(V)$ of V in $\mathbb{T}^2$ is dense (see Example 20.1.1). For applications in analysis, it is often necessary to know more, namely, that $\pi(V)$ is **uniformly distributed** in $\mathbb{T}^2$. Roughly speaking, this means that a long segment of $\pi(V)$ visits all parts of the torus equally often (see Exercise 1).

(20.3.2) **Definition.** Let

- μ be a probability measure on a topological space X, and
- $c: [0, \infty) \to X$ be a continuous curve in X.

We say that c is **uniformly distributed** in X (with respect to μ) if, for every continuous function $f: X \to \mathbb{R}$ with compact support, we have

$$\lim_{T \to \infty} \frac{1}{T} \int_0^T f(c(t))\, dt = \int_X f\, d\mu.$$

Ratner's Orbit-Closure Theorem tells us that if

- U is any one-parameter subgroup of G, and
- x is any point in G/Γ,

then the closure of the U-orbit Ux is a nice submanifold of G/Γ. The following theorem tells us that the U-orbit is uniformly distributed in this submanifold.

(20.3.3) Ratner's Equidistribution Theorem. *Let*

- $\{u^t\}$ *be any one-parameter unipotent subgroup of G,*
- $x \in G/\Gamma$, *and*
- $c(t) = u^t x$, *for $t \in [0, \infty)$.*

Then there is a connected, closed subgroup L of G, such that

1) *there is a (unique) L-invariant probability measure μ on Lx,*
2) *the curve c is uniformly distributed in Lx, with respect to μ,*
3) *the closure of $\{ c(t) \mid t \in [0, \infty) \}$ is Lx (so Lx is closed in G/Γ), and*
4) $\{u^t\} \subseteq L$.

In the special case where V is a one-parameter unipotent subgroup of G, the following theorem is a consequence of the above Equidistribution Theorem (see Exercise 3).

(20.3.4) Ratner's Classification of Invariant Measures. *Suppose*

- V *is a subgroup of G that is generated by unipotent elements, and*
- μ *is any ergodic V-invariant probability measure on G/Γ.*

Then there is a closed subgroup L of G, and some $x \in G/\Gamma$, such that μ is the unique L-invariant probability measure on Lx.

Furthermore, L can be chosen so that:

1) *L has only finitely many connected components,*
2) *L contains the identity component of V, and*
3) *Lx is closed in G/Γ.*

Here is a sample consequence of the Measure-Classification Theorem:

(20.3.5) Corollary. *Suppose*

- u_i *is a nontrivial unipotent element of G_i, for $i = 1, 2$,*
- $f: G_1/\Gamma_1 \to G_2/\Gamma_2$ *is a measurable map that intertwines the translation by u_1 with the translation by u_2; that is,*

$$f(u_1 x) = u_2 f(x), \quad \text{for a.e. } x \in G_1/\Gamma_1,$$

   and

- G_1 *is connected and almost simple.*

Then f *is a continuous function (a.e.).*

Proof. Let

- $G = G_1 \times G_2$,
- $\Gamma = \Gamma_1 \times \Gamma_2$,
- $u = (u_1, u_2) \in G$, and
- $\text{graph}(f) = \{ (x, f(x)) \mid x \in G_1/\Gamma_1 \} \subset G/\Gamma$.

The projection from G/Γ to G_1/Γ_1 defines a natural one-to-one correspondence between $\text{graph}(f)$ and G_1/Γ_1. In fact, since f intertwines u_1 with u_2, it is easy to see that the projection provides an isomorphism between the action of u_1 on G_1/Γ_1 and the action of u on $\text{graph}(f)$. In particular, the u_1-invariant probability measure μ_1 on $G_1\Gamma_1$ naturally corresponds to a u-invariant probability measure μ on $\text{graph}(f)$.

The Moore Ergodicity Theorem (14.2.4) tells us that μ_1 is ergodic for u_1, so μ is ergodic for u. Hence, Ratner's Measure-Classification Theorem (20.3.4) provides

- a closed subgroup L of G, and
- $x \in G/\Gamma$,

such that

- μ is the L-invariant measure on Lx, and
- Lx is closed.

Since the definition of μ implies that $\mu(\text{graph}(f)) = 1$, and the choice of L implies that the complement of Lx has measure 0, we may assume, by changing f on a set of measure 0, that

$$\text{graph}(f) \subseteq Lx.$$

Assume, for simplicity, that L is connected and G_1 is simply connected. Then the natural projection from L to G_1 is an isomorphism (see Exercise 4), so there is a (continuous) homomorphism $\rho\colon G_1 \to G_2$, such that

$$L = \text{graph}(\rho).$$

Assuming, for simplicity, that $x = (e, e)$, this implies that $f(g\Gamma) = \rho(g)\Gamma$ for all $g \in G$. So f, like ρ, is continuous. $\qquad\square$

As an example of the many important consequences of Ratner's Measure-Classification Theorem (20.3.4), we point out that it implies the Equidistribution Theorem (20.3.3). The proof is not at all obvious, and we will not attempt to explain it here, but the following simple example illustrates the important precept that knowing all of the invariant measures can lead to an equidistribution theorem.

(20.3.6) **Proposition.** *Let*
- *$\{g^t\}$ be a one-parameter subgroup of G,*
- *μ be the G-invariant probability measure on G/Γ,*
- *$x \in G/\Gamma$, and*
- *$c(t) = g^t x$.*

If
- *μ is the only g^t-invariant probability measure on G/Γ, and*
- *G/Γ is compact,*

then the curve c is uniformly distributed in G/Γ, with respect to μ.

Proof. Suppose c is not uniformly distributed. Then there is a sequence $T_k \to \infty$, and some continuous function $f_0 \in C(G/\Gamma)$, such that

$$\lim_{k \to \infty} \frac{1}{T_k} \int_0^{T_k} f_0(c(t)) \, dt \neq \mu(f_0) \qquad (20.3.7)$$

By passing to a subsequence, we may assume that

$$\lambda(f) = \lim_{k \to \infty} \frac{1}{T_k} \int_0^{T_k} f(c(t)) \, dt$$

exists for every $f \in C(G/\Gamma)$ (see Exercise 5). Then:
1) λ is a continuous linear functional on the space $C(G/\Gamma)$ of continuous functions on G/Γ, so the Riesz Representation Theorem (B6.10) tells us that λ is a measure on G/Γ.
2) $\lambda(1) = 1$, so λ is a probability measure.
3) From the definition of λ, it is not difficult to see that λ is g^t-invariant (see Exercise 6).

Since μ is the only g^t-invariant probability measure, we must have $\lambda = \mu$. However, (20.3.7) says $\lambda(f_0) \neq \mu(f_0)$, so this is a contradiction.   □

(20.3.8) *Remark.* Here is a rough outline of how the three theorems are proved:
1) Measure-Classification is proved in the case where V is unipotent.
   - The general case of Measure-Classification follows from this.
2) Equidistribution is a consequence of Measure-Classification.
3) Equidistribution easily implies Orbit-Closure in the special case where V is a one-parameter unipotent subgroup.
   - The general case of Orbit-Closure can be deduced from this.

(20.3.9) *Remarks.*
1) Ratner's Measure-Classification Theorem (20.3.4) remains valid if the lattice Γ is replaced with any closed subgroup of G. However, the other two theorems do not remain valid in this generality.

2) Ratner's Theorems assume that V is generated by unipotent elements, but N. Shah suggested that they might remain valid under the much weaker assumption that the Zariski closure of V is generated by unipotent elements. For the important case where the Zariski closure of V is semisimple (with no compact factors), this was recently proved by Y. Benoist and J.-F. Quint.

Exercises for §20.3.

#1. Suppose μ is a probability measure on a compact metric space X. Show that a curve $c: [0, \infty) \to X$ is uniformly distributed with respect to μ if and only if, for every open subset $\mathcal{O}$ of X, such that $\mu(\partial\mathcal{O}) = 0$, we have

$$\lim_{T \to \infty} \frac{1}{T} l\{t \in [0, T] \mid c(t) \in \mathcal{O}\} = \mu(\mathcal{O}).$$

[*Hint:* Bound the characteristic function of $\mathcal{O}$ above and below by continuous functions and apply Definition 20.3.2.]

#2. Show that if v is a nonzero vector of irrational slope in $\mathbb{R}^2$, then the curve $c(t) = \pi(tv)$ is uniformly distributed in $\mathbb{T}^2$ (with respect to the usual Lebesgue measure on the torus).

[*Hint:* Any continuous function on $\mathbb{T}^2$ can be approximated by a trigonometric polynomial $\sum a_{m,n}e^{2\pi imx+2\pi iny}$.]

#3. Show that if $V = \{u^t\}$ is a one-parameter unipotent subgroup of G, then the conclusions of Theorem 20.3.4 follow from Theorem 20.3.3.

[*Hint:* Pointwise Ergodic Theorem (see Exercise 14.3#8).]

#4. In the setting of the proof of Corollary 20.3.5, show that the projection $L \to G_1$ is a (surjective) covering map.

[*Hint:* Show $L \cap (\{e\} \cap G_2)$ is discrete, by using Fubini's Theorem and the fact that graph(f) has nonzero measure.]

#5. Suppose
 • $T_k \to \infty$,
 • G/Γ is compact, and
 • $c: [0, \infty) \to G/\Gamma$ is a continuous curve.
Show there is a subsequence $T_{k_i} \to \infty$, such that

$$\lambda(f) = \lim_{i \to \infty} \frac{1}{T_{k_i}} \int_0^{T_{k_i}} f(c(t))\, dt$$

exists for all $f \in C(G/\Gamma)$.

[*Hint:* It suffices to consider a countable subset of $C(G/\Gamma)$ that is dense in the topology of uniform convergence.]

#6. In the notation used in the proof of Proposition 20.3.6, show, for every $f \in C(G/\Gamma)$ and every $s \in \mathbb{R}$, that

$$\lambda(f) = \lim_{k \to \infty} \frac{1}{T_k} \int_0^{T_k} f(g^s c(t)) \, dt.$$

§20.4. Shearing — a key ingredient in the proof

The known proofs of any of the three variants of Ratner's Theorem are quite lengthy, so we will just illustrate one of the main ideas that is involved. To keep things simple, we will assume $G = SL(2, \mathbb{R})$.

(20.4.1) **Notation.** Throughout this section,

- $G = SL(2, \mathbb{R})$,
- $u^t = \begin{bmatrix} 1 & 0 \\ t & 1 \end{bmatrix}$ is a one-parameter unipotent subgroup of G,
- $a^t = \begin{bmatrix} e^t & 0 \\ 0 & e^{-t} \end{bmatrix}$ is a one-parameter diagonal subgroup of G, and
- $X = SL(2, \mathbb{R})/\Gamma$.

The proofs of Ratner's Theorems depend on an understanding of what happens to two nearby points of X as they are moved by the one-parameter subgroup u^t.

(20.4.2) **Definition.** If x and y are any two points of G/Γ, then there exists $q \in G$, such that $y = qx$. If x is close to y (which we denote $x \approx y$), then q may be chosen close to the identity. Therefore, we may define a metric d on G/Γ by

$$d(x, y) = \min \left\{ \|q - \mathrm{Id}\| \;\middle|\; \begin{array}{l} q \in G, \\ qx = y \end{array} \right\},$$

where

- Id is the identity matrix, and
- $\| \cdot \|$ is any (fixed) matrix norm on $\mathrm{Mat}_{2 \times 2}(\mathbb{R})$. For example, one may take

$$\left\| \begin{bmatrix} a & b \\ c & d \end{bmatrix} \right\| = \max\{|a|, |b|, |c|, |d|\}.$$

(Actually, this definition does not guarantee $d(x, y) = d(y, x)$, so it may not define a metric, but let us ignore this minor issue.)

Now, we consider two points x and qx, with $q \approx \mathrm{Id}$, and we wish to calculate $d(u^t x, u^t qx)$ (see Figure 20.4A).

- To get from x to qx, one multiplies by q; therefore

$$d(x, qx) = \|q - \mathrm{Id}\|.$$

FIGURE 20.4A. The u^t-orbits of two nearby orbits.

- To get from $u^t x$ to $u^t qx$, one multiplies by $u^t q u^{-t}$; therefore
$$d(u^t x, u^t qx) = \| u^t q u^{-t} - \mathrm{Id} \|.$$
(Actually, this equation only holds when the right-hand side is small — there are infinitely many elements g of G with $g u^t x = u^t qx$, and the distance is obtained by choosing the smallest one, which may not be $u^t q u^{-t}$ if t is large.)

Letting
$$q - \mathrm{Id} = \begin{bmatrix} a & b \\ c & d \end{bmatrix},$$
a simple matrix calculation shows that
$$u^t q u^{-t} - \mathrm{Id} = \begin{bmatrix} a - bt & b \\ c + (a - d)t - bt^2 & d + bt \end{bmatrix}. \tag{20.4.3}$$

(20.4.4) **Notation.** For convenience, let $x_t = u^t x$ and $y_t = u^t y$.

Consider the right-hand side of Equation (20.4.3), with a, b, c, and d very small. Indeed, let us say they are infinitesimal (too small to see). As t grows, it is the quadratic term in the bottom left corner that will be the first matrix entry to attain macroscopic size. Comparing with the definition of u^t (see Notation 20.4.1), we see that this is exactly the direction of the u^t-orbit. Therefore:

(20.4.5) **Proposition** (Shearing Property). *The fastest relative motion between two nearby points is parallel to the orbits of the flow.*

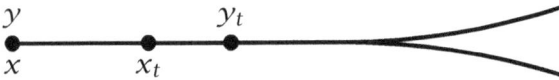

FIGURE 20.4B. Shearing: If two points start out so close together that we cannot tell them apart, then the first difference we see will be that one gets ahead of the other, but (apparently) following the same path. It is only much later that we will be able to detect any difference between their paths.

(20.4.6) *Remarks.*

1) The only exception to Proposition 20.4.5 is that if q is in the centralizer $C_G(u^t)$, then $u^t q u^{-t} = q$ for all t; in this case, the points x_t and y_t simply move along together at exactly the same speed, with no relative motion.

2) In contrast to the above discussion of u^t,
   • the matrix a^t is diagonal, but
   • the largest entry in

$$a^t q a^{-t} = \begin{bmatrix} a & e^{2t}b \\ e^{-2t}c & d \end{bmatrix}$$

is an off-diagonal entry,

so, under the action of the diagonal group, points move apart (at exponential speed) in a direction transverse to the orbits (see Figure 20.4C).

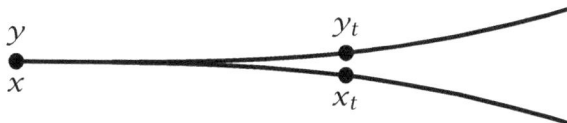

FIGURE 20.4C. Divergence under a diagonal subgroup: when two points start out so close together that we cannot tell them apart, the first difference we see will be in a direction transverse to the orbits.

The Shearing Property (20.4.5) shows that the direction of fastest relative motion is along u^t. However, in the proof of Ratner's Theorems, it turns out that we wish to ignore motion *along* the orbits, and consider, instead, only the component of the relative motion that is **transverse** (or perpendicular) to the orbits of u^t. This direction, by definition, does not belong to $\{u^t\}$.

(20.4.7) **Definition.** Suppose, as before, that x and y are two points in X with $x \approx y$. Then, by continuity, $x_t \approx y_t$ for a long time. Eventually, we will be able to see a difference between x_t and y_t. The Shearing Property (20.4.5) tells us that, when this first happens, y_t will be indistinguishable from some point on the orbit of x; that is, $y_t \approx x_{t'}$ for some t'. This will continue for another long time (with t' some function of t), but we can expect that y_t will eventually diverge from the orbit of x — this is **transverse divergence**. (Note that this transverse divergence is a second-order effect; it is only apparent after we mod out the relative motion along the orbit.) Letting $x_{t'}$ be the point on the orbit of x that is closest to y_t, we write $y_t = g x_{t'}$ for some $g \in G$. Then $g - \mathrm{Id}$ represents the transverse divergence. When this transverse divergence first becomes macroscopic, we wish to understand which of the matrix entries of $g -$ Id are macroscopic.

In the matrix on the right-hand side of Equation (20.4.3), we have already observed that the largest entry is in the bottom left corner, the direction of $\{u^t\}$. If we ignore that entry, then the two diagonal entries are the largest of the remaining entries. The diagonal corresponds to the subgroup $\{a^t\}$. Therefore, the fastest **transverse** divergence is in the direction of $\{a^t\}$. Notice that $\{a^t\}$ normalizes $\{u^t\}$.

(20.4.8) **Proposition.** *The fastest **transverse** motion is along some direction in the normalizer of u^t.*

More precisely, if $x, y \in X$, with $x \approx y$, and $r > 0$ is much smaller than the injectivity radius of X, then either:

1) *there exist large $t, t' \in \mathbb{R}$ and $g \in \mathcal{N}_G(\{u^t\})$ such that*
$$u^t y \approx g u^{t'} x \text{ and } \|g\| = d(g, \{u^t\}) = r,$$

   or

2) *for all $t \in \mathbb{R}$, there exists $t' \in \mathbb{R}$, such that $u^t y \approx u^{t'} x$ (i.e., there is no transverse motion, only shearing).*

To illustrate how understanding the transverse motion can be useful, let us prove a very special case of Ratner's Orbit-Closure Theorem (20.1.3).

(20.4.9) **Proposition.** *Let $C = \overline{\{u^t\}x}$, for some $x \in X$, and assume*

- *C is a **minimal** u^t-invariant closed subset of X (this means that no nonempty, proper, closed subset of C is $\{u^t\}$-invariant), and*

- *$\{g \in G \mid gC = C\} = \{u^t\}$.*

Then $C = \{u^t\}x$, so C is a submanifold of X.

Proof. We wish to show $C \subseteq \{u^t\}x$, but Exercise 1 implies that it suffices to prove only the weaker statement that $C \subseteq \mathcal{N}_G(\{u^t\}) x$.

Suppose $C \nsubseteq \mathcal{N}_G(\{u^t\}) x$. Then, since C is connected, there exists $y \in C$, with $y \approx x$, but $y \notin \mathcal{N}_G(\{u^t\}) x$. From Proposition 20.4.8, we see that there exist $t, t' \in \mathbb{R}$ and
$$g \in \mathcal{N}_G(\{u^t\}), \text{ with } g \notin \{u^t\}, \text{ such that } u^t y \approx g u^{t'} x.$$
For simplicity, let us pretend that
$$u^t y \text{ is equal to } g u^{t'} x,$$
rather than merely being approximately equal (see Exercise 2). Then we have $C \cap gC \neq \varnothing$ (because $u^t y \in C$ and $g u^{t'} x \in gC$). This contradicts Exercise 1. $\qquad \square$

Exercises for §20.4.

#1. Under the assumptions of Proposition 20.4.9, show:

if $g \in \mathcal{N}_G(\{u^t\})$, but $g \notin \{u^t\}$, then $C \cap gC = \varnothing$.

(In particular, $gx \notin C$.)

[*Hint:* gC is u^t-invariant (because $g \in \mathcal{N}_G(\{u^t\})$), so $C \cap gC$ is a u^t-invariant subset of the minimal set C.]

#2. Complete the proof of Proposition 20.4.9 by eliminating the pretense that $u^t y$ is *equal* to $gu^{t'} x$.

[*Hint:* The compact sets C and $\{g \in \mathcal{N}_G(\{u^t\}) \mid \|g\| = d(g, \{u^t\}) = r\} \cdot C$ are disjoint, so it is impossible for a point in one set to be arbitrarily close to a point in the other set.]

Notes

Theorem 20.1.3 is due to M. Ratner [13], under the assumption that V is either unipotent or connected. (A shorter proof can be found in [7].) This additional hypothesis was removed by N. Shah [15] (except for a technical problem involving Conclusions (2) and (3) that was resolved in [3, Cor. 3.5.4]).

See [9] for a more thorough introduction to Ratner's Theorems, their proofs, and some applications.

See [16, Lem. 2] for the construction of orbits whose closure is not a submanifold, demonstrating the pathology in Warning 20.1.5 and Example 20.2.1. However, it was conjectured by G. A. Margulis [6, §1.1] that certain analogues of Ratner's Theorems are valid in some situations where the subgroup V is a split torus of dimension > 1; see [3, §4.4c] and [8] for references on this open problem and its applications.

Corollary 20.2.2 was proved by N. Shah [14]. The generalization in Remark 20.2.3(1) is due to T. Payne [10].

Corollary 20.2.5 is due to G. A. Margulis [4]. See [5] for a survey of its history and later related developments.

Corollary 20.2.6 was discovered by N. Shah [15, Cor. 1.5]. This consequence of Ratner's Theorem played an important role in [17].

Theorem 20.3.3 is due to M. Ratner [13].

Theorem 20.3.4 was proved by M. Ratner [12] in the case where V is either unipotent or connected. (See [2] for a shorter and more self-contained proof in the case where $V \cong \mathrm{SL}(2, \mathbb{R})$.) The general case is due to N. Shah [15].

Corollary 20.3.5 was proved by M. Ratner [11] if $G_1 \cong G_2 \cong \mathrm{SL}(2, \mathbb{R})$. The general case is due to D. Witte [19].

Proposition 20.3.6 is a special case of a classical result in Ergodic Theory that can be found in textbooks such as [18, Thm. 6.19].

See [1] for the work of Y. Benoist and J.-F. Quint mentioned in Remark 20.3.9(2). Shah's suggestion about Zariski closures appears in [15, end of §1, p. 232].

The discussion of shearing in Section 20.4 is excerpted from [9, §1.5], except that Proposition 20.4.9 is a variant of [9, Prop. 1.6.10].

References

[1] Y. Benoist and J.-F. Quint: Stationary measures and invariant
    subsets of homogeneous spaces (III), *Ann. of Math.* (2) 178 (2013),
    no. 3, 1017–1059. MR 3092475,
    http://dx.doi.org/10.4007/annals.2013.178.3.5

[2] M. Einsiedler: Ratner's theorem on $SL(2, \mathbb{R})$-invariant measures,
    Jahresber. Deutsch. Math.-Verein. 108 (2006), no. 3, 143–164.
    MR 2265534

[3] D. Kleinbock, N. Shah, and A. Starkov: Dynamics of subgroup
    actions on homogeneous spaces of Lie groups and applications to
    number theory, in B. Hasselblatt and A. Katok, eds.: *Handbook of
    Dynamical Systems, Vol. 1A.* North-Holland, Amsterdam, 2002,
    pages 813–930. MR 1928528

[4] G. A. Margulis: Formes quadratriques indéfinies et flots unipotents
    sur les espaces homogènes, *C. R. Acad. Sci. Paris Sér. I Math.* 304
    (1987), no. 10, 249–253. MR 0882782

[5] G. A. Margulis: Oppenheim Conjecture, in M. Atiyah and
    D. Iagolnitzer, eds.: *Fields Medallists' Lectures.* World Sci. Publ.,
    River Edge, NJ, 1997, pp. 272–327. MR 1622909

[6] G. A. Margulis: Problems and conjectures in rigidity theory, in
    V. Arnold et al., eds.: *Mathematics: Frontiers and Perspectives.*
    Amer. Math. Soc., Providence, RI, 2000, pp. 161–174. MR 1754775

[7] G. A. Margulis and G. M. Tomanov: Invariant measures for actions of
    unipotent groups over local fields on homogeneous spaces, *Invent.
    Math.* 116 (1994) 347–392. MR 1253197,
    http://eudml.org/doc/144192

[8] F. Maucourant: A non-homogeneous orbit closure of a diagonal
    subgroup, *Ann. of Math.* (2) 171 (2010), no. 1, 557–570.
    MR 2630049, http://dx.doi.org/10.4007/annals.2010.171.557

[9] D. W. Morris: *Ratner's Theorems on Unipotent Flows.* University of
    Chicago Press, Chicago, IL, 2005. ISBN 0-226-53984-9, MR 2158954,
    http://arxiv.org/abs/math/0310402

[10] T. L. Payne: Closures of totally geodesic immersions into locally
    symmetric spaces of noncompact type, *Proc. Amer. Math. Soc.* 127

(1999), no. 3, 829–833. MR 1468202,
http://dx.doi.org/10.1090/S0002-9939-99-04552-9

[11] M. Ratner: Rigidity of horocycle flows, *Ann. of Math.* 115 (1982), no. 3, 597–614. MR 0657240, http://www.jstor.org/stable/2007014

[12] M. Ratner: On Raghunathan's measure conjecture, *Ann. of Math.* 134 (1991), no. 3, 545–607. MR 1135878, http://www.jstor.org/stable/2944357

[13] M. Ratner: Raghunathan's topological conjecture and distributions of unipotent flows, *Duke Math. J.* 63 (1991), no. 1, 235–280. MR 1106945, http://dx.doi.org/10.1215/S0012-7094-91-06311-8

[14] N. Shah: Closures of totally geodesic immersions in manifolds of constant negative curvature, in É. Ghys, A. Haefliger and A. Verjovsky, eds.: *Group Theory from a Geometrical Viewpoint (Trieste, 1990),* World Sci. Publ., River Edge, NJ, 1991, pp. 718–732. MR 1170382

[15] N. Shah: Invariant measures and orbit closures on homogeneous spaces for actions of subgroups generated by unipotent elements, in S. G. Dani, ed.: *Lie Groups and Ergodic Theory (Mumbai, 1996),* Tata Inst. Fund. Res., Bombay, 1998, pp. 229–271. MR 1699367

[16] A. N. Starkov: Structure of orbits of homogeneous flows and the Ragunatana conjecture, *Russian Math. Surveys* 45 (1990), no. 2, 227–228. (Translated from *Uspekhi Mat. Nauk* 45 (1990), no. 2(272), 219–220.) MR 1069361, http://dx.doi.org/10.1070/RM1990v045n02ABEH002338

[17] V. Vatsal: Uniform distribution of Heegner points, *Invent. Math.* 148 (2002) 1–46. MR 1892842, http://dx.doi.org/10.1007/s002220100183

[18] P. Walters: *An Introduction to Ergodic Theory.* Springer, New York, 1982 ISBN 0-387-90599-5, MR 0648108

[19] D. Witte: Rigidity of some translations on homogeneous spaces, *Invent. Math.* 81 (1985), no. 1, 1–27. MR 0796188, http://eudml.org/doc/143244

Appendices

Appendix A
Basic Facts about Semisimple Lie Groups

§A1. Definitions

We are interested in groups of matrices that are (topologically) closed:

(A1.1) **Definitions.**

1) Let $\mathrm{Mat}_{\ell \times \ell}(\mathbb{R})$ be the set of all $\ell \times \ell$ matrices with real entries. This has a natural topology, obtained by identifying it with the Euclidean space $\mathbb{R}^{\ell^2}$.

2) Let $\mathrm{SL}(\ell, \mathbb{R}) = \{\, g \in \mathrm{Mat}_{\ell \times \ell}(\mathbb{R}) \mid \det g = 1 \,\}$. This is a group under matrix multiplication (see Exercise 1), and it is a closed subset of $\mathrm{Mat}_{\ell \times \ell}(\mathbb{R})$ (see Exercise 2).

3) A *Lie group* is any (topologically) closed subgroup of some $\mathrm{SL}(\ell, \mathbb{R})$.

Recall that an abstract group is *simple* if it has no nontrivial, proper, normal subgroups. For Lie groups, we relax this to allow normal subgroups that are discrete (except that the one-dimensional abelian groups $\mathbb{R}$ and $\mathbb{T}$ are not considered to be simple).

(A1.2) **Definition.** A Lie group G is *simple* if it has no nontrivial, connected, closed, proper, normal subgroups, and G is not abelian.

(A1.3) **Example.** It can be shown that $G = \mathrm{SL}(\ell, \mathbb{R})$ is a simple Lie group (when $\ell > 1$). If ℓ is even, then $\{\pm \mathrm{Id}\}$ is a subgroup of G, and it is normal, but, because this subgroup is not connected, it does not disqualify G from being simple as a Lie group.

Recall: The Standing Assumptions (4.0.0 on page 41) are in effect, so, as always, Γ is a lattice in the semisimple Lie group $G \subseteq \mathrm{SL}(\ell, \mathbb{R})$.

(A1.4) *Remark.* Although Definition A1.2 only refers to *closed* normal subgroups, it turns out that, except for the center, there are no normal subgroups at all: if *G* is simple, then every proper, normal subgroup of *G* is contained in the center of *G*.

(A1.5) **Other terminology.** Some authors say that $SL(n, \mathbb{R})$ is **almost simple**, and reserve the term "simple" for groups that have no (closed) normal subgroups at all, not even finite ones.

A Lie group is said to be *semisimple* if it is a direct product of simple groups, modulo passing to a finite-index subgroup and/or modding out a finite group:

(A1.6) **Definitions.**

1) G_1 is **isogenous** to G_2 if there is a finite, normal subgroup N_i of a finite-index subgroup G'_i of G_i, for $i = 1, 2$, such that G'_1/N_1 is isomorphic to G'_2/N_2.

2) *G* is **semisimple** if it is isogenous to a direct product of simple Lie groups. That is, *G* is isogenous to $G_1 \times \cdots \times G_r$, where each G_i is simple.

(A1.7) **Example.** $SL(2, \mathbb{R}) \times SL(3, \mathbb{R})$ is a semisimple Lie group that is not simple (because $SL(2, \mathbb{R})$ and $SL(3, \mathbb{R})$ are normal subgroups).

(A1.8) *Remark* (see Exercise A4#9). If *G* is semisimple, then the center of *G* is finite.

(A1.9) **Assumption** (cf. the Standing Assumptions (4.0.0)). Now that we have the definition of a semisimple group, we will henceforth assume in this chapter that the symbol *G* always denotes a semisimple Lie group with only finitely many connected components (but the symbol Γ will never appear).

(A1.10) **Warning.** A *Lie group* is usually defined to be any group that is also a smooth manifold, such that the group operations are C^∞ functions. Proposition A6.2(1) below shows that every closed subgroup of $SL(\ell, \mathbb{R})$ is a Lie group in the usual sense. However, the converse is false: not every Lie group (in the usual sense) can be realized as a subgroup of some $SL(\ell, \mathbb{R})$. (In other words, not every Lie group is **linear**.) Therefore, our Definition A1.1(3) is more restrictive than the usual definition. (However, every connected Lie group is "locally isomorphic" to a linear Lie group.)

Exercises for §A1.

#1. Show that $SL(\ell, \mathbb{R})$ is a group under matrix multiplication.

[*Hint:* You may assume (without proof) basic facts of linear algebra, such as the fact that a square matrix is invertible if and only if its determinant is not 0.]

#2. Show that $SL(\ell, \mathbb{R})$ is a closed subset of $Mat_{\ell \times \ell}(\mathbb{R})$.

[*Hint:* For a continuous function, the inverse image of a closed set is closed.]

#3. Recall that $GL(\ell, \mathbb{R}) = \{ g \in Mat_{\ell \times \ell}(\mathbb{R}) \mid \det g \neq 0 \}$, and that this is a group under matrix multiplication. Show that it is (isomorphic to) a Lie group, by showing it is isomorphic to a closed subgroup of $SL(\ell + 1, \mathbb{R})$.

#4. Suppose $G = G_1 \times \cdots \times G_r$, where each G_i is simple, and N is a connected, closed, normal subgroup of G. Show there is a subset S of $\{1, \ldots, r\}$, such that $N = \prod_{i \in S} G_i$.

[*Hint:* If the projection of N to G_i is all of G_i, then $G_i = [G_i, G_i] = [N, G_i] \subseteq N$.]

#5. Show that if G is semisimple, and N is any closed, normal subgroup of G, then G/N is semisimple.

[*Hint:* Exercise 4.]

#6. Suppose N is a connected, closed, normal subgroup of G. Show that there is a connected, closed, normal subgroup H of G, such that G is isogenous to $N \times H$.

[*Hint:* Exercise 4.]

§A2. The simple Lie groups

It is clear from Definition A1.6(2) that the study of semisimple groups requires a good understanding of the simple groups. Probably the most elementary examples of simple Lie groups are special linear groups and orthogonal groups, but symplectic groups and unitary groups are also fundamental. A group of any of these types is called "classical." (The other simple groups are "exceptional," and are less easy to construct.)

(A2.1) **Definition.** G is a *classical group* if it is isogenous to the direct product of any collection of the groups constructed in Examples A2.3 and A2.4 below. That is, each simple factor of G is either a special linear group or the isometry group of a bilinear, Hermitian, or skew-Hermitian form, over $\mathbb{R}$, $\mathbb{C}$, or $\mathbb{H}$ (where $\mathbb{H}$ is the algebra of quaternions).

(A2.2) **Notation.** Let

- g^T denote the transpose of the matrix g,
- g^* denote the adjoint (that is, the conjugate-transpose) of g,
- G° denote the identity component of the Lie group G, and
- $I_{m,n} = \mathrm{diag}(1, 1, \ldots, 1, -1, -1, \ldots, -1) \in Mat_{(m+n) \times (m+n)}(\mathbb{R})$, where the number of 1's is m, and the number of -1's is n.

(A2.3) **Example.**

1) The *special linear group* $SL(n, \mathbb{R})$ is a simple Lie group (if $n \geq 2$).

2) **Special orthogonal group.** Let
$$SO(m,n) = \{\, g \in SL(m+n,\mathbb{R}) \mid g^T I_{m,n}\, g = I_{m,n} \,\}.$$
This is always semisimple (if $m+n \geq 3$). It may not be connected, but the identity component $SO(m,n)^\circ$ is simple if either $m+n = 3$ or $m+n \geq 5$. (Furthermore, the index of $SO(m,n)^\circ$ in $SO(m,n)$ is ≤ 2.)

We use $SO(n)$ to denote $SO(n,0)$ (or $SO(0,n)$, which is the same group).

3) **Special unitary group.** Let
$$SU(m,n) = \{\, g \in SL(m+n,\mathbb{C}) \mid g^* I_{m,n}\, g = I_{m,n} \,\}.$$
Then $SU(m,n)$ is simple if $m+n \geq 2$.

We use $SU(n)$ to denote $SU(n,0)$ (or $SU(0,n)$).

4) **Symplectic group.** Let
$$J_{2m} = \begin{pmatrix} 0 & \mathrm{Id}_{m \times m} \\ -\mathrm{Id}_{m \times m} & 0 \end{pmatrix} \in GL(2m,\mathbb{R})$$
(where $\mathrm{Id}_{m \times m}$ denotes the $m \times m$ identity matrix), and let
$$Sp(2m,\mathbb{R}) = \{\, g \in SL(2m,\mathbb{R}) \mid g^T J_{2m}\, g = J_{2m} \,\}.$$
Then $Sp(2m,\mathbb{R})$ is simple if $m \geq 1$.

(A2.4) Example. Additional simple groups can be constructed by replacing the field $\mathbb{R}$ with either the field $\mathbb{C}$ of complex numbers or the division ring $\mathbb{H}$ of quaternions:

1) **Complex and quaternionic special linear groups:** $SL(n,\mathbb{C})$ and $SL(n,\mathbb{H})$ are simple Lie groups (if $n \geq 2$).

   Note: The noncommutativity of $\mathbb{H}$ causes some difficulty in defining the determinant of a quaternionic matrix. To avoid this problem, we define the **reduced norm** of a quaternionic $n \times n$ matrix g to be the determinant of the $2n \times 2n$ complex matrix obtained by identifying $\mathbb{H}^n$ with $\mathbb{C}^{2n}$. Then, by definition, g belongs to $SL(n,\mathbb{H})$ if and only if its reduced norm is 1. It is not difficult to see that the reduced norm of a quaternionic matrix is always a (nonnegative) real number (see Exercise 1).

2) **Complex and quaternionic special orthogonal groups:**
$$SO(n,\mathbb{C}) = \{\, g \in SL(n,\mathbb{C}) \mid g^T \,\mathrm{Id}\, g = \mathrm{Id} \,\}$$
and
$$SO(n,\mathbb{H}) = \{\, g \in SL(n,\mathbb{H}) \mid \tau_r(g^T)\,\mathrm{Id}\, g = \mathrm{Id} \,\},$$
where τ_r is the **reversion** on $\mathbb{H}$ defined by
$$\tau_r(a_0 + a_1 i + a_2 j + a_3 k) = a_0 + a_1 i - a_2 j + a_3 k.$$
(Note that $\tau_r(ab) = \tau_r(b)\,\tau_r(a)$ (see Exercise 2); τ_r is included in the definition of $SO(n,\mathbb{H})$ in order to compensate for the noncommutativity of $\mathbb{H}$ (see Exercise 3).)

3) *Complex symplectic group*: Let
$$\text{Sp}(2m, \mathbb{C}) = \{\, g \in \text{SL}(2m, \mathbb{C}) \mid g^T J_{2m}\, g = J_{2m} \,\}.$$

4) *Symplectic unitary groups*: Let
$$\text{Sp}(m, n) = \{\, g \in \text{SL}(m + n, \mathbb{H}) \mid g^* I_{m,n}\, g = I_{m,n} \,\}.$$

Here, as usual, g^* denotes the conjugate-transpose of g; recall that the **conjugate** of a quaternion is defined by

$$\overline{a + bi + cj + dk} = a - bi - cj - dk$$

(and that $\overline{xy} = \overline{y}\,\overline{x}$). We use $\text{Sp}(n)$ to denote $\text{Sp}(n, 0)$ (or $\text{Sp}(0, n)$).

(A2.5) **Other terminology.** Some authors use

- $\text{SU}^*(2n)$ to denote $\text{SL}(n, \mathbb{H})$,
- $\text{SO}^*(2n)$ to denote $\text{SO}(n, \mathbb{H})$, or
- $\text{Sp}(n, \mathbb{R})$ to denote $\text{Sp}(2n, \mathbb{R})$.

(A2.6) *Remark.* $\text{SL}(2, \mathbb{R})$ is the smallest connected, noncompact, simple Lie group; it is contained (up to isogeny) in any other. For example:

1) If $\text{SL}(n, \mathbb{R})$, $\text{SL}(n, \mathbb{C})$, or $\text{SL}(n, \mathbb{H})$ is not compact, then $n \geq 2$, so the group contains $\text{SL}(2, \mathbb{R})$.

2) If $\text{SO}(m, n)$ is semisimple and not compact, then $\min\{m, n\} \geq 1$ and $\max\{m, n\} \geq 2$, so it contains $\text{SO}(1, 2)$, which is isogenous to $\text{SL}(2, \mathbb{R})$.

3) If $\text{SU}(m, n)$ or $\text{Sp}(m, n)$ is not compact, then $\min\{m, n\} \geq 1$, so the group contains $\text{SU}(1, 1)$, which is isogenous to $\text{SL}(2, \mathbb{R})$.

4) $\text{Sp}(2m, \mathbb{R})$ and $\text{Sp}(2m, \mathbb{C})$ both contain $\text{Sp}(2, \mathbb{R})$, which is equal to $\text{SL}(2, \mathbb{R})$.

5) If $\text{SO}(n, \mathbb{C})$ is semisimple and not compact, then $n \geq 3$, so the group contains $\text{SO}(1, 2)$, which is isogenous to $\text{SL}(2, \mathbb{R})$.

6) If $\text{SO}(n, \mathbb{H})$ is not compact, then $n \geq 2$, so it contains a subgroup conjugate to $\text{SU}(1, 1)$, which is isogenous to $\text{SL}(2, \mathbb{R})$.

The classical groups are just examples, so one would expect there to be many other (more exotic) simple Lie groups. Amazingly, that is not the case — there are only finitely many others:

(A2.7) **Theorem** (É. Cartan). *Every simple Lie group is isogenous to either*

1) *a classical group, or*

2) *one of the finitely many exceptional groups.*

See Sections 18.1 and 18.3 for an indication of the proof of Theorem A2.7.

Exercises for §A2.

#1. For all nonzero $g \in \mathrm{Mat}_{n \times n}(\mathbb{H})$, show that the reduced norm of g is a nonnegative real number.

[*Hint:* Use row and column operations in $\mathrm{Mat}_{n \times n}(\mathbb{H})$ to reduce to the case where g is upper triangular. For $n = 1$, the reduced norm of g is $g\,\overline{g}$.]

#2. In the notation of Example A2.4, show that $\tau_r(ab) = \tau_r(b)\,\tau_r(a)$ for all $a, b \in \mathbb{H}$.

[*Hint:* Calculate explicitly, or note that $\tau_r(x) = j\,\overline{x}\,j^{-1}$ (and $\overline{xy} = \overline{y}\,\overline{x}$).]

#3. For $g, h \in \mathrm{Mat}_{n \times n}(\mathbb{H})$, show that $\tau_r((gh)^T) = \tau_r(h^T)\,\tau_r(g^T)$.

#4. Show that $\mathrm{SO}(n, \mathbb{H})$ is a subgroup of $\mathrm{SL}(n, \mathbb{H})$.

§A3. Haar measure

Standard texts on real analysis construct a translation-invariant measure on $\mathbb{R}^n$. this is called **Lebesgue measure**, but the analogue for other Lie groups is called "Haar measure:"

(A3.1) **Proposition** (Existence and Uniqueness of Haar Measure). *If H is any Lie group, then there is a unique (up to a scalar multiple) σ-finite Borel measure μ on H, such that*

1) *$\mu(C)$ is finite, for every compact subset C of H, and*
2) *$\mu(hA) = \mu(A)$, for every Borel subset A of H, and every $h \in H$.*

(A3.2) **Definitions.**

1) The measure μ of Proposition A3.1 is called the **left Haar measure** on H. Analogously, there is a unique **right Haar measure** with $\mu(Ah) = \mu(A)$ (see Exercise 2).
2) H is **unimodular** if the left Haar measure is also a right Haar measure. (This means $\mu(hA) = \mu(Ah) = \mu(A)$.)

(A3.3) *Remark.* Haar measure is always **inner regular**: $\mu(A)$ is the supremum of the measures of the compact subsets of A.

(A3.4) **Proposition.** *There is a continuous homomorphism $\Delta \colon H \to \mathbb{R}^+$, such that, if μ is any (left or right) Haar measure on H, then*

$$\mu(hAh^{-1}) = \Delta(h)\,\mu(A), \text{ for all } h \in H \text{ and any Borel set } A \subseteq H.$$

Proof. Let μ be a left Haar measure. For each $h \in H$, define $\phi_h \colon H \to H$ by $\phi_h(x) = hxh^{-1}$. Then ϕ_h is an automorphism of H, so $(\phi_h)_* \mu$ is a left Haar measure. By uniqueness, we conclude that there exists $\Delta(h) \in \mathbb{R}^+$, such that $(\phi_h)_* \mu = \Delta(h)\,\mu$. It is easy to see that Δ is a continuous homomorphism. By using the construction of right Haar measure in Exercise 2, it is easy to verify that the same formula also applies to it. $\square$

(A3.5) **Definition.** The function Δ defined in Proposition A3.4 is called the *modular function* of H.

(A3.6) **Other terminology.** Some authors call $1/\Delta$ the modular function, because they use the conjugation $h^{-1}Ah$, instead of hAh^{-1}.

(A3.7) **Corollary.** *Let Δ be the modular function of H, and let A be a Borel subset of H.*

1) *If μ is a right Haar measure on H, then $\mu(hA) = \Delta(h)\,\mu(A)$, for all $h \in H$.*
2) *If μ is a left Haar measure on G, then $\mu(Ah) = \Delta(h^{-1})\,\mu(A)$, for all $h \in H$.*
3) *H is unimodular if and only if $\Delta(h) = 1$, for all $h \in H$.*
4) *$\Delta(h) = |\det(\mathrm{Ad}_H h)|$ for all $h \in H$ (see Notation A6.17).*

(A3.8) *Remark.* G is unimodular, because semisimple groups have no nontrivial (continuous) homomorphisms to $\mathbb{R}^+$ (see Exercise 3).

(A3.9) **Proposition.** *Let μ be a left Haar measure on a Lie group H. Then $\mu(H) < \infty$ if and only if H is compact.*

Proof. ($\Leftarrow$) See Proposition A3.1(1).

($\Rightarrow$) Since $\mu(H) < \infty$ (and the measure μ is inner regular), there is a compact subset C of H, such that $\mu(C) > \mu(H)/2$. Then, for any $h \in H$, we have

$$\mu(hC) + \mu(C) = \mu(C) + \mu(C) = 2\mu(C) > \mu(H),$$

so hC cannot be disjoint from C. This implies that h belongs to the set $C \cdot C^{-1}$, which is compact. Since h is an arbitrary element of H, we conclude that $H = C \cdot C^{-1}$ is compact. $\qquad\square$

Exercises for §A3.

#1. Prove the existence (but not uniqueness) of Haar measure on H, without using Proposition A3.1, under the additional assumption that the Lie group H is a C^∞ submanifold of $\mathrm{SL}(\ell, \mathbb{R})$ (cf. Proposition A6.2(1)).

[*Hint:* For $k = \dim H$, there is a differential k-form on H that is invariant under left translations.]

#2. Suppose μ is a left Haar measure on H, and define $\tilde{\mu}(A) = \mu(A^{-1})$. Show $\tilde{\mu}$ is a right Haar measure.

#3. Assume G is connected. Show that if $\phi\colon G \to A$ is a continuous homomorphism, and A is abelian, then ϕ is trivial.

[*Hint:* The kernel of a continuous homomorphism is a closed, normal subgroup.]

§A4. G is almost Zariski closed

(A4.1) **Definitions.**

1) We use $\mathbb{R}[x_{1,1}, \ldots, x_{\ell,\ell}]$ to denote the set of real polynomials in the ℓ^2 variables $\{x_{i,j} \mid 1 \le i, j \le \ell\}$.

2) For any $Q \in \mathbb{R}[x_{1,1}, \ldots, x_{\ell,\ell}]$, and any $g \in \mathrm{Mat}_{\ell \times \ell}(\mathbb{C})$, we use $Q(g)$ to denote the value obtained by substituting the matrix entries $g_{i,j}$ into the variables $x_{i,j}$. For example, if $Q = x_{1,1}x_{2,2} - x_{1,2}x_{2,1}$, then $Q(g)$ is the determinant of the first principal 2×2 minor of g.

3) For any subset Q of $\mathbb{R}[x_{1,1}, \ldots, x_{\ell,\ell}]$, let
$$\mathrm{Var}(Q) = \{g \in \mathrm{SL}(\ell, \mathbb{R}) \mid Q(g) = 0, \ \forall Q \in Q\}.$$
This is the **variety** associated to Q.

4) A subset H of $\mathrm{SL}(\ell, \mathbb{R})$ is **Zariski closed** if there exists a subset Q of $\mathbb{R}[x_{1,1}, \ldots, x_{\ell,\ell}]$, such that $H = \mathrm{Var}(Q)$. (In the special case where H is a subgroup of $\mathrm{SL}(\ell, \mathbb{R})$, we may also say that H is a **real algebraic group** or an **algebraic group that is defined over** $\mathbb{R}$.)

5) The **Zariski closure** of a subset H of $\mathrm{SL}(\ell, \mathbb{R})$ is the (unique) smallest Zariski closed subset of $\mathrm{SL}(\ell, \mathbb{R})$ that contains H. This is sometimes denoted $\overline{\overline{H}}$. (It can also be denoted $\overline{H}$, if this will not lead to confusion with the closure of H in the ordinary topology.)

(A4.2) **Example.**

1) $\mathrm{SL}(\ell, \mathbb{R})$ is Zariski closed. Let $Q = \varnothing$.

2) The group of diagonal matrices in $\mathrm{SL}(\ell, \mathbb{R})$ is Zariski closed. Let $Q = \{x_{i,j} \mid i \ne j\}$.

3) For any $A \in \mathrm{GL}(\ell, \mathbb{R})$, the centralizer of A is Zariski closed. Let
$$Q = \left\{ \left. \sum_{k=1}^{\ell} (x_{i,k}A_{k,j} - A_{i,k}x_{k,j}) \ \right| \ 1 \le i, j \le \ell \right\}.$$

4) If we identify $\mathrm{SL}(n, \mathbb{C})$ with a subgroup of $\mathrm{SL}(2n, \mathbb{R})$, by identifying $\mathbb{C}$ with $\mathbb{R}^2$, then $\mathrm{SL}(n, \mathbb{C})$ is Zariski closed, because it is the centralizer of T_i, the linear transformation in $\mathrm{GL}(2n, \mathbb{R})$ that corresponds to scalar multiplication by i.

5) The classical groups of Examples A2.3 and A2.4 are Zariski closed (if we identify $\mathbb{C}$ with $\mathbb{R}^2$ and $\mathbb{H}$ with $\mathbb{R}^4$ where necessary).

(A4.3) **Other terminology.**

- Other authors use $\mathrm{GL}(\ell, \mathbb{R})$ in the definition of $\mathrm{Var}(Q)$, instead of $\mathrm{SL}(\ell, \mathbb{R})$. Our choice leads to no loss of generality, and simplifies the theory slightly. (In the GL theory, one should, for technical reasons, stipulate that the function $1/\det(g)$ is considered to be a polynomial. In our setting, $\det g$ is the constant function 1, so this is not an issue.)

- What we call Var(Q) is actually only the *real* points of the variety. Algebraic geometers usually consider the solutions in $\mathbb{C}$, rather than $\mathbb{R}$, but our preoccupation with real Lie groups leads to our emphasis on real points.

(A4.4) **Example.** Let

$$
H = \left\{ \begin{pmatrix} e^t & 0 & 0 & 0 \\ 0 & e^{-t} & 0 & 0 \\ 0 & 0 & 1 & t \\ 0 & 0 & 0 & 1 \end{pmatrix} \;\middle|\; t \in \mathbb{R} \right\} \subset \mathrm{SL}(4, \mathbb{R}).
$$

Then H is a 1-dimensional subgroup that is not Zariski closed. Its Zariski closure is

$$
\overline{\overline{H}} = \left\{ \begin{pmatrix} a & 0 & 0 & 0 \\ 0 & 1/a & 0 & 0 \\ 0 & 0 & 1 & t \\ 0 & 0 & 0 & 1 \end{pmatrix} \;\middle|\; \begin{matrix} a \in \mathbb{R} \smallsetminus \{0\}, \\ t \in \mathbb{R} \end{matrix} \right\} \subset \mathrm{SL}(4, \mathbb{R}).
$$

The point here is that the exponential function is transcendental, not polynomial, so no polynomial can capture the relation that ties the diagonal entries to the off-diagonal entry in H. Therefore, as far as polynomials are concerned, the diagonal entries in the upper left are independent of the off-diagonal entry, as we see in the Zariski closure.

(A4.5) *Remark.* If H is Zariski closed, then the set Q of Definition A4.1 can be chosen to be finite (because the ring $\mathbb{R}[x_{1,1}, \ldots, x_{\ell,\ell}]$ is Noetherian).

Everyone knows that a (nonzero) polynomial in one variable has only finitely many roots. The following important fact generalizes this observation to any collection of polynomials in any number of variables.

(A4.6) **Theorem.** *Every Zariski closed subset of* $\mathrm{SL}(\ell, \mathbb{R})$ *has only finitely many connected components.*

(A4.7) **Definition.** A closed subgroup H of $\mathrm{SL}(\ell, \mathbb{R})$ is *almost Zariski closed* if it has only finitely many components, and there is a Zariski closed subgroup H_1 of $\mathrm{SL}(\ell, \mathbb{R})$, such that $H^\circ = H_1^\circ$. In other words, in the terminology of Definition 4.2.1, H is **commensurable** to a Zariski closed subgroup.

(A4.8) **Examples.**

1) Let H be the group of diagonal matrices in $\mathrm{SL}(2, \mathbb{R})$. Then H is Zariski closed (see Example A4.2(2)), but H° is not: any polynomial that vanishes on the diagonal matrices with positive entries will also vanish on the diagonal matrices with negative entries. So H° is almost Zariski closed, but it is not Zariski closed.

2) Let $G = \mathrm{SO}(1,2)^\circ$. Then G is almost Zariski closed (because $\mathrm{SO}(1,2)$ is Zariski closed), but G is not Zariski closed (see Exercise 1).

These examples are typical: a connected Lie group is almost Zariski closed if and only if it is the identity component of a group that is Zariski closed.

The following fact gives the Zariski closure a central role in the study of semisimple Lie groups.

(A4.9) Theorem. *If $G \subseteq SL(\ell, \mathbb{R})$, then G is almost Zariski closed.*

Proof. Let $\overline{\overline{G}}$ be the Zariski closure of G. Then $\overline{\overline{G}}$ is semisimple. (For example, if G is irreducible in $SL(\ell, \mathbb{C})$, then $\overline{\overline{G}}$ is also irreducible, so Corollary A7.7 below implies that $\overline{\overline{G}}^{\circ}$ is semisimple.)

Since G has only finitely many connected components (see Assumption A1.9), we may assume, by passing to a subgroup of finite index, that it is connected. This implies that the normalizer $\mathcal{N}_{SL(\ell,\mathbb{R})}(G)$ is Zariski closed (see Exercise 2). Therefore $\overline{\overline{G}}$ is contained in the normalizer, which means that G is a normal subgroup of $\overline{\overline{G}}$.

Hence (up to isogeny), we have $\overline{\overline{G}} = G \times H$, for some closed, normal subgroup H of $\overline{\overline{G}}$ (see Exercise A1#6). So $G = C_{\overline{\overline{G}}}(H)^{\circ}$ is almost Zariski closed (see Example A4.2(3)).   □

(A4.10) Warning. Theorem A4.9 relies on our standing assumption that G is semisimple (see Example A4.4). (Actually, it suffices to know that, besides being connected, G is perfect; that is, $G = [G, G]$ is equal to its commutator subgroup.)

Exercises for §A4.

#1. Show that $SO(1,2)^{\circ}$ is not Zariski closed.

[*Hint:* We have

$$\frac{1}{2}\begin{pmatrix} s + \frac{1}{s} & s - \frac{1}{s} & 0 \\ s - \frac{1}{s} & s + \frac{1}{s} & 0 \\ 0 & 0 & 2 \end{pmatrix} \in SO(1,2)^{\circ} \qquad \Leftrightarrow \qquad s > 0.$$

If a rational function $f: \mathbb{R} \smallsetminus \{0\} \to \mathbb{R}$ vanishes on $\mathbb{R}^{+}$, then it also vanishes on $\mathbb{R}^{-}$.]

#2. Show that if H is a connected Lie subgroup of $SL(\ell, \mathbb{R})$, then the normalizer $\mathcal{N}_{SL(\ell,\mathbb{R})}(H)$ is Zariski closed.

[*Hint:* $g \in \mathcal{N}(H)$ if and only if $g\mathfrak{h}g^{-1} = \mathfrak{h}$, where $\mathfrak{h} \subseteq Mat_{\ell \times \ell}(\mathbb{R})$ is the Lie algebra of H.]

#3. Show that if $\overline{\overline{H}}$ is the Zariski closure of a subgroup H of G, then $g\overline{\overline{H}}g^{-1}$ is the Zariski closure of gHg^{-1}, for any $g \in G$.

#4. Suppose G is a connected subgroup of $SL(\ell, \mathbb{R})$ that is almost Zariski closed, and that $\mathcal{Q} \subset \mathbb{R}[x_{1,1}, \ldots, x_{\ell,\ell}]$.
   a) Show that $G \cap \text{Var}(\mathcal{Q})$ is a closed subset of G.

b) Show that if $G \not\subseteq \mathrm{Var}(\mathcal{Q})$, then $G \cap \mathrm{Var}(\mathcal{Q})$ does not contain any nonempty open subset of G.

c) Show that if $G \not\subseteq \mathrm{Var}(\mathcal{Q})$, then $G \cap \mathrm{Var}(\mathcal{Q})$ has measure zero, with respect to the Haar measure on G.

[*Hint:* For (b) and (c), you may assume, without proof, that, for some d, there exist

$$\varnothing = \mathrm{Var}(\mathcal{Q}_{-1}) \subseteq \mathrm{Var}(\mathcal{Q}_0) \subseteq \mathrm{Var}(\mathcal{Q}_1) \subseteq \cdots \subseteq \mathrm{Var}(\mathcal{Q}_d) = \mathrm{Var}(\mathcal{Q}),$$

such that $G \cap (\mathrm{Var}(\mathcal{Q}_k) \smallsetminus \mathrm{Var}(\mathcal{Q}_{k-1}))$ is a (possibly empty) k-dimensional C^∞ submanifold of G, for $0 \le k \le d$. $(G \cap \mathrm{Var}(\mathcal{Q}_{k-1})$ is called the **singular set** of the variety $G \cap \mathrm{Var}(\mathcal{Q}_k)$.)]

#5. Show, for any subspace V of $\mathbb{R}^\ell$, that
$$\mathrm{Stab}_{\mathrm{SL}(\ell,\mathbb{R})}(V) = \{\, g \in \mathrm{SL}(\ell, \mathbb{R}) \mid gV = V \,\}$$
is Zariski closed.

#6. A Zariski-closed subset of $\mathrm{SL}(\ell, \mathbb{R})$ is **irreducible** if it cannot be written as the union of two Zariski-closed, proper subsets. Show that every Zariski-closed subset A of $\mathrm{SL}(\ell, \mathbb{R})$ has a unique decomposition as an irredundant, finite union of irreducible, Zariski-closed subsets. (By irredundant, we mean that no one of the sets is contained in the union of the others.)

[*Hint:* The ascending chain condition on ideals of $\mathbb{R}[x_{1,1}, \ldots, x_{\ell,\ell}]$ implies the descending chain condition on Zariski-closed subsets, so A can be written as a finite union of irreducibles. To make the union irredundant, the irreducible subsets must be maximal.]

#7. Let H be a connected subgroup of $\mathrm{SL}(\ell, \mathbb{R})$. Show that if $H \subseteq A_1 \cup A_2$, where A_1 and A_2 are Zariski-closed subsets of $\mathrm{SL}(\ell, \mathbb{R})$, then either $H \subseteq A_1$ or $H \subseteq A_2$.

[*Hint:* The Zariski closure $\overline{\overline{H}} = B_1 \cup \cdots \cup B_r$ is an irredundant union of irreducible, Zariski-closed subsets (see Exercise 6). For $h \in H$, we have $\overline{\overline{H}} = hB_1 \cup \cdots \cup hB_r$, so uniqueness implies that h acts as a permutation of $\{B_j\}$. Because H is connected, conclude that $\overline{\overline{H}} = B_1$ is irreducible.]

#8. Assume G is connected, and $G \subseteq \mathrm{SL}(\ell, \mathbb{R})$. Show there exist
- a finite-dimensional real vector space V,
- a vector v in V, and
- a continuous homomorphism $\rho \colon \mathrm{SL}(\ell, \mathbb{R}) \to \mathrm{SL}(V)$,

such that $G = \mathrm{Stab}_{\mathrm{SL}(\ell,\mathbb{R})}(v)^\circ$.

[*Hint:* Let V_n be the vector space of polynomial functions on $\mathrm{SL}(\ell, \mathbb{R})$, and let W_n be the subspace consisting of polynomials that vanish on G. Then $\mathrm{SL}(\ell, \mathbb{R})$ acts on V_n by translation, and W_n is G-invariant. For n sufficiently large, W_n contains generators of the ideal of all polynomials vanishing on G, so $G = \mathrm{Stab}_{\mathrm{SL}(\ell,\mathbb{R})}(W_n)^\circ$. Now let V be the exterior power $\bigwedge^d V_n$, where $d = \dim W_n$, and let v be a nonzero vector in $\bigwedge^d W_n$.]

#9. Show that the center of G is finite.

[*Hint:* The identity component of the Zariski closure of $Z(G)$ is a connected, normal subgroup of G.]

§A5. Three useful theorems

§A5(i). Real Jordan decomposition.

(A5.1) **Definition.** Let $g \in \mathrm{GL}(n, \mathbb{R})$. We say that g is

1) *semisimple* if g is diagonalizable (over $\mathbb{C}$),

2) *hyperbolic* if
   - g is semisimple, and
   - every eigenvalue of g is real and positive,

3) *elliptic* if
   - g is semisimple, and
   - every eigenvalue of g is on the unit circle in $\mathbb{C}$,

4) *unipotent* (or *parabolic*) if 1 is the only eigenvalue of g over $\mathbb{C}$.

(A5.2) *Remark.* A matrix g is semisimple if and only if the minimal polynomial of g has no repeated factors.

1) Because its eigenvalues are real, any hyperbolic g element is diagonalizable over $\mathbb{R}$. That is, there is some $h \in \mathrm{GL}(\ell, \mathbb{R})$, such that $h^{-1}gh$ is a diagonal matrix.

2) An element is elliptic if and only if it is contained in some compact subgroup of $\mathrm{GL}(\ell, \mathbb{R})$. In particular, if g has finite order (that is, if $g^n = \mathrm{Id}$ for some $n > 0$), then g is elliptic.

3) A matrix $g \in \mathrm{GL}(\ell, \mathbb{R})$ is unipotent if and only if the characteristic polynomial of g is $(x - 1)^\ell$. (That is, 1 is the only root of the characteristic polynomial, with multiplicity ℓ.) Another way of saying this is that g is unipotent if and only if $g - \mathrm{Id}$ is nilpotent (that is, if and only if $(g - \mathrm{Id})^n = 0$ for some $n \in \mathbb{N}$).

(A5.3) *Remark.* Remark A2.6 implies that if G is not compact, then it contains nontrivial hyperbolic elements, nontrivial elliptic elements, and nontrivial unipotent elements.

(A5.4) **Proposition** (Real Jordan Decomposition). *Any element g of G can be written uniquely as the product $g = aku$ of three **commuting** elements a, k, u of G, such that a is hyperbolic, k is elliptic, and u is unipotent.*

§A5(ii). Engel's Theorem on unipotent subgroups.

(A5.5) **Definition.** A subgroup U of $\mathrm{SL}(\ell, \mathbb{R})$ is said to be *unipotent* if all of its elements are unipotent.

(A5.6) **Example.** Let N be the group of upper-triangular matrices with 1's on the diagonal; that is,

$$N = \left\{ \begin{bmatrix} 1 & & & \\ & 1 & & \text{\Large *} \\ & & \ddots & \\ \text{\Large 0} & & & 1 \end{bmatrix} \right\} \subseteq SL(\ell, \mathbb{R}).$$

It is obvious that N is unipotent.

Therefore, it is obvious that every subgroup of N is unipotent. Conversely:

(A5.7) **Theorem** (Engel's Theorem). *Every unipotent subgroup of* $SL(\ell, \mathbb{R})$ *is conjugate to a subgroup of the group N of Example A5.6.*

§A5(iii). Jacobson-Morosov Lemma.

(A5.8) **Theorem** (Jacobson-Morosov Lemma). *For every unipotent element u of G, there is a subgroup H of G isogenous to* $SL(2, \mathbb{R})$, *such that* $u \in H$.

Exercises for §A5.

#1. Show that an element of $SL(\ell, \mathbb{R})$ is unipotent if and only if it is conjugate to an element of the subgroup N of Example A5.6.
   [*Hint:* If g is unipotent, then all of its eigenvalues are real, so it can be triangularized over $\mathbb{R}$.]

#2. Show that the Zariski closure of every unipotent subgroup is unipotent.

§A6. The Lie algebra of a Lie group

(A6.1) **Definition.** A map ρ from one Lie group to another is a ***homomorphism*** if

- it is a homomorphism of abstract groups (i.e., $\rho(ab) = \rho(a)\,\rho(b)$), and
- it is continuous.

(Hence, an ***isomorphism*** of Lie groups is a continuous isomorphism of abstract groups, whose inverse is also continuous.)

Although the definition only requires homomorphisms to be continuous, it turns out that they are always infinitely differentiable:

(A6.2) **Proposition.** *Suppose H_1 and H_2 are closed subgroups of* $GL(\ell_i, \mathbb{R})$, *for $i = 1, 2$. Then*

1) H_i *is a C^∞ submanifold of* $GL(\ell_i, \mathbb{R})$,

2) *every (continuous) homomorphism from H_1 to H_2 is C^∞ and*

3) *if $H_1 \subseteq H_2$, then the coset space H_2/H_1 is a C^∞ manifold.*

(A6.3) *Remark.* In fact, the submanifolds and homomorphisms are real analytic, not just C^∞, but we will have no need for this stronger statement.

(A6.4) *Remark.* If H is any Lie group, then conjugation by any element h of H is an automorphism. That is, if we define a map $\varphi_h \colon H \to H$ by $\varphi_h(x) = h^{-1}xh$, then φ_h is a continuous automorphism of H. Any such automorphism is said to be "***inner***." The group of all inner automorphisms is isomorphic to $H/Z(H)$, where $Z(H)$ is the center of H. For some groups, there are many other automorphisms. For example, every inner automorphism of an abelian group is trivial, but the automorphism group of $\mathbb{R}^n$ is $GL(n, \mathbb{R})$, which is quite large. In contrast, it can be shown that the group of inner automorphisms of G has finite index in $\operatorname{Aut}(G)$ (since G is semisimple).

(A6.5) **Definitions.**

1) For $A, B \in \operatorname{Mat}_{\ell \times \ell}(\mathbb{R})$, the ***commutator*** (or ***Lie bracket***) of A and B is the matrix $[A, B] = AB - BA$.

2) A vector subspace $\mathfrak{h}$ of $\operatorname{Mat}_{\ell \times \ell}(\mathbb{R})$ is a ***Lie algebra*** if it is closed under the Lie bracket. That is, for all $A, B \in \mathfrak{h}$, we have $[A, B] \in \mathfrak{h}$.

3) A map ρ from one Lie algebra to another is a ***homomorphism*** if
   - it is a linear transformation, and
   - it preserves brackets (that is, $[\rho(A), \rho(B)] = \rho([A, B])$).

4) Suppose H is a closed subgroup of $GL(\ell, \mathbb{R})$. Then H is a C^∞ manifold, so it has a tangent space at every point; the tangent space at the identity element e is called the ***Lie algebra*** of H. Note that, since H is contained in the vector space $\operatorname{Mat}_{\ell \times \ell}(\mathbb{R})$, its Lie algebra can be identified with a vector subspace of $\operatorname{Mat}_{\ell \times \ell}(\mathbb{R})$.

(A6.6) **Notation.** Lie algebras are usually denoted by lowercase German letters: the Lie algebras of G and H are $\mathfrak{g}$ and $\mathfrak{h}$, respectively.

(A6.7) **Examples.**

1) The Lie algebra $\mathfrak{sl}(\ell, \mathbb{R})$ of $SL(\ell, \mathbb{R})$ is the set of matrices whose trace is 0 (see Exercise 2).

2) The Lie algebra $\mathfrak{so}(n)$ of $SO(n)$ is the set of $n \times n$ skew-symmetric matrices of trace 0 (see Exercise 3).

It is an important fact that the Lie algebra of H is indeed a Lie algebra:

(A6.8) **Proposition.** *If H is a closed subgroup of $SL(\ell, \mathbb{R})$, then the Lie algebra of H is closed under the Lie bracket.*

Here is a very useful reformulation:

(A6.9) **Corollary.** *Suppose H_1 and H_2 are Lie groups, with Lie algebras $\mathfrak{h}_1$ and $\mathfrak{h}_2$. If H_1 is a subgroup of H_2, then $\mathfrak{h}_1$ is a Lie subalgebra of $\mathfrak{h}_2$.*

Hence, for every closed subgroup of H, there is a corresponding Lie subalgebra of $\mathfrak{h}$. Unfortunately, the converse may not be true: although every Lie subalgebra corresponds to a subgroup, the subgroup might not be closed.

(A6.10) **Example.** The 2-torus $\mathbb{T}^2 = \mathbb{R}^2 / \mathbb{Z}^2$ can be identified with the Lie group $\mathrm{SO}(2) \times \mathrm{SO}(2)$. For any line through the origin in $\mathbb{R}^2$, there is a corresponding 1-dimensional subgroup of $\mathbb{T}^2$. However, if the slope of the line is irrational, then the corresponding subgroup of $\mathbb{T}^2$ is dense, not closed.

Therefore, in order to obtain a subgroup corresponding to each Lie subalgebra, we need to allow subgroups that are not closed:

(A6.11) **Definition.** Suppose H_1 and H_2 are Lie groups, and $\rho \colon H_1 \to H_2$ is a homomorphism. Then $\rho(H_1)$ is a **Lie subgroup** of H_2.

(A6.12) **Proposition.** *If H is a Lie group with Lie algebra $\mathfrak{h}$, then there is a one-to-one correspondence between the connected Lie subgroups of H and the Lie subalgebras of $\mathfrak{h}$.*

(A6.13) **Definitions.** Let H be a Lie group in $\mathrm{SL}(\ell, \mathbb{R})$.

1) If $h \colon \mathbb{R} \to H$ is any (continuous) homomorphism, we call h a **one-parameter subgroup** of H, and we usually write h^t, instead of $h(t)$.

2) We define $\exp \colon \mathrm{Mat}_{\ell \times \ell}(\mathbb{R}) \to \mathrm{GL}(\ell, \mathbb{R})$ by

$$\exp X = \sum_{k=0}^{\infty} \frac{1}{k!} X^k.$$

This is called the **exponential map**.

(A6.14) **Proposition.** *Let $\mathfrak{h}$ be the Lie algebra of a Lie group $H \subseteq \mathrm{SL}(\ell, \mathbb{R})$.*

1) *For any $X \in \mathfrak{h}$, the function $x^t = \exp(tX)$ is a one-parameter subgroup of H.*

2) *Conversely, every one-parameter subgroup of H is of this form, for some unique $X \in \mathfrak{h}$.*

Furthermore, for $X \in \mathrm{Mat}_{\ell \times \ell}(\mathbb{R})$, we have

$$X \in \mathfrak{h} \iff \forall t \in \mathbb{R}, \ \exp(tX) \in H.$$

(A6.15) **Definition.** Lie groups H_1 and H_2 are **locally isomorphic** if there is a connected Lie group H and homomorphisms $\rho_i \colon H \to H_i^\circ$, for $i = 1, 2$, such that each ρ_i is a covering map.

(A6.16) **Proposition.** *Two Lie groups are locally isomorphic if and only if their Lie algebras are isomorphic.*

(A6.17) **Notation** (adjoint representation). Suppose $\mathfrak{h}$ is the Lie algebra of a closed subgroup H of $\mathrm{SL}(\ell, \mathbb{R})$. For $h \in H$ and $x \in \mathfrak{h} \subseteq \mathrm{Mat}_{\ell \times \ell}(\mathbb{R})$, we define

$$(\mathrm{Ad}_H h)(x) = hxh^{-1} \in \mathfrak{h}.$$

Then $\mathrm{Ad}_H \colon H \to \mathrm{GL}(\mathfrak{h})$ is a (continuous) homomorphism. It is called the *adjoint representation* of H.

Exercises for §A6.

#1. Suppose $\rho \colon \mathbb{R}^m \to \mathbb{R}^n$ is a continuous map that preserves addition. (That is, we have $\rho(x + y) = \rho(x) + \rho(y)$.) Show (without using Proposition A6.2) that ρ is a linear transformation (and is therefore C^∞). This is a very special case of Proposition A6.2.

[*Hint:* By assumption, we have $\rho(kx) = k\rho(x)$ for all $k \in \mathbb{Z}$, so $\rho(tx) = t\rho(x)$ for all $t \in \mathbb{Q}$ (why?). Then continuity implies this is true for all $t \in \mathbb{R}$.]

#2. Verify Example A6.7(1).

[*Hint:* $A \in \mathfrak{sl}(\ell, \mathbb{R})$ iff $\frac{d}{dt} \det(\mathrm{Id} + tA)|_{t=0} = 0$, and, letting $\lambda = 1/t$, we have $\det(\mathrm{Id} + tA) = t^\ell \det(\lambda I + A) = t^\ell (\lambda^\ell + (\mathrm{trace}\, A)\lambda^{\ell-1} + \cdots = 1 + (\mathrm{trace}\, A)t + \cdots$.]

#3. Verify Example A6.7(2).

[*Hint:* A matrix A of trace 0 is in $\mathfrak{so}(n)$ iff $\frac{d}{dt}(\mathrm{Id} + tA)^T(\mathrm{Id} + tA)|_{t=0} = 0$. Calculate the derivative by using the Product Rule.]

#4. In the notation of Notation A6.17, show, for all $h \in H$, that $\mathrm{Ad}_H h$ is an automorphism of the Lie algebra $\mathfrak{h}$. (In particular, it is an invertible linear transformation, so it is in $\mathrm{GL}(\mathfrak{h})$.)

[*Hint:* The map $a \mapsto hah^{-1}$ is a diffeomorphism of H that fixes e, so its derivative is a linear transformation of the tangent space at e.]

§A7. How to show a group is semisimple

A semisimple group $G = G_1 \times \cdots G_r$ will often have connected, normal subgroups (such as the simple factors G_i). However, these normal subgroups cannot be abelian (see Exercise 1). The converse is a major theorem in the structure theory of Lie groups:

(A7.1) **Theorem.** *A connected Lie group H is semisimple if and only if it has no nontrivial, connected, abelian, normal subgroups.*

(A7.2) *Remark.* A connected Lie group R is *solvable* if every nontrivial quotient of R has a nontrivial, connected, abelian, normal subgroup. (For example, abelian groups are solvable.) It can be shown that every connected Lie group H has a unique maximal connected, closed, solvable, normal subgroup. This subgroup is called the *radical* of H, and is denoted $\mathrm{Rad}\, H$. Our statement of Theorem A7.1 is equivalent to the more usual statement that H is semisimple if and only if $\mathrm{Rad}\, H$ is trivial (see Exercise 2).

The following result makes it easy to see that the classical groups, such as $SL(n, \mathbb{R})$, $SO(m, n)$, and $SU(m, n)$, are semisimple (except a few abelian groups in small dimensions).

(A7.3) Definition. A subgroup H of $GL(\ell, \mathbb{R})$ (or $GL(\ell, \mathbb{C})$) is **irreducible** if there are no nontrivial, proper, H-invariant subspaces of $\mathbb{R}^\ell$ (or $\mathbb{C}^\ell$, respectively).

(A7.4) Example. $SL(\ell, \mathbb{R})$ is an irreducible subgroup of $SL(\ell, \mathbb{C})$ (see Exercise 3).

(A7.5) Warning. In a different context, the adjective "irreducible" can have a completely different meaning when it is applied to a group. For example, saying that a lattice is irreducible (as in Definition 4.3.1) has nothing to do with Definition A7.3.

(A7.6) *Remark.* If H is a subgroup of $GL(\ell, \mathbb{C})$ that is *not* irreducible (that is, if H is **reducible**), then, after a change of basis, we have

$$H \subseteq \begin{pmatrix} GL(k, \mathbb{C}) & * \\ 0 & GL(n - k, \mathbb{C}) \end{pmatrix},$$

for some k with $1 \leq k \leq n - 1$.
   Similarly for $GL(\ell, \mathbb{R})$.

(A7.7) Corollary. *If H is a nonabelian, closed, connected, irreducible subgroup of $SL(\ell, \mathbb{C})$, then H is semisimple.*

Proof. Suppose A is a connected, abelian, normal subgroup of H. For each function $w: A \to \mathbb{C}^\times$, let

$$V_w = \{ v \in \mathbb{C}^\ell \mid \forall a \in A, \ a(v) = w(a) v \}.$$

That is, a nonzero vector v belongs to V_w if

- v is an eigenvector for every element of A, and
- the corresponding eigenvalue for each element of a is the number that is specified by the function w.

Of course, $0 \in V_w$ for every function w; let $W = \{ w \mid V_w \neq 0 \}$. (This is called the set of **weights** of A on $\mathbb{C}^\ell$.)
   Each element of a has an eigenvector (because $\mathbb{C}$ is algebraically closed), and the elements of A all commute with each other, so there is a common eigenvector for the elements of A. Therefore, $W \neq \emptyset$. From the usual argument that the eigenspaces of any linear transformation are linearly independent, one can show that the subspaces $\{ V_w \mid w \in W \}$ are linearly independent. Hence, W is finite.
   For $w \in W$ and $h \in H$, a straightforward calculation shows that $hV_w = V_{h(w)}$, where $(h(w))(a) = w(h^{-1}ah)$. That is, H permutes the subspaces $\{V_w\}_{w \in W}$. Because H is connected and W is finite, this implies

$hV_w = V_w$ for each w; that is, V_w is an H-invariant subspace of $\mathbb{C}^\ell$. Since H is irreducible, we conclude that $V_w = \mathbb{C}^\ell$.

Now, for any $a \in A$, the conclusion of the preceding paragraph implies that $a(v) = w(a)\, v$, for all $v \in \mathbb{C}^\ell$. Therefore, a is a scalar matrix.

Since $\det a = 1$, this scalar is an ℓth root of unity. So A is a subgroup of the group of ℓth roots of unity, which is finite. Since A is connected, we conclude that $A = \{e\}$, as desired. $\qquad\qquad\qquad\qquad\quad\square$

Here is another useful characterization of semisimple groups.

(A7.8) Corollary. *Let H be a closed, connected subgroup of* $\mathrm{SL}(\ell, \mathbb{C})$. *If*

- *the center $Z(H)$ is finite, and*
- *$H^* = H$ (where $*$ denotes the "adjoint," or conjugate-transpose),*

then H is semisimple.

Proof. Because $H^* = H$, it is not difficult to show that H is **completely reducible**: there is a direct sum decomposition $\mathbb{C}^\ell = \bigoplus_{j=1}^r V_j$, such that the restriction $H|_{V_j}$ is irreducible, for each j (see Exercise 6).

Let A be a connected, normal subgroup of H. The proof of Corollary A7.7 (omitting the final paragraph) shows that $A|_{V_j}$ consists of scalar multiples of the identity, for each j. Hence $A \subset Z(H)$. Since A is connected, but (by assumption) $Z(H)$ is finite, we conclude that A is trivial. $\qquad\qquad\qquad\qquad\qquad\qquad\qquad\qquad\qquad\qquad\quad\square$

(A7.9) *Remark.* There is a converse: if G is semisimple (and connected), then G is conjugate to a subgroup H, such that $H^* = H$. However, this is more difficult to prove.

Exercises for §A7.

#1. Prove ($\Rightarrow$) of Theorem A7.1.

#2. Show that a connected Lie group H is semisimple if and only if H has no nontrivial, connected, solvable, normal subgroups.

[*Hint:* If R is a solvable, normal subgroup of H, then $[R, R]$ is also normal in H. Repeating this eventually yields an abelian, normal subgroup.]

#3. Show that no nontrivial, proper $\mathbb{C}$-subspace of $\mathbb{C}^\ell$ is invariant under $\mathrm{SL}(\ell, \mathbb{R})$.

[*Hint:* Suppose $v, w \in \mathbb{R}^\ell$, not both 0. If they are linearly independent, then there exists $g \in \mathrm{SL}(\ell, \mathbb{R})$ with $g(v + iw) = v - iw$. Otherwise, there exists nonzero $\lambda \in \mathbb{C}$ with $\lambda(v + iw) \in \mathbb{R}^\ell$.]

#4. Give an example of a nonabelian, closed, connected, irreducible subgroup H of $\mathrm{SL}(\ell, \mathbb{R})$, such that H is not semisimple.

[*Hint:* $\mathrm{U}(2)$ is an irreducible subgroup of $\mathrm{SO}(4)$.]

#5. Suppose $H \subseteq \mathrm{SL}(\ell, \mathbb{C})$. Show that H is completely reducible if and only if, for every H-invariant subspace W of $\mathbb{C}^\ell$, there is an H-invariant subspace W' of $\mathbb{C}^\ell$, such that $\mathbb{C}^\ell = W \oplus W'$.

[*Hint:* ($\Rightarrow$) If $W' = V_1 \oplus \cdots \oplus V_s$, and $W' \cap W = \{0\}$, but $(W' \oplus V_j) \cap W \neq \{0\}$ for every $j > s$, then $W' + W = \mathbb{C}^\ell$. ($\Leftarrow$) Let W be maximal among the subspaces that are direct sums of irreducibles, and let V be a minimal H-invariant subspace of W'. Then $W \oplus V$ contradicts the maximality of W.]

#6. Suppose $H = H^* \subseteq \mathrm{SL}(\ell, \mathbb{C})$.
   a) Show that if W is an H-invariant subspace of $\mathbb{C}^\ell$, then the orthogonal complement $W^\perp$ is also H-invariant.
   b) Show that H is completely reducible.

Notes

See [6] for a very brief introduction to Lie groups, compatible with Definition A1.1(3). Similar elementary approaches are taken in the books [1] and [3].

Almost all of the material in this appendix (other than §A4) can be found in Helgason's book [4]. However, we do not follow Helgason's notation for some of the classical groups (see Terminology A2.5).

Theorem A4.9 is proved in [5, Thm. 8.3.2, p. 112].

Proposition A5.4 can be found in [4, Lem. IX.7.1, p. 430].

See [2, Prop. 2 in §11.2 of Chapter 8, p. 166] or [7, Thm. 3.17, p. 100] for a proof of the Jacobson-Morosov Lemma (A5.8).

See [9, Thm. 2.7.5, p. 71] for a proof of Proposition A6.16.

Remark A7.9 is due to Mostow [8].

References

[1] A. Baker: *Matrix Groups.* Springer, London, 2002. ISBN 1-85233-470-3, MR 1869885

[2] N. Bourbaki: Lie Groups and Lie Algebras, Chapters 7–9. Springer, Berlin, 2005. ISBN 3-540-43405-4, MR 2109105

[3] B. Hall: Lie Groups, Lie Algebras, and Representations. Springer, New York, 2003. ISBN 0-387-40122-9, MR 1997306

[4] S. Helgason: *Differential Geometry, Lie Groups, and Symmetric Spaces.* Academic Press, New York, 1978. ISBN 0-12-338460-5, MR 0514561

[5] G. P. Hochschild: *Basic Theory of Algebraic Groups and Lie Algebras.* Springer, New York, 1981. ISBN 0-387-90541-3, MR 0620024

[6] R. Howe: Very basic Lie theory, *Amer. Math. Monthly* 90 (1983), no. 9, 600–623. Correction 91 (1984), no. 4, 247. MR 0719752, http://dx.doi.org/10.2307/2323277

[7] N. Jacobson: *Lie Algebras.* Dover, New York, 1979. ISBN
0-486-63832-4, MR 0559927

[8] G. D. Mostow: Self adjoint groups, *Ann. Math.* 62 (1955) 44–55.
MR 0069830, http://www.jstor.org/stable/2007099

[9] V. S. Varadarajan: *Lie Groups, Lie Algebras, and their Representations.*
Springer, Berlin Heidelberg New York, 1984. ISBN 0-387-90969-9,
MR 0746308

Appendix B
Assumed Background

Since the target audience of this book includes mathematicians from a variety of backgrounds (and because very different theorems sometimes have names that are similar, or even identical), this chapter lists (without proof or discussion) specific notations, definitions, and theorems of graduate-level mathematics that are assumed in the main text. (Undergraduate-level concepts, such as the definitions of groups, metric spaces, and continuous functions, are generally not included.) All of this material is standard, so proofs can be found in graduate textbooks (and on the internet).

§B1. Groups and group actions

(B1.1) **Notation.** Let H be a group, and let K be a subgroup.

1) We usually use e to denote the identity element.

2) $Z(H) = \{ z \in H \mid hz = zh \text{ for all } h \in H \}$ is the ***center*** of H.

3) $C_H(K) = \{ h \in H \mid hk = kh \text{ for all } k \in K \}$ is the ***centralizer*** of K in H.

4) $\mathcal{N}_H(K) = \{ h \in H \mid hKh^{-1} = K \}$ is the ***normalizer*** of K in H.

(B1.2) **Definition.** An ***action*** of a Lie group H on a topological space X is a continuous function $\alpha \colon H \times X \to X$, such that

- $\alpha(e, x) = x$ for all $x \in X$, and
- $\alpha(g, \alpha(h, x)) = \alpha(gh, x)$ for $g, h \in H$ and $x \in X$.

(B1.3) **Definitions.** Let a (discrete) group Λ act on a topological space M.

1) The action is ***free*** if no nonidentity element of Λ has a fixed point.

Recall: The Standing Assumptions (4.0.0 on page 41) are in effect, so, as always, Γ is a lattice in the semisimple Lie group $G \subseteq \mathrm{SL}(\ell, \mathbb{R})$.

2) It is ***properly discontinuous*** if, for every compact subset C of M, the set $\{\lambda \in \Lambda \mid C \cap (\lambda C) \neq \varnothing\}$ is finite.

3) For any $p \in M$, we define $\mathrm{Stab}_\Lambda(p) = \{\lambda \in \Lambda \mid \lambda p = p\}$. This is a subgroup of Λ called the ***stabilizer*** of p in Λ.

4) M is ***connected*** if it is **not** the union of two nonempty, disjoint, proper, open subsets.

5) M is ***locally connected*** if every neighborhood of every $p \in M$ contains a connected neighborhood of p.

(B1.4) **Proposition.** *If Λ acts freely and properly discontinuously on a topological space M, then the natural map $\pi \colon M \to \Lambda \backslash M$ is a **covering map**.*

Under the simplifying assumption that M is locally connected, this means that every $p \in \Lambda \backslash M$ has a connected neighborhood U, such that the restriction of π to each connected component of $\pi^{-1}(U)$ is a homeomorphism onto U.

§B2. Galois theory and field extensions

(B2.1) **Theorem** (Fundamental Theorem of Algebra)**.** *The field $\mathbb{C}$ of complex numbers is algebraically closed; that is, every nonconstant polynomial $f(x) \in \mathbb{C}[x]$ has a root in $\mathbb{C}$.*

(B2.2) **Proposition.** *Let F be a subfield of $\mathbb{C}$, and let $\sigma \colon F \to \mathbb{C}$ be any embedding. Then σ extends to an automorphism $\hat{\sigma}$ of $\mathbb{C}$.*

(B2.3) **Notation.** If F is a subfield of a field L, then $|L : F|$ denotes $\dim_F L$, the dimension of L as a vector space over F. This is called the ***degree*** of L over F.

(B2.4) **Proposition.** *If F and L are subfields of $\mathbb{C}$, such that $F \subseteq L$, then $|L : F|$ is equal to the number of embeddings σ of L in $\mathbb{C}$, such that $\sigma|_F = \mathrm{Id}$.*

(B2.5) **Definition.** An extension L of a field F (of characteristic zero) is ***Galois*** if, for every irreducible polynomial $f(x) \in F[x]$, such that $f(x)$ has a root in L, there exist $\alpha_1, \ldots, \alpha_n \in L$, such that
$$f(x) = (x - \alpha_1) \cdots (x - \alpha_n).$$
That is, if an irreducible polynomial in $F[x]$ has a root in L, then all of its roots are in L.

(B2.6) **Definition.** Let L be a Galois extension of a field F. Then
$$\mathrm{Gal}(L/F) = \{\sigma \in \mathrm{Aut}(L) \mid \sigma|_F = \mathrm{Id}\}.$$
This is the ***Galois group*** of L over F.

(B2.7) **Proposition.** *If L is a Galois extension of a field F of characteristic 0, then $|\mathrm{Gal}(L/F)| = |L : F|$.*

(B2.8) **Corollary.** *If L is a Galois extension of a field F of characteristic 0, then there is a one-to-one correspondence between*
- *the subfields K of L, such that $F \subseteq K$, and*
- *the subgroups H of* $\mathrm{Gal}(L/F)$.

Specifically, the subgroup of $\mathrm{Gal}(L/F)$ *corresponding to the subfield K is* $\mathrm{Gal}(L/K)$.

§B3. Algebraic numbers and transcendental numbers

(B3.1) **Definitions.**
1) A complex number z is **algebraic** if there is a nonzero polynomial $f(x) \in \mathbb{Z}[x]$, such that $f(z) = 0$.
2) A complex number is **transcendental** if it is not algebraic.
3) A (nonzero) polynomial is **monic** if its leading coefficient is 1; that is, we may write $f(x) = \sum_{k=0}^{n} a_k x^k$ with $a_n = 1$.
4) A complex number z is an **algebraic integer** if there is a *monic* polynomial $f(x) \in \mathbb{Z}[x]$, such that $f(z) = 0$.

(B3.2) **Proposition.** *If α is an algebraic number, then there is some nonzero $m \in \mathbb{Z}$, such that $m\alpha$ is an algebraic integer.*

(B3.3) **Proposition.** *The set of algebraic integers is a subring of $\mathbb{C}$.*

(B3.4) **Proposition.** *Fix some $n \in \mathbb{N}^+$. Let*
- *ω be a primitive n^{th} root of unity, and*
- *$\mathbb{Z}_n^\times$ be the multiplicative group of units modulo n.*

Then there is an isomorphism
$$f \colon \mathbb{Z}_n^\times \to \mathrm{Gal}(\mathbb{Q}[\omega]/\mathbb{Q}) \colon k \mapsto f_k,$$
such that $f_k(\omega) = \omega^k$, for all $k \in \mathbb{Z}_n^\times$.

§B4. Polynomial rings

(B4.1) **Definition.** A commutative ring R is **Noetherian** if the following equivalent conditions hold:
1) Every ideal of R is finitely generated.
2) If $I_1 \subseteq I_2 \subseteq \cdots$ is any increasing chain of ideals of R, then there is some m, such that $I_m = I_{m+1} = I_{m+2} = \cdots$.

(B4.2) **Proposition** (Hilbert Basis Theorem). *For any field F, the polynomial ring $F[x_1, \ldots, x_s]$ (in any number of variables) is Noetherian.*

(B4.3) **Theorem.** *Let F be a subfield of a field L. If L is finitely generated as an F-algebra (that is, if there exist $c_1, \ldots, c_r \in L$, such that $L = F[c_1, \ldots, c_r]$), then L is algebraic over F.*

(B4.4) **Proposition** (Nullstellensatz). *Let*
- *F be an algebraically closed field,*
- *$F[x_1, \ldots, x_r]$ be a polynomial ring over F, and*
- *I be any proper ideal of $F[x_1, \ldots, x_r]$.*

Then there exist $a_1, \ldots, a_r \in F$, such that $f(a_1, \ldots, a_r) = 0$ for all $f(x_1, \ldots, x_r) \in I$.

(B4.5) **Corollary.** *If B is any finitely generated subring of $\mathbb{C}$, then there is a nontrivial homomorphism from B to the algebraic closure $\overline{\mathbb{Q}}$ of $\mathbb{Q}$.*

(B4.6) **Lemma** (Eisenstein Criterion). *Let $f(x) \in \mathbb{Z}[x]$. If there is a prime number p, and some $a \in \mathbb{Z}_p \smallsetminus \{0\}$, such that*
- *$f(x) \equiv ax^n \pmod{p}$, where $n = \deg f(x)$, and*
- *$f(0) \not\equiv 0 \pmod{p^2}$,*

then $f(x)$ is irreducible over $\mathbb{Q}$.

§B5. General topology

(B5.1) **Definitions.** Let X be a topological space.

1) A subset C of X is **precompact** (or **relatively compact**) if the closure of C is compact.

2) X is **locally compact** if every point of X is contained in a precompact, open subset.

3) X is **separable** if it has a countable, dense subset.

4) If I is an index set (of any cardinality), and X_i is a topological space, for each $i \in I$, then the Cartesian product $\times_{i \in I} X_i$ has a natural "**product topology**," in which a set is open if and only if it is a union (possibly infinite) of sets of the form $\times_{i \in I} U_i$, where each U_i is an open subset of X_i, and we have $U_i = X_i$ for all but finitely many i.

(B5.2) **Theorem** (Tychonoff's Theorem). *If X_i is a compact topological space, for each $i \in I$, then the Cartesian product $\times_{i \in I} X_i$ is also compact (with respect to the product topology).*

(B5.3) **Proposition** (Zorn's Lemma). *Suppose $\leq$ is a binary relation on a set $\mathcal{P}$, such that:*
- *If $a \leq b$ and $b \leq c$, then $a \leq c$.*
- *If $a \leq b$ and $b \leq a$, then $a = b$.*
- *$a \leq a$ for all a.*
- *If $C \subseteq \mathcal{P}$, such that, for all $c_1, c_2 \in C$, either $c_1 \leq c_2$ or $c_2 \leq c_1$, then there exists $b \in \mathcal{P}$, such that $c \leq b$, for all $c \in C$.*

Then there exists $a \in \mathcal{P}$, such that $a \not\leq b$, for all $b \in \mathcal{P}$.

§B6. Measure theory

(B6.1) **Assumption.** Throughout this section, X and Y are complete, separable metric spaces. (Recall that **complete** means all Cauchy sequences converge.)

(B6.2) **Definitions.**

1) The **Borel σ-algebra** $\mathcal{B}(X)$ of X is the smallest collection of subsets of X that:
   - contains every open set,
   - is closed under countable unions (that is, if $A_1, A_2, \ldots \in \mathcal{B}$, then $\bigcup_{i=1}^{\infty} A_i \in \mathcal{B}$), and
   - is closed under complements (that is, if $A \in \mathcal{B}$, then $X \setminus A \in \mathcal{B}$).

2) Each element of $\mathcal{B}(X)$ is called a **Borel set**.

3) A function $f \colon X \to Y$ is **Borel measurable** if $f^{-1}(A)$ is a Borel set in X, for every Borel set A in Y.

4) A function $\mu \colon \mathcal{B}(X) \to [0, \infty]$ is called a **measure** if it is **countably additive**. This means that if $A_1, A_2, \ldots$ are pairwise disjoint, then

$$\mu\left(\bigcup_{i=1}^{\infty} A_i\right) = \sum_{i=1}^{\infty} \mu(A_i).$$

5) A measure μ on X is **Radon** if $\mu(C) < \infty$, for every compact subset C of X.

6) A measure μ on X is **σ-finite** if X is the union of countably many sets of finite measure. This means $X = \bigcup_{i=1}^{\infty} A_i$, with $\mu(A_i) < \infty$ for each i.

(B6.3) **Proposition.** *If μ is a measure on X, and f is a measurable function on X, such that $f \geq 0$, then the integral $\int_X f\, d\mu$ is a well-defined element of $[0, \infty]$, such that:*

1) *$\int_X \chi_A\, d\mu = \mu(A)$ if χ_A is the characteristic function of A.*

2) *$\int_X (a_1 f_1 + a_2 f_2)\, d\mu = a_1 \int_X f_1\, d\mu + a_2 \int_X f_2\, d\mu$ for $a_1, a_2 \in [0, \infty)$.*

3) *if $\{f_n\}$ is a sequence of measurable functions on X, such that we have $0 \leq f_1 \leq f_2 \leq \cdots$, then*

$$\int_X \lim_{n \to \infty} f_n\, d\mu = \lim_{n \to \infty} \int_X f_n\, d\mu.$$

(B6.4) **Corollary** (Fatou's Lemma). *If $\{f_n\}_{n=1}^{\infty}$ is a sequence of measurable functions on X, with $f_n \geq 0$ for all n, and μ is a measure on X, then*

$$\int_X \liminf_{n \to \infty} f_n\, d\mu \leq \liminf_{n \to \infty} \int_X f_n\, d\mu.$$

(B6.5) **Proposition.** *If X is locally compact and separable, then every Radon measure μ on X is **inner regular**. This means*

$\mu(E) = \sup\{\mu(C) \mid C \text{ is a compact subset of } E\}$, *for every Borel set E.*

(B6.6) **Proposition** (Lusin's Theorem). *Assume μ is a Radon measure on X, and X is locally compact. Then, for every measurable function $f\colon X \to \mathbb{R}$, and every $\epsilon > 0$, there is a continuous function $g\colon X \to \mathbb{R}$, such that*

$$\mu(\{\, x \in X \mid f(x) \neq g(x)\,\}) < \epsilon.$$

(B6.7) **Definition.** If μ is a measure on X, and $f\colon X \to Y$ is measurable, then the **push-forward** of μ is the measure $f_*\mu$ on Y that is defined by

$$(f_*\mu)(A) = \mu(f^{-1}(A)) \quad \text{for } A \subseteq Y.$$

(B6.8) **Proposition** (Fubini's Theorem). *Assume*

- *X_1 and X_2 are complete, separable metric spaces, and*
- *μ_i is a σ-finite measure on X_i, for $i = 1, 2$.*

Then there is a measure $\nu = \mu_1 \times \mu_2$ on $X_1 \times X_2$, such that:

1) *$\nu(E_1 \times E_2) = \nu(E_1) \cdot \nu(E_2)$ when E_i is a Borel subset of X_i for $i = 1, 2$, and*

2) *$\int_{X_1 \times X_2} f \, d\nu = \int_{X_1} \int_{X_2} f(x_1, x_2) \, d\mu_2(x_2) \, d\mu_1(x_1)$ when the function $f\colon X_1 \times X_2 \to [0, \infty]$ is Borel measurable.*

(In particular, $\int_{X_2} f(x_1, x_2) \, d\mu_2(x_2)$ is a measurable function of x_1.)

(B6.9) **Definitions.**

1) The **support** of a function $f\colon X \to \mathbb{C}$ is defined to be the closure of $\{\, x \in X \mid f(x) \neq 0 \,\}$.

2) $C_c(X) = \{\,$ continuous functions $f\colon X \to \mathbb{C}$ with compact support $\}$.

3) $\lambda\colon C_c(X) \to \mathbb{C}$ is a **positive linear functional** on $C_c(X)$ if:
   - it is linear (that is, $\lambda(a_1 f_1 + a_2 f_2) = a_1 \lambda(f_1) + a_2 \lambda(f_2)$ for $a_1, a_2 \in \mathbb{C}$ and $f_1, f_2 \in C(X)$), and
   - it is positive (that is, if $f(x) \geq 0$ for all x, then $\lambda(f) \geq 0$).

(B6.10) **Theorem** (Riesz Representation Theorem). *Assume X is locally compact and separable. If λ is any positive linear functional on $C_c(X)$, then there is a Radon measure μ on X, such that*

$$\lambda(f) = \int_X f \, d\mu \quad \text{for all } f \in C_c(X).$$

(B6.11) **Definitions.** Assume μ is a measure on X, and the function $\varphi\colon X \to \mathbb{C}$ is measurable.

1) For $1 \leq p < \infty$, the $\mathscr{L}^p$-**norm** of φ is

$$\|\varphi\|_p = \left(\int_X |\varphi(x)|^p \, d\mu(x) \right)^{1/p}.$$

2) An assertion $P(x)$ is said to be true for **almost all** $x \in X$ (or to be true **almost everywhere**, which is usually abbreviated to **a.e.**), if $\mu(\{\, x \mid P(x) \text{ is false} \,\}) = 0$.

3) In particular, two functions φ_1 and φ_2 are equal (a.e.) if
$$\mu(\{\, x \mid \varphi_1(x) \neq \varphi_2(x) \,\}) = 0.$$
This defines an equivalence relation on the set of (measurable) functions on X.

4) The $\mathscr{L}^\infty$-*norm* (or *essential supremum*) of φ is
$$\|\varphi\|_\infty = \min\{\, a \in (-\infty, \infty] \mid \varphi(x) \leq a \text{ for a.e. } x \,\}.$$

5) $\mathscr{L}^p(X, \mu) = \{\, \varphi \colon X \to \mathbb{C} \mid \|\varphi\|_p < \infty \,\}$, for $1 \leq p \leq \infty$. An element of $\mathscr{L}^p(X, \mu)$ is called an $\mathscr{L}^p$-*function* on X. Actually, two functions in $\mathscr{L}^p(X, \mu)$ are identified if they are equal almost everywhere, so, technically, $\mathscr{L}^p(X, \mu)$ should be defined to be a set of equivalence classes, instead of a set of functions.

(B6.12) **Definition.** Two measures μ and ν on X are in the same *measure class* if they have exactly the same sets of measure 0:
$$\mu(A) = 0 \iff \nu(A) = 0.$$
(This defines an equivalence relation.)

(B6.13) **Theorem** (Radon-Nikodym Theorem). *Two σ-finite measures μ and ν on X are in the same class if and only if there is a measurable function $D \colon X \to \mathbb{R}^+$, such that $\mu = D\nu$. That is, for every measurable subset A of X, we have $\mu(A) = \int_A D \, d\nu$.*

The function D is called the **Radon-Nikodym derivative** $d\mu/d\nu$.

§B7. Functional analysis

(B7.1) **Definitions.** Let $\mathbb{F}$ be either $\mathbb{R}$ or $\mathbb{C}$, and let V be a vector space over $\mathbb{F}$.

1) A *topological vector space* is a vector space V, with a topology, such that the operations of scalar multiplication and vector addition are continuous (that is, the natural maps $\mathbb{F} \times V \to V$ and $V \times V \to V$ are continuous).

2) A subset C of V is *convex* if, for all $v, w \in C$ and $0 \leq t \leq 1$, we have $tv + (1 - t)w \in C$.

3) A topological vector space V is *locally convex* if every neighborhood of 0 contains a convex neighborhood of 0.

4) A locally convex topological vector space V is *Fréchet* if its topology can be given by a metric that is *complete* (that is, such that every Cauchy sequence converges to a limit point).

5) A *norm* on V is a function $\| \ \| \colon V \to [0, \infty)$, such that:
   (a) $\|v + w\| \leq \|v\| + \|w\|$ for all $v, w \in V$,
   (b) $\|av\| = |a| \, \|v\|$ for $a \in \mathbb{F}$ and $v \in V$, and
   (c) $\|v\| = 0$ if and only if $v = 0$.

Note that any norm $\| \ \|$ on V provides a metric that is defined by $d(v, w) = \|v - w\|$. Thus, the norm determines a topology on V.

6) A **Banach space** is a vector space $\mathcal{B}$, together with a norm $\| \ \|$, such that the resulting metric is complete. (Banach spaces are Fréchet.)

7) An **inner product** on V is a function $\langle \ | \ \rangle \colon V \times V \to \mathbb{F}$, such that
   (a) $\langle av + bw \mid x \rangle = a\langle v|x \rangle + b\langle w|x \rangle$ for $a, b \in \mathbb{F}$ and $v, w, x \in V$,
   (b) $\langle v \mid w \rangle = \overline{\langle w \mid v \rangle}$ for $v, w \in V$, where $\bar{a}$ denotes the complex conjugate of a, and
   (c) $\langle v \mid v \rangle \geq 0$ for all $v \in V$, with equality iff $v = 0$.
   Note that if $\langle \ | \ \rangle$ is an inner product on V, then a norm on V is defined by the formula $\|v\| = \sqrt{\langle v \mid v \rangle}$.

8) A **Hilbert space** is a vector space $\mathcal{H}$, together with an inner product $\langle \ | \ \rangle$, such that the resulting normed vector space is complete. (Hence, every Hilbert space is a Banach space.)

9) An **isomorphism** between Hilbert spaces $(\mathcal{H}_1, \langle \ | \ \rangle_1)$ and $(\mathcal{H}_2, \langle \ | \ \rangle_2)$ is an invertible linear transformation $T \colon \mathcal{H}_1 \to \mathcal{H}_2$, such that
$$\langle Tv \mid Tw \rangle_2 = \langle v \mid w \rangle_1 \ \text{ for all } v, w \in \mathcal{H}_1.$$
An isomorphism from $\mathcal{H}$ to itself is called a **unitary operator** on $\mathcal{H}$.

(B7.2) **Example.** If μ is a measure on X, then the $\mathcal{L}^p$-norm makes $\mathcal{L}^p(X, \mu)$ into a Banach space (for $1 \leq p \leq \infty$). Furthermore, $\mathcal{L}^2(X, \mu)$ is a Hilbert space, with the inner product
$$\langle \varphi \mid \psi \rangle = \int_X \varphi(x) \, \overline{\psi(x)} \, d\mu(x).$$

(B7.3) **Definitions.** Let $\mathcal{B}$ be a Banach space (over $\mathbb{F} \in \{\mathbb{R}, \mathbb{C}\}$).

1) A **continuous linear functional** on $\mathcal{B}$ is a continuous function $\lambda \colon \mathcal{B} \to \mathbb{F}$ that is linear (which means $\lambda(av + bw) = a\lambda(v) + b\lambda(w)$ for $a, b \in \mathbb{F}$ and $v, w \in \mathcal{B}$).

2) $\mathcal{B}^* = \{$ continuous linear functionals on $\mathcal{B}\}$ is the **dual** of $\mathcal{B}$. This is a Banach space: the norm of a linear functional λ is
$$\|\lambda\| = \sup\{ |\lambda(v)| \mid v \in \mathcal{B}, \ \|v\| \leq 1 \}.$$

3) For each $v \in \mathcal{B}$, there is a linear function $e_v \colon \mathcal{B}^* \to \mathbb{F}$, defined by $e_v(\lambda) = \lambda(v)$. The **weak* topology** on $\mathcal{B}^*$ is the coarsest topology for which every e_v is continuous.
   In other words, the basic open sets in the weak* topology are of the form $\{ \lambda \in \mathcal{B}^* \mid \lambda(v) \in U \}$, for some $v \in \mathcal{B}$ and some open subset U of $\mathbb{F}$. A set in $\mathcal{B}^*$ is open if and only if it is a union of sets that are finite intersections of basic open sets.

4) Any continuous, linear transformation from $\mathcal{B}$ to itself is called a **bounded operator** on $\mathcal{B}$.

5) The set of bounded operators on $\mathcal{B}$ is itself a Banach space, with the **operator norm**

$$\|T\| = \sup\{\,\|T(v)\| \mid \|v\| \le 1\,\}.$$

(B7.4) **Proposition** (Banach-Alaoglu Theorem). *If $\mathcal{B}$ is any Banach space, then the closed unit ball in $\mathcal{B}^*$ is compact in the weak* topology.*

(B7.5) **Proposition** (Hahn-Banach Theorem). *Suppose*

- *$\mathcal{B}$ is a Banach space over $\mathbb{F}$,*
- *W is a subspace of $\mathcal{B}$ (not necessarily closed), and*
- *$\lambda\colon W \to \mathbb{F}$ is linear.*

If $|\lambda(w)| \le \|w\|$ for all $w \in W$, then λ extends to a linear functional $\hat{\lambda}\colon \mathcal{B} \to \mathbb{F}$, such that $|\hat{\lambda}(v)| \le \|v\|$ for all $v \in \mathcal{B}$.

(B7.6) **Proposition** (Open Mapping Theorem). *Assume X and Y are Fréchet spaces, and $f\colon X \to Y$ is a continuous, linear map.*

1) *If f is surjective, and $\mathcal{O}$ is any open subset of X, then $f(\mathcal{O})$ is open.*
2) *If f is bijective, then the inverse $f^{-1}\colon Y \to X$ is continuous.*

(B7.7) **Assumption.** Hilbert spaces are always assumed to be separable.

This has the following consequence:

(B7.8) **Proposition.** *There is only one infinite-dimensional Hilbert space (up to isomorphism). In other words, every infinite-dimensional Hilbert space is isomorphic to $\mathcal{L}^2(\mathbb{R}, \mu)$, where μ is Lebesgue measure.*

(B7.9) **Definitions.**

1) If $\mathcal{H}_1$ and $\mathcal{H}_2$ are Hilbert spaces, then the **direct sum** $\mathcal{H}_1 \oplus \mathcal{H}_2$ is a Hilbert space, under the inner product

$$\langle (\varphi_1, \varphi_2) \mid (\psi_1, \psi_2) \rangle = \langle \varphi_1 \mid \psi_1 \rangle + \langle \varphi_2 \mid \psi_2 \rangle.$$

   By induction, this determines the direct sum of any finite number of Hilbert spaces; see Definition 11.6.1 for the direct sum of infinitely many.

2) We use "$\perp$" as an abbreviation for "is orthogonal to." Therefore, if $\varphi, \psi \in \mathcal{H}$, then $\varphi \perp \psi$ means $\langle \varphi \mid \psi \rangle = 0$. For subspaces $\mathcal{K}, \mathcal{K}'$ of $\mathcal{H}$, we write $\mathcal{K} \perp \mathcal{K}'$ if $\varphi \perp \varphi'$ for all $\varphi \in \mathcal{K}$ and $\varphi' \in \mathcal{K}'$.

3) The **orthogonal complement** of a subspace $\mathcal{K}$ of $\mathcal{H}$ is

$$\mathcal{K}^\perp = \{\, \varphi \in \mathcal{H} \mid \varphi \perp \mathcal{K} \,\}.$$

   This is a closed subspace of $\mathcal{H}$. We have $\mathcal{H} = \mathcal{K} + \mathcal{K}^\perp$ and $\mathcal{K} \perp \mathcal{K}^\perp$, so $\mathcal{H} = \mathcal{K} \oplus \mathcal{K}^\perp$.

4) The **orthogonal projection** onto a closed subspace $\mathcal{K}$ of $\mathcal{H}$ is the (unique) bounded operator $P\colon \mathcal{H} \to \mathcal{K}$, such that
   - $P(\varphi) = \varphi$ for all $\varphi \in \mathcal{K}$, and

- $P(\psi) = 0$ for all $\varphi \in \mathcal{K}^\perp$.

(B7.10) **Definitions.** Let $T: \mathcal{H} \to \mathcal{H}$ be a bounded operator on a Hilbert space $\mathcal{H}$.

1) The **adjoint** of T is the bounded operator T^* on $\mathcal{H}$, such that
$$\langle T\varphi \mid \psi \rangle = \langle \varphi \mid T^*\psi \rangle \text{ for all } \varphi, \psi \in \mathcal{H}.$$
   It does not always exist, but T^* is unique if it does exist.

2) T is **self-adjoint** (or **Hermitian**) if $T = T^*$.

3) T is **normal** if $TT^* = T^*T$.

4) T is **compact** if there is a nonempty, open subset $\mathcal{O}$ of $\mathcal{H}$, such that $T(\mathcal{O})$ is precompact.

(B7.11) **Proposition.** *Let T be a bounded operator on a Hilbert space $\mathcal{H}$.*

1) *If $T(\mathcal{H})$ is finite-dimensional, then T is compact.*

2) *The set of compact operators on $\mathcal{H}$ is closed (in the topology defined by the operator norm).*

(B7.12) **Proposition** (Spectral Theorem). *If T is any bounded, normal operator on any Hilbert space $\mathcal{H}$, then there exist*

- *a finite measure μ on $[0,1]$,*
- *a bounded, measurable function $f: [0,1] \to \mathbb{C}$, and*
- *an isomorphism $U: \mathcal{H} \to \mathcal{L}^2([0,1], \mu)$,*

such that $U(T\varphi) = f\, U(\varphi)$, for all $\varphi \in \mathcal{H}$ (where $f\, U(\varphi)$ denotes the pointwise multiplication of the functions f and $U(\varphi)$).

   Furthermore:

1) *T is unitary if and only if $|f(x)| = 1$ for a.e. $x \in [0,1]$.*

2) *T is self-adjoint if and only if $f(x) \in \mathbb{R}$ for a.e. $x \in [0,1]$.*

(B7.13) **Definition.** In the situation of Proposition B7.12, the **spectral measure** of T is $f_*\mu$.

(B7.14) **Corollary** (Spectral Theorem for compact, self-adjoint operators). *Let T be a bounded operator on any Hilbert space $\mathcal{H}$. Then T is both self-adjoint and compact if and only if there exists an orthonormal basis $\{e_n\}$ of $\mathcal{H}$, such that*

1) *each e_n is an eigenvector of T, with eigenvalue λ_n,*

2) *$\lambda_n \in \mathbb{R}$, and*

3) *$\lim_{n \to \infty} \lambda_n = 0$.*

(B7.15) **Proposition** (Fréchet-Riesz Theorem). *If λ is any continuous linear functional on a Hilbert space $\mathcal{H}$, then there exists $\psi \in \mathcal{H}$, such that $\lambda(\varphi) = \langle \varphi \mid \psi \rangle$ for all $\varphi \in \mathcal{H}$.*

Appendix C
A Quick Look at *S*-Arithmetic Groups

Classically, and in the main text of this book, the Lie groups under consideration were manifolds over the field $\mathbb{R}$ of real numbers. However, in some areas of modern mathematics, especially Number Theory and Geometric Group Theory, it is important to understand the lattices in Lie groups not only over the classical field $\mathbb{R}$ (or $\mathbb{C}$), but also over "nonarchimedean" fields of p-adic numbers. The natural analogues of arithmetic groups in this setting are called "*S*-arithmetic groups." Roughly speaking, this generalization is obtained by replacing the ring $\mathbb{Z}$ with a slightly larger ring.

(C0.1) **Definition.** For any finite set $S = \{p_1, p_2, \ldots, p_n\}$ of prime numbers, let

$$\mathbb{Z}_S = \left\{ \frac{p}{q} \in \mathbb{Q} \;\middle|\; \begin{array}{c} \text{every prime factor} \\ \text{of } q \text{ is in } S \end{array} \right\} = \mathbb{Z}[1/p_1, 1/p_2, \ldots, 1/p_n].$$

This is called the ring of *S-integers*.

(C0.2) **Example.**

1) The prototypical example of an arithmetic group is $\mathrm{SL}(\ell, \mathbb{Z})$.
2) The corresponding example of an *S*-arithmetic group is $\mathrm{SL}(\ell, \mathbb{Z}_S)$ (where S is a finite set of prime numbers).

That is, while arithmetic groups do not allow their matrix entries to have denominators, *S*-arithmetic groups allow their matrix entries to have denominators that are products of certain specified primes.

Most of the results in this book can be generalized in a natural way to *S*-arithmetic groups. (The monographs [5] and [8] treat *S*-arithmetic groups alongside arithmetic groups throughout.) We will now give a very brief description of these more general results.

(C0.3) *Remark.* The monograph [5] of Margulis deals with a more general class of S-arithmetic groups that allows $\mathbb{R}$ to be replaced with certain "local" fields of characteristic p, but we discuss only the fields of characteristic 0.

§C1. Introduction to S-arithmetic groups

Most of the theory in this book (and much of the importance of the theory of arithmetic groups) arises from the fundamental fact that $G_{\mathbb{Z}}$ is a lattice in G. Since the ring $\mathbb{Z}_S$ is not discrete (unless $S = \varnothing$), the group $G_{\mathbb{Z}_S}$ is usually not discrete, so it is usually not a lattice in G. Instead, it is a lattice in a group G_S that will be defined in this section.

The construction of $\mathbb{R}$ as the completion of $\mathbb{Q}$ can be generalized as follows:

(C1.1) **Definition** (*p-adic numbers*). Let p be a prime number.

1) If x is any nonzero rational number, then there is a unique integer $v = v_p(x)$, such that we may write
$$x = p^v \frac{a}{b},$$
where a and b are relatively prime to p. (We let $v_p(0) = \infty$.) Then $v_p(x)$ is called the ***p-adic valuation*** of x.

2) Let
$$d_p(x,y) = p^{-v_p(x-y)}.$$
It is easy to verify that d_p is a metric on $\mathbb{Q}$. It is called the ***p-adic metric***.

3) Let $\mathbb{Q}_p$ be the completion of $\mathbb{Q}$ with respect to this metric. (That is, $\mathbb{Q}_p$ is the set of equivalence classes of convergent Cauchy sequences.) This is a field that naturally contains $\mathbb{Q}$. It is called the ***field of p-adic numbers***.

4) If **G** is an algebraic group over $\mathbb{Q}$, we can define the group $\mathbf{G}(\mathbb{Q}_p)$ of $\mathbb{Q}_p$-points of **G**.

(C1.2) **Notation.** To discuss real numbers and p-adic numbers uniformly, it is helpful to let $\mathbb{Q}_\infty = \mathbb{R}$.

The construction of arithmetic subgroups by restriction of scalars (see Section 5.5) is based on the fact that the ring $\mathcal{O}$ of integers in a number field F embeds as a cocompact, discrete subring in $\bigoplus_{v \in S_\infty} F_v$. Using this fact, it was shown that $\mathbf{G}(\mathcal{O})$ is a lattice in $\times_{v \in S_\infty} \mathbf{G}(F_v)$.

Similarly, to obtain a lattice in a p-adic group $\mathbf{G}(\mathbb{Z}_p)$, or, more generally, in a product $\times_{v \in S \cup \{\infty\}} \mathbb{F}_v$ of p-adic groups and real groups, we note that

$$\mathbb{Z}_S \text{ embeds as a cocompact, discrete subring in } \bigoplus_{p \in S \cup \{\infty\}} \mathbb{Q}_p.$$

Using this fact, it can be shown that
$$\mathbf{G}(\mathcal{O}_S) \text{ is a lattice in } \mathbf{G}_S = \underset{p \in S \cup \{\infty\}}{\times} \mathbf{G}(\mathbb{Q}_p).$$
We call $\mathbf{G}(\mathcal{O}_S)$ an *S-arithmetic subgroup*.

(C1.3) **Example.** Let $\mathbf{G}$ be the special linear group SL_n.
1) Letting $S = \emptyset$, we have $\mathbb{Z}_S = \mathbb{Z}$ and $G_S = \mathrm{SL}(n, \mathbb{R})$. So $\mathrm{SL}(n, \mathbb{Z})$ is an S-arithmetic lattice in $\mathrm{SL}(n, \mathbb{R})$. This is a special case of the fact that every arithmetic lattice is an S-arithmetic lattice (with $S = \emptyset$).
2) Letting $S = \{p\}$, where p is a prime, we see that $\mathrm{SL}(n, \mathbb{Z}[1/p])$ is a lattice in $\mathrm{SL}(n, \mathbb{R}) \times \mathrm{SL}(n, \mathbb{Q}_p)$.
3) More generally, letting $S = \{p_1, p_2, \dots, p_r\}$, where $p_1, \dots, p_r$ are primes, we see that $\mathrm{SL}(n, \mathbb{Z}_S)$ is a lattice in
$$\mathrm{SL}(n, \mathbb{R}) \times \mathrm{SL}(n, \mathbb{Q}_{p_1}) \times \mathrm{SL}(n, \mathbb{Q}_{p_2}) \times \cdots \times \mathrm{SL}(n, \mathbb{Q}_{p_r}).$$
This is an elaboration of our previous comment that $\mathrm{SL}(\ell, \mathbb{Z}_S)$ is the prototypical example of an S-arithmetic group (see Example C0.2).

(C1.4) *Remark* [2, Chap. 7]. In the study of arithmetic subgroups of a Lie group G, the symmetric space G/K is a very important tool. In the theory of S-arithmetic subgroups of $\mathbf{G}_S$, this role is taken over by a space called the *Bruhat-Tits building* of $\mathbf{G}_S$. It is a Cartesian product
$$X_S = (G/K) \times \underset{p \in S}{\times} X_p,$$
where X_p is a contractible simplicial complex on which $\mathbf{G}(\mathbb{Q}_p)$ acts properly (but not transitively).

Optional: Readers familiar with the basic facts of Algebraic Number Theory will realize that the above discussion has the following natural generalization:

(C1.5) **Definition** ([5, p. 61], [8, p. 267]). Let
- $\mathcal{O}$ be the ring of integers of an algebraic number field F,
- S be a finite set of finite places of F, and
- $\mathbf{G}$ be a semisimple algebraic group over F, and
- $\mathbf{G}_S = \times_{v \in S \cup S_\infty} \mathbf{G}(F_v)$.

Then $\mathbf{G}(\mathcal{O}_S)$ is an *S-arithmetic subgroup* of $\mathbf{G}_S$.

(C1.6) *Remark.* More generally, much as in Definition 5.1.19, if
- Γ' is an S-arithmetic subgroup of $\mathbf{G}'_S$, and
- $\varphi : \mathbf{G}'_S \to \mathbf{G}_S$ is a surjective, continuous homomorphism, with compact kernel,

then any subgroup of $\mathbf{G}_S$ that is commensurable to $\varphi(\Gamma')$ may be called an S-arithmetic subgroup of $\mathbf{G}_S$.

(C1.7) **Theorem** [8, Thm. 5.7, p. 268]. *Every S-arithmetic subgroup of $\mathbf{G}_S$ is a lattice in $\mathbf{G}_S$.*

§C2. List of results on S-arithmetic groups

(C2.1) **Warning.** The Standing Assumptions (4.0.0) do **not** apply in this appendix, because G is not assumed to be a *real* Lie group.

Instead:

(C2.2) **Assumption.** Throughout the remainder of this appendix:
- G is a semisimple algebraic group over $\mathbb{Q}$,
- S is a finite set of prime numbers, and
- Γ is an S-arithmetic lattice in $\mathbf{G}_S = \times_{p \in S \cup \{\infty\}} \mathbf{G}(\mathbb{Q}_p)$.

To avoid trivialities, we assume $\mathbf{G}_S$ is not compact, so Γ is infinite.

(C2.3) **Definition.** As a substitute for real rank in this setting, let
$$S\text{-rank}\,\mathbf{G} = \sum_{p \in S \cup \{\infty\}} \text{rank}_{\mathbb{Q}_p}\,\mathbf{G}.$$

(C2.4) *Remark.* All of these results generalize to the setting of Definition C1.5, but we restrict our discussion to $\mathbb{Q}$ for simplicity.

The following theorems on S-arithmetic groups are all stated without proof, but each result is provided with a reference for further reading. The reader should be aware that these references are almost always secondary sources, not the original appearance of the result in the literature.

Results related to Chapter 4 (Basic Properties of Lattices).

(4.4.3$_S$) $\Gamma \backslash \mathbf{G}_S$ is compact if and only if the identity element e is **not** an accumulation point of $\Gamma^{\mathbf{G}_S}$ [9, Thm. 1.12, p. 22].

(4.4.4$_S$) If Γ has a nontrivial, unipotent element, then $\Gamma \backslash \mathbf{G}_S$ is not compact. In fact, Godement's Criterion (5.3.1$_S$) tells us that the converse is also true.

(4.5.1$_S$) The Borel Density Theorem holds, for any continuous homomorphism $\rho \colon \mathbf{G}_S \to \mathrm{GL}(V)$, where V is a vector space over $\mathbb{R}$, $\mathbb{C}$, or any p-adic field $\mathbb{Q}_p$ [5, Thm. II.2.5 (and Lem. II.2.3), p. 84].

(4.7.10$_S$) Γ is finitely presented [8, Thm. 5.11, p. 272].

(4.8.2$_S$) (Selberg Lemma) Γ has a torsion-free subgroup of finite index [9, Thm. 6.11, p. 93].

(4.9.2$_S$) (Tits Alternative) Γ has a nonabelian free subgroup [5, App. B, pp. 351–353].

Results related to Chapter 5 (What is an Arithmetic Group?).

(5.2.1$_S$) (Margulis Arithmeticity Theorem) If S-rank $\mathbf{G} \geq 2$, then every irreducible lattice in $\mathbf{G}_S$ is S-arithmetic [5, Thm. IX.1.10, p. 298, and Rem. (vi) on p. 290]. (Note that our definition of irreducibility is stronger than the one used in [5].)

(5.3.1$_S$) (Godement Criterion) $\Gamma \backslash \mathbf{G}_S$ is compact if and only if Γ has no nontrivial unipotent elements [8, Thm. 5.7(2), p. 268].

Remark. Exercise 4.4#10 (easily) implies:

1) [13, Thm. 1] If v is any nonarchimedean place of F, then every lattice in $\mathbf{G}(F_v)$ is cocompact.

2) If $\mathbf{G}(S_\infty)$ is compact, then every lattice in G is cocompact.

Warning. We know that if $\mathcal{O}$ is the ring of integers of F, then $\mathbf{G}(\mathcal{O})$ embeds as an arithmetic lattice in $\times_{v \in S_\infty} \mathbf{G}(F_v)$, but that restriction of scalars allows us to realize this same lattice as the $\mathbb{Z}$-points of an algebraic group defined over $\mathbb{Q}$ (cf. Proposition 5.5.8). This means that all arithmetic groups can be found by using only algebraic groups that are defined over $\mathbb{Q}$, not other number fields. It is important to realize that the same cannot be said for S-arithmetic groups: most extensions of $\mathbb{Q}$ provide many S-arithmetic groups that cannot be obtained from $\mathbb{Q}$.

For example, suppose p is a prime in $\mathbb{Z}$, but p factors in the integers $\mathcal{O}$ of an extension field, and a is a prime factor of p in $\mathcal{O}$. Then the subgroup $SL(2, \mathcal{O}[1/p])$ can be obtained by restriction of scalars, but $SL(2, \mathcal{O}[1/a])$ is an $\{a\}$-arithmetic subgroup that cannot be obtained by this method.

A result related to Chapter 12 (Amenable Groups).

(12.4.5$_S$) For $v \in S$, if $\mathbf{G}(F_v)$ is not compact, then $\mathbf{G}(F_v)$ is not amenable [12, Rem. 8.7.11, p. 260].

Results related to Chapter 13 (Kazhdan's Property (T)).

(13.2.4$_S$) If rank$_{F_v} G \geq 2$, for every simple factor G of $\mathbf{G}(F_v)$, and every $v \in S$, then $\mathbf{G}_S$ has Kazhdan's property [5, Cor. III.5.4, p. 130].

(13.4.1$_S$) If $\mathbf{G}_S$ has Kazhdan's property, then Γ also has Kazhdan's property [5, Thm. III.2.12, p. 117].

(13.4.3$_S$) If Γ has Kazhdan's property, then $\Gamma/[\Gamma, \Gamma]$ is finite [5, Thm. III.2.5, p. 115].

A result related to Chapter 16 (Margulis Superrigidity Theorem).

Assumption. Assume

- S-rank $G \geq 2$,
- Γ is irreducible, and
- w is a place of some algebraic number field F'.

(16.1.6$_S$) (Margulis Superrigidity Theorem [5, Prop. VII.5.3, p. 225]) If

- G' is a Zariski-connected, noncompact, simple algebraic group over F'_w, with trivial center, and
- $\varphi: \Gamma \to G'(F'_w)$ is a homomorphism, such that $\varphi(\Gamma)$ is:
  - Zariski dense in G', and
  - not contained in any compact subgroup of $G'(F'_w)$,

then φ extends to a continuous homomorphism $\hat{\varphi}: G_S \to G'(F'_w)$.

Furthermore, there is some $v \in S$, such that F_v is isomorphic to a subfield of a finite extension of F'_w.

Warning. Exercise 16.4#1 does not extend to the setting of S-arithmetic groups: for example, the lattice $SL(n, \mathbb{Z})$ is not cocompact, but the image of the natural inclusion $SL(n, \mathbb{Z}) \hookrightarrow SL(n, \mathbb{Q}_p)$ is precompact.

Results related to Chapter 17 (Normal Subgroups of Γ).

(17.1.1$_S$) (Margulis Normal Subgroups Theorem [5, Thm. VIII.2.6, p. 265]) Assume

- S-rank $\mathbf{G} \geq 2$,
- Γ is irreducible, and
- N is a normal subgroup of Γ.

Then either N is finite, or Γ/N is finite.

(17.2.1$_S$) If S-rank $\mathbf{G} = 1$, then Γ has (many) normal subgroups N, such that neither N nor Γ/N is finite [4, Cor. 7.6].

Results related to Chapter 18 (Arithmetic Subgroups of Classical Groups).

(18.1.1$_S$) Let $\overline{\mathbb{Q}_p}$ be the algebraic closure of $\mathbb{Q}_p$. Then all but finitely many of the simple Lie groups over $\overline{\mathbb{Q}_p}$ are isogenous to either $SL(n, \overline{\mathbb{Q}_p})$, $SO(n, \overline{\mathbb{Q}_p})$, or $Sp(2n, \overline{\mathbb{Q}_p})$, for some n [3, Thm. 11.4, pp. 57–58, and Thm. 18.4, p. 101].

(18.1.7$_S$), (18.5.3$_S$) Every $\mathbb{Q}$-form or $\mathbb{Q}_p$-form of $SL(n, \overline{\mathbb{Q}_p})$, $SO(n, \overline{\mathbb{Q}_p})$, or $Sp(n, \overline{\mathbb{Q}_p})$ is of classical type, except for some "triality" forms of $SO(8, \overline{\mathbb{Q}_p})$ (cf. Remark 18.5.10) [8, §2.3].

(18.5.10$_S$) Unlike $\mathbb{R}$, the field $\mathbb{Q}_p$ has extensions of degree 3, so some $\mathbb{Q}_p$-forms of $SO(8, \overline{\mathbb{Q}_p})$ are triality groups, even though there are no such $\mathbb{R}$-forms of $SO(8, \mathbb{C})$.

(18.7.1$_S$) $\mathbf{G}_S$ has a cocompact, S-arithmetic lattice [1].

(18.7.4$_S$) If $\mathbf{G}$ is isotypic, then $\mathbf{G}_S$ has a cocompact, irreducible lattice that is S-arithmetic [1].

(18.7.5$_S$) If $\mathbf{G}_S$ has an irreducible, S-arithmetic lattice, then $\mathbf{G}$ is isotypic.

A result related to Chapter 19 (Construction of a Coarse Fundamental Domain). If $\mathcal{F}$ is any coarse fundamental domain for $\mathbf{G}(\mathbb{Z})$ in $\mathbf{G}(\mathbb{R})$, then there is a compact subset C of $\times_{p \in S} \mathbf{G}(\mathbb{Q}_p)$, such that $\mathcal{F} \times C$ is a coarse fundamental domain for $\mathbf{G}(\mathbb{Z}_S)$ in $\mathbf{G}_S$ [8, Prop. 5.11, p. 267].

This implies that every S-arithmetic subgroup of $\mathbf{G}_S$ is a lattice [8, Thm. 5.7, p. 268], but the short proof outlined in Section 7.4 does not seem to generalize to this setting.

Results related to Chapter 20 (Ratner's Theorems on Unipotent Flows). Ratner's three main theorems (20.1.3, 20.3.3, and 20.3.4) have all been generalized to the S-arithmetic setting by Ratner [10, 11] and Margulis-Tomanov [6, 7] (independently).

References

[1] A. Borel and G. Harder: Existence of discrete cocompact subgroups of reductive groups over local fields, *J. Reine Angew. Math.* 298 (1978) 53–64. MR 0483367, http://eudml.org/doc/151965

[2] K. Brown: *Buildings.* Springer, New York, 1989. ISBN 0-387-96876-8, MR 0969123.
http://www.math.cornell.edu/~kbrown/buildings/brown1988.pdf

[3] J. E. Humphreys: *Introduction to Lie Algebras and Representation Theory.* Springer, Berlin Heidelberg New York, 1972. MR 0323842

[4] A. Lubotzky: Lattices in rank one Lie groups over local fields, *Geom. Funct. Anal.* 1 (1991), no. 4, 406–431. MR 1132296,
http://dx.doi.org/10.1007/BF01895641

[5] G. A. Margulis: *Discrete Subgroups of Semisimple Lie Groups.* Springer, Berlin Heidelberg New York, 1991. ISBN 3-540-12179-X, MR 1090825

[6] G. A. Margulis and G. M. Tomanov: Invariant measures for actions of unipotent groups over local fields on homogeneous spaces, *Invent. Math.* 116 (1994), no. 1–3, 347–392. MR 1253197,
http://eudml.org/doc/144192

[7] G. A. Margulis and G. M. Tomanov: Measure rigidity for almost linear groups and its applications, *J. Anal. Math.* 69 (1996), 25–54. MR 1428093, http://dx.doi.org/10.1007/BF02787100

[8] V. Platonov and A. Rapinchuk: *Algebraic Groups and Number Theory.* Academic Press, Boston, 1994. ISBN 0-12-558180-7, MR 1278263

[9] M. S. Raghunathan: *Discrete Subgroups of Lie Groups.* Springer, New York, 1972. ISBN 0-387-05749-8, MR 0507234

[10] M. Ratner: Raghunathan's conjectures for Cartesian products of real and *p*-adic Lie groups, *Duke Math. J.* 77 (1995), no. 2, 275–382. MR 1321062, http://dx.doi.org/10.1215/S0012-7094-95-07710-2

[11] M. Ratner: On the *p*-adic and *S*-arithmetic generalizations of Raghunathan's conjectures, in S. G. Dani, ed.: *Lie Groups and Ergodic Theory (Mumbai, 1996).* Tata Inst. Fund. Res., Bombay, 1998, pp. 167–202. MR 1699365

[12] H. Reiter and J. D. Stegeman: *Classical Harmonic Analysis and Locally Compact Groups.* Oxford University Press, New York, 2000. ISBN 0-19-851189-2, MR 1802924

[13] T. Tamagawa: On discrete subgroups of *p*-adic algebraic groups, in O. F. G. Schilling, ed.: *Arithmetical Algebraic Geometry (Proc. Conf. Purdue Univ., 1963).* Harper & Row, New York, 1965, pp. 11–17. MR 0195864

List of Notation

Chapter 1. What is a Locally Symmetric Space?

$G°$ = identity component of G, page 4

Chapter 2. Geometric Meaning of $\mathbb{R}$-rank and $\mathbb{Q}$-rank

rank X = dimension of maximal flat, page 21
$\text{rank}_{\mathbb{R}}\, G$ = maximal dimension of closed, simply connected flat, page 22

Chapter 3. Brief Summary

Chapter 4. Basic Properties of Lattices

$^b a = bab^{-1}$ for $a, b \in G$, page 50
$^G\Gamma = \{ {}^g\gamma \mid \gamma \in \Gamma,\ g \in G \}$, page 50
$\mathbb{P}(V)$ = projectivization of vector space V, page 59
$\mathcal{F}$ = (coarse) fundamental domain, page 61

Chapter 5. What is an Arithmetic Group?

$\text{Var}(\mathcal{Q}) = \{ g \in \text{SL}(\ell, \mathbb{R}) \mid Q(g) = 0,\ \forall Q \in \mathcal{Q} \}$, page 85
$G_A = G \cap \text{SL}(n, A)$, page 86
$\text{Comm}_G(\Gamma)$ = commensurator of Γ in G, page 91
$\mathcal{L} = \mathbb{Z}$-lattice in a vector space, page 96

Chapter 6. Examples of Arithmetic Groups

$\mathbb{H}_F^{a,b}$ = quaternion algebra, page 116
$\text{PO}(1, n) = \text{O}(1, n) / \{ \pm \text{Id} \}$, page 127
τ_c = conjugation on quaternion algebra, page 146
τ_r = reversion anti-involution on quaternion algebra, page 146

Chapter 7. SL(n, $\mathbb{Z}$) is a lattice in SL(n, $\mathbb{R}$)

$\mathfrak{S}_{c_1,c_2,c_3} = N_{c_1,c_2} A_{c_3} K$ is a Siegel set for SL($n,\mathbb{Z}$), page 156

l = Lebesgue measure on $\mathbb{R}$, page 164

$\widehat{W}$ = {primitive vectors in discrete subgroup W of $\mathbb{R}^n$}, page 164

$\widehat{W}^+$ = set of representatives of $\widehat{W}/\{\pm 1\}$, page 164

$d(W) = \|w_1 \wedge \cdots \wedge w_k\|$, page 169

$\mathcal{S}(\mathcal{L})$ = {full, nontrivial subgroups of $\mathcal{L}$}, page 170

Chapter 8. Real Rank

$\operatorname{rank}_{\mathbb{R}} G$ = real rank of G, page 178

Chapter 9. $\mathbb{Q}$-Rank

$\operatorname{rank}_{\mathbb{Q}} \Gamma$ = $\mathbb{Q}$-rank of Γ, page 192

Chapter 10. Quasi-Isometries

Chapter 11. Unitary Representations

$\mathcal{H}$ = Hilbert space with inner product $\langle \mid \rangle$, page 211

$\mathcal{U}(\mathcal{H})$ = group of unitary operators on $\mathcal{H}$, page 211

π = unitary representation, page 211

$\mathbb{1}$ = trivial representation, page 212

π_{reg} = regular representation, page 212

$\mathcal{L}_\Gamma(G;\mathcal{H})$ = { Γ-equivariant functions from G to $\mathcal{H}$ }, page 218

$\mathcal{L}_\Gamma^2(G;\mathcal{H})$ = { square-integrable functions in $\mathcal{L}_\Gamma(G;\mathcal{H})$ }, page 218

$\operatorname{Ind}_\Gamma^G(\pi)$ = induced representation, page 219

A^* = {characters of abelian group A}, page 221

Chapter 12. Amenable Groups

$\operatorname{Prob}(X)$ = {probability measures on X}, page 236

$\mathcal{L}^\infty(H;C)$ = {bounded functions from H to C}, page 250

$\mathcal{L}_\Lambda^\infty(H;C)$ = {essentially Λ-equivariant maps in $\mathcal{L}^\infty(H;C)$}, page 250

Chapter 13. Kazhdan's Property (T)

$\mathcal{B}(\mathcal{H})$ = {bounded operators on $\mathcal{H}$}, page 266

$\sigma \le \pi$ means σ is a subrepresentation of π, page 268

$\sigma \prec \pi$ means σ is weakly contained in π, page 269

Chapter 14. Ergodic Theory

$\mathcal{B}(X)$ = {Borel subsets}, modulo sets of measure 0, page 294

$A \bigtriangleup B$ = symmetric difference of A and B, page 294

ext C = {extreme points of C}, page 295

$\Sigma C = \{\, tc \mid t \in [0, \infty),\ c \in C \,\}$, page 296

Chapter 15. Mostow Rigidity Theorem

Chapter 16. Margulis Superrigidity Theorem

$\mathcal{E}_\varphi$ = flat vector bundle over G/Γ, page 323

Chapter 17. Normal Subgroups of Γ

Chapter 18. Arithmetic Subgroups of Classical Groups

$\mathcal{H}^1(X; M)$ = 1st cohomology of X, page 370

Chapter 19. Construction of a Coarse Fundamental Domain

Chapter 20. Ratner's Theorems on Unipotent Flows

$x \approx y$ means x is close to y, page 417

Appendix A. Basic Facts about Semisimple Lie Groups

$\mathrm{Mat}_{\ell \times \ell}(\mathbb{R})$ = {$\ell \times \ell$ matrices with real entries}, page 427

$\mathrm{SL}(\ell, \mathbb{R})$ = {$\ell \times \ell$ matrices of determinant 1}, page 427

$\mathrm{GL}(\ell, \mathbb{R})$ = {invertible $\ell \times \ell$ matrices}, page 429

g^T = transpose of the matrix g, page 429

$\mathrm{diag}(a_1, a_2 \ldots, a_n)$ = diagonal matrix with entries $a_1, \ldots, a_n$, page 429

$\mathrm{SL}(n, \mathbb{R})$, $\mathrm{SO}(m, n)$, $\mathrm{SU}(m, n)$, $\mathrm{Sp}(2m, \mathbb{R})$: classical Lie groups, page 429

$\mathrm{SL}(n, \mathbb{C})$, $\mathrm{SL}(n, \mathbb{H})$, $\mathrm{SO}(n, \mathbb{C})$, $\mathrm{SO}(n, \mathbb{H})$, $\mathrm{Sp}(2m, \mathbb{C})$, $\mathrm{Sp}(m, n)$: more classical groups, page 430

$\overline{x}$ = conjugate of the quaternion x, page 431

Δ = modular function of H, page 433

$\overline{\overline{H}}$ = Zariski closure of H, page 434

$\mathfrak{g}$, $\mathfrak{h}$ = Lie algebras of G and H, page 440

Ad_H = adjoint representation of H, page 442

Appendix B. Assumed Background

e = identity element of a group, page 447

$Z(H)$ = center of the group H, page 447

$C_H(K)$ = centralizer of K in H, page 447

$\mathcal{N}_H(K)$ = normalizer of K in H, page 447

$\mathrm{Stab}_\Lambda(p)$ = stabilizer of p, page 448

$|L : F|$ = degree of field extension, page 448

$\mathrm{Gal}(L/F)$ = Galois group, page 448

$\mathcal{B}(X)$ = {Borel subsets of X}, page 451

$f_*\mu$ = push-forward of measure μ, page 452

$C_c(X)$ = {continuous functions with compact support}, page 452

$\mathscr{L}^p(X,\mu)$ = {$\mathscr{L}^p$-functions on X}, page 453

$\mathcal{B}$ = a Banach space, page 454

$\mathcal{H}$ = a Hilbert space, page 454

$\mathcal{B}^*$ = dual of $\mathcal{B}$, page 454

$\|T\|$ = operator norm, page 455

$\perp$ = "is orthogonal to", page 455

$\mathcal{K}^\perp$ = orthogonal complement of the subspace $\mathcal{K}$, page 455

Appendix C. A Quick Look at S-Arithmetic Groups

S = a finite set of prime numbers, page 457

$\mathbb{Z}_S = \mathbb{Z}[1/p_1, 1/p_2, \ldots, 1/p_n]$ is the ring of S-integers, page 457

$\mathbb{Q}_p$ = field of p-adic numbers, page 458

$\mathbb{Q}_\infty = \mathbb{R}$, page 458

$\mathbf{G}_S = \times_{p \in S \cup \{\infty\}} \mathbf{G}(\mathbb{Q}_p)$, page 459

S-rank $\mathbf{G} = \sum_{p \in S \cup \{\infty\}} \mathrm{rank}_{\mathbb{Q}_p} \mathbf{G}$, page 460

Index

List of Named Theorems

www.ingramcontent.com/pod-product-compliance
Lightning Source LLC
Chambersburg PA
CBHW060315200326

41519CB00011BA/1731